Lecture Notes in Computer Science 1334
Edited by G. Goos, J. Hartmanis and J. van Leeuwen

Springer
Berlin
Heidelberg
New York
Barcelona
Budapest
Hong Kong
London
Milan
Paris
Santa Clara
Singapore
Tokyo

Yongfei Han Tatsuaki Okamoto
Sihan Qing (Eds.)

Information and Communications Security

First International Conference, ICIS '97
Beijing, China, November 11-14, 1997
Proceedings

Springer

Series Editors

Gerhard Goos, Karlsruhe University, Germany

Juris Hartmanis, Cornell University, NY, USA

Jan van Leeuwen, Utrecht University, The Netherlands

Volume Editors

Yongfei Han, Program Co-Chair
Gemplus Technologies Asia PTE LTD
89, Science Park Drive #04-01/05
The Rutherford, Singapore Science Park, Singapore 118261
E-mail: YongFei.HAN@ccmail.edt.fr

Tatsuaki Okamoto, Program Co-Chair
NTT Labs, Room 609A
1-1 Hikarinooka, Yokosuka-shi, 239 Japan
E-mail: okamoto@sucaba.isl.ntt.co.jp

Sihan Qing, Director
Engineering Research Center for Information Security Technology
Chinese Academy of Sciences
Beijing 100080, P.R. China
E-mail: qsh@sun.ihep.ac.cn

Cataloging-in-Publication data applied for

Die Deutsche Bibliothek - CIP-Einheitsaufnahme

Information and communications security : first international conference ; proceedings / ICICS '97, Beijing, China, November 11 - 13, 1997. Yongfei Han ... (ed.). - Berlin ; Heidelberg ; New York ; Barcelona ; Budapest ; Hong Kong ; London ; Milan ; Paris ; Santa Clara ; Singapore ; Tokyo : Springer, 1997
(Lecture notes in computer science ; Vol. 1334)
ISBN 3-540-63696-X

CR Subject Classification (1991): E.3-4, G.2.1, D.4.6,F.2.1-2, C.2, J.1, K.6.5

ISSN 0302-9743
ISBN 3-540-63696-X Springer-Verlag Berlin Heidelberg New York

Typesetting: Camera-ready by author
SPIN 10647862 06/3142 – 5 4 3 2 1 0 Printed on acid-free paper

Preface

The International Conference on Information and Communications Security (ICICS) represents international research and development in the area of information and communications security. The conference is held every two years (during years when Asiacrypt is not held) in different Asian countries, and it attracts an audience from the academic, commercial, and industrial communities. ICICS '97 is held in the Chinese Academy of Sciences in Beijing, P. R. China, November 11-14 1997.

ICICS '97 was sponsored by the Chinese Academy of Sciences, the National Natural Science Foundation of China, and the China Computer Federation.

The conference was organized by the Engineering Research Center for Information Security Technology in the Chinese Academy of Sciences in co-operation with the International Association for Cryptologic Research (IACR), IEICE, and the Asiacrypt Steering Committee.

There were 87 papers submitted for inclusion, from an international authorship, 37 papers among them have been accepted as regular papers, and 11 as short papers. We would like to thank the authors of all papers submitted, both those whose work is included in these proceedings, and those whose work could not be accommodated. Without their research and writing up there would be no conference.

The number of submitted papers put an additional strain on the program committee members. We are grateful to them all for their work in reviewing the papers in a short time and for freely giving us the benefit of their experience and support in a variety of ways. We also would like to thank all the other reviewers for their reviewing the papers.

We wish to thank all the participants, organizers, and contributors of the ICICS '97 conference.

August, 1997

Yongfei Han
Tatsuaki Okamoto
Sihan Qing

ICICS Steering Committee and ICICS '97 Program Committee

1 STEERING COMMITTEE

Chin-Chen Chang
James W. Gray, III
Yongfei Han
Kwangjo Kim
Tatsuaki Okamoto
Sihan Qing
Vijay Varadharajan

2 PROGRAM COMMITTEE

Co-Chairs
Yongfei Han (GemPlus)
Tatsuaki Okamoto (NTT, Japan)
Sihan Qing (ERCIST, China)
Member
Elisa Bertino (University of Milano, Italy)
Chin-Chen Chang (NCCU, Taiwan)
Robert Deng (NUS, Singapore)
Dieter Gollmann (University of London, UK)
James W. Gray, III (UST, Hong Kong)
Erland Jonsson (Chalmers University of Technology, Sweden)
Kwangjo Kim (ETRI, S. Korea)
Wenbo Mao (HP, UK)
Tsutomu Matsumoto (Yokohama National University, Japan)
Mitsuru Matsui (Mitsubishi, Japan)
Ueli Maurer (ETH, Switzerland)
Catherline Meadows (Naval Research Lab. USA)
Kazuo Ohta (NTT, Japan)
Eiji Okamoto (JAIST, Japan)
Jean-Jacques Quisquater (UCL, Belgium)
Phil Rogaway (UC Davis, USA)
Ravi Sandhu (George Mason University, USA)
Yiqun Lisa Yin (RSA, USA)
Moti Yung (CertCo, USA)
Vijay Varadharajan (Western Sydney University, Australia)
General Chair and Organizing Chair
General Chair: Sihan Qing
Organizing Chair: Xizhen Ni

CONTENTS

Session 13: Key Recovery/ Fair Cryptosystem

Session 14: Intellectual Property Protection

Session 15: Protocols

Session 16: Electronic Commerce

Minimizing the Use of Random Oracles in Authenticated Encryption Schemes

Mihir Bellare[1] and Phillip Rogaway[2]

[1] Dept. of Computer Science & Engineering, University of California at San Diego, 9500 Gilman Drive, La Jolla, California 92093, USA. E-Mail: mihir@watson.ibm.com. URL: http://www-cse.ucsd.edu/users/mihir.

[2] Dept. of Computer Science, Engineering II Bldg., University of California at Davis, Davis, CA 95616, USA. E-mail: rogaway@cs.ucdavis.edu. URL: http://wwwcsif.cs.ucdavis.edu/~rogaway.

Abstract. A cryptographic scheme is "provably secure" if an attack on the scheme implies an attack on the underlying primitives it employs. A cryptographic scheme is "provably secure in the random-oracle model" if it uses a cryptographic hash function F and is provably secure when F is modeled by a public random function. Demonstrating that a cryptographic scheme is provably secure in the random-oracle model engenders much assurance in the scheme's correctness. But there may remain some lingering fear that the concrete hash function which instantiates the random oracle differs from a random function in some significant way. So it is good to limit reliance on random oracles. Here we describe two encryption schemes which use their random oracles in a rather limited way. The schemes achieve semantic security and plaintext awareness under specified assumptions. One scheme uses the RSA primitive; another uses Diffie-Hellman. In either case messages longer than the modulus length can be safely and directly encrypted without relying on the hash functions modeled as random-oracles to be good for private-key encryption.

1 Introduction

1.1 Provable security and random oracles

A cryptographic scheme S based on a primitive P is said to be *provably secure* if the security of P has been demonstrated to imply the security of S. More precisely, we use this phrase when someone has formally defined the goal G_P for some primitive P; someone has formally defined and the goal G_S for some scheme S; and then someone has proven that the existence of an adversary A_S who breaks scheme S, in the sense of violating G_S, implies the existence of an adversary A_P who breaks protocol P, in the sense of violating G_P.

What provable security means is that as long as we are ready to believe that P is secure, then there are no attacks on S. This obviates the need to consider any specific cryptanalytic attacks on S. Provable security can vastly increase assurance in a cryptographic scheme. For this reason it is a highly desirable goal.

For many cryptographic goals there are protocols known which establish provable security. But achieving provable security is often quite difficult, and schemes which achieve it are usually more complex and less efficient than their not-provably-secure counterparts. To address this problem the current authors suggested a few years back that the *random oracle model* could provide an effective tool to simultaneously achieve efficiency and something which is "close to" provable security [3]. The idea is to assume during algorithm design and analysis that all parties have access to a *public random oracle*— that is, a publicly-know "black box" which, on input of a string x, returns a *random* string $R(x)$ of some appropriate length. The *random-oracle paradigm* is to do provable security in this enriched model of computation and then, after a protocol and proof have been worked out, to *instantiate* the random oracle with an SHA-like hash function. The *thesis* underlying the random oracle paradigm is that substantial assurance remains despite the not-theoretically-justified instantiation step. For more details on this approach, see [3].

The buying of provable-security-style assurance without loss of efficiency has made the random oracle model an attractive choice for doing rigorous yet practical work in several cryptographic domains. In particular, the approach has been followed for asymmetric encryption [3,4] and digital signatures [3,5,19]. It is a particularly attractive approach for designing the sort of simple, efficient, as-high-assurance-as-possible schemes one wants for cryptographic standards.

One such standards effort is currently going on. The IEEE working group known as "P1363" has been drafting a *Standard for Public-Key Cryptography* [11]. This will be the first document owned by any standard-setting authority which provides general-purpose, bit-level specification for doing public key encryption, digital signatures, and key agreement using public-key techniques.

The P1363 committee has been considering several schemes (for both encryption and digital signatures) which are provably secure in the random-oracle model. This is a major gain in assurance over *ad. hoc.* design. All the same, some concerns have been voiced about the use of random oracles in P1363 schemes. The concerns are of two types: general questions about what a proof in the the random oracle model really means; and specific concerns about the way in which random oracles have been used in particular candidate schemes. In the next two subsections we address each of type of concern, in turn.

1.2 The security guarantee from proofs in the random-oracle model

It is important to neither over-estimate nor under-estimate what the random-oracle paradigm buys you in terms of security guarantees. Here we explain some of the issues and guarantees. See also [3].

Provable security in the random-oracle model is significantly different from (and fundamentally weaker than) provable-security in the standard model. At issue is the fact that when a scheme is designed assuming a random oracle R, and then this oracle is replaced by a concrete hash function F, there is no "standard" assumption on F which is adequate to ensure that F is a good-enough instantiation of R to cause no problems for the particular scheme. So what can a proof of security in the random-oracle model really mean? Some researchers have suggested that the answer is *nothing.* But this is not true.

Schemes proven correct in the random-oracle model typically involve both number theoretic primitives (RSA, discrete exponentiation, etc.) and the use of some sort of confusion/diffusion hash function, which is being modeled as a random oracle. The analysis makes only *standard* and *weak* assumptions on the number-theoretic function.

In practice, attacks on schemes which intertwine both a number-theoretic primitive (like RSA) and a hashing step F (like SHA) invariably treat F as a random function. In other words, cryptanalysis of "mixed" schemes is invariably done by assuming that F is random. But then the random oracle analysis guarantees that this line of attack *cannot succeed* unless the underlying number-theoretic primitive gets broken, and in a strong way.

In other words, let us call an attack *generic* if the attack works for random choice of the function F. Attacks on mixed schemes have always been generic. And if the scheme is secure in the random-oracle model, all generic attacks fail, short of the underlying number-theoretic tool getting completely broken.

In other words, a proof in the random-oracle model buys a significant and unquestionable guarantee in the standard model: proven security against generic attacks. We also claim, this time heuristically, that in "natural" schemes in which number-theory and hashing are closely intertwined, it is difficult to imagine anyone ever coming up with a non-generic attack.

We comment that while it is possible to find "contrived" schemes that are provably secure in the random oracle model but fail when a SHA-like function is put in place of the random oracle, such examples choose the details of the scheme after one has chosen the function F which instantiates the random oracle. If the scheme is chosen before the hash function then no example, however artificial, has been suggested in which the scheme is provably secure in the random-oracle model but insecure when the random oracle is conservatively instantiated. Thus it is unlikely that a real-world scheme should turn out to have such a characteristic. In contrast to this, published schemes which lack reductions (ie. proofs of any kind) are quite routinely found to be in error.

All in all, a scheme which can be proven secure in the random-oracle model has much better assurance than one which cannot.

1.3 Use of random oracles in P1363 encryption schemes

ENCRYPTION BASED ON RSA. In a (non-current) draft of the P1363 document (Aug 22, 1996) a formatting procedure called "OAEP" [4] was used in a way that permitted strings longer than the modulus length to be directly encrypted using the RSA primitive.[1] OAEP uses two hash functions, G and H, and in the security proof for RSA-using-OAEP encryption [4], G and H are modeled

[1] When we say the RSA *primitive* we mean the function $f_{N,e}(x)$ defined by $x \mapsto x^e \bmod N$, for appropriate choice of numbers N, e, x. When we speak of an RSA encryption *scheme* we mean a particular way to use the RSA primitive to encrypt a string. In general, P1363 schemes (which manipulate on strings) are

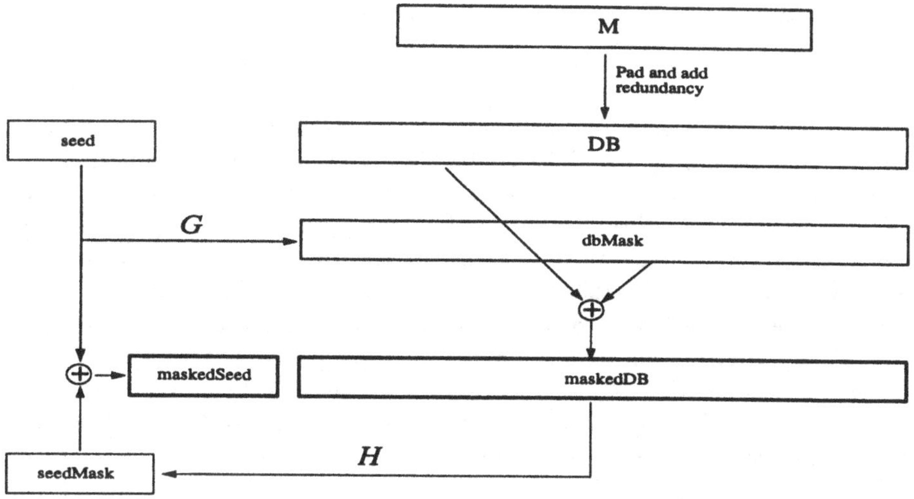

Fig. 1. *The generic OAEP method. Some redundancy (e.g. 0-bits or the hash of a known value) is concatenated to plaintext* M *to make the data block* DB. *A randomly chosen string* seed *is selected. The ciphertext is obtained by encrypting an arbitrary piece of* U = maskedSeed||maskedDB. *In the proof of security, functions* G *and* H *are modeled as random functions.*

as random oracles. Referring now to Figure 1, it is proven in [4] that if *any* part of $U = maskedSeed||maskedDB$ is encrypted using the RSA primitive $x \mapsto x^e \bmod N$, then, in the random-oracle model, the RSA-using-OAEP encryption scheme is secure. (Here one assumes that the RSA primitive is hard to invert on random strings.) In the P1363 draft the RSA primitive was applied to as many bits of U as would fit (just less than the length of the modulus N). The remaining bits of U, if any, are simply dropped into the ciphertext.

About this method Carl Ellison observed: "if we let the OAEP-masked PK [public key] payload [U] get bigger than the modulus, then that part of the payload which overhangs the PK modulus is being symmetrically encrypted by the [function G in the] OAEP operation" [9]. This is true. What happens in this case is that a part of the payload is not RSA-encapsulated by the RSA primitive and so, in this case, there is no interaction between the RSA primitive and (part of the output of) the function G. The lack of such interaction is unfortunate, because, as explained in the introduction, such interaction might serve to cover up any defects in G's behaving like a random oracle. In particular, the claim that non-generic attacks are highly unlikely breaks down.

We emphasize that, formally, none of this is a problem. In the random-oracle model, the overhanging text is protected just as well as the text covered by the

built on top of primitives (which implement number-theoretic functions). The "scheme"/"primitive" distinction is a nice choice of language introduced by the P1363 effort.

RSA primitive. But in passing to an instantiation of the random oracle, the concern does make sense.

ENCRYPTION BASED ON DISCRETE LOG AND ELLIPTIC CURVES. The issue described above is not specific to data formatting long messages under RSA encryption. In encryption schemes which P1363 has considered based on the discrete log or elliptic curves, the underlying hash function is effectively used for symmetric encryption regardless of the length of the plaintext. Here again there is a legitimate desire to avoid such direct use for encryption of a function likely to be derived from a cryptographic hash and modeled by a random oracle.

The current authors agreed to serve on that subcommittee. This paper represents their contribution-to-date on the question at hand. We describe two encryption schemes. One is just a way to realize the RSA-based scheme in the current P1363 draft. The second encryption is based on discrete log (DL). It could just as easily be based on elliptic curves.

1.4 A suggestions for RSA-based encryption

We keep the structure of OAEP exactly in tact and simply refine the interpretation of the functions G and H. In doing this, we preserve plaintext-awareness [4] as well as semantic security [10,4]. The underlying idea springs from the fact that the proof of security for RSA-using-OAEP [4] did not really need G and H to consist *entirely* of random oracles. Instead, it suffices that G be made up of two parts: a function modeled by a random oracle, GG, composed with a pseudorandom generator prg; and H may likewise be partitioned into two parts: a collision-resistant hash function h composed, in a particular way, with a function modeled by a random oracle, HH. A more exact description of this way to realize G and H is given in Section 2 and in Figures 2—3.

One might think it pointless to realize G and H of OAEP in a way that continues to require the use of random oracles! But no; nice dividends are obtained by drawing this prg/GG and h/HH division of labor. Let us describe some of the benefits.

The length of the input that we must feed GG and HH is independent of the length of the plaintext M. Same for the length of the output that GG and HH must generate. This is an advantage in itself, as it is easier to believe in the goodness of hash functions over a smaller domain and range. Next, when the plaintext M is longer than the RSA modulus nothing is "private-key encrypted" using a function modeled by a random oracle. Instead, such work is done by a function which need only be a pseudorandom generator (PRG) and which can, therefore, be implemented by a standard usage of a block cipher. Thus the part of OAEP which "looks like" a symmetric encryption can be be implemented using tools designed for that purpose. Finally, the output-values of GG and HH are kept entirely within the scope of the RSA-primitive. So even if GG and HH should be found to have some behavior which makes them seem different from a random oracle, this is only going to be relevant if that non-random behavior is something that interacts badly with the RSA primitive.

1.5 A suggestions for DL-based encryption

Fix a large prime p. In this description all arithmetic operations are understood to be done mod p. Where appropriate, regard points in $\mathbf{Z}_p^*$ as binary strings.

To encrypt a message M under the recipient's public key g^v choose a random $u \in [0, p-2]$ and compute g^u and g^{uv}. The ciphertext will include g^u. The rest of the ciphertext is formed from M using symmetric keys obtained as $macKey \| encKey = GG(g^u \| g^{uv})$. Here GG is a hash function, to be modeled as a random oracle in the security analysis. Now symmetrically encrypt the message M using encryption key *encKey*, and authenticate the resulting ciphertext using MAC key *macKey*. Full details can be found in Section 3 and Figure 4.

The properties one needs of the symmetric encryption and the message authentication code are extremely weak: we require semantic security with respect to an adversary who can obtain only *one* ciphertext or can ask for the MAC of only *one* message. We call these goals a "one-time symmetric encryption" and a "one-time MAC." Nice features of the scheme described are analogous to those for the RSA-based scheme. The function modeled as a random oracle, GG, is applied to strings of length independent of the length of the plaintext, and the output length needed from it is, in effect, a small constant. We never use a function modeled by a random oracle to symmetrically encrypt; we do that with a primitive intended for encryption.

1.6 Related work

See Appendix A for a discussion of related schemes. See [2] for a fuller version of this paper (due to page limitations, most material on the security of our schemes has been cut).

2 Encrypting with RSA

2.1 The idea

The P1363 "data format" known as "DFE-OAEP" is diagramed in Figure 2. The encryption scheme which uses this formatting method is known as "RSAAES." We describe here a nice way to realize the functions G and H of DFE-OAEP:

$$G(x) = prg(GG(x)) \quad \text{and}$$
$$H(x, y) = HH(x, h(x, y)) \quad \text{where}$$

- *prg* is an *ordinary* pseudorandom generator. It is *not* a random oracle.
- h is an *ordinary* collision-resistant hash function. It is not a random oracle.
- GG and HH are random oracles.

In the formula for HH, x is the part of the message that is covered by the RSA encryption and y is the part of the message which "overhangs" and is not so covered. Of course y may be empty.

We comment that, using DFE-OAEP within RSAAES:

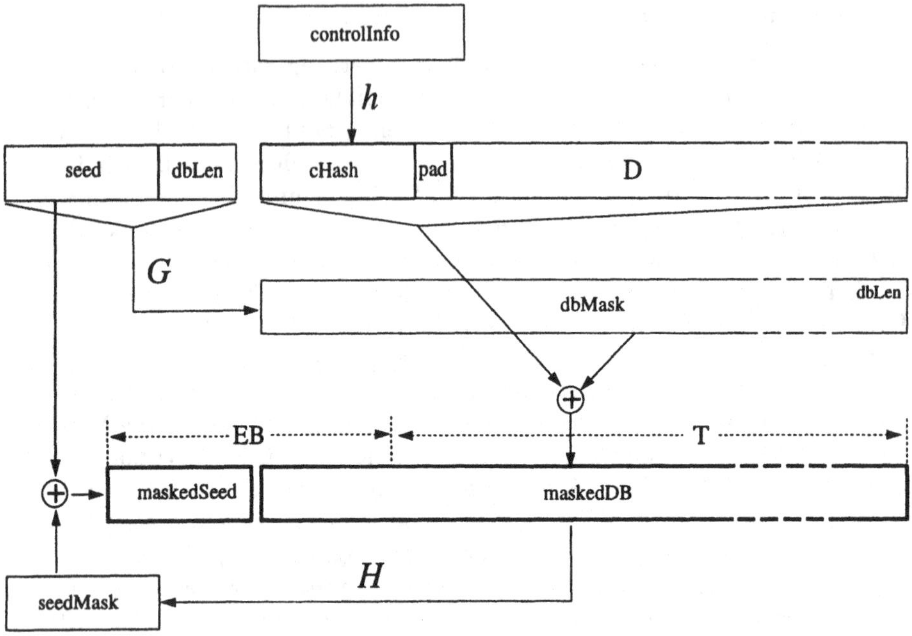

Fig. 2. *Details on the data formatting method DFE-OAEP of P1363. The output has been relabeled (as in an earlier P1363 draft) to allow for the encryption of plaintexts longer than the modulus.*

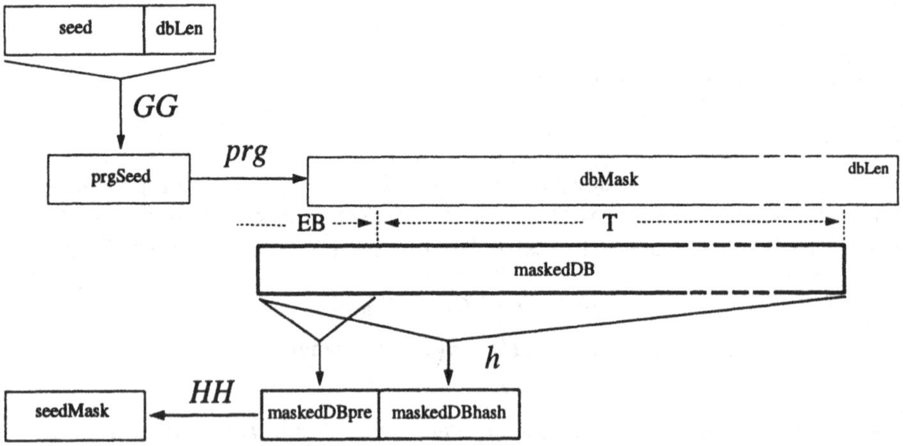

Fig. 3. *A suggestion for how to realize the functions G and H in DFE-OAEP:* $G(x) = prg(GG(x))$, *where prg is a pseudorandom generator and GG is modeled as a random oracle; and* $H(x, y) = HH(x, h(x, y))$, *where h is a collision-resistant hash function and HH is modeled as a random oracle.*

- Random oracles GG and HH are asked to produce only a fixed number of output octets (bytes).
- Random oracles GG is applied to a string of fixed length.
- Random oracle HH is applied to a string whose length depends only on the RSA blocksize, not the length of the message which is being encrypted.
- The output of GG is not directly manifest in an RSAAES-encrypted string. Instead, it is further concealed by the RSA primitive and, in the case of long plaintexts, by the pseudorandom generator *prg*.
- The output of HH is not directly manifest in an RSAAES-encrypted string. Instead, it is further concealed by the RSA primitive.

2.2 Auxiliary functions

We begin by describing the new set of auxiliary functions. For simplicity, we have used a single input and output length, *hLen*, whenever this is sufficient. We have also allowed for a multiplicity of arguments wherever this has proven convenient.

- **Psuedorandom Generator.** This is a function *prg* that takes a secret input r of length *hLen* and a positive integer ℓ and produces an ℓ-octet (ie., ℓ-byte) string $prg(r)$. (In the notation $prg(r)$ we omit mention of ℓ.) For the purposes of algorithm analysis, the assumption on *prg* is that it be a "pseudorandom generator." Roughly said, $prg(r)$ should appear to be uniformly distributed when r is a random and uniformly distributed secret, from the point of view of an adversary who is not given r. (What distinguishes a *prg* from a random oracle is that its security relies on secret coins, namely that r is not available to the attacker.)
- **Idealized Hash Function.** This is a function F that maps a sequence of octet strings, $x_1, x_2, \ldots, x_n$, into a string $F(x_1, x_2, \ldots, x_n)$ of length *hLen*. For the purpose of algorithm analysis, the assumption on F is that it behaves like a random oracle. That is, F is intended for instantiating what, during protocol design and analysis, is regarded as a random oracle. We use this type of primitive as little as we can, while still preserving efficiency and provability.
- **Mask Generation Function** A mask generation function, G, takes as input a string s of arbitrary length and a positive integer, ℓ. It produces an ℓ-octet string $G(s)$. (In the notation $G(s)$ we omit mention of ℓ.) A mask generation function, G, should be defined using some pseudorandom generator, *prg*, and some idealized hash function, GG. The definition is $G(s) = prg(GG(s))$. The pseudorandom generator is used to produce ℓ-octets of output. See Figure 3.
- **Collision-Resistant Hash Function** This is a function h that takes a sequence of arbitrary-length strings, $x_1, \ldots, x_n$, and produces an *hLen*-octet string $h(x_1, \ldots, x_n)$. For the purpose of algorithm analysis, the assumption on h is that it be "collision intractable". Roughly said, no one should be unable to find $(x_1, \ldots, x_n) \neq (y_1, \ldots, y_n)$ such that $h(x_1, \ldots, x_n) = h(y_1, \ldots, y_n)$.

- **Awareness-Preserving Hash Function** An awareness-preserving hash function, H, takes as input a pair of arbitrary-length strings, a and b, and outputs a string, $H(a,b)$, of *hLen* octets. An awareness-preserving hash function, H, should be defined using some collision-resistant hash function, h, and some idealized hash function, *HH*. The definition is $H(a,b) = HH(a, h(a,b))$. See Figure 3.

2.3 DFE-OAEP — Data formatting

Some notational adjustments are useful for our exposition so that one can see how the new auxiliary functions come into play. Again, see Figures 2 and 3.

Input:

- an octet string D of length *dLen* octets, the ***data***;
- a positive integer *ebLen*, the ***encryption block length***;
- an octet string *controlInfo*, the ***control information*** (may be empty).

Output:

- an octet string *EB* of length *ebLen* octets, the ***encryption block***;
- an octet string T of length $\min\{0, dLen - ebLen - 2hLen\}$ octets, the ***extension***.

Auxiliary functions:

- A ***mask generation function***, G;
- An ***awareness-preserving hash function***, H;
- A ***collision-resistant hash function***, h.

Algorithm:

(1) If $dLen < ebLen - 2hLen$ then prepend D with octets $00||00||\cdots||01$ so that the resulting string D' has length $dLen' = ebLen - 2hLen$. Else prepend D with a single octet 01 so the that resulting string D' has length $dLen' = dLen + 1$.

(2) Apply the collision-resistant hash function h to *controlInfo* to produce an *hLen*-octet string *cHash*: $cHash = h(controlInfo)$.

(3) Form the data block *DB* by setting $DB = cHash||D'$.

(4) Generate a fresh, random, *hLen*-octet string *seed*.

(5) Set $seedG = seed||dbLen'$ where $dbLen' = \text{I2OSP}(dbLen)$ (which function converts an Integer to an Octet String). Apply the mask generation function G to *seedG* to produce a *dbLen*-octet string *dbMask*. Then exclusive-or *dbMask* with *DB* to produce a *dbLen*-octet string *maskedDB*:

$$\begin{aligned} seedG &= seed||dbLen', \\ dbMask &= G(seedG), \\ maskedDB &= DB \oplus dbMask. \end{aligned}$$

Let *maskedDBprefix* be the first $dbLen - hLen$ octets of *maskedDB*.

(6) Apply the awareness-preserving hash function H to *maskedDBprefix* and *maskedDB* to produce an *hLen*-octet string *seedMask*. Then exclusive-or *seedMask* and *seed* to produce an *hLen*-octet string *maskedSeed*.

$$\begin{aligned} \mathit{seedMask} &= H(\mathit{maskedDBprefix}, \mathit{maskedDB}), \\ \mathit{maskedSeed} &= \mathit{seed} \oplus \mathit{seedMask}. \end{aligned}$$

(7) Let $U = \mathit{maskedSeed} \| \mathit{maskedDB}$. Let the encryption block (what we will apply the RSA primitive to) *EB* be the first *ebLen* octets of U, and let T be the rest of U.

3 Encrypting with Discrete Log

3.1 Auxiliary functions

We begin by describing the new set of auxiliary functions. The scheme we will call "DLAES" uses three types of auxiliary functions. Only one of these, an idealized hash function, has already been described. The other two are:

- **One-Time MAC.** This is a function *OneTimeMAC* that takes a string *macKey* of some length *macKeyLen*, and a sequence $x = (x_1, \ldots, x_n)$ of arbitrary length strings. It returns $\mathit{tag} = \mathit{OneTimeMAC}_{\mathit{macKey}}(x)$. For the purpose of algorithm analysis, the assumption on *OneTimeMAC* is that it be a "one-message-secure message authentication code." Roughly said, this means that an adversary, given a *single* authenticated message $(x, \mathit{OneTimeMAC}_{\mathit{macKey}}(x))$, should be unable to come up with any *second* message x' and a correct tag $\mathit{OneTimeMAC}_{\mathit{macKey}}(x')$ for it. This is a much *weaker* assumption than the standard one for a secure message authentication code. A one-time MAC can be achieved "cryptographically" (eg., by HMAC [1]) or "non-cryptographically" (eg., by an almost-univeral-2 family of hash functions).
- **One-Time Encryption** This is a function *OneTimeENC* that takes a string *encKey* of some length *encKeyLen*, and a string M of arbitrary length, and produces a string $\mathit{encM} = \mathit{OneTimeENC}_{\mathit{encKey}}(M)$. There must be a corresponding decryption function which, when presented with key *encKey*, recovers M from *encM*. For the purpose of algorithm analysis, the assumption on *OneTimeENC* is that it be a "one-message-secure symmetric encryption scheme." Roughly said, this means that for any equal length messages M_0, M_1, an adversary, given a *single* set of strings $(M_0, M_1, \mathit{OneTimeENC}_{\mathit{encKey}}(M_b))$, for b a random and unknown bit, should be unable to tell if the encrypted message is an encryption of M_0 or M_1. Just as a one-time MAC is weaker than a standard MAC, a one-time encryption is a weaker assumption than a standard notion for secure encryption. It only allows one to securely encrypt one string.

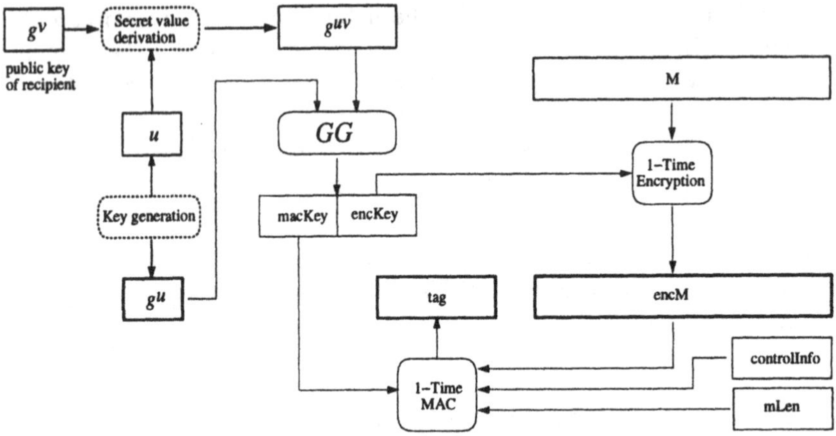

Fig. 4. *A suggestion for how to encrypt with discrete log.*

3.2 DLAES encryption operation

Refer to Figure 4.

Input:

- an octet string M of length *mLen* octets, the ***message***;
- an octet string *controlInfo*, the ***control information*** (may be empty);
- a Diffie-Hellman public key, g^v, the ***recipient's public key***, and associated discrete log parameters.

Output:

- an octet string *EM* of length l+*hLen*+*mLen* octets, the ***encrypted message***.

Auxiliary functions:

- A ***one-time MAC***, *OneTimeMAC*, using a key of length *macKeyLen* octets and generates tags of length *macLen* octets;
- A ***one-time encryption***, *OneTimeENC*, using a key of length *encKeyLen* octets;
- An ***idealized hash function***, *GG*.

Algorithm:
For x a point in some finite field, we will write "$\_x$" to denote the string which encodes x (according to some fixed formatting conventions).

(1) Generate a "one-time key pair" (g^u, u). (In [11] this is done with a primitive called "DLKGP.") Convert g^u to an octet string $\_g^u$. (In [11] this is done with a primitive called "I2OSP.")

(2) Derive a field element g^{uv}, the secret value, from the public key g^v and the one-time private key u. (In [11], the primitive which does this is called "DLSVDP.") Convert g^{uv} to an octet string $\_g^{uv}$ (using I2OSP).

(3) Apply the idealized hash function *GG* to $\_g^u$ and $\_g^{uv}$ to obtain a string of *macKeyLen*+*encKeyLen* octets *ggOutput*. Let the first *macKeyLen* octets of *ggOutput* be denoted *macKey*. Let the remaining octets of *ggOutput* be denoted *encKey*.

$$ggOutput = GG(\_g^u, \_g^{uv})$$
$$macKey = ggOutput[1..macKeyLen]$$
$$encKey = ggOutput[macKeyLen + 1..macKeyLen + encKeyLen]$$

(4) Apply *OneTimeENC* to message *M* and key *encKey* to generate ciphertext *encM*: $encM = OneTimeENC_{encKey}(M)$.

(5) Apply *OneTimeMAC* to the octet strings *encM*, *controlInfo*, and $mLen' =$ I2OSP(*mLen*) using key *macKey* to obtain the octet string *tag*: $tag = OneTimeMAC_{macKey}(encM, controlInfo, mLen')$.

(6) Concatenate the octet strings $\_u$, *encM* and *tag* to produce the encrypted message *EM*: $EM = \_u \| encM \| tag$.

(7) Output *EM* as the encrypted message.

Acknowledgments

The first author was supported by NSF CAREER Award CCR-9624439 and a Packard Foundation Fellowship in Science and Engineering. The second author was supported by NSF CAREER Award CCR-9624560 and a grant from Certicom Corporation.

References

1. M. Bellare, R. Canetti and H. Krawczyk, "Keying hash functions for message authentication," *Advances in Cryptology – Crypto 96 Proceedings*, Lecture Notes in Computer Science Vol. 1109, N. Koblitz ed., Springer-Verlag, 1996.
2. M. Bellare and P. Rogaway, "Minimizing the use of random oracles in authenticated encryption schemes." More complete version of this paper, http://wwwcsif.cs.ucdavis.edu/˜rogaway.
3. M. Bellare and P. Rogaway, "Random oracles are practical: A paradigm for designing efficient protocols," *Proceedings of the First Annual Conference on Computer and Communications Security*, ACM, 1993.
4. M. Bellare and P. Rogaway, "Optimal asymmetric encryption– How to encrypt with RSA". Current version available at URL of either author. Earlier version in *Advances in Cryptology – Eurocrypt 94 Proceedings*, Lecture Notes in Computer Science Vol. 950, A. De Santis ed., Springer-Verlag, 1994.
5. M. Bellare and P. Rogaway, "The exact security of digital signatures– How to sign with RSA and Rabin." Current version available at URL of either author. Earlier version in *Advances in Cryptology – Eurocrypt 96 Proceedings*, Lecture Notes in Computer Science Vol. 1070, U. Maurer ed., Springer-Verlag, 1996.
6. M. Blum and S. Goldwasser, "An efficient probabilistic public-key encryption scheme which hides all partial information," *Advances in Cryptology – Crypto 84 Proceedings*, Lecture Notes in Computer Science Vol. 196, R. Blakely ed., Springer-Verlag, 1984.

7. I. Damgård, "Towards practical public key cryptosystems secure against chosen ciphertext attacks," *Advances in Cryptology – Crypto* 91 *Proceedings*, Lecture Notes in Computer Science Vol. 576, J. Feigenbaum ed., Springer-Verlag, 1991.
8. D. Dolev, C. Dwork and M. Naor, "Non-malleable cryptography," *Proceedings of the* 23rd *Annual Symposium on Theory of Computing*, ACM, 1991.
9. C. Ellison, personal communication, September 1996.
10. S. Goldwasser and S. Micali, "Probabilistic Encryption," *Journal of Computer and System Sciences* **28**, 270-299, April 1984.
11. IEEE P1363 Committee, *IEEE P1363 Working Draft*, February 6, 1997. Lisa Yiquin, editor. Current draft in http://stdsbbs.ieee.org/groups/1363/index.html.
12. D. Johnson and S. Matyas. "Asymmetric encryption: evolution and enhancements," CryptoBytes, Vol. 2, No. 1, Spring 1996.
13. D. Johnson, A. Lee, W. Martin, S. Matyas and J. Wilkins, "Hybrid key distribution scheme giving key record recovery," IBM Technical Disclosure Bulletin, 37(2A), 5–16, February 1994.
14. D. Johnson, S. Matyas, M. Peyravian, "Encryption of long blocks using a short-block encryption procedure." November 1996. Available in http://stdsbbs.ieee.org/groups/1363/index.html.
15. IEEE P1363 Committee, Burt Kaliski, chair. Unpublished e-mail memorandum with subject heading: "Technical Subcommittee on Encryption Schemes." September 10, 1996.
16. M. Luby and C. Rackoff, "How to construct pseudorandom permutations from pseudorandom functions." *SIAM J. Computation*, Vol. 17, No. 2, April 1988.
17. S. Matyas, M. Peyravian, A. Roginsky, "Security analysis of Feistel ladder formatting procedure." March 1997. Available in http://stdsbbs.ieee.org/groups/1363/index.html.
18. M. Naor and M. Yung, "Public-key cryptosystems provably secure against chosen ciphertext attacks," *Proceedings of the* 22nd *Annual Symposium on Theory of Computing*, ACM, 1990.
19. D. Pointcheval and J. Stern, "Security proofs for signatures," *Advances in Cryptology – Eurocrypt* 96 *Proceedings*, Lecture Notes in Computer Science Vol. 1070, U. Maurer ed., Springer-Verlag, 1996.
20. R. Rivest, A. Shamir and L. Adleman, "A method for obtaining digital signatures and public key cryptosystems," CACM 21 (1978).
21. RSA Data Security, Inc., "PKCS #1: RSA Encryption Standard," June 1991.
22. Y. Zheng, "Public key authenticated encryption schemes using universal hashing," (15 Oct 96). Unpublished contribution to P1363. ftp://stdsbbs.ieee.org/pub/p1363/contributions/aes-uhf.ps
23. Y. Zheng and J. Seberry, "Immunizing public key cryptosystems against chosen ciphertext attack." *IEEE Journal on Selected Areas in Communications*, vol. 11, no. 5, 715–724 (1993).

Appendix A — Related work

Raw encryption. In the beginning [20], people did not seem to notice that there was a "gap" between using the RSA primitive and encrypting a string with RSA. That is, to encrypt a string x using RSA exponent e and modulus N one would insist that x, regarded as a binary number, was a point in $\mathbf{Z}_N^*$, and then the encryption of x would be $x^e \bmod N$. One might call this *raw* RSA encryption. One will find plenty of books and articles which imply that raw RSA encryption is a reasonable way to encrypt. But it is not.

Goldwasser and Micali were the first to explain [10] that public-key encryption, to be generally useful, had better probabilistic. (If the encryption is deterministic then, for example, it is trivial to decrypt messages drawn from a small and known message space.) In the same paper that introduced provable security, they showed one way to achieve it for public-key encryption: probabilistically encrypt each bit of the message x using a "hardcore bit" of the underlying trapdoor permutation.

Of course the [10] method is very inefficient While there were some subsequent improvements in efficiency for provably-secure public-key encryption schemes based on certain number theoretic assumptions (e.g., [6]), practical encryption schemes remained *ad. hoc.*, without any sort of provable-security properties.

SIMPLE-EMBEDDING SCHEMES. Still, the message that "deterministic encryption is bad" was appreciated and did filter into practice. The paradigm that emerged for RSA encryption was this: probabilistically, invertibly embed the plaintext message x into a string $r_x \in Z_N^*$; then take the encryption of x to be $f(r_x) \stackrel{\text{def}}{=} (r_x)^e \bmod N$. In [4] we called this approach a *simple-embedding scheme*.

Here is the simple-embedding scheme from RSA PKCS #1 [21]: r_x is a zero byte, 00; followed by the byte 02; followed by at least 8 random non-zero bytes, r; followed by the byte 00; followed by the message x. We use enough non-zero random bytes to make the entire message come out to $|N|$ bits. So the the encryption of x is then $\mathcal{E}_{\text{PKCS\#1}}(x) = f(00\|02\|r\|00\|x)$.
No attacks on the above method are know. But a group at IBM objected (among other things) to the fact that x appears directly in the scope of the RSA primitive: what if the RSA primitive should leak some function of those (unmasked) bits? A slightly simplified version of their approach is $\mathcal{E}_{\text{IBM}}^{G}(x) = f(x0^{k_1} \oplus G(r) \parallel r)$. Of concern with both of these schemes is that there is no compelling reason to believe that x is as hard to compute from $f(r_x)$ as r_x is hard to compute from $f(r_x)$—let alone that all interesting properties of x are well-hidden. To get over this, the current authors suggested the scheme: $\mathcal{E}_{\text{BR}}^{G}(x) = f(r) \parallel G(r) \oplus x$ and proved it semantically secure in the random oracle model.

BEYOND SEMANTIC SECURITY. Chosen-ciphertext security was provably achieved by [18], but the scheme is extremely inefficient. More practical encryption schemes which aimed at achieving chosen ciphertext security were proposed by Damgård [7] and Zheng and Seberry [23], but they are not proven secure under any assumption, even a random oracle one. Non-malleability is provably achieved by [8], but the scheme is extremely inefficient. An efficient scheme proven in [3] to achieve both non-malleability and chosen-ciphertext security in the random-oracle model is $\overline{\mathcal{E}}_{\text{BR}}^{G,H}(x) = f(r)\|G(r) \oplus x\|H(rx)$.
SUGGESTION OF ZHENG. Also responding to P1363, Zheng suggested [22] the encryption scheme shown in Figure 5, based on ideas of [23]. The box labeled "universal hash" represents a family of almost strongly universal-2 hash func-

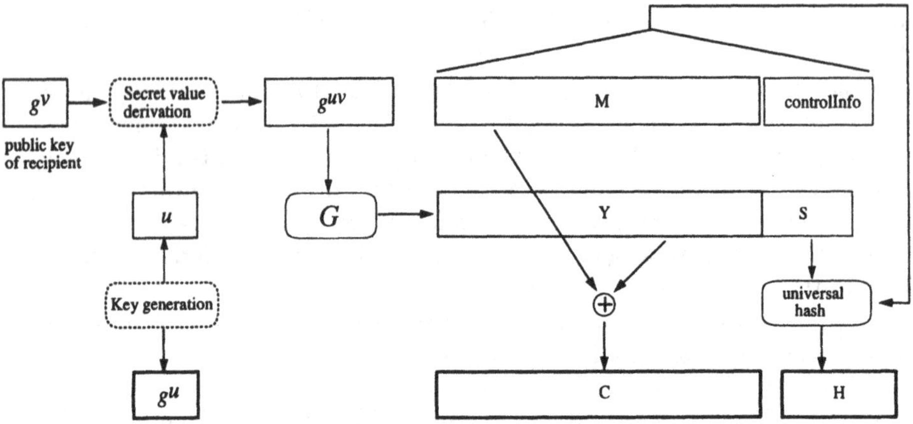

Fig. 5. *A suggestion by Zheng for how to encrypt with discrete log. The box labeled "universal hash" is a family of almost strong universal-2 hash functions; the particular function selected is indexed by* S *and evaluated at* M‖controlInfo*. It is not clear what type of object G is supposed to be.*

tions.[2] Zheng asserts that "as no one-way hash function is used by these schemes, their security does NOT rely on the random hash oracle assumption." Yet the scheme in [22] is not secure (even for semantic security) under standard assumptions. For what exactly would one assume about G? Not that it is a PRG: just imagine what happens if G happens to leak some function λ of its seed, g^{uv}, and it so happens that λ is easy to compute from g^u and g^v. This can not be ruled out under the Diffie-Hellman assumption combined with the assumption that G is a PRG.

Let us emphasize that Zheng's suggestion may work well, in practice, for realistic choice of G. We are simply asserting that the scheme is not going to be provably secure under standard assumptions. It is possible that something could be proven if G is a random oracle, but such a claim has not been made, and certainly not been proved.

Relation of the scheme of Section 3 to others. Our scheme of Figure 4 bears some resemblance to the schemes of [22,23]. Namely, if you implement the one-time MAC with a universal-2 hash function, and implement the one-time encryption by a Vernam cipher, then the schemes start to look quite similar. Our point has not been to come up with a different-looking scheme, but to understand what assumptions are needed where, so that we could minimize our use of random oracles.

Suggestion of Johnson, Matyas, Peyravian. Another contemporaneous suggestion was put forth by Johnson, Matyas and Peyravian [14]. Assume that

[2] Recall that a set of functions $\mathcal{H} = \{h : A \to B\}$ is almost strongly universal-2 if for all $x \neq y$ the random variable $h(x)\|h(y)$ is nearly uniformly distributed (as h ranges over $\mathcal{H}$).

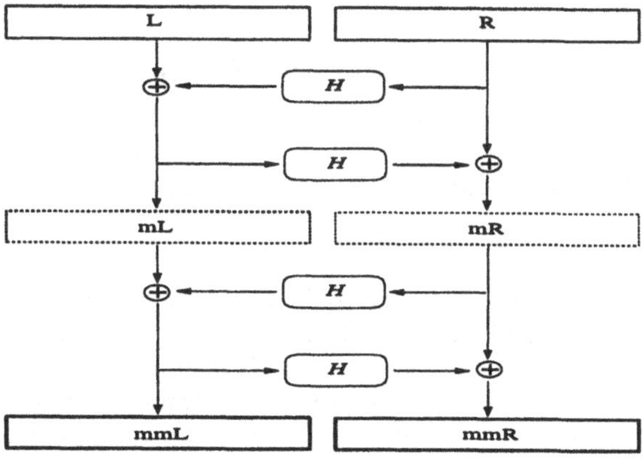

Fig. 6. *Data formatting method suggested by Johnson, Matyas and Peyravian. After adding redundancy and randomness to the plaintext message (if necessary), the resulting data block,* X = L||R, *is subjected to 4 rounds of a Feistel network, using a round function H. Any piece of the resulting message* Y = mmL||mmR *may then be encrypted using any desired primitive.*

the message X already has "sufficient" redundancy in it (e.g., 0-bits or the hash of some known value are embedded in it), and assume that X already has "sufficient" unpredictability (e.g., random bits have been embedded in it). Then to asymmetrically encrypt X subject it to 4 rounds of a Feistel network, as shown in Figure 6, to obtain Y. Encrypt, using an arbitrary encryption primitive, an arbitrary piece of Y.

What can we assume about H to get an authenticated encryption scheme using the method just described? The authors suggest that H should be a "strong collision-resistant one-way hash function," but standard assumptions won't be enough. It is easy to verify, for example, that if H preserves the least significant bit of its input (ie., $\text{lsb}(H(x)) = \text{lsb}(x)$) then the entire procedure is unsound. Such a property of a hash function is no way excluded by the standard assumptions which are made on a hash function.

Again, we are not saying that a realistic embodiment of the [14] scheme will be bad. In fact, we think just the opposite, and we believe that the scheme can be proven secure in the random oracle model. Generalizing the method and applying a result of [16] may help in this direction.

The approach in [14] does not seem to do anything to help reduce the assumptions one needs on hash functions to gain some type of provable security. But the method has the potential of giving rise to something more versatile than own own suggestions, because the randomness and redundancy are inserted arbitrarily into the plaintext X and then an arbitrary piece of the masked plaintext Y is encrypted with an arbitrary primitive.

Zero-Knowledge Proofs of Decision Power: New Protocols and Optimal Round-Complexity

GIOVANNI DI CRESCENZO* KOUICHI SAKURAI** MOTI YUNG***

Abstract. We consider perfect zero-knowledge proof systems for "proving the power to decide whether a membership in a language is true or not". We extend the definition of the model, and then extend the class of languages in it; (so far only the language of quadratic residuosity modulo a Blum integer was known to be applicable to this model). More precisely, we present a protocol for all known random self-reducible languages (i.e., graph isomorphism, quadratic residuosity, discrete log). This protocol can be executed with only 4 rounds of communication. Finally we extend a well-known lower bound for the number of rounds of zero-knowledge proofs of membership to our "decision power model". This shows that (under some technical restrictions) our protocol is round-optimal unless the considered language is in BPP (which seems unlikely).

1 Introduction

Zero-knowledge proofs, introduced by Goldwasser, Micali and Rackoff [GMR], are a method for a prover to convince a polynomial-time bounded verifier that a certain assertion (as membership of a string x to a language L) is true, without revealing any additional information. Since their introduction, the importance of zero-knowledge has been soon realized and applied to many areas in cryptography, such as identification schemes [FFS], two-party and multi-party protocols [Ya, GMW2], to name a few. Moreover, related and more elaborated notions have been investigated, such as proofs of decision power, introduced in [FFS] (under a different name). A proof of decision power is a method for a prover to convince a polynomial-time bounded verifier that he knows whether a certain assertion (as membership of a string x to a language L) is true or false, without revealing which is the case, nor any additional information. Proof systems in this model have been shown to have application to identification schemes and have started the important field of proofs of knowledge [FFS, FS, TW, BG], proof systems in which the prover, on input a string x, convinces a verifier that he knows a string y such that a certain relation $R(x, y)$ holds.

Since the introduction of proof of decision power, few languages have been shown to have a zero-knowledge proof system in this model. Furthermore, in the case of *perfect zero-knowledge*, i.e., when the property of the protocol being zero-knowledge is proved without resorting to unproven assumptions, only one language, namely, quadratic residuosity modulo Blum integers, was shown to have such a protocol (also assuming that

* Computer Science and Engineering Dep., University of California San Diego, La Jolla, CA, 92093-0114. E-mail: giovanni@cs.ucsd.edu

** Dept. of Computer Science and Comm. Eng., Kyushu Univ., Fukuoka 812-81, Japan. E-mail:sakurai@csce.kyushu-u.ac.jp

*** CertCo, New York NY, USA. E-mail: moti@certco.com, moti@cs.columbia.edu

the integer was previously verified). In particular, the mentioned language belongs to NP ∩ co-NP and the current state of the art gives the impression that perfect zero-knowledge proofs of decision power, because of their definition, might be obtained only for languages in this class. Notice that this does not happen in the model of proofs of membership, where also languages believed not to be in NP ∩ co-NP, as graph isomorphism, have such protocols [GMW1].

Our results.

In this paper we first give an more general definition of "perfect zero-knowledge interactive proof of decision power." We then extend the class of languages that are known to have such a proof. We present a protocol for proving statements in any of the known random self-reducible languages (e.g., graph isomorphism, quadratic residuosity modulo composite integers and discrete log related languages). When combined with techniques from [DDPY], this protocol extends to the languages of monotone formulae over statements of membership in the above random self-reducible languages.

An important and perhaps unexpected feature of our protocol is that it has only 4 rounds of communication (notice that perfect zero-knowledge proofs of membership for the same languages currently require 5 rounds [BMO], and only some special languages are known to have 4-rounds such proofs [DP, DDP]). This also motivates us to consider whether the obtained number of rounds is optimal. A lower bound on the number of rounds in zero-knowledge proofs of membership has been given in [GoKr], who showed that such protocols (under some technical restriction) can be obtained in at most 3 rounds only for languages in BPP. In this paper we extend this lower bound to the model of proofs of decision power. This shows that our protocol is round-optimal unless the languages considered are in BPP (which is not believed to be true). Therefore, we obtain a tight result (which is therefore more powerful than what is known in the model of proofs of membership).

2 Notations and definitions

We start with some basic definitions about interactive protocols and then define perfect zero-knowledge proof systems of decision power.

Basic definitions. If A and B are two interactive probabilistic Turing machine, by pair (A,B) we denote an interactive protocol. Let x be an input common to A and B. By $\mathrm{tr}_{(A,B)}(x)$ we denote the transcript of the interaction between A and B on input x, that is, the messages written on B's communication tape during such interaction. By $\mathrm{OUT}_B(\mathrm{tr}_{(A,B)}(x))$ we denote B's output given the transcript $\mathrm{tr}_{(A,B)}(x)$. We define $View_B(x)$, B's view of the interaction with A on input x, as the probability space that assigns to pairs $(R; \mathrm{tr}_{(A,B(R))}(x))$ the probability that R is the content of B's random tape and that $\mathrm{tr}_{(A,B(R))}(x)$ is the transcript of an execution of protocol (A,B) on input x given that R is B's random tape. If L is a language, by $\chi_L : \{0,1\}^* \rightarrow \{0,1\}$ we denote the indicator function for the language L (i.e., $\chi_L(x) = 1$ if and only if $x \in L$).

A definition for Zero-Knowledge Proofs of Decision Power. A zero-knowledge proof of decision power is an interactive protocol in which a prover convinces a poly-bounded verifier that he knows whether a string x belongs to a language L or not, without revealing which is the case, or any other information. The formal definition for zero-knowledge proofs of decision power has three requirements: regularity, validity, and zero-knowledge, which we first informally describe. The regularity requirement states that the verifier accepts with high probability for any input x, in the language or not. The validity requirement states that there exists an extractor that, for any input x, and interacting with any prover that forces the verifier to accept with 'sufficiently high' probability, is able to decide whether $x \in L$ or not, within a 'properly bounded' expected time. This differs from previous work on proofs of knowledge in which the extractor existed only for input in the language and was required to output a string satisfying a polynomial relation with the input. Our approach allows to consider even languages above NP. Finally, the zero-knowledge requirement states that for all input x, in the language or not, the conversation between prover and verifier is efficiently simulatable, in the usual sense of [GMR].

The model for proving whether a membership statement for a certain language was true or not was first introduced in [FFS]. The definition of the validity property has later been fully formalized in [BG] for the case of proofs of knowledge. In this paper we adapt the approach used in [BG] to the case of proofs of decision power. We note that the notion of computational power has been introduced in [Yu].

Definition 1. Let P be an interactive probabilistic Turing machine and V an interactive probabilistic polynomial-time Turing machine that share the same input. Let L be a language and $err : \{0,1\}^* \rightarrow [0,1]$ be a function. We say that a pair (P,V) is a PERFECT ZERO-KNOWLEDGE INTERACTIVE PROOF SYSTEM OF DECISION POWER with decision error err for L if

Regularity. For all x, $\text{Prob}(\text{OUT}_V(\text{tr}_{(P,V)}(x)) = \text{ACCEPT}) \geq 1 - 2^{-|x|}$.

Validity. There exists a probabilistic oracle machine E (called the extractor) such that for all x and any Turing machine P$'$, and letting $acc_{P'}(x) = \text{Prob}(\text{OUT}_V(\text{tr}_{(P',V)}(x)) = \text{ACCEPT})$, the following holds: if $acc_{P'}(x) > err(x)$ then,
- $\text{Prob}(\text{OUT}_E(\text{tr}_{(P',E)}(x)) = \chi_L(x)) \geq 1 - 2^{-|x|}$.
- The machine E halts within expected time bounded by $\frac{poly(|x|)}{(acc_{P'}(x) - err(x))}$.

Perfect Zero-Knowledge. For any Turing machine V$'$, there exists a probabilistic Turing machine $S_{V'}$ (called the simulator) running in expected polynomial-time such that for all x, the probability spaces $View_{V'}(x)$ and $S_{V'}(x)$ are equal.

We remark that algorithm E is allowed to use the prover P' as an oracle. Using an old protocol from [FFS] one obtains the result:

Theorem 2. *[FFS] There exists a perfect zero-knowledge proof system of decision power for the language of Quadratic Residuosity modulo Blum integers.*

3 Perfect zero-knowledge proofs of decision power

In this section we investigate proofs of decision power that are perfect zero-knowledge. The emphasis is both on extending the class of languages that are known to have such protocols, and on finding protocols with minimum number of rounds of communication. Our main result is a protocol for all random self-reducible languages, which is round-optimal unless the given language is in BPP. In Subsection 3.1 we give some insight about the subtlety of constructing proofs of decision power. In Subsection 3.2 we show the protocol in the case of graph isomorphism language (for ease of exposition). In Subsection 3.3 we prove that the presented protocol is correct. In Subsection 3.4 we present extensions and generalizations.

3.1 The subtlety of constructing proofs of decision power

In principle it might be possible to directly use interactive proof systems of membership in order to construct proof systems of decision power. In particular, consider the following protocol transformation: Given a proof of membership (A,B) for the language OR(L,$\overline{L}$) defined as the set of pairs (x_1, x_2) such that $(x_1 \in L) \vee (x_2 \notin L)$ - [DDPY] have such proofs for Graph Isomorphism - derive a protocol (P,V) as (A,B) executed on input (x, x). One would observe that such transformation might be a reasonable approach to construct a proof system of decision power for L. Nevertheless, it turns out that this approach in general may fail; that is, the obtained (P,V) may fail to be a proof of decision power, and we show an example for this. That is, we consider a known proof of membership for a language OR(L,$\overline{L}$) which, when transformed as described above, does not constitute a proof of decision power. To see this, we show that there exists a cheating prover not knowing whether the input x is in the language L or not, that forces the verifier to accept with probability 1. Clearly, no valid extractor can be given and thus the protocol is not a proof of decision power. Here is our example.

We describe the proof system of membership given in [DDPY] for the language GI-OR-GNI of quadruples (G_0, G_1, H_0, H_1) of graphs such that G_0 is isomorphic to G_1 or H_0 is not isomorphic to H_1. We will see that the same protocol executed on input (G_0, G_1, G_0, G_1) does not give a proof of decision power or a transfer of decision for the graph isomorphism language.

On input the quadruple (G_0, G_1, H_0, H_1) of n-node graphs, the protocol in [DDPY] is constituted of the parallelization of n independent atomic protocols (A,B), and each iteration is as follows. B uniformly chooses a bit b and a permutation ψ, computes the graph $H = \psi(H_b)$, and sends H to V. Now, if $H_0 \not\approx H_1$ then A computes b. Then, A computes a graph G isomorphic to G_c, where c is bit b if $H_0 \not\approx H_1$ or a uniformly chosen bit otherwise. Then A sends G to B, which reveals his bit b and the permutation ψ. Finally, B sends an isomorphism between the graph G and G_d and B verifies that the isomorphism is correct and $b = d$. After all iterations, B accepts if and only if all his verifications have been satisfied.

The above protocol is a proof of membership for the language GI-OR-GNI. Consider the protocol (P,V) as (A,B) executed on input (G_0, G_1, G_0, G_1); now we see why (P,V) is not a proof of decision power or a deciding proof for the language GI. To this purpose, consider the following (dishonest) prover P$'$ not knowing whether $(G_0 \approx G_1)$ or $(G_0 \not\approx G_1)$: after receiving the graph H from V, (s)he computes the graph G as a uniformly chosen graph isomorphic to H, say, as $G = \pi(H)$, where π is a uniformly chosen permutation. We notice that G is still isomorphic to one of G_0, G_1; in fact, G is isomorphic to G_b. However, the difference here with an execution of the same protocol by the honest prover P, is that the dishonest prover P$'$ may not know the value of b. Nonetheless, after receiving the permutation ψ such that $H = \psi(G_b)$, P$'$ can compute a permutation ϕ such that $G = \phi(G_b)$, by just setting $\phi = \pi \circ \psi$, thus making V accept. Notice that V accepts with probability 1, even if P$'$ has no idea of whether $(G_0 \approx G_1)$ or $(G_0 \not\approx G_1)$. Clearly this implies that no extractor can be given for (P,V).

3.2 A proof of decision power for GI

An informal description. The common input to prover P and verifier V is a pair of graphs (G_0, G_1). We can divide (P,V) into three basic steps. The first step is done by

V; he randomly chooses a bit b and a graph G isomorphic to G_b, and sends it to P. In the second step, V proves to P that graph G has been correctly constructed, in a witness-indistinguishable fashion, that is, without revealing any information about bit b and the permutation used. In the third step, P checks that V's proof is accepting and then proves to V that he knows an isomorphism between graph G and one of the two input graphs G_0, G_1. V accepts if and only if this proof is convincing.

Implementation. The implementation of the first step is simple. We observe that the second and the third step can be implemented in various ways; perhaps, the simpler is to use the same protocol for both steps. Specifically, P and V will run twice a witness-indistinguishable subprotocol (from [GMR]), where in the first execution (second step of (P,V)) V acts as a prover and P as a verifier, and in the second execution (third step of (P,V)) the roles are reversed. By carefully interleaving such executions, we obtain only 4 rounds of communication between P and V. Let n be an integer and $m = n \log n$; a formal description of (P,V) is in Figure 1. We obtain the following

Theorem 3. *The protocol (P, V) is a 4-round perfect zero-knowledge proof of decision power (with decision error 0) for GI.*

3.3 Proof of main theorem

Clearly, V's program can be performed in polynomial time. Now we prove the three requirements in Definition 1: regularity, validity and perfect zero-knowledge.

Regularity. First of all notice that if P and V behave honestly, then P's verifications in his last step are satisfied with probability 1. This implies that with probability 1 the graph G sent by V in his first step is isomorphic to at least one of G_0, G_1. Now, observe that regardless of whether $G_0 \approx G_1$ or not, the prover can compute an isomorphism between G and one of G_0, G_1 and then meet V's verification in the third step of the protocol. Specifically, if $G_0 \approx G_1$ then G is isomorphic to both, and, say, the permutation between G and G_0 can be used to run his program in the third step of the protocol. Instead, if $G_0 \not\approx G_1$ then G is isomorphic only to G_b, and then the permutation between G and G_b can be computed by P and used to run his program in the third step of the protocol. Thus, in both cases, V accepts with probability 1.

Validity. We show an extractor E which satisfies Definition 1. Recall that E uses as an oracle prover P′ which may deviate arbitrarily from P's program and makes V accept with a certain probability $acc_{P'}(G_0, G_1)$.

The extractor E. On input (G_0, G_1), E starts by running m times a procedure, called Iso-ext, which we now describe.

The procedure Iso-ext takes as input a bit b and returns either a bit v or a special string *fail*. Precisely, each time the procedure is executed, it takes as input a uniformly and independently chosen bit b_i. The procedure starts by repeatedly running the program of the verifier V interacting with P′ until an accepting conversation is obtained. In this conversation P′ has received a graph G chosen by the procedure as isomorphic to G_b; also, P′ has sent some pairs of graphs (D_{i0}, D_{i1}) and answered correctly to V's questions represented by bits e_i. Then the procedure Iso-ext rewinds P′ until after his first step. Now, V's second round is run again by sending some uniformly chosen e_i' instead of the bits e_i sent before (here the procedure also makes sure that the sequence $(e_1', \ldots, e_m')$ is distinct from all previously chosen, including $(e_1, \ldots, e_m)$). This step is

The Protocol (P,V)

Input to P and V: (G_0, G_1), where G_0, G_1 are n-node graphs.

$V1$: Uniformly choose bit b and a permutation π and compute $G = \pi(G_b)$;
for $i = 1, \ldots, m$,
uniformly choose bit a_i and two permutations η_{i0}, η_{i1};
compute graphs $A_{i0} = \eta_{i0}(G_{a_i})$ and $A_{i1} = \eta_{i1}(G_{1-a_i})$;

$P \leftarrow V$: $(G, (A_{10}, A_{11}), \ldots, (A_{m0}, A_{m1}))$.

$P1$: For $i = 1, \ldots, m$,
uniformly choose bits c_i, d_i and permutations ψ_{i0}, ψ_{i1};
compute graphs $D_{i0} = \psi_{i0}(G_{d_i})$ and $D_{i1} = \psi_{i1}(G_{1-d_i})$.

$P \rightarrow V$: $((c_1, \ldots, c_m), (D_{10}, D_{11}), \ldots, (D_{m0}, D_{m1}))$.

$V2$: For $i = 1, \ldots, m$,
uniformly choose a bit e_i;
if $c_i = 0$ then set $\sigma_i = (\eta_{i0}, \eta_{i1})$;
if $c_i = 1$ then set $\sigma_i = \pi \circ \eta_{i,b\oplus a_i}^{-1}$;

$P \leftarrow V$: $((e_1, \ldots, e_m), (\sigma_1, \ldots, \sigma_m))$.

$P2$: For $i = 1, \ldots, m$,
if $c_i = 0$ then
let $\sigma_i = (\eta_{i0}, \eta_{i1})$;
check that $A_{i0} = \eta_{i0}(G_{a_i})$ and $A_{i1} = \eta_{i1}(G_{1-a_i})$, for some bit a_i;
if $c_i = 1$ then check that $G = \sigma_i(A_{i0})$ or $G = \sigma_i(G_{i1})$;
if the above verifications are not satisfied then halt;
if $e_i = 0$ then set $\tau_i = (\psi_{i0}, \psi_{i1})$;
if $e_i = 1$ then
if $G_0 \approx G_1$ then
randomly choose a bit g_i;
compute a permutation τ_i such that $G = \tau_i(G_{g_i})$;
if $G_0 \not\approx G_1$ then
compute bit b and permutation π such that $G = \pi(G_b)$;
set $\tau_i = \pi \circ \psi_{b\oplus d_i}$.

$P \rightarrow V$: $((\tau_1, \ldots, \tau_m))$.

$V3$: For $i = 1, \ldots, m$,
if $e_i = 0$ then
let $\tau_i = (\psi_{i0}, \psi_{i1})$;
check that $D_{i0} = \psi_{i0}(G_{d_i})$ and $D_{i1} = \psi_{i1}(G_{1-d_i})$, for some bit d_i;
if $e_i = 1$ then check that $G = \tau_i(D_{i0})$ or $G = \tau_i(D_{i1})$.
If all verifications are successful then output: ACCEPT else output: REJECT. Halt.

Figure 1: A proof system of decision power for GI

repeated until another accepting conversation is obtained. Now, in the case the procedure never finds a second (or even a first) accepting conversation, then it outputs *fail*. If this does not happen, then this implies that P′ has given answers to bit e_i and bit e_i' corresponding to the same pair of graphs (D_{i0}, D_{i1}), for $i = 1, \ldots, m$. Since there exists an i such that $e_i \neq e_i'$, from the answers to such two distinct bits, the procedure can easily compute an isomorphism ϕ between G and one of G_0, G_1. In this case the output of procedure Iso-ext will be a bit v such that the isomorphism ϕ obtained by P′ is such that $G = \phi(G_v)$.

Now, if procedure Iso-ext has ever output *fail* then E runs a search procedure to find a permutation π such that $G_0 = \pi(G_1)$, or a proof that no such permutation exists; if such a permutation is found, then E outputs 1; if not, E outputs 0.

Instead, consider the case procedure Iso-ext never outputs *fail*. As mentioned above, E runs m times the procedure Iso-ext, each time on input a uniformly chosen bit b_i. Then, let v_i be the bit output by the procedure Iso-ext, when given b_i as input, for $i = 1, \ldots, m$. Then E outputs 0 (meaning that the graphs G_0, G_1 are not isomorphic) if $b_i = v_i$, for $i = 1, \ldots, m$, and 1 (meaning the contrary) otherwise.

The next two lemmas show that the extractor E indeed satisfies Definition 1.

Lemma 4. *For all pairs* (G_0, G_1) *of* n*-node graphs, for all sufficiently large* n, *and all* P', *if* $acc_{P'}(x) > 0$ *then* $\text{Prob}(\text{OUT}_E(z, \text{tr}_{(P',E)}(G_0, G_1)) = \chi_{GI}(G_0, G_1)) \geq 1 - 2^{-n}$.

Proof. First of all we observe that if the extractor E outputs because of the search procedure then clearly its output is correct with probability 1. Now we consider the case in which the extractor E outputs after running n times the procedure Iso-ext. First, assume that $G_0 \not\approx G_1$. In this case, in each execution of procedure Iso-ext, E sends a graph G isomorphic to G_{b_i} to P′; also, procedure Iso-ext finds an isomorphism between G and exactly one of G_0, G_1, which can only be G_{b_i}. Thus, it holds that $v_i = b_i$, for $i = 1, \ldots, m$, and thus E's output is correct with probability 1. Now, assume that $G_0 \approx G_1$. In this case, in each execution of procedure Iso-ext, E sends a graph G isomorphic to G_b to P′, and proves that he knows an isomorphism between G and one of G_0, G_1. Since this proof is witness-hiding (see [FS]), no information is revealed about bit b to any P′, and thus the probability that $v_i = b_i$ is exactly 1/2. This means that the probability that there exists a j such that $b_j \neq v_j$, from which it follows that E's output is correct, is at least $1 - 2^{-m} \geq 1 - 2^{-n}$. □

Lemma 5. *For all pairs* (G_0, G_1) *of* n*-node graphs, for all sufficiently large* n, *and all* P', *if* $acc_{P'}(x) > 0$ *then* E *halts within expected time bounded by* $\frac{|x|^c}{(acc_{P'}(x) - err(x))}$, *for some constant* $c > 0$.

Proof. The first reason E can output is because of the result of the procedure Iso-ext; in this case the expected running time of E is properly bounded, for the following two reasons: 1) at each iteration such procedure essentially runs the program of verifier V, which is strict polynomial time; 2) the expected number of iterations is at most $2/acc_{P'}(G_0, G_1)$. It follows that E's expected time is at most $poly(n)/acc_{P'}(G_0, G_1)$. Now consider the other case, that is, when E outputs because the result of the search procedure; clearly, this procedure may take exponential time. However, this happens when prover P′ makes V accept only in correspondence to one of the sequences $(e_1, \ldots, e_m)$. This implies that in this case the probability $acc_{P'}(G_0, G_1)$ is at most 2^{-m}, and E's expected running time is then $poly(n) \cdot n! \cdot 2^{-m} \leq poly(n)$. □

Perfect zero-knowledge. We show a simulator S which satisfies Definition 1. Recall that S interacts with a verifier V′ (treated as a black box) which may deviate arbitrarily from V's program.

The simulator S. On input (G_0, G_1), S will first of all feed V′ with a random string of appropriate length. Then S obtains the first message from V′ and runs P's program to simulate the first message by P. Then he obtains the second message from V′, which terminates the proof of knowledge from V′. Now, if this proof is convincing, then S uses this proof to extract the knowledge communicated by V′ through this proof. That is, S runs the extractor for the proof of knowledge by V′ and obtains a permutation between G and one of G_0, G_1. Then S simulates the last message by P by running P's program, and using the obtained permutation as auxiliary input. Finally S outputs the conversation thus obtained.

To show that the simulator S indeed satisfies Definition 1, we need to show two properties: first, S's output is distributed exactly as the output of the protocol; second, S's running time is expected polynomial time.

To see that the first property is satisfied, we start by observing that the messages from V′ are clearly equally distributed in both spaces, since they are computed in the same way. The first message from the prover is equally distributed in both spaces since S runs algorithm P to compute it. The second message of the prover is also computed by S using algorithm P; here S uses the permutation extracted from the proof of knowledge by V′ as his auxiliary-input. Although this auxiliary-input may be different from the one used by P during the protocol, the second message by P has the same distribution, no matter which auxiliary-input is used by V, since P is running a witness-indistinguishable proof of knowledge.

To see that the second property is satisfied, we observe that the simulator computes the first message from the prover, by running P's program which is polynomial time here. Then S runs the extractor for the proof of knowledge by V′, which, by properties of proofs of knowledge (see [BG]) we know to run in expected polynomial time. Finally, he uses the witness obtained from this extraction to run P's program in polynomial time and simulate the last step of the protocol.

3.4 Extensions

Random self-reducible languages (RSR) [AFK] are well-known for their use in cryptography and have received a lot of attention also in the context of perfect zero-knowledge proof systems [GMR, GMW1, TW, DDPY]. They include graph isomorphism, quadratic residuosity, subgroup membership modulo a prime and the decision Diffie-Hellman problem. The protocol presented for GI extends to all other known random self-reducible languages. Let L be any of the mentioned RSR languages. We have the following:

Theorem 6. *There exists a 4-round perfect zero-knowledge proof system of decision power with error 0 for L.*

We omit the simple proof of this theorem. Also, combining the technique used in Section 3.2 with constructions in [DDPY], we obtain protocols for the language of monotone formulae over all known random self-reducible languages.

Theorem 7. *There exists a 4-round perfect zero-knowledge proof system of decision power with error 0 for the language $\Phi(L)$ of monotone formulae over membership statements to L.*

4 On the round-complexity of zk proofs of decision power

In this section we investigate the round-complexity of zero-knowledge proofs of decision power. We show that if a language L has a 3-round zero-knowledge proof of decision power with 'sufficiently small' decision error, then L is in BPP. This result is obtained as an extension of the technique in [GoKr], which proved the analogue impossibility result for zero-knowledge proofs of membership. We obtain the following

Theorem 8. *Let L be a language and let (P,V) be a 3-round zero-knowledge proof system of decision power with decision error $err(|x|) \leq 1/q(|x|)$ for L, for any super-polynomial function $q(\cdot)$. Then L is in BPP.*

We observe that the decision error obtained is the best possible since there exist 3-round proofs of decision power with decision error $1/poly(|x|)$ (an example is a parallelization of $O(\log n)$ copies of the atomic protocol in [FFS] for quadratic residuosity). As a particular case, this theorem holds for proof systems of decision power with decision error 0, as the ones obtained in previous section. Thus, we derive the following, perhaps unexpected result, which is is currently not known in the model of proofs of membership.

Theorem 9. *The proof systems of decision power with decision error 0 presented in Section 3.2 for the languages GI, for any known language in RSR, and language $\Phi(L)$ (for any L known in RSR) are round-optimal unless the given language is in BPP.*

4.1 Proof of Theorem 8

Let (P,V) be a 3-round zero-knowledge proof system of decision power with decision error $1/q(\cdot)$, for any super-polynomial function $q(\cdot)$, for L. Assume that, on common input x, P first sends to V a message α, then V responds to P with a message β, and finally P sends to V a message γ. Note that V's message β is constructed from the P's message α and V's (private) random coin tosses r. After the above interactions, V checks whether or not $\rho_V(x, r, \alpha, \gamma) =$ "accept," where ρ_V is a polynomial (in $|x|$) time computable predicate. We assume that $|r| = l(|x|)$, for some polynomial l, and denote by M the simulator associated to (P,V).

Our goal is to construct a BPP algorithm A which decides L. We will accomplish this into two steps. First, we will construct a prover P' which makes the verifier accept with probability at least $1/p(n)$, for some polynomial p, and such that P' runs in probabilistic polynomial time. Then, we using such a prover and the extractor of the assumed proof of decision power to construct algorithm A.

Constructing a poly-time prover. We use the simulator M as a subroutine for constructing the prover P'. Recall the whole simulation process is completely determined by the input x to the protocol, the content of M's random tape R_M, and the responses by the verifier. Let $t(\cdot)$ be a polynomial. Define the following procedure F_M that uses M as a sub-procedure. On input $x \in \{0,1\}^*$, the procedure F_M chooses $R_M \in_R \{0,1\}^{q(|x|)}$ and $r^i \in_R \{0,1\}^{\ell(|x|)}$ $(1 \leq i \leq t)$, then runs M on input x and R_M. If M first generates α^1, then the procedure F_M responds with $\beta^1 = \beta_V(x, r^1, \alpha^1)$. For the *i-th* different string α^i generated by M, F_M responds with $\beta^i = V(x, r^i, \alpha^i)$, and if the same string α^j $(1 \leq j \leq i)$ is generated by M, then F_M responds with β^j as before. For all x, $|x| = n$, all $R_M \in \{0,1\}^{r(n)}$ and all $r_i \in \{0,1\}^{l(n)}$, define a vector $(x, R_M, r_1, \ldots, r_t)$ to be *M-good* if $M(x, R_M, r_1, \ldots, r_t) = (x, r, \alpha, \gamma)$ is an accepting conversation for a verifier V, i.e., $\rho_V(x, r, \alpha, \gamma) =$ "accept," and a vector $(x, R_M, r_1, r_2, \ldots, r_t)$ to be *(M,i)-good* if it is good for M and $\alpha = \alpha_i$, $r = r_i$.

We will use the following lemma from [GoKr] for proofs of membership.

Lemma 10. *[GoKr] Let (P, V) be a 3-round zero-knowledge proof system of membership for language L, and let M be the associated simulator. Then there exists a polynomial $t(\cdot)$ such that*

1. *If $x \notin L$ then a negligible fraction of vectors $(x, R_M, r_1, \ldots, r_t)$ are M-good.*
2. *If $x \in L$ then at least half of the vectors $(x, R_M, r_1, \ldots, r_t)$ are (M, i)-good, for some $i \in \{1, \ldots, t(|x|)\}$.*

The proof of this lemma can be easily adapted to the case of proofs of decision power with decision error $1/q(\cdot)$, for any super-polynomial function $q(\cdot)$, where we obtain:

Lemma 11. *Let (P, V) be a 3-round zero-knowledge proof system of decision power with error $1/q(\cdot)$ for language L, for any super-polynomial function $q(\cdot)$, and let M be the associated simulator. Then there exists a polynomial $t(\cdot)$ such that for all x, at least half of the vectors $(x, R_M, r_1, \ldots, r_t)$ are (M, i)-good, for some $i \in \{1, \ldots, t(|x|)\}$.*

Using this lemma, we construct a prover P' which runs in probabilistic polynomial time and makes the verifier accept with non-negligible probability. We observe that the index i such that a vector $(x, R_M, r_1, \ldots, r_t)$ is *(M,i)-good* may depend on the answers from the (possibly dishonest) verifier. Therefore, the prover P' will randomly choose an index i among the polynomially many possible, and hope that the message received from the verifier corresponds to an (M, i)-good vector (this happens with non-negligible probability). Formally, consider the following interactive proof system (P^*, V):

Protocol (P^*, V)

Input: $x \in \{0,1\}^*$.

P1-1: P^* chooses $R_M \in_R \{0,1\}^{q(|x|)}$ and $r_j \in_R \{0,1\}^{\ell(|x|)}$ $(1 \le j \le t)$.

P1-2: P^* computes $(x, r, \alpha, \gamma) = F_M(x, R_M, r_1, r_2, \ldots, r_t)$.

P1-3: If $\rho_V(x, r, \alpha, \gamma) =$ "reject," then P^* halts; otherwise P^* continues.

P1-4: P^* uniformly chooses an index i such that $1 \le i \le t(|x|)$.

P→ V: $\alpha = \alpha^{(i)}$.

V1: V chooses $r \in_R \{0,1\}^{\ell(|x|)}$, and computes $\beta = \beta_V(x, r, \alpha^{(i)})$.

V → P: β

P2-1: P^* chooses $(t - i)$ strings $\delta_j \in_R \{0,1\}^{\ell(|x|)}$ $(i + 1 \le j \le t)$;
if $F_M(x, R_M, r_1, \ldots, r_{i-1}, r, \delta_{i+1}, \ldots, \delta_t) = (x, r, \alpha, \gamma)$
and $\rho_V(x, r, \alpha, \beta, \gamma) =$ "reject,", then stop.

P → V: γ.

V2: V checks whether or not $\rho_V(x, r, \alpha, \beta, \gamma) =$ "accept."

Note that P^* runs in probabilistic polynomial time since so does procedure F_M. Also, the probability that V accepts is at least the probability that the vector $(x, R_M, r_1, \ldots, r_{i-1},$

$r, \delta_{i+1}, \ldots, \delta_t$) computed in step P2-3 is (M, i)-good. Since i is uniformly chosen by P^*, this probability is at least $1/t(|x|)$, and thus non-negligible.

Construction of the BPP algorithm A **for** L. It follows from the definition of validity that there exists an algorithm E which runs in time $poly(|x|)/(acc_{P^*}(x) - err(x))$ with black-box access to P^* (see the protocol above) that on input x, outputs b such that $\chi_L(x) = b$ with overwhelming probability. Then, on input x, algorithm A just runs algorithm E using prover P^* as a black box.

In order to see that algorithm A runs in probabilistic polynomial time, we observe that the running time of algorithm E is $poly(|x|)/(acc_{P^*}(x) - err(x))$, which is equal to $t(|x|) \cdot poly(|x|)$, and thus polynomial, since $err(x) = 0$ and the acceptance probability of P^* was showed to be at least $1/t(|x|)$.

References

[AFK] M. Abadi, J. Feigenbaum, and J. Kilian, *On Hiding Information from an Oracle*, STOC 87.

[BG] M. Bellare and O. Goldreich, *On Defining Proofs of Knowledge*, CRYPTO '92.

[BMO] M. Bellare, S. Micali, and R. Ostrovsky, *Perfect Zero Knowledge in Constant Rounds*, STOC 90.

[DP] G. Di Crescenzo and G. Persiano, *Round-Optimal Perfect Zero-Knowledge Proofs*, Information Processing Letters, vol. 50, (1994), pag. 93–99.

[DDP] A. De Santis, G. Di Crescenzo, G. Persiano, *The Knowledge Complexity of Quadratic Residuosity Languages*, Theoretical Computer Science, vol. 132, (1994), pag. 291–317.

[DDPY] A. De Santis, G. Di Crescenzo, G. Persiano and M. Yung, *On Monotone Formula Closure of SZK*, FOCS 94.

[FFS] U. Feige, A. Fiat, and A. Shamir, *Zero-Knowledge Proofs of Identity*, Journal of Cryptology, vol. 1, 1988, pp. 77–94.

[FS] U. Feige and A. Shamir, *Witness-Indistinguishable and Witness-Hiding Protocols*, STOC 90.

[Fo] L. Fortnow, *The Complexity of Perfect Zero Knowledge*, STOC 87.

[GHY] Z. Galil, S. Haber, and M. Yung, *Minimum-Knowledge Interactive Proofs for Decision Problems*, SIAM Journal on Computing, vol. 18, n.4, pp. 711–739 (previous version in FOCS 85).

[GoKr] O. Goldreich and H. Krawczyk, *On the Composition of Zero-Knowledge Proof Systems*, ICALP 1990.

[GMW1] O. Goldreich, S. Micali, and A. Wigderson, *Proofs that Yield Nothing but their Validity or All Languages in NP Have Zero-Knowledge Proof Systems*, Journal of the ACM, vol. 38, n. 1, 1991, pp. 691–729.

[GMW2] O. Goldreich, S. Micali, and A. Wigderson, *How to play any mental game*, STOC 88.

[GMR] S. Goldwasser, S. Micali, and C. Rackoff, *The Knowledge Complexity of Interactive Proof-Systems*, SIAM Journal on Computing, vol. 18, n. 1, February 1989.

[Sa] K. Sakurai, *A hidden cryptographic assumption in no-transferable identification schemes*, Asiacrypt 96.

[TW] M. Tompa and H. Woll, *Random Self-Reducibility and Zero-Knowledge Interactive Proofs of Possession of Information*, FOCS 87.

[Ya] A. Yao, *Theory and Applications of Trapdoor Functions*, FOCS 85.

[Yu] M. Yung, *Zero-Knowledge Proofs of Computational Power*, Eurocrypt 89.

Computational Learning Theoreitc Cryptanalysis of Language Theoretic Cryptosystems

Takeshi Koshiba

High Performance Computing Research Center,
Fujitsu Laboratories Ltd.,
4-1-1 Kamikodanaka, Nakahara-ku, Kawasaki-shi, Kanagawa 211-88, Japan.
Email: koshiba@flab.fujitsu.co.jp

Abstract. From the point of view of computational learning theory, we analyze the security of Richelieu cryptosystems, which are based on language theory. Richelieu cryptosystems use generating systems of slender languages as key generators. We consider the problem of learning slender languages using queries and additional information. We show that families of slender languages are not polynomial-time learnable by only using membership queries, but certain subfamilies of slender languages are polynomial-time learnable from representative samples and membership queries. Based on these results, we discuss the security of Richelieu cryptosystems.

1 Introduction

Security of recent cryptosystems such as RSA and ElGamal ones is based on the difficulty of problems on number theory and group theory. The difficulty is defined in terms of computational aspects in number theory and algebra. On the other hand, language theoretic or semigroup theoretic notions can be used as bases for cryptosystems. Security of cryptosystems based on language theory is measured with computational aspects of language theory. *Computational learnability* or *inferability* of languages is a means to measure the security of the cryptosystems.

Let us consider the relation between security of cryptosystems and learnability of languages. Let $\mathcal{E} = \{\mathcal{E}_k\}_{k\geq 1}$ be a cryptosystem, where each $\mathcal{E}_k$ consists of a key space $\mathcal{K}_k$, a plaintext space $\mathcal{P}_k$ and a ciphertext space $\mathcal{C}_k$. Each key $K \in \mathcal{K}_k$ represents a subset of $\mathcal{P}_k \times \mathcal{C}_k$. We denote by $L(K)$ this subset. Roughly speaking,

> $\mathcal{E}$ is secure if there is no (randomized) polynomial time algorithm A such that for any $K \in \mathcal{K}_k$ and for any small finite subset S of $L(K)$ that is drawn according to some fashion, the probability
>
> $$\Pr[\, A(S) = K' \in \mathcal{K}_k \text{ and } K \sim K' \,]$$
>
> is high, where $K \sim K'$ means, for almost all plaintext P, $(P, C) \in L(K)$ if and only if $(P, C) \in L(K')$.

Attacks such as "known plaintext attack" and "chosen ciphertext attack" can be defined by specifying fashions of drawing S from $L(K)$.

On the other hand, we informally define notions of learnability of languages. Usually, the notion of learnability of languages is defined in terms of their representation. Languages L, sets of strings over a finite alphabet Σ, are generated by generating systems such as term rewriting systems or grammar systems. Representation of languages means the description of term rewriting systems or grammar systems of the languages. Let G be a generating system and $L(G)$ be a language generated by G. Let $\{\mathcal{G}\}_{k\geq 1}$ be a family of generating systems and $\mathcal{L}$ be the corresponding language class. That is, $\mathcal{L} = \{L(G) : G \in \mathcal{G}\}$. Sometimes, we find it convenient to view languages as a subset $\bar{L}$ of $\Sigma^* \times \{0,1\}$, where $(w,1) \in \bar{L}$ if $w \in L$, $(w,0) \in \bar{L}$ otherwise. Roughly speaking,

> $\mathcal{L}$ is learnable in terms of $\mathcal{G}$ if there is a (randomized) polynomial time algorithm A such that for any $G \in \mathcal{G}_k$ and for any small finite subset S of $L(G)$ that is drawn according to some fashion, the probability
>
> $$\Pr[\ A(S) = G' \in \mathcal{G}_k \text{ and } G \sim G'\ ]$$
>
> is high, where $G \sim G'$ means, for almost all strings w, $(w,b) \in \bar{L}(G)$ if and only if $(w,b) \in \bar{L}(G')$.

For any cryptosystem $\mathcal{E}$ and any family of generating systems $\mathcal{G}$, if, for each k, $L(\mathcal{K}_k)$ and $L(\mathcal{G}_k)$ are very related to each other, a learning algorithm for $\mathcal{G}$ may break the cryptosystem $\mathcal{E}$, where $L(\mathcal{K}_k)$ (resp., $L(\mathcal{G}_k)$) denotes $\{L(K) : K \in \mathcal{K}_k\}$ (resp., $\{L(G) : G \in \mathcal{G}_k\}$). In such a situation, cryptosystems $\mathcal{E}$ must be secure against any type of learning algorithms for $\mathcal{G}$.

Let us consider some concrete cryptosystems. Formal language theoretic study of the classic Richelieu cryptosystem (see, e.g., [12]) has arisen from [1]. It is found that slender languages are useful to the key management in the cryptographic frame. Here, a language is said to be m-thin if the cardinality of the set of all strings with length n is bounded by a constant positive integer m for every $n \geq 0$. A language is said to be slender if it is m-thin for some positive integer m. In a recent series of papers [7, 10, 11], several interesting properties of slender languages have been investigated. These properties enable us to analyze the security of the cryptosystem from the point of view of computational learning theory.

Before we see the relation between security of Richelieu cryptosystems and learnability of slender languages, wee see how Richelieu cryptosystems work. In order to decrypt the message from a ciphertext, a key of the same length as the ciphertext is used in Richelieu cryptosystems. If the set of keys forms a slender language then the legal receiver has only to know its generating system such as a grammar system. That is, by checking at most k strings of a given length, the receiver can find the corresponding key and decrypt the ciphertext. Further details about Richelieu cryptosystems will be discussed in Section 3.

If an eavesdropper could use an efficient learning algorithm for slender languages then he might attack such cryptosystems easily. If a grammar system which generates keys is identified then all ciphertext strings can be decrypted by the eavesdropper. Secure cryptosystems must be robust against various kinds of attacks. We can say that a paradigm of learnability, "identifiability in the limit" [4] of languages, corresponds with "insecureness" of the cryptosystem against "known plaintext attack" by the following reason. The opponent who uses known plaintext attack possesses a string of plaintext and the corresponding ciphertext. From a string of plaintext and the corresponding ciphertext, he can construct the corresponding key. If he can possess a sequence of plaintext strings and the corresponding sequence of ciphertext strings, he can also obtain the corresponding sequence of keys. If the key generating system is efficiently learnable from any presentation of target key K, he might forge the cryptosystem by using the learning algorithm. (Presentation means an infinite sequence of elements in $L(K)$ and learning algorithms use initial segments of presentation.) Moreover, we can say that another paradigm of learnability, "identifiability by using membership queries" (see, e.g., [2]) corresponds with "insecureness" of the cryptosystem against "chosen ciphertext attack" by the following reason. The opponent who uses chosen ciphertext attack has obtained temporary access to the decryption machinery. Hence he can choose a ciphertext string and construct the corresponding plaintext string. Since the corresponding key is obtained from a ciphertext string and the corresponding plaintext string, he can construct a set of keys. If the key generating system is efficiently learnable by using membership queries, he might forge the cryptosystem by using the learning algorithm and membership queries. In such a situation, the study of learning problems plays an important role.

In this paper, we consider the learning problem of slender context-free languages using queries. In other words, we consider the security of Richelieu cryptosystems equipped with generating systems that generate slender context-free languages. The results in case of slender context-free languages are applicable to the case of slender regular languages or slender context-sensitive languages. We show that the family of slender context-free languages is not polynomial-time learnable by only using membership queries, but a certain subfamily of slender context-free languages is polynomial-time learnable from representative samples and membership queries. We show that the results of learnability of slender context-free languages can be extended to slender context-sensitive languages of some properties. This implies that Richelieu cryptosystems equipped with generating systems that generate slender context-free languages or slender context-sensitive languages of the properties are not always be secure.

2 Preliminaries

Here, we give preliminary notations and provide necessary properties. Let Σ denote an alphabet, Σ^* the set of all strings over Σ including the empty string λ and Σ^n the set of all strings over Σ of length n. For each string s, $|s|$ denotes the length of s. For each set S, $card(S)$ denotes the cardinality of S.

For any integer $m \geq 1$, a language L is said to be *m-thin*, if $card(L \cap \Sigma^n) \leq m$

for each n. A language is said to be *slender*, if it is m-thin for some integer m. A language is said to be *thin*, if it is 1-thin.

A language L is called a *finite union of paired loops* if, for some integer $k \geq 1$ and strings u_i, v_i, w_i, x_i, y_i with $1 \leq i \leq k$,

$$L = \bigcup_{i=1}^{k} \{u_i v_i^n w_i x_i^n y_i \mid n \geq 0\}.$$

A finite union of paired loop is said to be *disjoint* if any two sets in the union are pairwise disjoint.

Proposition 1. [7, 11] *A language L is slender context-free if and only if L is a finite union of paired loops.*

Proposition 2. [10] *A language L is a finite union of paired loops if and only if L is a disjoint finite union of paired loops.*

Since, in this paper, we mainly consider the security of Richelieu cryptosystems equipped with generating systems that generate slender context-free languages, we simply note properties of slender languages other than context-free languages. A language L is called a *finite union of single loops* if, for some integer $k \geq 1$ and strings u_i, v_i, w_i with $1 \leq i \leq k$, $L = \bigcup_{i=1}^{k} \{u_i v_i^n w_i \mid n \geq 0\}$. In [10], it has been shown that a language L is slender regular if and only if L is a finite union of single loops. Some slender context-sensitive languages can be described by a finite union of terms of the form $x_1 y_1^i x_2 y_2^i \cdots x_n y_n^i x_{n+1}$ and are generated by simple regulated rewriting systems, say, linear grammar systems with slender regular control sets [8]. If, for some integers $k, m \geq 1$ and strings $x_{(i,1)}, y_{(i,1)}, x_{(i,2)}, y_{(i,2)}, \ldots, x_{(i,m)}, y_{(i,m)}, x_{(i,m+1)}$ with $1 \leq i \leq k$, $L = \bigcup_{i=1}^{k} \{x_{(i,1)} y_{(i,1)}^n x_{(i,2)} y_{(i,2)}^n \cdots x_{(i,m)} y_{(i,m)}^n x_{(i,m+1)} \mid n \geq 0\}$, then L is called an *m-tuple loops language*. Note that any m-tuple loops languages are slender and context-sensitive.

A string s is said to be *primitive* if s is not a power of another string, that is, $s \neq \lambda$ and $s \in \{t^n \mid n \geq 0\}$ for some string $t \in \Sigma^*$ implies $s = t$. If a nonempty string t is not primitive then there exists a primitive string s such that $t = s^k$ for some $k \geq 2$. In this case, s is called a *primitive substring* of t and denoted by $\rho(t)$. Moreover, $|t|/|\rho(t)|$ is denoted by $\alpha(t)$. We set $\rho(\lambda) = \lambda$ and $\alpha(\lambda) = 1$ for the sake of simplicity. Two strings s and t are said to be *conjugate* if there exist strings $u, v \in \Sigma^*$ such that $s = uv$ and $t = vu$.

Proposition 3. [6] *Let s and t be primitive strings. Assume that*

$$us^n w_1 = vt^m w_2, \quad |u| \geq |v|, \quad n|s| \geq |st|, \quad m|t| \geq |t^2 s| + |u| - |v|.$$

Then s and t are conjugate and, for any integer $k \geq 0$, $us^{n+k} w_1 = vt^{m+k} w_2$.

3 Formal Language Theoretic Cryptosystem

In the cryptosystem customarily attributed to Richelieu, both the sender and the legal receiver have identical sheets of cardboard with holes. When the sheet is positioned on top of the ciphertext, the plaintext becomes visible through the holes. This is a special case of the cryptographic method referred to as "garbage-in-between" [12]. Some letters of the ciphertext, specified according to their position, are significant and the other letters are meaningless.

The decryption in Richelieu cryptosystems can be formulated in terms of the operation of *guided filtering*. Let x and y be strings of equal length n over Σ and $\{0,1\}$, respectively. Denote by $s_i(x)$ and $N_a(x)$ the ith letter of x and the number of occurrences of a letter a in x, respectively. Guided filtering of x through y is defined by $gf(x,y) = a_1a_2\cdots a_n$, where, for $1 \le i \le n$,

$$a_i = \begin{cases} s_i(x) \text{ if } s_i(y) = 1, \\ \quad\lambda \quad \text{if } s_i(y) = 0. \end{cases}$$

Intuitively, x is the ciphertext and y is the sheet, where the holes are at the positions indicated by occurrences of the letter 1.

The string y can be viewed as the decryption key: plaintext results when guided filtering is applied to the ciphertext and the key. As usual in classical cryptography, in addition to security considerations, one has to deal with problems of key management. While it depends on the encryption method whether the plaintext can be recovered from the ciphertext without a filtering key, the effect of shuffling should be reversed even without the exact knowledge of a key.

Assume that the message x is encrypted to the ciphertext z by scrambling x with the string y. A desirable situation would be that x is easily recovered from z without the exact knowledge of y.

Consider strings over Σ, and let Γ be a subalphabet of Σ. Extend the notation $N_a(x)$ to concern subalphabets by

$$N_\Gamma(x) = \sum_{a\in\Gamma} N_a(x).$$

The operation of *guided sparse substitution* is defined for strings x and y satisfying $N_\Gamma(y) \ge |x| = m$ as follows:

$$gss(y,x) = y_1a_1y_2a_2\cdots y_ma_my_{m+1},$$

where

$$y = y_1b_1y_2b_2\cdots y_mb_my_{m+1}, \; x = a_1a_2\cdots a_m,$$
$$y_i \in (\Sigma - \Gamma)^*, \; b_i \in \Gamma, \; a_i \in \Sigma \text{ for all } i.$$

Assume now that the strings y used as encryption keys come from a language L over Σ such that L contains at most one string, or some constant k strings, of any given length n with finitely many exceptions. Assume further that the legal receiver knows L and the designated subset Γ. Given a ciphertext $gss(y,x) = z$, the legal receiver can find y since $|y| = |z|$. If $|z|$ is one of the exceptional values of n, the legal receiver has to try finitely many possible y's. Of course, the exceptional values of n may also be avoided in the encryption.

The occurrences of the letters of Γ in y indicate the "position of the holes in the cardboard." The plaintext can be read through the holes. Since we have only $N_\Gamma(y) \geq |x|$ and not necessarily $N_\Gamma(y) = |x|$, we obtain in this fashion a string xw, where the original plaintext occurs as a prefix. The prefix which is actually intended has to be found out in some other fashion, such as considering meaningfulness. Note that it would be too dangerous to require $N_\Gamma(y) = |x|$ from the point of view of secrecy.

We can see that languages discussed in this section are thin languages or slender languages defined in the previous section and that these languages play central roles in cryptosystems such as Richelieu cryptosystems [1].

4 Learning Slender Context-free Languages Using Queries

In this section, we mainly consider the polynomial-time identifiability of slender context-free languages by using queries. If slender context-free languages are learnable in polynomial-time, then Richelieu cryptosystems equipped with generating systems that generate slender context-free languages are not secure. Since slender context-free languages are finite unions of paired loops, i.e., 2-tuple loops languages, all the results are extendable to m-tuple loops languages for any integer $m \geq 1$. That is, polynomial-time learnability of m-tuple loops languages implies that Richelieu cryptosystems equipped with generating systems that generate m-tuple loops languages are not secure.

Let L be a slender context-free language. A *representative sample* S of L is a finite set of strings $\{u_1v_1w_1x_1y_1, u_2v_2w_2x_2y_2, \ldots, u_kv_kw_kx_ky_k\}$ such that $L = \bigcup_{i=1}^{k}\{u_iv_i^jw_ix_i^jy_i \mid j \geq 0\}$. If the sets in the union are pairwise disjoint, S is called a δ-representative sample of L. From Proposition 2, there exists a δ-representative sample for each slender context-free language. Although it is usual to consider that slender context-free languages are generated by context-free grammar systems, we use sets of quintuples of strings as generating systems for slender context-free languages. If T is a set of quintuples of strings, $L(T)$ denotes $\{uv^iwx^iy \mid (u,v,w,x,y) \in T, i \geq 0\}$. The *size* of the generating system T of L is the sum of the length of each quintuples in T. Queries we consider are (1) membership queries and (2) membership queries and additional information, that is, representative samples. We denote by $memQ(w)$ an oracle that answers whether a string w belongs to the target language or not.

Theorem 4. *There is no algorithm A such that, for any n and for any slender context-free language generated by a generating system of size n, when A runs with an oracle to answer membership queries for any strings, A halts and outputs a hypothesis T such that $L = L(T)$ and the time used by A is bounded by a polynomial in n.*

Proof. Since any singleton $L = \{w\}$ is slender context-free, we have only to consider the identifiability of the family of singletons using membership queries.

Even if an algorithm knows the length of w, it asks $card(\Sigma)^{|w|} - 1$ membership queries in the worst case to identify L.

Here, we consider the polynomial-time learnability of slender context-free languages by using membership queries and additional information, since the family of slender context-free languages is not polynomial-time learnable by using membership queries only.

The following observation shows that the family of slender languages is not polynomial-time learnable by using membership queries and any representative samples.

Example 1. Let $p_1, p_2, \ldots, p_k$ be pairwise prime integers. Let

$$T = \left(\bigcup_{j=1}^{k} \{(a^i, a^{p_j}, \lambda, b^{p_j}, b^i) \mid i = 0, 1, \ldots, p_j - 2\} \right) \cup \{(a, \lambda, \lambda, \lambda, b)\}.$$

Then $L = L(T)$ is a slender context-free language and $S = \{uvwxy \mid (u, v, w, x, y) \in T\}$ is a representative sample for L. We can easily show that the shortest string of the form $a^n b^n$ that is not in L is $a^{p_1 p_2 \cdots p_k - 1} b^{p_1 p_2 \cdots p_k - 1}$. The length of this string is not bounded by a polynomial in the sum of the length of all strings in S. This is actually bounded by a sub-exponential (see, e.g., [3]).

In this case, if representative samples are restricted to δ-representative samples, the difficulty may be solved. In general, this restriction may not bring solutions to overcome the difficulty. The following observation shows that the family of slender languages may not be polynomial-time learnable by using membership queries and any δ-representative samples.

Example 2. Let $T = \{(\lambda, a^6, \lambda, b^6, \lambda),\ (a, a^{10}, \lambda, b^{10}, b),\ (a^3, a^{10}, \lambda, b^{10}, b^3),\ (a^5, a^{10}, \lambda, b^{10}, b^5),\ (a^2, a^{15}, \lambda, b^{15}, b^2),\ (a^4, a^{15}, \lambda, b^{15}, b^4),\ (a^7, a^{15}, \lambda, b^{15}, b^7),\ (a^{14}, a^{15}, \lambda, b^{15}, b^{14})\}$. We can easily show that $S = \{uvwxy \mid (u, v, w, x, y) \in T\}$ is a δ-representative sample of $L(T)$. We assume that $t = (a^p, a^q, \lambda, b^q, b^p)$ is a quintuple such that $S \cup \{a^{p+q} b^{p+q}\}$ is a δ-representative sample of a slender context-free language. Then q must be equal to 30, which is the least common multiple of 6, 10 and 15. This means that, for some p, any strings of the form $a^{p+30} b^{p+30}$ do not occur in $L(T)$. As stated in Example 1, the least common multiple of numbers is not bounded by any polynomial in the sum of the numbers.

Here, we consider subfamilies of slender context-free languages that are polynomial-time learnable by using membership queries and representative samples.

Let T be a set of quintuples of strings. T is said to be *pairwise unconjugate* if for any distinct quintuples $(u_1, v_1, w_1, x_1, y_1)$ and $(u_2, v_2, w_2, x_2, y_2)$ in T, none of the following cases holds:

case 1. $\rho(v_1)$ and $\rho(v_2)$ are conjugate, $\rho(x_1)$ and $\rho(x_2)$ are conjugate, and $\rho(v_1)\rho(x_1) \neq \lambda$ or $\rho(v_2)\rho(x_2) \neq \lambda$,

case 2. some three of $\rho(v_1)$, $\rho(v_2)$, $\rho(x_1)$ and $\rho(x_2)$ are conjugate and the other is empty,
case 3. $\rho(v_1)$ and $\rho(x_2)$ are conjugate and $\rho(v_2)\rho(x_1) = \lambda$,
case 3'. $\rho(v_2)$ and $\rho(x_1)$ are conjugate and $\rho(v_1)\rho(x_2) = \lambda$.

If a slender context-free language L has a representative sample S such that $S = \{uvwxy \mid (u, v, w, x, y) \in T\}$ and T is a pairwise unconjugate set of quintuples of strings, S is called a *pairwise unconjugate representative sample* of L and L is said to be *pairwise unconjugate.*

Example 3. Let $L = \{(abb)^n c(dd)^n \mid n \geq 0\} \cup \{(bab)^n cdd^n \mid n \geq 0\} \cup \{e^n fe^n \mid n \geq 0\}$. From Proposition 1, L is slender context-free. Let $T = \{(\lambda, abb, c, dd, \lambda), (\lambda, bab, cd, d, \lambda), (\lambda, e, f, e, \lambda)\}$ be a set of quintuples. $\{uvwxy \mid (u, v, w, x, y) \in T\}$ is a representative sample of L, but T is not a pairwise unconjugate set of quintuples, because $\rho(abb) = abb$, $\rho(bab) = bab$, $\rho(dd) = \rho(d) = d$ and abb and bab are conjugate. It is not hard to see that any set T' of quintuples such that $\{uvwxy \mid (u, v, w, x, y) \in T'\}$ is a representative sample of L is not pairwise unconjugate. This implies that L is not pairwise unconjugate.

Pairwise unconjugate slender context-free languages seem to be somewhat artificial and tailored for a special purpose. Since almost all languages in the family of slender context-free languages are pairwise unconjugate, and since languages which are not pairwise unconjugate are unsuitable for cryptographic purpose from the point of view of secrecy, the following theorem is meaningful from the point of view of analysis of the real problem.

Theorem 5. *There is an algorithm A such that, for any n and for any pairwise unconjugate slender context-free language L generated by a generating system of size n, when A runs with an oracle to answer membership queries for strings, A is given a pairwise unconjugate representative sample as an input, halts and outputs a hypothesis T such that $L = L(T)$ and the time used by A is bounded by a polynomial in n.*

Proof. Consider the algorithm LPUSCF in Figure 1. Let L be a target pairwise unconjugate slender context-free language. Assume that $S = \{u_1v_1w_1x_1y_1, \ldots, u_kv_kw_kx_ky_k\}$ is given as a pairwise unconjugate representative sample such that $L = \bigcup_{i=1}^{k}\{u_iv_i^jw_ix_i^jy_i \mid j \geq 0\}$. Let l be the size of the representative sample. For each string s in the representative sample and for each factorization (u, v, w, x, y) of s such that $s = uvwxy$, the algorithm asks membership queries. First, we show the following claim.

Claim. If $\{uv^jwx^jy \mid 5l < j \leq 5l + l^2\} \subseteq L$ then $\{uv^jwx^jy \mid j > 5l\} \subseteq L$.

Since the claim holds for the case $|vx| = 0$, we assume that $|vx| \geq 1$. For each m ($5l < m \leq 5l + l^2$), there exist integers i_m and n such that $uv^mwx^my = u_{i_m}v_{i_m}^nw_{i_m}x_{i_m}^ny_{i_m}$. It is easy to verify that the assumption in Proposition 3 holds for $m > 5l$. In the case where $|v| = 0$, from Proposition 3, one of the following statements holds:

- $\rho(x)$,$\rho(v_{i_m})$ and $\rho(x_{i_m})$ are conjugate,
- $\rho(x)$ and $\rho(v_{i_m})$ are conjugate and $\rho(x_{i_m}) = \lambda$,
- $\rho(x)$ and $\rho(x_{i_m})$ are conjugate and $\rho(v_{i_m}) = \lambda$.

In the case where $|x| = 0$, we obtain similar results as above. In the case where $|v| \geq 1$, $|x| \geq 1$ and $\rho(v)$ and $\rho(x)$ are conjugate, one of the following statements holds:

- $\rho(v)$, $\rho(x)$, $\rho(v_i)$ and $\rho(x_{i_m})$ are conjugate,
- $\rho(v)$, $\rho(x)$ and $\rho(v_{i_m})$ are conjugate and $\rho(x_{i_m}) = \lambda$,
- $\rho(v)$, $\rho(x)$ and $\rho(x_{i_m})$ are conjugate and $\rho(v_{i_m}) = \lambda$.

In the case where $|v| \geq 1$, $|x| \geq 1$, and $\rho(v)$ and $\rho(x)$ are not conjugate, we obtain that $\rho(v)$ and $\rho(v_{i_m})$ are conjugate and $\rho(x)$ and $\rho(x_{i_m})$ are conjugate. In any case, from Proposition 3, we can say that, for each m $(5l < m \leq 5l + l^2)$,

$$L_m = \{uv^j wx^j y \mid j = m + j' \cdot \mathrm{lcm}\{\alpha(z), \alpha(z_{i_m})\}/\alpha(z),\ j' \geq 0\} \subseteq L,$$

where z (resp., z_{i_m}) is a nonempty string and equal to either v or x (resp., either v_{i_m} or x_{i_m}). Note that if both v and x (resp., v_{i_m} and x_{i_m}) are nonempty then $\alpha(v) = \alpha(x)$ (resp., $\alpha(v_{i_m}) = \alpha(x_{i_m})$). Since S is pairwise unconjugate, $\alpha(z_{i_m})$ is constant for $m = 5l+1, \ldots, 5l+l^2$. This implies that $\mathrm{lcm}\{\alpha(z), \alpha(z_{i_{5l+1}}), \ldots, \alpha(z_{i_{5l+l^2}})\} \leq l^2$. Since consecutive l^2 strings, which are from $uv^{5l+1}wx^{5l+1}y$ to $uv^{5l+l^2}wx^{5l+l^2}y$, are in $\bigcup_{m=5l+1}^{5l+l^2} L_m$, we can say that

$$\{uv^j wx^j y \mid j > 5l\} = \bigcup_{m=5l+1}^{5l+l^2} L_m \subseteq L.$$

This completes the proof of the claim.

From the above claim, the algorithm does not output unnecessary quintuples of strings. On the other hand, the algorithm enumerates all possible quintuples of strings. Therefore, the algorithm identifies the target language using $O(n^2)$ membership queries and a given representative sample whose size is $O(n)$. This completes the proof.

We generalize the notion of pairwise unconjugateness. Let T be a set of quintuples of strings. T is said to be *c-conjugate*, if $c = \max\{card(T') \mid T' \subseteq T$ and T' is not pairwise unconjugate$\}$. Note that T is pairwise unconjugate if and only if T is 1-conjugate. If a slender context-free language L has a representative sample S such that $S = \{uvwxy \mid (u, v, w, x, y) \in T\}$ and T is a c-conjugate set of quintuples of strings, S is called a *c-conjugate representative sample* of L and L is said to be *c-conjugate*.

Theorem 6. *There is an algorithm A such that, for any n and for any c-conjugate slender context-free language L generated by a generating system of size n, when A runs with an oracle to answer membership queries for strings, A is given a c-conjugate representative sample as an input, halts and outputs a hypothesis T such that $L = L(T)$ and the time used by A is bounded by a polynomial in n.*

```
Procedure LPUSCF
input: a representative sample {s_1, s_2, ..., s_k}
output: a set of quintuples of strings

begin
l := Σ_{i=1}^{k} |s_i|;
foreach s in {s_1, s_2, ..., s_k}
   foreach (u, v, w, x, y) is a factorization of s such that s = uvwxy
      if |vx| = 0 then output (u, v, w, x, y)
      else
         begin
            i := 5l + l^2;
            while (memQ(uv^i wx^i y) = "yes" and i > 5l + 1) do i := i - 1;
            if memQ(uv^i wx^i y) = "no" then continue;
            while (memQ(uv^i wx^i y) = "yes" and i > 0) do i := i - 1;
            if memQ(uv^i wx^i y) = "no" then output (uv^{i+1}, v, wx^{i+1}, x, y);
            while (i ≥ 0) do begin
               if memQ(uv^i wx^i y) = "yes" then output (uv^i, λ, wx^i, λ, y);
               i := i - 1;
            end;
         end;
end.
```

Fig. 1. Algorithm LPUSCF

Proof. Similar discussion to the proof of Theorem 5 completes the proof.

We have considered slender context-free languages. For m-tuple loops languages, representative samples and similar notion as pairwise unconjugateness can be defined. All the results about learnability can be extended to the family of all m-tuple loops languages for any fixed m. Due to limitations of space, we omit the details.

5 Concluding Remarks

LPUSCF is an algorithm to learn a subfamily of slender context-free languages. The algorithm is also applicable to the family of slender context-free languages. Let L be a target slender context-free language. In this case, LPUSCF always outputs a hypothesis T' such that $L(T') \supseteq L$. So, we may say that LPUSCF is an approximate learning algorithm for the family of slender context-free languages. From the point of view of cryptanalysis, even this naive approximate algorithm is one evidence of insecureness of the cryptosystems.

Observe Example 1 again. If we make LPUSCF run with S as its input, LPUSCF outputs a hypothesis which represents $\{a^i b^i \mid i \geq 0\}$. This hypothesis is practically a good approximation for L.

We have shown that a subfamily of slender context-free languages can be identified using queries and from additional information in polynomial time. This reveals a weakness of slender context-free languages (or, m-tuple languages) as cryptosystems. As one of ways to make the Richelieu cryptosystems robust, we have to consider another family of slender languages.

Acknowledgment

The author is grateful to Yuji Takada, Kazuhiro Yokoyama, Yasubumi Sakakibara and Hirokazu Anai for their valuable comments and suggestions.

References

1. M. Andraşiu, G. Păun, J. Dassow, and A. Salomaa. Language-theoretic problems arising from Richelieu cryptosystems. *Theoretical Computer Science*, 116(2):339–357, 1993.
2. D. Angluin. Queries and concept learning. *Machine Learning*, 2(4):319–342, 1988.
3. J. Berstel and M. Mignotte. Deux propriétés décidables des suites récurrentes linéaires. *Bulletin de la Société Mathématique de France*, 104(2):173–184, 1976.
4. E. M. Gold. Language identification in the limit. *Information and Control*, 10(5):447–474, 1967.
5. M. A. Harrison. *Introduction to Formal Language Theory*. Addison-Wesley, Reading:Massachusetts, 1978.
6. J. I. Hmelevskii. Equations in free semigroups. *Proceedings of the Steklov Institute of Mathematics*, 107:1–270, 1976.
7. L. Ilie. On a conjecture about slender context-free languages. *Theoretical Computer Science*, 132(1-2):427–434, 1994.
8. T. Koshiba. On a hierarchy of slender languages based on control sets. To appear in *Fundamenta Informaticae*.
9. M. Lothaire. *Combinatorics on Words*. Addison-Wesley, Reading:Massachusetts, 1983.
10. G. Păun and A. Salomaa. Thin and slender languages. *Discrete Applied Mathematics*, 61(3):257–270, 1995.
11. D. Raz. On slender context-free languages. In *Lecture Notes in Computer Science (STACS'95)*, Vol. 900, pp. 445–454. Springer-Verlag, 1995.
12. A. Salomaa. *Public-Key Cryptography*. Springer-Verlag, Berlin, 1990.

A Language for Specifying Sequences of Authorization Transformations and Its Applications

Yun Bai and Vijay Varadharajan

Distributed System and Network Security Research Unit
Department of Computing
University of Western Sydney, Nepean
P.O.Box 10, Kingswood 2747, Australia
Email: {ybai,vijay}@st.nepean.uws.edu.au

Abstract. A formal language to specify authorization policies and their transformations has been proposed in [1]. The authorization policy was specified using a policy base which consisted of a finite set of facts and a finite set of access constraints. In this paper, we modify the language to consider a sequence of authorization policy transformations. The syntax and semantics of the modified authorization policy language is presented. The central issue addressed in this paper is as follows: given a policy base and a sequence of transformations, what is the resulting policy base after performing the sequence of transformations? The language is able to represent incomplete information and allows denials to be expressed explicitly. We also use the proposed language to specify a variety of well known access control policies such as static separation of duty, dynamic separation of duty and Chinese wall security policy.

Key words: Authorization Policies, Formal Language, Example Policy Specifications, Policy Transformations

1 Introduction

A formal language to specify authorization policies and their transformations has been proposed in [1]. Authorization policy provides the ability to limit and control access to system, applications and information, and to limit what entities can do with the information and resources. The language proposed was based on a first order language. The authorization policy was specified using a policy base which consisted of a finite set of facts and a finite set of access constraints. The facts represented explicitly the access rights the subjects held for the objects. The access constraints, on the other hand, are rules that should be satisfied by the authorization policy. The concept of model(s) of the policy base was used for checking the consistency of the policy base and for performing transformations of authorizations under the model-based semantics [4]. In this paper, we modify the language to consider a sequence of transformations. In particular, we address the following issue: given a policy base and a sequence of transformations, what is the resulting policy base after performing a sequence of transformations? Both the syntax and the semantics of the modified language are given. Then we use these new constructs to specify a variety of well known authorization policies such as separation of duty, Chinese wall security policy and dynamic separation of duty.

To simplify our presentation, in this paper, we assume the existence of a system security officer administering the authorization transformations. This assumption enables us to concentrate on a single administering agent system and hence avoid the problem of coordination among multiagents.

The paper is organized as follows. Section 2 describes the modified language $\mathcal{L}$ by outlining both its syntax and semantics and gives some well known authorization policies such as separation of duty, Chinese wall security policy and dynamic separation of duty specified by the language. Section 3 illustrates examples of transformations of authorizations. Section 4 discusses some important properties of the language $\mathcal{L}$. Finally, section 5 concludes the paper with some final remarks.

2 Authorization Policy Language $\mathcal{L}$

This section specifies the syntax and semantics of the modified language $\mathcal{L}$.

2.1 Syntax of $\mathcal{L}$

Let $\mathcal{L}$ be a sorted language with six disjoint sorts for *subject, group-subject, access-right, group-access-right* and *object, group-object* respectively together with some other predicate symbols and connectives:

1. Sort *subject* with subject constants $S, S_1, S_2, \cdots$, and subject variables $s, s_1, s_2, \cdots$.
2. Sort *group-subject* with group subject constants $G, G_1, G_2, \cdots$, and group subject variables $g, g_1, g_2, \cdots$.
3. Sort *access-right* with access right constants $A, A_1, A_2, \cdots$, and access right variables $a, a_1, a_2, \cdots$.
4. Sort *group-access-right* with group access right constants $GA, GA_1, GA_2, \cdots$, and group access right variables $ga, ga_1, ga_2, \cdots$.
5. Sort *object* with object constants $O, O_1, O_2, \cdots$, and object variables $o, o_1, o_2, \cdots$.
6. Sort *group-object* with group object constants $GO, GO_1, GO_2, \cdots$, and group object variables $go, go_1, go_2, \cdots$.
7. A ternary predicate symbol *holds* which takes arguments as *subject, object or group-object* and *access-right or group-access-right* respectively.
8. A ternary predicate symbol *g-holds* which takes arguments as *group-subject, object or group-object* and *access-right or group-access-right.* respectively.
9. A binary predicate symbol $\in$ which takes arguments as *subject* and *group-subject* or *access-right* and *group-access-right* or *object* and *group-object* respectively.
10. A binary predicate symbol $\subseteq$ whose both arguments are *group-subjects, group-access-rights* or *group-objects.*

11. Logical connectives $\wedge$ and $\neg$.

In language $\mathcal{L}$, the fact that a subject S has access right R for object O is represented using a ground atom $holds(S, R, O)$. On the other hand, the fact that a subject S is a member of G is represented by $S \in G$. Similarly, we represent inclusion relationships between subject groups such as $G_1 \subseteq G_2$ or between access right groups such as $GA_1 \subseteq GA_2$.

In general, we define a *fact* F to be an atomic formula of $\mathcal{L}$ or its negation, while a *ground fact* is a fact without variable occurrence. We view $\neg\neg F$ as F. *Fact expressions* of $\mathcal{L}$ are defined as follows: (i) each fact is a fact expression; (ii) if ϕ and ψ are fact expressions, then $\phi \wedge \psi$ is also a fact expression. A *ground fact expression* is a fact expression without variable occurrence. A ground fact expression is called a *ground instance* of a fact expression if this ground fact expression is obtained from the fact expression by replacing each of its variable occurrence with the same sort constant. Now we are ready to formally define propositions of $\mathcal{L}$.

A *policy proposition* of $\mathcal{L}$ is an expression of the form

$$\phi \textbf{ after } T_1, \cdots, T_m, \tag{1}$$

where ϕ is a ground fact expression and $T_1, \cdots, T_m$ $(m \geq 0)$ are transformation names. Intuitively, this proposition means that after performing transformations $T_1, \cdots, T_m$ sequentially, the ground fact expression ϕ holds. If $m = 0$, we will rewrite (1) as

$$\textbf{initially } \phi, \tag{2}$$

which is called *initial policy proposition.*

A *transformation proposition* is an expression of the form

$$T \textbf{ causes } \phi \textbf{ if } \psi, \tag{3}$$

where T is a transformation name, ϕ and ψ are ground fact expressions. Intuitively, a transformation proposition expresses the following meaning: at a given state, if the pre-condition ψ is true[1], then after performing the transformation T at this state, the ground fact expression ϕ will be true in the resulting state.

A *policy domain description* D in $\mathcal{L}$ is a finite set of initial policy propositions and transformation propositions.

Example 1. The following is a domain description:

$$\textbf{initially } \; holds(S, Read, O) \wedge holds(S, Write, O),$$
$$Delete\text{-}write(S, Write, O) \;\; \textbf{causes} \;\; \neg holds(S, Write, O).$$

This domain description expresses the following information: initially subject S has *Read* and *Write* rights on object O. A transformation named *Delete-write*$(S, Write, O)$ is available in this domain; if this transformation occurs, then the subject S will no longer have the *Write* right for object O.

[1] We will formally define a state and the semantics of a transformation proposition in the next subsection.

Example 2. Let us now consider the static separation of duty access policy. This policy refers to the fact that a certain set of access rights cannot be allowed for the same subject. For example, consider the operations Submit, Evaluate and Approve associated with a Budget object. A static separation of duty policy requires that the same subject cannot be authorized for all of the three operations.

Let S be the subject, B be the budget object. The domain description is specified as follows:

initially $\neg(holds(S, Submit, B) \wedge holds(S, Evaluate, B) \wedge holds(S, Approve, B))$.

Example 3. Let us now consider the more general access policy on dynamic separation of duty.

In this case, a subject can potentially execute any operation in a given set, though s/he cannot execute all of them. By executing some, s/he will automatically rule out the possibility of executing the others. The policy is referred to as dynamic in the sense that which actions a user can execute is determined by the user. For instance, consider the following simple example. Let a group *officer* be represented using a group-subject *G-Officer*. Let this group have access rights to submit, evaluate and approve a budget. Let the budget be represented using an object B. Now if a subject S belongs to *G-Officer*, that is, $S \in$ *G-Officer*, then $holds(S, Submit, B)$, $holds(S, Evaluate, B)$ and $holds(S, Approve, B)$. Let the transformations be $Rqst(S, Submit, B)$, $Rqst(S, Evaluate, B)$ and $Rqst(S, Approve, B)$. The domain description D can be represented as follows:

initially $S \in$ *G-Officer*,
initially $holds(S, Submit, B)$,
initially $holds(S, Evaluate, B)$,
initially $holds(S, Approve, B)$,

$Rqst(S, Evaluate, B)$ **causes**
$\neg holds(S, Approve, B)$
$\neg holds(S, Submit, B)$

$Rqst(S, Approve, B)$ **causes**
$\neg holds(S, Evaluate, B)$
$\neg holds(S, Submit, B)$

$Rqst(S, Submit, B)$ **causes**
$\neg holds(S, Evaluate, B)$
$\neg holds(S, Approve, B)$

Example 4. Let us now consider the specification of the Chinese wall access policy [3] using our domain description. The Chinese wall access policy can be viewed as a special kind of dynamic separation of duty. In Chinese wall policy, objects are grouped into *company datasets*, for instance Company-1 and Company-2.

Company datasets whose organizations are in competition are then grouped together into *conflict of interest classes*. If a subject accesses an object in a company dataset 1, it cannot be allowed anymore to access any object in a company dataset that appear in a conflict of interest class with dataset 1. In our language, company datasets can be represented by a *group-object*. For instance, if $Company_1$ and $Company_2$ are in the same conflict of interest class, a subject who has accessed an object of $Company_1$ will not be allowed to access any object in $Company_2$ and vice versa.

Suppose O_1 is an object of $Company_1$ and O_2 is an object of $Company_2$ and S is a subject. Initially, let us assume that S can access both O_1 and O_2. Let us assume that we have the following transformations: $Rqst(S, Access, O_1)$ and $Rqst(S, Access, O_2)$. The domain description D is specified as follows:

initially $O_1 \in Company_1$,
initially $O_2 \in Company_2$,
initially $holds(S, Access, O_1)$,
initially $holds(S, Access, O_2)$,

$Rqst(S, Access, O_1)$ **causes**
$\neg holds(S, Access, O_2)$
if $O_1 \in Company_1 \wedge O_2 \in Company_2$,

$Rqst(S, Access, O_2)$ **causes**
$\neg holds(S, Access, O_1)$
if $O_1 \in Company_1 \wedge O_2 \in Company_2$,

That is, if S accesses O_1, then it will not be able to access O_2 due to the transformation $Rqst(S, Access, O_1)$. Similarly, if S accesses O_2, it will not be able to access O_1.

Note that the full language described in [2] has constraints and quantifiers for the general statements such as "for any subject S" and "for any object O". Here we only use the simplified specification to illustrate the examples specified by the language.

2.2 Semantics of $\mathcal{L}$

Now we define the semantics of language $\mathcal{L}$. A *state* is a set of ground facts. Given a ground fact F (i.e. F is $holds(S, R, O)$ or $\neg holds(S, R, O)$) and a state σ, we say F is *true* in σ iff $F \in \sigma$, and F is *false* in σ iff $\neg F \in \sigma$. A ground fact expression $\phi \equiv F_1 \wedge \cdots \wedge F_k$, where each F_i $(1 \leq i \leq k)$ is a ground fact, is *true* in σ iff each F_i $(1 \leq i \leq k)$ is in σ. Furthermore, a fact expression with variables is true in σ iff each of its ground instances is true in σ. A state σ is *complete* if for any ground fact F of $\mathcal{L}$, F or $\neg F$ is in σ. Otherwise σ is called a *partial state*. An *inconsistent state* σ is a state containing a pair of complementary ground facts F and $\neg F$.

A *transition function* ρ maps a set of (T, σ) into a set of states, where T is a transformation name and σ is a state. Intuitively, $\rho(T, \sigma)$ denotes the resulting state caused by performing transformation T in σ. A *structure* M is a pair (σ, ρ), where σ is a state, and ρ is a transition function. For any structure M and any set of transformations $T_1, \cdots, T_m$, the notation $M^{T_1,\cdots,T_m}$ denotes the state

$$\rho(T_m, \rho(T_{m-1}, \cdots, \rho(T_1, \sigma) \cdots)),$$

where ρ is the transition function of M, and σ is the state of M.

We denote a policy proposition (given by 1) *is satisfied* in a structure M as $M \models_{\mathcal{L}} \phi$ **after** $T_1, \cdots, T_m$. This is true iff ϕ is true in the state $M^{T_1,\cdots,T_m}$. Given a domain description D, we say that a state σ_0 is the *initial state* of D iff (i) for each initial policy proposition, **initially** ϕ of D, ϕ is true in σ_0; (ii) if there is another state σ satisfying condition (i), then $\sigma_0 \subseteq \sigma$ (i.e. σ_0 is the least state satisfying all initial policy propositions of D).

Definition 1. A structure (σ_0, ρ) is a *model* of a domain description D iff σ_0 is a consistent initial state of D (i.e. $\sigma_0 \neq |\mathcal{L}|$), and for any transformation T and state σ, the following conditions hold:

(i) if D includes a transformation proposition T **causes** ϕ **if** ψ, and ψ is true in σ, then ϕ is true in $\rho(T, \sigma)$;
(ii) if D does not include such transformation proposition, then for each ground fact F, $F \in \rho(T, \sigma)$ iff $F \in \sigma$.

We say that a domain description D is *consistent* if D has a model. A policy proposition ϕ **after** $T_1, \cdots, T_m$ *is entailed* by D, denoted as $D \models_{\mathcal{L}} \phi$ **after** $T_1, \cdots, T_m$, iff it is true in each model of D. A model M of D is *complete* if for any policy proposition ϕ **after** $T_1, \cdots, T_m$, either $M \models_{\mathcal{L}} \phi$ **after** $T_1, \cdots, T_m$ or $M \models_{\mathcal{L}} \neg\phi$ **after** $T_1, \cdots, T_m$.

Proposition 2. *The following results hold:*

(i) A consistent domain description D has a unique model;
(ii) D has a complete model iff D has a complete initial state.

Example 5. Continuation of Example 3. For the dynamic separation of duty example, the initial state of D is:

$$\sigma_0 = \{S \in \textit{G-Officer}, holds(S, Submit, B), holds(S, Evaluate, B), \\ holds(S, Approve, B)\}.$$

It can be easily shown that the following results hold:

$$D \models_{\mathcal{L}} S \in \textit{G-Officer} \wedge holds(S, Submit, B) \wedge \neg holds(S, Evaluate, B) \wedge \\ \neg holds(S, Approve, B) \\ \textbf{after } Rqst(S, Submit, B),$$
$$D \models_{\mathcal{L}} S \in \textit{G-Officer} \wedge holds(S, Evaluate, B) \wedge \neg holds(S, Submit, B) \wedge \\ \neg holds(S, Approve, B)$$

$$\begin{aligned} &\textbf{after } Rqst(S, Evaluate, B),\\ D \models_{\mathcal{L}}\ & S \in G\text{-}Officer \wedge holds(S, Approve, B) \wedge \neg holds(S, Submit, B) \wedge \\ &\neg holds(S, Evaluate, B)\\ &\textbf{after } Rqst(S, Approve, B). \end{aligned}$$

Example 6. Continuation of Example 4. For the Chinese wall policy, the initial state of D is:

$$\begin{aligned} \sigma_0 = \{&O_1 \in Company_1, O_2 \in Company_2, holds(S, Access, O_1)\\ &holds(S, Access, O_2)\}. \end{aligned}$$

From the above description, it is not difficult to show that

$$\begin{aligned} D \models_{\mathcal{L}}\ & O_1 \in Company_1 \wedge O_2 \in Company_2 \wedge holds(S, Access, O_1) \wedge \\ &\neg holds(S, Access, O_2)\\ &\textbf{after } Rqst(S, Access, O_1),\\ D \models_{\mathcal{L}}\ & O_1 \in Company_1 \wedge O_2 \in Company_2 \wedge holds(S, Access, O_2) \wedge \\ &\neg holds(S, Access, O_1)\\ &\textbf{after } Rqst(S, Access, O_2). \end{aligned}$$

3 Document Release Example Revisited

In this section, we consider a slightly modified version of the well-known document release example[5] and specify the authorization transformations using our language $\mathcal{L}$.

Example 7. The following is cited from [5]:

> "$\cdots$ *a scientist creates a document and hence gets* own, read *and* write *access rights to it. After preparing the document for publication, the scientist asks for a review from a patent officer. In the process, the scientist loses the* write *right to the document, since it is clearly undesirable for a document to be edited during or after a (successful) review. After review of the document, the patent officer grants the scientist an approval. It is reasonable to disallow further attempts to review the document after an approval is granted. Thus the* review *right for the document is lost as approval is granted. After obtaining approval from the patent officer, the scientist can publish the document by getting a* release *right for the document.* $\cdots$"

We use subject constant *Sci* to denote the scientist, subject constant *PO* to denote the patent officer , object constant *Doc* to denote the document, access right constants *Own, Read, Write, Review, Pat-ok, Pat-reject, Release* to denote the rights *own, read, write, review, patent-ok, patent-reject* and *release* respectively. We also have the following transformations $Rqst(Sci, Doc, PO)$, $Get\text{-}approval(Sci, Doc, PO)$, $Get\text{-}rejection(Sci, Doc, PO)$, $Release\text{-}doc(Sci, Doc)$ and $Revise\text{-}doc(Sci, Doc)$. The domain description D expressing the access policy within our framework is given as follows:

initially $holds(Sci, Own, Doc)$,
initially $holds(Sci, Read, Doc)$,
initially $holds(Sci, Write, Doc)$,

$Rqst(Sci, Doc, PO)$ **causes**
$holds(PO, Review, Doc) \wedge \neg holds(Sci, Write, Doc)$
if $holds(Sci, Own, Doc) \wedge holds(Sci, Write, Doc)$,

$Get\text{-}approval(Sci, Doc, PO)$ **causes**
$holds(Sci, Pat\text{-}ok, Doc) \wedge \neg holds(PO, Review, Doc)$
if $holds(PO, Review, Doc) \wedge holds(Sci, Own, Doc)$,

$Get\text{-}rejection(Sci, Doc, PO)$ **causes**
$holds(Sci, Pat\text{-}reject, Doc) \wedge \neg holds(PO, Review, Doc)$,
if $holds(PO, Review, Doc) \wedge holds(Sci, Own, Doc)$,

$Release\text{-}doc(Sci, Doc)$ **causes**
$holds(Sci, Release, Doc) \wedge \neg holds(Sci, Pat\text{-}ok, Doc)$
if $holds(Sci, Pat\text{-}ok, Doc)$,

$Revise\text{-}doc(Sci, Doc)$ **causes** $holds(Sci, Write, Doc)$
if $holds(Sci, Pat\text{-}reject, Doc)$.

The initial state of D is

$$\sigma_0 = \{holds(Sci, Own, Doc), holds(Sci, Read, Doc), holds(Sci, Write, Doc)\}.$$

Let us now consider the policy propositions that are entailed from D. From the semantics presented in section 2.2, we can prove the following theorem.

Theorem 3. *The following results hold.*

$D \models_{\mathcal{L}} holds(PO, Review, Doc) \wedge \neg holds(Sci, Write, Doc)$
after $Rqst(Sci, Doc, PO)$,
$D \models_{\mathcal{L}} holds(Sci, Pat\text{-}ok, Doc) \wedge \neg holds(PO, Review, Doc)$
after $Rqst(Sci, Doc, PO)$, $Get\text{-}approval(Sci, Doc, PO)$,
$D \models_{\mathcal{L}} holds(Sci, Pat\text{-}reject, Doc) \wedge \neg holds(PO, Review, Doc)$
after $Rqst(Sci, Doc, PO)$, $Get\text{-}rejection(Sci, Doc, PO)$,
$D \models_{\mathcal{L}} holds(Sci, Release, Doc) \wedge holds(Sci, Pat\text{-}ok, Doc)$
after $Rqst(Sci, Doc, PO)$, $Get\text{-}approval(Sci, Doc, PO)$, $Release\text{-}doc(Sci, Doc)$,
$D \models_{\mathcal{L}} holds(Sci, Write, Doc)$

after *Rqst*(*Sci*, *Doc*, *PO*), *Get-rejection*(*Sci*, *Doc*, *PO*),
Revise-doc(*Sci*, *Doc*).

The results presented in Theorem 1 describe the expected solutions with respect to the execution of different sequences of transformations.

Furthermore the following theorem shows that the performance of sequence of transformations will not affect the scientist's own right for the document.

Theorem 4. *The following results hold.*

$D \models_{\mathcal{L}}$ *holds*(*Sci*, *Own*, *Doc*)
after *Rqst*(*Sci*, *Doc*, *PO*),
$D \models_{\mathcal{L}}$ *holds*(*Sci*, *Own*, *Doc*)
after *Rqst*(*Sci*, *Doc*, *PO*), *Get-approval*(*Sci*, *Doc*, *PO*),
$D \models_{\mathcal{L}}$ *holds*(*Sci*, *Own*, *Doc*)
after *Rqst*(*Sci*, *Doc*, *PO*), *Get-rejection*(*Sci*, *Doc*, *PO*),
$D \models_{\mathcal{L}}$ *holds*(*Sci*, *Own*, *Doc*)
after *Rqst*(*Sci*, *Doc*, *PO*), *Get-approval*(*Sci*, *Doc*, *PO*),
Release-doc(*Sci*, *Doc*),
$D \models_{\mathcal{L}}$ *holds*(*Sci*, *Own*, *Doc*)
after *Rqst*(*Sci*, *Doc*, *PO*), *Get-rejection*(*Sci*, *Doc*, *PO*),
Revise-doc(*Sci*, *Doc*).

4 Properties of Language $\mathcal{L}$

Let us now consider some of the important properties of the language $\mathcal{L}$.

- *Incomplete information is allowed.* In $\mathcal{L}$, the state of the authorization policies can be specified to be *incomplete* in the sense that some authorization facts may not be represented in the state. For instance, in Example 7 presented in section 3, the initial state σ_0 is incomplete because it does not include the facts *holds*(*PO*, *Review*, *Doc*) or ¬*holds*(*PO*, *Review*, *Doc*).
- *Denials are expressed explicitly.* As incomplete information is allowed in the state of authorization policies, denials (negations of authorization policies) must be explicitly represented in the state.
- *The entailment relation* $\models_{\mathcal{L}}$ *of* $\mathcal{L}$ *is nonmonotonic* with respect to transformation propositions. Recall that a domain description D is a finite set of policy propositions and transformation propositions. The nonmonotonicity of $\models_{\mathcal{L}}$ with respect to transformation propositions states that adding more transformation propositions into D may result in a policy proposition being no longer entailed in the domain description. This is because a new transformation proposition may change an authorization policy to become negative. For example, consider a domain description D consisting of the following policy and transformation propositions:
  initially *holds*(*S*, *Read*, *File*),
  Assign-write(*S*, *Write*, *File*) **causes** *holds*(*S*, *Write*, *File*).

Clearly, we have $D \models_{\mathcal{L}}$ *holds*$(S, Write, File)$ **after** *Assign-write*$(S, Write, File)$. However, if we add another transformation proposition into D:

Delete-write$(S, Write, File)$ **causes** $\neg$*holds*$(S, Write, File)$
if *holds*$(S, Write, File)$,

Then we have $D' \models_{\mathcal{L}} \neg$*holds*$(S, Write, File)$ **after**
Assign-write$(S, Write, File)$,
Delete-write$(S, Write, File)$.

where D' is the new domain description with the above transformation proposition added into D.

– $\models_{\mathcal{L}}$ *is also nonmonotonic with respect to policy propositions.* This can be observed from the following example. Suppose a domain description D consists of the following policy and transformation propositions:

initial *holds*$(S, Own, File)$,
Delete-own$(S, Own, File)$ **causes** $\neg$*holds*$(S, Own, File)$ **if** $S \in G$.

As the pre-condition $S \in G$ of *Delete-own*$(S, Own, File)$ does not hold in the initial state, we have $D \models_{\mathcal{L}}$ *holds*$(S, Own, File)$ **after** *Delete-own*$(S, Own, File)$. However, if we add a policy proposition $S \in G$ into D to form a new domain description D', then we have

$D' \models_{\mathcal{L}} \neg$*holds*$(S, Own, File)$ **after** *Delete-own*$(S, Own, File)$.

5 Concluding remarks

In this paper, we have proposed a modified higher level language $\mathcal{L}$ to specify sequences of authorization transformations. We have shown that language $\mathcal{L}$ has a simple syntax and semantics, but is powerful enough to represent some well known access policy examples involving sequences of authorization transformations. Using the definition of policy proposition, we are able to compute both the final state after performing a sequence of transformations as well as any intermediate state within the sequence of transformations. The language can represent incomplete information and allow denials to be represented explicitly. The entailment relation $\models_{\mathcal{L}}$ of $\mathcal{L}$ has a nonmonotonic property with respect to both policy propositions and transformation propositions.

Though the language is powerful, it can be improved still further. We now briefly mention some of the additional features of the language that we are currently working on. In the full version of our manuscript [2], we have extended language $\mathcal{L}$ to a new high level language $\mathcal{L}^d$ which has additional *default propositions.* An example of a default proposition is as follows: if ϕ is true in a state σ and it can not be derived that γ is true in σ, then we can infer that ψ is true in σ. The *default proposition* has the form:

$$\phi \textbf{ implies } \psi \textbf{ with absence } \gamma, \tag{4}$$

When γ is missing from the proposition, we rewrite (4) as

$$\phi \textbf{ implies } \psi, \tag{5}$$

which is a special case of default proposition and is viewed as a *causal* or *inheritance* relation between ϕ and ψ. When both γ and ϕ are missing from the proposition, we rewrite (4) as

$$\textbf{always } \psi, \tag{6}$$

which represents a *constraint* that should be satisfied by any state of the domain. So the domain description D in $\mathcal{L}^d$ is a finite set of initial policy propositions, transformation propositions and default propositions.

References

1. Y. Bai and V. Varadharajan, A logic for state transformations in authorization policies. In *the Proceedings of the 10th IEEE Computer Security Foundations Workshop*, pp 173-182, Massachusetts, June, 1997.
2. Y. Bai and V. Varadharajan, An Authorization Policy Language : Syntax and Semantics, Department of Computing, University of Western Sydney, Nepean, May 1997.
3. D.F.C.Brewer and M.J.Nash, The Chinese wall security policy. In *Proceedings of IEEE Symposium on Security and Privacy*, pp 215-228, Oakland, May 1989.
4. T.S-C. Chou, M. Winslett, Immortal: a Model-based Belief Revision System, *The 2nd International Conference on Principles of Knowledge Representation and Reasoning*, Morgan Kaufman Publishers Inc. pp 99–110, 1991.
5. R.S. Sandhu and S. Ganta, On the Minimality of Testing for Rights in Transformation Models. In *Proceedings of IEEE Symposium on Research in Security and Privacy*, pp 230-241, 1994.

On the Decomposition Constructions for Perfect Secret Sharing Schemes *

Hung-Min Sun[1] and Bor-Liang Chen[2]

[1] Department of Information Management, Chaoyang University of Technology, Wufeng, Taichung County, Taiwan 413, e-mail: hmsun@dec8.cyut.edu.tw
[2] Department of Mathematics, Tung-Hai University, Taichung, Taiwan 407, e-mail: blchen@s867.thu.edu.tw

Abstract. We propose the concept of weight-decomposition construction for perfect secret sharing schemes. This construction is more general than previous constructions. Based on the weight-decomposition construction, we improve the information rate in 4 cases of the left unsolved 18 cases of secret sharing schemes for connected graphs on six vertices. In addition, we also propose some efficient decomposition constructions for perfect secret sharing schemes with access structures of constant rank. Compared with the best previous constructions, our constructions have some improved lower bounds on the information rate.

1 Introduction

A secret sharing scheme is a method which allows a secret K to be shared among a set of participants $\boldsymbol{P}$ in such a way that only *qualified* subsets of participants can recover the secret [1,9]. The information kept by each participant is called share. The collection of subsets of participants that can reconstruct the secret in this way is called *access structure*, denoted by Γ. It is natural to require Γ to be monotone, that is, if $X \in \Gamma$ and $X \subseteq X' \subseteq \boldsymbol{P}$, then $X' \in \Gamma$. A minimal qualified subset $Y \in \Gamma$ is a subset of participants such that $Y' \notin \Gamma$ for all $Y' \subset Y$, $Y' \neq Y$. The basis of Γ, denoted by Γ_0, is the family of all minimal qualified subsets. For any $\Gamma_0 \subseteq 2^P$, the *closure* of Γ_0 is defined to be $cl(\Gamma_0) = \{ X' : \exists X \in \Gamma_0, X \subseteq X' \subseteq \boldsymbol{P} \}$. Therefore, an access structure Γ is the same as the closure of its basis Γ_0, $cl(\Gamma_0)$. A secret sharing scheme is called *perfect* if unqualified subsets of participants obtain no information regarding the secret [5,7]. It means that the prior probability $p(K = K_0)$ equals the conditional probability $p(K = K_0 \mid$ given any shares of an unqualified set). By using the entropy function H from [8], we can state the requirements for a perfect secret sharing scheme as follows:

(1) any qualified subset can reconstruct the secret:

$$\forall_{X \in \Gamma} \; H(K|X) = 0, \text{ and}$$

* This research was supported in part by the National Science Council of the Republic of China under grant number NSC-86-2213-E-324-002.

(2) any unqualified subset has no information on the secret:

$$\forall_{X \notin \Gamma}\ H(K|\,X) = H(K).$$

An important issue in the implementation of perfect secret sharing schemes is the size of shares. Let K be the secret space and **S** be the maximum share space. The information rate for the secret sharing scheme is defined as $\rho = \log_2 |\mathbf{K}| / \log_2 |\mathbf{S}|$ [5]. The information rate for share S_i is defined as $\rho_i = \log_2 |\mathbf{K}| / \log_2 |\mathbf{S}_i|$ where $\mathbf{S}_i$ is the share space for S_i. We will use the notation $PS(\Gamma_0, \rho, q)$ to denote a perfect secret sharing scheme with access structure $cl(\Gamma_0)$ and information rate ρ for a set of q keys. Given any access structure Γ, Ito et al. [9,10] showed that there exists a perfect secret sharing scheme to realize the structure. Benaloh and Leichter [1] proposed a different algorithm to realize secret sharing schemes for any given monotone access structure. In both constructions, the information rate decreases exponentially as a function of n, the number of participants. After that, many researchers focused on studying the perfect secret sharing scheme for graph-based access structure Γ having basis Γ_0, where Γ_0 is the collection of the pairs of participants corresponding to edges [2-7,11-12]. Among these constructions, Stinson [12] proposed the idea of decomposition construction which is more general than previous constructions [2-6,11]. In addition, he proved that, for any graph G with n vertices having maximum degree d, there exists a perfect secret sharing scheme for the access structure based on G in which the information rate is at least $2/(d+1)$. Recently, van Dijk [7] showed that Stinson's lower bound is tight because he proved that there exists graphs having maximum degree d such that the optimal information rate is at most $2/(d+1-\varepsilon)$ for all $d \geq 3$ and $\varepsilon > 0$.

The optimal information rates of secret sharing schemes for all graphs on at most five vertices were determined in recent years [3,11]. In [7], van Dijk studied the information rate of secret sharing schemes for connected graphs on six vertices. In 94 of the 112 connected graphs on six vertices he determined the exact values of the optimal information rate.

The *rank* of an access structure Γ is the maximum cardinality of a minimal qualified subset. An access structure is *uniform* if every minimal qualified subset has the same cardinality. Therefore, the graph-based access structure is the case of access structure with rank two. Perfect secret sharing schemes with access structures of constant rank were studied by Stinson [11]. He applied Steiner systems to construct perfect secret sharing schemes with access structures of rank three. The constructed secret sharing scheme has the information rate $\rho \geq \dfrac{4}{(n-1)(n-2)}$ if Γ is non-uniform and $n \equiv 2, 4 \pmod 6$, or $\rho \geq \dfrac{6}{(n-1)(n-2)}$ if Γ is uniform and $n \equiv 2, 4 \pmod 6$, where n is the number of participants [11]. Note that if n doesn't satisfy the condition: $n \equiv 2, 4 \pmod 6$, it is necessary to find an $n' > n$ such that $n' \equiv 2, 4 \pmod 6$. Based on the edge-colourings of bipartite graphs, Stinson also studied the construction of secret sharing schemes with access structures of rank m. The constructed secret sharing schemes have the information rate $\rho \geq \dfrac{m}{(2m-1)\cdot\binom{n-1}{m-2}+d}$, where n is the number of participants and d is the maximum degree of any participant [11].

In this paper, we propose the weight-decomposition construction which is more general than previous constructions. Based on the proposed weight-decomposition construction, we can improve the information rate in 4 cases of the left unsolved 18 cases of secret sharing schemes for connected graphs on six vertices. In addition, we also propose some efficient decomposition constructions for perfect secret sharing schemes with access structures of constant rank. If Γ is an access structure (either uniform or non-uniform) of rank three on n participants, we show that there exists a secret sharing scheme with information rate $\rho \geq \dfrac{6}{(n-1)^2+2}$, for $n \geq 5$. If Γ is an access structure of rank m on n participants, we show that there exists a secret sharing scheme with information rate $\rho \geq \dfrac{n-m+1}{\binom{n}{m}}$. Compared with the best previous constructions [11], our constructions have some improved lower bounds on the information rate.

2 Preliminaries

Suppose Γ is an access structure having basis Γ_0. A λ-decomposition of Γ_0 consists of a collection $\{\Gamma_1, \ldots, \Gamma_t\}$ such that the following requirements are satisfied.

(1) $\Gamma_h \subseteq \Gamma_0$ for $1 \leq h \leq t$.

(2) For each $X \in \Gamma_0$, there exist at least λ indexes $i_1 < \cdots < i_\lambda$ such that $X \in \Gamma_{i_j}$ for $1 \leq j \leq \lambda$.

Let P_h be the set of participants in a scheme with access structure $cl(\Gamma_h)$. Stinson [12] proposed the decomposition construction (DC) for secret sharing schemes. The proposed construction is more general than other well-known constructions [2-6].

Theorem 2.1 [12]: (*Decomposition Construction*, DC) Let Γ be an access structure on n participants, having basis Γ_0, and suppose that $\{\Gamma_1, \ldots, \Gamma_t\}$ is a λ-decomposition of Γ_0. Assume that for each access structure $cl(\Gamma_h)$, there exists a perfect secret sharing scheme with information rate ρ_{ih} for each $p_i \in P_h$, and a set of q keys. Then there exists a $PS(\Gamma_0, \rho, q^\lambda)$, where

$$\rho = \min\left\{\frac{\lambda}{\sum_{\{h: p_i \in P_h\}} (1/\rho_{ih})} : 1 \leq i \leq n\right\}.$$

Let DC_1 and DC_2 be two different decomposition constructions of an access structure. We say that DC_1 dominates DC_2 if $\rho_i^1 \geq \rho_i^2$ for all participant p_i, where ρ_i^1 is the information rate for p_i in DC_1 and ρ_i^2 is the information rate for p_i of DC_2.

Let's consider the case when the basis of an access structure is a graph and Γ_i's are complete multipartite graphs. Because there exists a *PS*(G, ρ=1, q) for any complete multipartite graph [5], we can obtain the following theorem.

Theorem 2.2[7,12]: Suppose access structure G is a graph with vertex set V and edge set E for which a complete multipartite covering exists, say $\Pi=\{G_1, \ldots, G_t\}$. For each vertex $v\in V$ define $R_v = |\{i: v\in V_i\}|$, where V_i denotes the vertex set of G_i. For each edge $e\in E$ define $T_e = |\{i: e\in E_i\}|$, where E_i denotes the edge set of G_i. Let $R = \max\{R_v : v\in V\}$ and $T = \min\{T_e : e\in E\}$. Then there exists a $PS(G, \rho, q^T)$, where q is a prime power and $\rho \geq T/R$.

By decomposing graph into stars, Stinson [12] showed that for any graph G with n vertices having maximum degree d, there exists a perfect secret sharing scheme for the access structure in which the information rate is at least $2/(d+1)$. In the following, we propose a construction which is similar to the one proposed by Stinson [12].

We assume that $\boldsymbol{P} = \{p_1, p_2, \cdots, p_n\}$ is the set of participants corresponding to the vertices of the graph G, and the secret $K = (K_1, K_2)$ is taken randomly from GF(q)× GF(q), where q is a prime and $q > n$. Let $f(x) = K_2 x + K_1 \pmod q$. y_i is computed from $f(x)$ as follows: $y_i = f(i) \pmod q$, for $i = 1, \ldots, n$.

Obviously, given y_i and y_j, for $i \neq j$, $f(x)$ can be determined uniquely. Therefore, one who gets two or more y_i's can recover the secret K. However, one without knowledge of any y_i obtains no information about the secret. Note that one who gets one y_i can obtain partial information about the secret.

The dealer selects n random numbers, $r_1, \ldots, r_n$, over GF(q). The share of participant p_i is given by

$S_i = < a_{i,1}, \cdots, a_{i,t}, \cdots, a_{i,n} >$, where $1 \leq t \leq n$,

$a_{i,t} = r_i \pmod q$ if $t = i$,

$a_{i,t} = r_t + y_t \pmod q$ if $\overline{p_i p_t}$ is an edge of G, and

$a_{i,t}$ is empty if $t \neq i$ and $\overline{p_i p_t}$ is not an edge of G.

The share of participant p_i is an n-dimensional vector. Except that $a_{i,j}$'s (for all j, $\overline{p_i p_j} \notin E(G)$) are empty, every $a_{i,j}$ is over GF(q). Therefore, the size of share S_i is $\log(q^{d_i+1})$, where d_i is the degree of vertex p_i of G. The maximal size of the shares is $\log(q^{d+1})$, where d is the maximum degree of G. The size of the secret is $\log(q^2)$. Thus, the information rate of the secret sharing scheme is $\rho = \dfrac{2\cdot\log q}{(d+1)\cdot\log q} = \dfrac{2}{d+1}$.

3 Weight-decomposition Construction

A weighted graph is a graph with weights associated with the edges. We use the notation $W_e(G)$ to denote the weight of edge e in graph G, and $W(G)$ to denote the maximum weight of all weights of edges in graph G. We define that a perfect secret sharing scheme for weighted graph $G=(V, E)$ with integer weight $W_e(G)$ ($W_e(G) \geq 1$) for each $e \in E$, is a secret sharing scheme which satisfies the following requirements:

(1) pair of participants corresponding to an edge e of G obtains $W_e(G)/W(G)$ information on the secret, i.e.,
$H(K|X) = (1 - W_e(G)/W(G)) \cdot H(K)$ for X corresponding to edge $e \in E$, and

(2) pair of participants corresponding to a non-edge of G obtains no information on the secret, i.e.,
$H(K|X) = H(K)$ for X corresponding to a non-edge of G.

Note that the secret sharing scheme for graph structure is the special case when $W_e(G)=1$ for all $e \in E$. It is also remarked that the size of shares may be smaller than the size of secret for a perfect secret sharing scheme with weighted graph.

A secret sharing scheme for weighted graph can be realized by the same concept of decomposition construction. As an example, we demonstrate the construction of the secret sharing scheme for the weighted graph G in Figure 1 by decomposition construction technique. By examining all DCs, we find that there exist three DCs which are not dominated by other DCs. These three DCs are listed in Figure 2, Figure 3, and Figure 4. It is clear that the information rates for shares in these DCs are:

Case I: $\rho_1=2,\ \rho_2=1,\ \rho_3=1,\ \rho_4=2,\ \rho_5=1,\ \rho_6=1$,

Case II: $\rho_1=2,\ \rho_2=1,\ \rho_3=2/3,\ \rho_4=2,\ \rho_5=2,\ \rho_6=2$, and

Case III: $\rho_1=2,\ \rho_2=2/3,\ \rho_3=1,\ \rho_4=2,\ \rho_5=2,\ \rho_6=2$.

In the following, we show that there exists other construction such that the constructed secret sharing scheme for weighted graph has higher information rate for shares than that constructed by using DC technique.

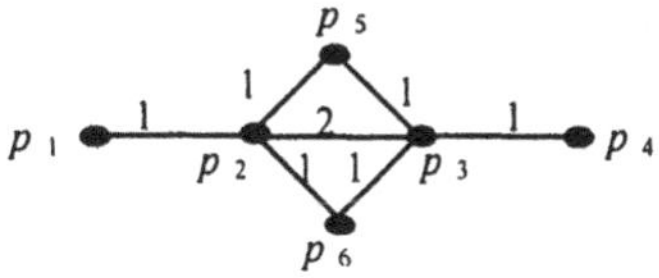

Fig. 1. Graph G

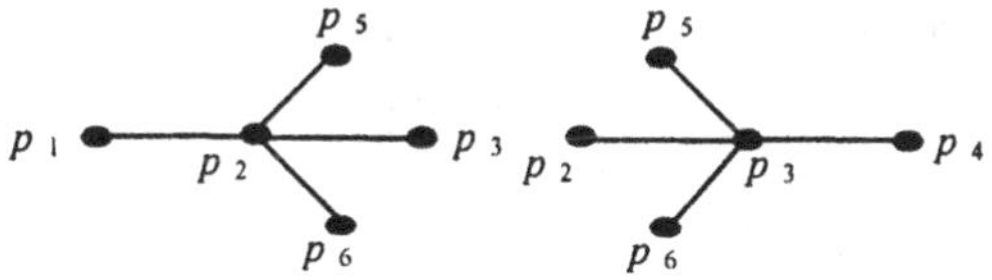

Fig. 2. DC for G (Case I)

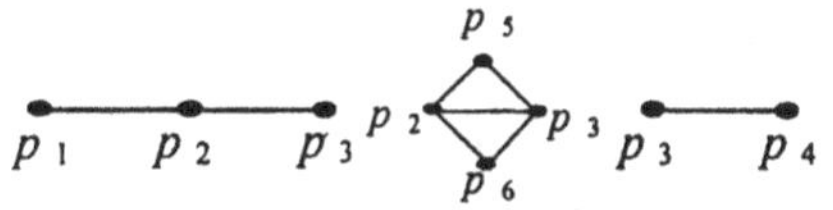

Fig. 3. DC for G (Case II)

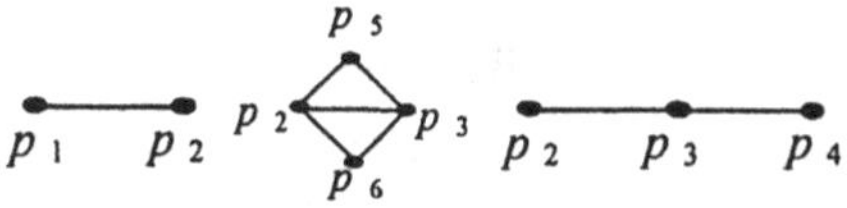

Fig. 4. DC for G (Case III)

We define that a double-star graph is a weighted graph which satisfies the following conditions:

(1) the graph can be decomposed into two stars exactly,
(2) the centers of the stars are two adjacent vertices, say p_i and p_j for $i \neq j$,
(3) the weight of $\overline{p_i p_j}$ is 2,
(4) except for $\overline{p_i p_j}$, each edge of the graph has the weight 1.

It is clear that the graph G in Figure 1 is a double-star graph. Given any double-star graph, we have the following theorem.

Theorem 3.1: Suppose G is a double-star graph with centers of the stars, p_i and p_j. Then there exists a secret sharing scheme for weighted graph G such that $\rho_i = \rho_j = 1$ and $\rho_k = 2$ for $k \neq i \neq j$.

Proof: We construct the secret sharing scheme in the following method. Assume that the secret $K = (K_1, K_2)$ is taken randomly from GF(q)× GF(q). The share of p_i is given by $S_i = (r_1 + K_1,\ r_2)$ (mod q) and the share of p_j is given by $S_j = (r_1,\ r_2 + K_2)$ (mod q). The share of p_k is given by

$S_k = r_1$ (mod q) if $\overline{p_i p_k} \in E(G)$ and $\overline{p_j p_k} \notin E(G)$,

$S_k = r_2$ (mod q) if $\overline{p_i p_k} \notin E(G)$ and $\overline{p_j p_k} \in E(G)$, and

$S_k = r_1 + r_2$ (mod q) if $\overline{p_i p_k} \in E(G)$ and $\overline{p_j p_k} \in E(G)$.

Here we ommit the proof of perfect secrecy and the computation of the information rates for shares. □

As an example of the construction in Theorem 3.1, we can obtain a secret sharing scheme for the weighted graph G in Figure 1, and the information rates for these shares are $\rho_1 = 2$, $\rho_2 = 1$, $\rho_3 = 1$, $\rho_4 = 2$, $\rho_5 = 2$, and $\rho_6 = 2$. It is clear that the construction dominates the decomposition constructions in Figure 2, Figure 3, and Figure4.

We generalize the concept of decomposition construction as follows:
Suppose $G=(V, E)$ be a weighted graph. A λ-weight-decomposition of G consists of a collection $\{G_1, \ldots, G_t\}$ such that the following requirements are satisfied.

(1) $G_h \subseteq G$ is a weighted graph, for $1 \le h \le t$.

(2) For each $e \in E$, there exist some indexes, say $i_1 < \cdots < i_k$, such that $\sum_{i_j} W_e(G_{i_j}) = \lambda$.

Let P_h be the set of participants in a scheme with access structure based on G_h. We have the following theorem.

Theorem 3.2: (*Weight-Decomposition Construction*, WDC) Let G be a graph of access structure on n participants, and suppose that $\{G_1, \ldots, G_t\}$ is a λ-weight-decomposition of G. Assume that for each weighted graph G_h, $1 \le h \le t$, there exists a perfect secret sharing scheme with information rate ρ_{ih} for each $p_i \in P_h$. Then there exists a perfect secret sharing scheme for G with information rate:

$$\rho = \min\left\{ \frac{\lambda}{\sum_{\{h: p_i \in P_h\}} (W(G_h)/\rho_{ih})} : 1 \le i \le n \right\}.$$

As an example of WDC, we consider that each G_h is either a complete multipartite graph or a double-star graph, for $1 \le h \le t$. Thus we can improve the information rate in 4 cases of the left unsolved 18 cases of secret sharing schemes for connected graphs on six vertices [7]. For graph G_9, we can improve the lower bound on the information rate from 1/2 to 4/7. For graph G_{22}, we can improve the lower bound on the information rate from 5/9 to 4/7. For graph G_{40}, we can improve the lower bound on the information rate from 5/9 to 4/7. For graph G_{61}, we can improve the lower bound on the information rate from 1/2 to 9/16. In Appendix we list the weight-decomposition constructions of these improvements.

4 Secret Sharing Schemes with Access Structures of Rank Three

In the section, we propose an efficient decomposition construction of perfect secret sharing schemes with access structures of rank three, and evaluate the information rate of the constructed scheme. For an access structure of rank three, with basis Γ_0, we can decompose Γ_0 into $\{\Gamma_1, \Gamma_2\}$ such that $\Gamma_0 = \Gamma_1 \cup \Gamma_2$ where $cl(\Gamma_1)$ is a uniform access structure of rank two and $cl(\Gamma_2)$ is a uniform access structure of rank three.

Assume that $\boldsymbol{P} = \{p_1, p_2, \cdots, p_n\}$ is the set of participants and the secret $K = (K_1, K_2, K_3, K_4, K_5, K_6)$ is taken randomly from $(\mathrm{GF}(q))^6$, where q is a prime and $q > 2n+2$. Let $f(x) = K_6x^5 + K_5x^4 + K_4x^3 + K_3x^2 + K_2x + K_1 \pmod q$. y_i is computed from $f(x)$ as follows: $y_i = f(i) \pmod q$, for $i = 1, \ldots, 2n+2$.

Thus one who gets six or more y_i's can recover $f(x)$ and then the secret K. However, one without knowledge of any y_i obtains no information about the secret.

We use G to denote the access structure $cl(\Gamma_1)$ whose rank is two. From section 2, we know that there exists a graph-based secret sharing scheme realizing $cl(\Gamma_1)$ in which the secret is (y_{2n+1}, y_{2n+2}) and the share of participant p_i is $S_i(G)$.

In addition, we define G_i, for $1 \le i \le n$, is the graph with vertices $V(G_i)$ and edges $E(G_i)$, where $V(G_i) = \{p_j \mid \text{for all } p_j, \text{ where } \{p_i, p_j, p_k\} \in \Gamma_2\}$ and $E(G_i) = \{\overline{p_jp_k} \mid \text{for all } \overline{p_jp_k}, \text{ where } \{p_i, p_j, p_k\} \in \Gamma_2\}$. The dealer selects $2n$ random numbers, $r_1, \ldots, r_{2n}$, over GF(q). As the construction in section 2, there exists a secret sharing scheme realizing G_i in which the secret is $(r_i + y_i, r_{n+i} + y_{n+i})$ and the share of participant p_j is $S_j(G_i)$ for $p_j \in V(G_i)$.

The share of participant p_i is given by

$S_i = < r_i,\ r_{n+i},\ a_{i,1},\ \cdots,\ a_{i,t},\ \cdots,\ a_{i,n}\ ,\ S_i(G) >$, where $1 \le t \le n$,

$a_{i,t} = S_i(G_t)$ if $p_i \in V(G_t)$,

$a_{i,t} = (r_t + y_t, r_{n+t} + y_{n+t})$ if $\overline{p_i p_t} \in E(G)$, and

$a_{i,t}$ is empty otherwise.

Theorem 4.1. The constructed secret sharing scheme is perfect.

The share of participant p_i is $S_i = < r_i,\ r_{n+i},\ a_{i,1},\ \cdots,\ a_{i,t},\ \cdots,\ a_{i,n},\ S_i(G) >$. The size of $a_{i,t}$ is equal to $\log(q^{d_i(G_t)+1})$ if $p_i \in V(G_t)$ or $\log(q^2)$ if $p_i \in V(G)$. The the size of $S_i(G)$ is equal to 0 if $p_i \notin V(G)$, or is equal to $\log(q^{d_i(G)+1})$ if $p_i \in V(G)$, where $d_i(G)$ is the degree of vertex p_i in G. Hence the size of share S_i is equal to $\log(q^{\sum_{t:p_i \in G_t}(d_i(G_t)+1)\ +\ 2})$ if $p_i \notin V(G)$, or $\log(q^{\sum_{t:p_i \in G_t}(d_i(G_t)+1)\ +\ d_i(G)\ +\ 3})$ if $p_i \in V(G)$. Because the size of the secret is equal to $\log(q^6)$, the information rate of the share S_i, ρ_i, is equal to $\dfrac{6}{\sum_{t:p_i \in G_t}(d_i(G_t)+1)\ +2}$ if $p_i \notin V(G)$ or is equal to

$$\frac{6}{\sum_{t:p_i \in G_t}(d_i(G_t)+1)\ + d_i(G)+3} \text{ if } p_i \in V(G).$$

Theorem 4.2. If Γ is an access structure (either uniform or non-uniform) of rank three on n participants, then there exists a $PS(\Gamma, \rho,\ q^6)$, where $\rho \ge \dfrac{6}{(n-1)^2+2}$.

Compared with the lower bound studied by Stinson [11], our lower bound is better than Stinson's lower bound in some cases. The comparison can be seen in Table 1.

Table 1. Bounds on the information rate for uniform access structures of rank three on n participants for $n \ge 5$, where * denotes the method which has the better bound.

	Uniform Structure		Non-uniform Structure	
n	Stinson's Method	Our Method	Stinson's Method	Our Method
$n \equiv 0 \pmod 6$	$\frac{6}{n(n+1)}$	$\frac{6}{(n-1)^2+2}$ *	$\frac{4}{n(n+1)}$	$\frac{6}{(n-1)^2+2}$ *
$n \equiv 1,3 \pmod 6$	$\frac{6}{n(n-1)}$	$\frac{6}{(n-1)^2+2}$ *	$\frac{4}{n(n-1)}$	$\frac{6}{(n-1)^2+2}$ *
$n \equiv 2,4 \pmod 6$	$\frac{6}{(n-1)(n-2)}$ *	$\frac{6}{(n-1)^2+2}$	$\frac{4}{(n-1)(n-2)}$	$\frac{6}{(n-1)^2+2}$ *
$n \equiv 5 \pmod 6$	$\frac{6}{(n+1)(n+2)}$	$\frac{6}{(n-1)^2+2}$ *	$\frac{4}{(n+1)(n+2)}$	$\frac{6}{(n-1)^2+2}$ *

5 Secret Sharing Schemes with Uniform Access Structures of Rank *m*

In the section, we propose an efficient decomposition construction of secret sharing schemes with uniform access structures of rank m. Let Γ be a uniform access structure of rank m on n participants. Assume that $\boldsymbol{P} = \{p_1, p_2, \cdots, p_n\}$ is the set of participants and the basis of Γ is Γ_0. We can decompose Γ_0 into the union of Γ_i's, for $1 \le i \le n$, where $\Gamma_i = \{X : X \in \Gamma_0$ and X contains participant $p_i\}$. Thus $\Gamma = cl(\Gamma_0) = cl(\Gamma_1) \cup \ldots \cup cl(\Gamma_n)$. We define $\Gamma_i{}^* = \{X : X \cup \{p_i\} \in \Gamma_i\}$, i.e., $\Gamma_i{}^*$ is the set of Γ_i which participant p_i is removed from each element in Γ_i. Therefore, each $cl(\Gamma_i{}^*)$ is a uniform access structure of rank m-1. We assume that the secret $K = (K_1, K_2, \ldots, K_m)$, where each K_i, for $1 \le i \le m$, is taken randomly from $(\mathrm{GF}(q))^{h(m-1)}$ which is the secret space of the secret sharing schemes with uniform access structures of rank m-1. Note that $h(i)$ is a function which indicates the secret space of the secret sharing schemes with uniform access structures of rank i to be $(\mathrm{GF}(q))^{h(i)}$. The dealer selects a polynomial $f(x)$ of degree $m \cdot h(m-1)-1$ with coefficients K and computes y_i as follows: $y_i = f(i)$ (mod q), for $i = 1, \ldots, n \cdot h(m-1)$.

Thus one who gets $m \cdot h(m-1)$ or more y_i's can recover $f(x)$ and then the secret K. However, one without knowledge of any y_i obtains no information about the secret. We use $Y_1, Y_2, \ldots, Y_n$ over $(\mathrm{GF}(q))^{h(m-1)}$ to denote these $n \cdot h(m-1)$ y_i's. The dealer selects n random numbers $R_1, R_2, \ldots, R_n$ over $(\mathrm{GF}(q))^{h(m-1)}$. We assume that there exists a secret sharing scheme realizing $cl(\Gamma_i{}^*)$ in which the secret is $R_i + Y_i$ and the share of participant p_j is $S_j(\Gamma_i{}^*)$.

The share of participant p_i is given by

$S_i = \langle R_i, S_i(\Gamma_1{}^*), \ldots, S_i(\Gamma_{i-1}{}^*), S_i(\Gamma_{i+1}{}^*), \ldots, S_i(\Gamma_n{}^*)\rangle$.

Thus, the constructed secret sharing scheme is a perfect secret sharing scheme with access structure Γ. Summaring, we have the following theorem.

Theorem 5.1. The constructed secret sharing scheme is perfect.

It is interesting to see the secret space and the lower bound of the information rate for the constructed secret sharing scheme. The secret space, $(\mathrm{GF}(q))^{h(m)}$, of the constructed secret sharing scheme is equal to $(\mathrm{GF}(q))^{m \cdot h(m-1)}$. Therefore, $h(m) = m \cdot h(m-1)$. From section 2, we know that there exist secret sharing schemes with unique access structure of rank two in which $h(2)$ is equal to 2. Therefore, we can obtain $h(m) = m!$. That is, the secret space of the constructed secret sharing scheme is equal to $(\mathrm{GF}(q))^{m!}$.

To evaluate the lower bound of the information rate of the constructed secret sharing schemes, we define $\rho(m, n)$ to be the lower bound of the information rate of secret sharing schemes with uniform access structures of rank m on n participants. Therefore, $\rho(m, n) = \dfrac{m}{(n-1)\cdot\dfrac{1}{\rho(m-1, n-1)}+1}$

Because $0 \le \rho(m-1, n-1) \le 1$, $\dfrac{\rho(m, n)}{\rho(m-1, n-1)} = \dfrac{m}{(n-1)+\rho(m-1, n-1)} \ge \dfrac{m}{n}$.

We can obtain

$\rho(m, n) \ge \dfrac{m}{n}\cdot\rho(m-1, n-1) \ge \dfrac{m\cdot(m-1)\cdot\ldots\cdot 3}{n\cdot(n-1)\cdot\ldots\cdot(n-k+3)}\cdot\rho(2, n-k+2)$. From section 2, we know that $\rho(2, n-k+2) \ge \dfrac{2}{n-k+2}$. Therefore, $\rho(m, n) \ge \dfrac{m!\cdot(n-m+1)!}{n!} = \dfrac{n-m+1}{\binom{n}{m}}$. Hence, we have the following theorem.

Theorem 5.2. Let Γ be a uniform access structure of rank m on n participants. Then there exists a $PS(\Gamma_0, \dfrac{n-m+1}{\binom{n}{m}}, q^{m!})$ for $q > n\cdot(m-1)!$.

Compared with the best previous lower bound of $\rho(m, n)$, studied by Stinson [11], which is $\dfrac{m}{(2m-1)\cdot\binom{n-1}{m-2}+d}$ where d is the maximum degree of any participant, our lower bound is better than Stinson's lower bound when $m \ge \dfrac{3+\sqrt{8n+1}}{4}$.

References

[1] J. Benaloh and J. Leichter, "Generalized secret sharing and monotone functions," in *Advances in Cryptology-Crypto'88 Proceedings*, Lecture Notes in Computer Science, Vol. 403, Springer-Verlag, Berlin, pp. 27-35, 1990.

[2] C. Blundo, A. De Santis, L. Gargano, and U. Vaccaro, "On the information rate of secret sharing schemes," in *Advance in Cryptology-CRYPTO'92, Lecture Notes in Comput. Sci.*, Vol. 740, pp. 148-167, 1993.

[3] C. Blundo, A. De Santis, D.R. Stinson and U. Vaccaro, "Graph decompositions and secret sharing schemes," in *Advance in Cryptology-Proceedings of Eurocrypt'92, Lecture Notes in Comput. Sci.*, Vol. 658, pp. 1-24, 1993.

[4] C. Blundo, A. De Santis, D.R. Stinson and U. Vaccaro, "Graph decompositions and secret sharing schemes," *Journal of Cryptology*, Vol. 8, pp. 39-63, 1995.

[5] E.F. Brickell and D.R. Stinson, "Some improved bounds on the information rate of perfect secret sharing schemes," *Journal of Cryptology*, Vol. 5, pp. 153-166, 1992.

[6] R. M. Capocelli, A. De Santis, L. Gargano, and U. Vaccaro, "On the size of shares for secret sharing schemes," *Journal of Cryptology*, Vol. 6, pp. 157-167, 1993.

[7] M. van Dijk, "On the information rate of perfect secret sharing schemes," Designs, Codes and Cryptography, Vol. 6, pp. 143-169, 1995.

[8] R.W. Hamming, *Coding and Information Theory*, Englewood Cliffs, Reading, NJ:Prentice-Hall, 1986.

[9] M. Ito, A. Saito and T. Nishizeki, "Secret sharing scheme realizing general access structure," in *Proc. IEEE Globecom'87*, Tokyo, pp. 99-102, 1987 .

[10] M. Ito, A. Saito and T. Nishizeki, "Multiple assignment scheme for sharing secret," *Journal of Cryptology*, Vol. 6, pp. 15-20, 1993.

[11] D.R. Stinson, "New general lower bounds on the information rate of secret sharing schemes," in *Advance in Cryptology-CRYPTO'92, Lecture Notes in Comput. Sci.*, Vol. 740, pp. 168-182, 1993.

[12] D.R. Stinson, "Decomposition constructions for secret sharing schemes," *IEEE Trans. Inform. Theory*, Vol. IT-40, pp. 118-125, 1994.

Appendix: Weight-Decomposition Construction

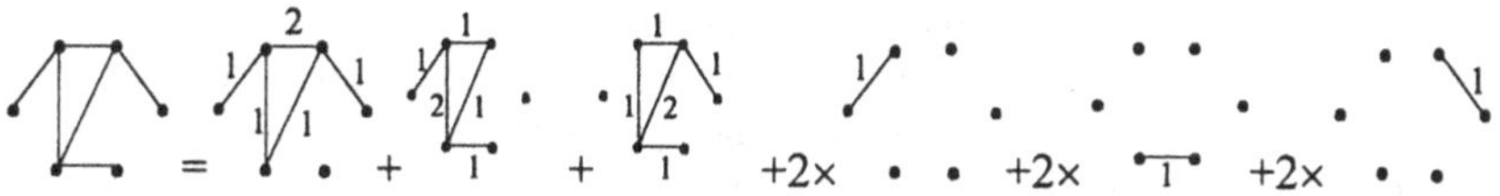

Graph G_9

Graph G_{22}

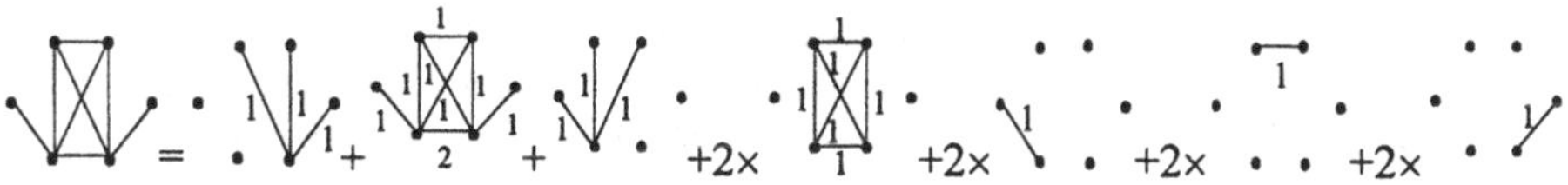

Graph G_{40}

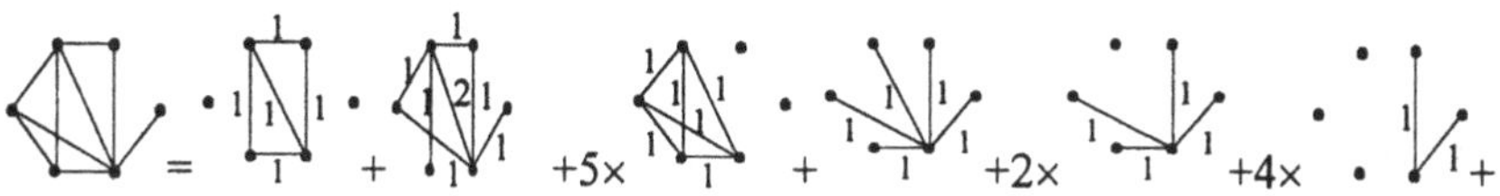

Graph G_{61}

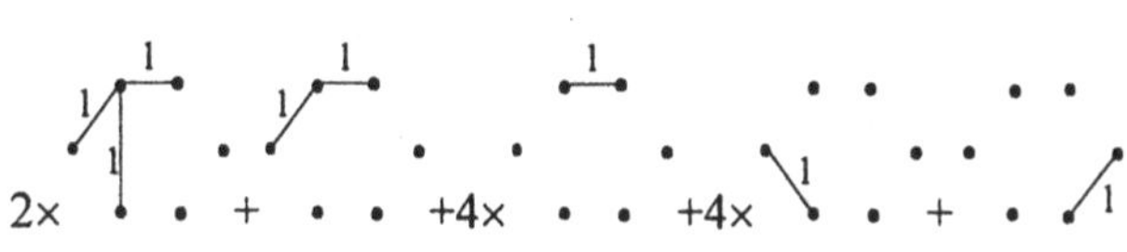

Traceable Visual Cryptography

(Extended Abstract)

Ingrid Biehl and Susanne Wetzel*

Universität des Saarlandes
Fachbereich 14 - Informatik
Postfach 151150
Phone number: +49 681 302 5607
Fax number: +49 681 302 4164
email: ingi@cs.uni-sb.de, wetzel@cs.uni-sb.de

Abstract. In this paper we present a new k out of n visual cryptography scheme which does not only meet the requirements of a basic visual cryptography scheme defined by Naor and Shamir [5] but is also traceable. A k out of n visual cryptography scheme is a special instance of a k out of n threshold secret sharing scheme [6]. Thus, no information about the original secret can be revealed if less than k share-holders combine their shares. In those systems it is inherently assumed that even if there are k or more share-holders with an interest in the abuse of the secret, then it is almost impossible that they can meet up as an entirety (e.g. because they are to cautious to inform too many others about their intentions) and combine their shares to misuse the secret. But in real scenarios it might not be too unlikely that the betrayers find together in small groups. Even though each one of these groups is too small to compute the original secret, the betrayers of such a group can impose a major security risk on the system by publishing the information about their shares. Suppose for example that $k-1$ betrayers find each other and do the publishing. Then all the other $n-k+1$ share-holders can potentially reveal the secret without ever meeting up with at least $k-1$ other share-holders as is intended by the system. In order to cope with this lack of security, we present in this paper the idea of traceable visual cryptography schemes which allows to track down the publishing saboteurs.

Keywords: Visual cryptography schemes, traceability

1 Introduction

In 1994 the basic problem of visual cryptography was introduced by Naor and Shamir [5]. In visual cryptography we are dealing with the problem of encrypting

* This research was done while the author was a member of the Graduiertenkolleg Informatik at the University of Saarbruecken, a fellowship program of the DFG (Deutsche Forschungsgemeinschaft).

pictures in a secure way such that the decryption can be done by the human visual system. The encryption of a secret image is achieved by encoding the information into several shadow images, called shares. The decoding is done by printing the shares on transparencies and stacking them. The system can be used by everyone without any knowledge of cryptography and without performing cryptographic computations.

Naor and Shamir have described a k out of n (with $k \leq n$) system where the secret is encoded in n shares and the decoding can be done by stacking k or more transparencies. Using less than k transparencies won't reveal the secret not even to an infinitely powerful cryptanalyst. Thus, the system is a special k out of n threshold secret sharing schème [6]. In practical scenarios it is assumed that even if there are k or more share-holders who are inclined to be involved in an act of sabotage, then it is almost impossible that they can meet up as an entirety and misuse the secret, e.g. because they might be too careful to tell (too many) others about their intentions. But it is not too unlikely that there might be many small saboteur groups, each one too small to compute the original secret. Nevertheless, such a small saboteur group can do a lot of damage by publishing the information about the shares of the group members. Thus, others can use the published information, possibly even for revealing the original secret without having at least k share-holders ever meet as a group.

A practical scenario is given by a key escrowing system, realized as a k out of n threshold secret sharing scheme with a large number n of escrow agencies (which is discussed by the cryptographic research community nowadays) where the agencies share the secret keys for encryption or electronic signatures for participating parties. Then, if for instance $k-1$ agencies are traitors and combine their shares of the parties' keys and publish this information, every other single escrow agent can retrieve the secret keys and illegally eavesdrop the communication of the parties using the system. This is a major lack of security and definetely will not help to build up some confidence of the general public in the system.

Obviously, it is important to provide mechanisms which allow the tracing of the publishing saboteurs so that they can appropriately be punished. In this paper, we present a traceable visual cryptography scheme which enables a trusted authority to track down the saboteurs by knowing only the so-called marking information but not the actual shares. In the following we will give a definition for a traceable visual cryptography scheme and illustrate the new model with an example.

2 Visual Cryptography Schemes

In the sequel we assume that the pictures are black-and-white. Each pixel is handled separately and is represented by m subpixels and appears in n modified versions (shares). Therefore, the resulting structure of a k out of n system can be described by an $n \times m$ Boolean matrix $B = (b_{ij})$. For a set $I \subseteq \{1, \ldots, n\}$, $\text{Shares}_B(I) = \{b_{i\cdot} : i \in I\}$ is the set of all rows $b_{i\cdot}$ of B whose row numbers i

are in I. With $v = \mathrm{OR}(s_1, \ldots, s_l)$ we denote the OR of a set of Boolean vectors s_i ($1 \leq i \leq l$, $l \in \mathbb{N}$) consisting of m components.

For $B = (b_{ij}) \in \{0,1\}^{n \times m}$, $b_{ij} = 1$ if and only if the j-th subpixel of the i-th transparency is black. Printing the m subpixels in close proximity, the human visual system averages the black and white contributions. The Hamming weight of the OR of k corresponding rows of B determines the greyness of the stack of k transparencies. The contrast is determined by the difference of the number of black subpixels in original white and black pixels.

Definition 1 [5]. Two collections of $n \times m$ Boolean matrices C_0 and C_1 are called a k out of n visual cryptography scheme ((k,n)-VCS), if there are constants $\alpha \geq \frac{1}{m}$ and $b \in \{1, \ldots, m\}$ such that the following conditions hold:

1. For any B in C_0, the OR v of any k of the n rows of B has a Hamming weight $H(v) \leq b - \alpha * m$.
2. For any B in C_1, the OR v of any k of the n rows of B has a Hamming weight $H(v) \geq b$.
3. For any subset $\{i_1, i_2, \ldots, i_q\}$ of $\{1, 2, \ldots, n\}$ with $q < k$, the two collections of $q \times m$ matrices B_j for $j \in \{0,1\}$ obtained by restricting each $n \times m$ matrix in C_j ($j \in \{0,1\}$) to the rows $\{i_1, i_2, \ldots, i_q\}$ contain the same matrices with the same frequencies.

The n shares (transparencies) are generated from the original secret picture by choosing a matrix of the set C_0 or C_1 equally distributed and independently for every pixel, depending only on the color of the pixel.

The first two conditions are called *contrast* ensuring that stacking k transparencies will reveal the original color of the pixel. The last condition is called *security*, implying that inspecting less than k transparencies will not give any information on the original pixel. The value m is the loss in resolution from the original picture to the shared one and should be as small as possible. α is the *relative contrast* and determines how well k transparencies will reveal the secret. The *minimal greyness* of a black pixel is determined by the parameter b. An example for a visual cryptography scheme will be presented in Sect. 4 where we will use this scheme for the construction of a traceable visual cryptography scheme.

3 Definition of Traceable Visual Cryptography Schemes

Based on the definition of a visual cryptography scheme (see Definition 1) we will now present the traitor model. We assume that $0 < t < k$ share-holders try to sabotage the system by stacking their shares and publishing the resulting information. In the following we are not interested in keeping other coalitions from stacking their shares on top of the published information and therefore revealing the secret illegaly. We are solely interested in tracing the traitors who have started the sabotage act by publishing the information. In order to guarantee that the traitors can be traced, it is necessary to insert some markings in the picture.

Prior to defining a traceable (k,n)-VCS we shall first introduce some notations: Given a visual cryptography scheme with $\mathcal{C} = (\mathcal{C}_0, \mathcal{C}_1)$ (where $\mathcal{C}_0$ and $\mathcal{C}_1$ are collections of $n \times m$ matrices) with threshold constants α and b and a vector v, the predicate $\text{ThresholdDecision}_{\mathcal{C}_0,\mathcal{C}_1}(v)$ is 0 if the Hamming weight of v is at most $b - \alpha m$ and is 1 if the Hamming weight of v is at least b. Let $\mathcal{K}$ be a collection of Boolean $n \times m$ matrices B, $t \in \mathbb{N}$ and S be an $t \times m$ matrix. Then $\mathcal{B}_{\mathcal{K}}(S)$ is the set of all matrices $B \in \mathcal{K}$ such that $\{s_{i\cdot} \mid 1 \le i \le t\}$ (set of rows of S) is a subset of $\{b_{i\cdot} : 1 \le i \le n\}$ (set of rows of B).

We can now give a formal definition of a *traceable (k,n)-VCS with (ε,δ)-security*:

Definition 2. A *traceable (k,n)-VCS with (ε,δ)-security* is a set of three collections of Boolean $n \times m$ matrices $\mathcal{C}_0, \mathcal{C}_1$ and $\mathcal{C}_M$, denoted by $\mathcal{C} = (\mathcal{C}_0, \mathcal{C}_1, \mathcal{C}_M)$. The matrices of the set $\mathcal{C}_M$ are called *marking matrices* and for each marking matrix $B \in \mathcal{C}_M$ there is a special row $1 \le r = r(B) \le n$, called *marking row*, which can be used for tracing. The collections $\mathcal{C}_0, \mathcal{C}_1$ and $\mathcal{C}_M$ have to satisfy the following properties:

1. $(\mathcal{C}_0, \mathcal{C}_1)$ is a (k,n)-VCS
2. For any subset $\{i_1, i_2, \ldots, i_q\}$ of $\{1, 2, \ldots, n\}$ with $q < k$, the three collections of $q \times m$ matrices B_j for $j \in \{0, 1, M\}$ obtained by restricting each $n \times m$ matrix in $\mathcal{C}_j$ ($j \in \{0,1,M\}$) to the rows $\{i_1, i_2, \ldots, i_q\}$ contain the same matrices with the same frequencies.
3. There is a *tracing algorithm Trace* such that the following holds: Take the information which is published by $t < k$ saboteurs (holding the shares $S = \{s_1, \ldots, s_t\}$) as the description of a Turing machine $\mathcal{A}$. If for $\mathcal{A}$ there is an integer u such that for all $U \subseteq \{1, \ldots, n\}$ with $|U| = u$ and $U \cap S = \emptyset$ there is a subset $S' = \{s'_1, \ldots, s'_{k-u}\} \subseteq S$, such that for all $B \in \mathcal{B}_{(\mathcal{C}_0 \cup \mathcal{C}_1 \cup \mathcal{C}_M)}(S)$

   $$\mathcal{A}(\text{Shares}_B(U)) = \text{ThresholdDecision}_{\mathcal{C}_0,\mathcal{C}_1}(\text{OR}(s'_1, \ldots s'_{k-u}, \text{Shares}_B(U))) \ .$$

   then the following is true:
   (a) (Protection against saboteurs)

   $$\sum_{B \in \mathcal{B}_M(S)} Pr\{B\} \cdot Pr\{Shares_B(\{r\}) \in S \Rightarrow \text{Trace}(\mathcal{A}, B) = r\} \ge 1 - \varepsilon$$

   (b) and (Security for innocent share-holders)

   $$\sum_{B \in \mathcal{B}_M(S)} Pr\{B\} \cdot Pr\{Shares_B(\{r\}) \notin S \text{ and } \text{Trace}(\mathcal{A}, B) = r\} \le \delta \ .$$

The first condition of a traceable visual cryptography scheme guarantees that it is also a visual cryptography scheme. The second one ensures that a coalition of less than k share-holders can not decide whether they have gotten their shares from a matrix corresponding to a white or a black pixels or even from a marking matrix. The third condition describes the traceability property. The motivation of the chosen formalisation is as follows: In our attack scenario we assume that the saboteurs publish some information $\mathcal{A}$ which can be interpreted as the description of a Turing machine. Note that the information $\mathcal{A}$ does not necessarily consist of $v = \mathrm{OR}(s_1, \ldots, s_t)$. E.g. the saboteurs might somehow combine their t shares to some kind of information which corresponds to less than t shares. For any sufficiently large subset of share-holders U with $|U| = u$ those share-holders can use the published information $\mathcal{A}$ in combination with their shares to reveal the correct information about the shared pixel. The *correct* information about the shared pixel is the value which the share-holders get if they would obtain $k-u$ shares from the saboteurs. Property 3.(a) guarantees that a saboteur holding the marking row can be traced with high probability and 3.(b) ensures that it is very unlikely that an innocent share-holder will be found guilty.

4 Construction of a Traceable (k,n)-VCS

In the following we will present the construction of a traceable (k,n)-VCS with (ε, δ)-security which is based on so-called *normal* visual cryptography schemes. In those schemes the sets $\mathcal{C}_0$ and $\mathcal{C}_1$ consist of all matrices obtained by permuting the columns of two Boolean *base matrices* S^0 and S^1, respectively.

4.1 A (k,k)-VCS

In [1, 2] Ateniese et al. have presented a (k,k)-VCS where K^0 is the matrix whose columns are all the Boolean k-vectors having an even number of 1's and K^1 is the matrix where the columns are the Boolean k-vectors with an odd number of 1's. Note, that the Hamming weight of each row in K^0 and K^1 is $m/2 = 2^{k-2}$. The collections $\mathcal{C}_0$ and $\mathcal{C}_1$ consist of all possible permutations of the columns of the corresponding base matrix. Ateniese et al. show that this construction results in a (k,k)-VCS with $|\mathcal{C}_0| = |\mathcal{C}_1| = 2^{k-1}!$ and parameters $m = 2^{k-1}$, $\alpha = 1/2^{k-1}$.

With K^0 as the matrix whose columns are all the Boolean k-vectors having an even number of 1's, K^1 can be chosen such that $k-1$ rows $a_{i_1}, \ldots, a_{i_{k-1}}$ are identical in both matrices and the i_k-th row of K^1 is the complement of the corresponding row of K^0. In the sequel, the complement of a_{i_k} will be denoted by $\overline{a_{i_k}}$. K^0 and K^1 differing in row $1 \le i \le k$ will simply be described with the notation K_i^0 and K_i^1 where $K_i^0 = K^0$ for all $1 \le i \le k$.

For the (k,k)-VCS with collections $\mathcal{C}_0$ and $\mathcal{C}_1$ obtained from K^0 and K^1 one can show the following lemma:

Lemma 3. *For every $p \le k-1$, $P \subset \{1, \ldots, k\}$, $|p| = P$ and every matrix $B \in \mathcal{C}_0 \cup \mathcal{C}_1$, the Hamming weight of $v = OR(Shares_B(P))$ is*

$$H(v) = 2^{k-2} + \sum_{j=1}^{p-1} 2^{k-2-j}$$

and therefore depends only on p.

Example 1.

$$K^0 = \begin{bmatrix} 0\,0\,0\,0\,1\,1\,1\,1 \\ 0\,0\,1\,1\,0\,0\,1\,1 \\ 0\,1\,0\,1\,0\,1\,0\,1 \\ 0\,1\,1\,0\,1\,0\,0\,1 \end{bmatrix} \quad \text{and } K_4^1 = \begin{bmatrix} 0\,0\,0\,0\,1\,1\,1\,1 \\ 0\,0\,1\,1\,0\,0\,1\,1 \\ 0\,1\,0\,1\,0\,1\,0\,1 \\ 1\,0\,0\,1\,0\,1\,1\,0 \end{bmatrix}$$

are the black and white base matrices for a (4, 4)-VCS and

$\mathcal{C}_0 = \{M : M$ is a matrix obtained by permuting the columns of $K^0\}$ and
$\mathcal{C}_1 = \{M : M$ is a matrix obtained by permuting the columns of $K^1\}$.

4.2 Normal (k, n)-VCS

Using the (k, k)-VCSs of Sect. 4.1, we can now construct normal (k, n)-VCSs with $k \leq n$. As in [1, 2, 5] we use an $n \times \ell$ matrix $SH(n, \ell, k)$ whose entries are elements of a ground set $\{a_1, \ldots, a_k\}$. The $n \times \ell$ matrix $SH(n, \ell, k)$ has the property that for any subset of k rows there exists at least one column such that the entries in the k given rows of that column are all distinct. One constructs the base matrices S^0 and S^1 for normal (k, n)-VCS by replacing the symbols $a_1, \ldots, a_k$ with the 1-st, ..., k-th rows of the corresponding base matrices K^0 and K_i^1 of the (k, k)-VCS (see Sect. 4.1) for some arbitrarily chosen $1 \leq i \leq k$. The collections $\mathcal{C}_0$ (respectively $\mathcal{C}_1$) which are obtained by permuting the columns of the corresponding base matrix S^0 (respectively S^1) in all possible ways, form a normal (k, n)-VCS with $m = \ell \times 2^{k-1}$.

In our construction of normal (k, n)-VCS, the $n \times \ell$ matrix SH is a representation of a *Hash Family H* which is a collection of ℓ k-wise independent hash functions (see [3, 5, 7]) with the following properties:

1. For all $h \in \mathcal{H}$ we have $h : \{1, \ldots, n\} \to \{1, \ldots, k\}$.
2. There is an $x \in \{1, \ldots, n\}$ such that for all subsets $X \subseteq \{1, \ldots, n\}$ of size k with $x \in X$ there is at least one $h \in \mathcal{H}$ such that $|h(X)| = k$ and

$$|\{j : 1 \leq j \leq n \wedge h(x) = h(j)\}| \leq \frac{n}{2} .$$

In general, the second condition is not necessary for constructing a (k, n)-VCS, but is required for the construction of a tracable visual cryptography scheme (see Sect. 4.3).

The property that the hash functions are k-wise independent means, that for any k distinct elements $x_1, \ldots, x_k \in \{1, \ldots, n\}$ and any k (not necessarily distinct) elements $y_1, \ldots, y_k \in \{1, \ldots, k\}$ the probability that for a randomly chosen $h \in \mathcal{H}$

$$h(x_j) = y_j \text{ for all } 1 \leq j \leq k$$

is the same. It follows that for all subsets $X \subseteq \{1, \ldots, n\}$ of size k and for all $1 \leq q \leq k$ the probability that a randomly chosen $h \in \mathcal{H}$ yields q different values on X is the same. Applying Lemma 3 we can therefore conclude:

Lemma 4. *If one uses the rows of the base matrices K^0 and K_i^1 for some arbitrarily chosen $1 \leq i \leq k$ to fill the SH matrix, then for every $p \leq k-1$, $P \subset \{1, \ldots, n\}$, $|P| = p$ and every matrix $B \in C_0 \cup C_1$, the Hamming weight of $v = OR(Shares_B(P))$ depends only on p.*

Moreover we need the following lemma:

Lemma 5. *Let B be some matrix in $C_0 \cup C_1$, $p < k$ and P_0, P_1 two subsets of p rows of B. Then there is a permutation of the columns which applied to P_0 results in P_1.*

Proof. It is sufficient to look at the base matrices. Each p-tuple of values as entries of p rows in a single column of the SH matrix appears in each set of p rows with the same frequency. This is guaranteed by the k-wise independence of the hash functions. □

In the following example we will show a part of a SH matrix and the corresponding base matrix S^0 of a (3, 4)-VCS.

Example 2. For

$$SH = \begin{bmatrix} \cdots a_1 \; a_2 \cdots \\ \cdots a_1 \; a_3 \cdots \\ \cdots a_2 \; a_1 \cdots \\ \cdots a_2 \; a_3 \cdots \\ \cdots a_3 \; a_1 \cdots \\ \cdots a_3 \; a_2 \cdots \end{bmatrix} \quad \text{and } K^0 = \begin{bmatrix} 0\,0\,1\,1 \\ 0\,1\,0\,1 \\ 0\,1\,1\,0 \end{bmatrix}$$

the white base matrix is constructed as:

$$S^0 = \begin{bmatrix} \cdots & 0011 & 0101 & \cdots \\ \cdots & 0011 & 0110 & \cdots \\ \cdots & 0101 & 0011 & \cdots \\ \cdots & 0101 & 0110 & \cdots \\ \cdots & 0110 & 0011 & \cdots \\ \cdots & 0110 & 0101 & \cdots \end{bmatrix}$$

4.3 Construction of a Traceable Scheme

Based on the construction of a (k, n)-VCS with $C = (C_0, C_1)$ in Sect. 4.2, we can now present a traceable $(k, \frac{n}{2})$-VCS with $C' = (C'_0, C'_1, C'_M)$. The construction is as follows:

Let $x \in \{1, \ldots, n\}$ be such that for all subsets $X \subseteq \{1, \ldots, n\}$ of size k with $x \in X$ there is at least one $h \in \mathcal{H}$ such that $|h(X)| = k$ and $|\{j : 1 \leq j \leq n \wedge h(x) = h(j)\}| \leq \frac{n}{2}$. Then $\mathcal{H}'$ is defined as the subset of $\mathcal{H}$ of all functions with $|\{j : 1 \leq j \leq n \wedge h(x) = h(j)\}| \leq n/2$. Note, that the choice of $\mathcal{H}$ implies that for each subset $X \subseteq \{1, \ldots, n\}$ of size k and $x \in X$ the set $\mathcal{H}'$ contains at least one function h' such that $|h'(X)| = k$.

Lemma 4 implies that the Hamming weight for white as well as black pixels is fixed. Let w_C be the Hamming weight in case of C_0 matrices, b_C the one for C_1 matrices and $d_C = b_C - w_C$ where $C = (C_0, C_1)$.

Exchange the first row and the x-th row with each other. Then construct for each $h' \in \mathcal{H}'$ a submatrix of the SH matrix by deleting all rows $j \neq x$ with $h'(x) = h'(j)$. According to the properties of the functions in $\mathcal{H}'$, at least $n/2$ rows will remain. If there are more than $n/2$ rows left, delete the rows whose index is greater than $n/2$. Concatenating these $|\mathcal{H}'|$ submatrices yields the $\frac{n}{2} \times l$ matrix $\widetilde{SH}$.

One obtains the *elementary white matrix* by inserting the rows of K^0 into the values of $\widetilde{SH}$. Replacing the values in the first submatrix by the rows of K^1 and the values in the other $|\mathcal{H}'| - 1$ copies by the rows of K^0 yields the *elementary black matrix*. Thus the OR of k different shares of the elementary white matrix has Hamming weight $|\mathcal{H}'| \cdot w_C$ and $b_C + (|\mathcal{H}'| - 1) \cdot w_C$ for the elementary black matrix.

The *elementary marking matrix* is created by modifying the elementary white matrix as follows: In each copy let a_j be the entry in the first row and in the column of the function h' in $\mathcal{H}'$ corresponding to this copy. Instead of using the j-th row of K^0 we insert the j-th row of K_j^1 that is the Boolean complement of the j-th row of K^0. Note, that in this column it is the only occurence of a_j. Take k shares which are pairwise distinct in this column of the copy of $\widetilde{SH}$. Then, the Hamming weight of the OR of the fragment of the shares corresponding to this column in the copy of $\widetilde{SH}$ is 2^{k-1} according to the properties of the (k,k)-VCS. In the elementary white matrix the OR of the same fragment is $2^{k-1} - 1$. Since the elementary marking matrix contains such a fragment for each subset of k different shares, we achieve that the OR of k different shares of the elementary marking matrix is at least $|\mathcal{H}'| \cdot w_C + 1$ if the first row is in the set and $|\mathcal{H}'| \cdot w_C$ otherwise.

The set C'_0 is constructed by permuting the rows of the *white base matrix* S'^0 which is the concatenation of d_C identical copies of the elementary white matrix. The set C'_1 is constructed by permuting the rows of the *black base matrix* S'^1 which is the concatenation of the elementary black matrix and $d_C - 1$ identical copies of the elementary white matrix. Thus, the Hamming weight of the OR of k different rows of any matrix in C'_0 is $w_{C'} = d_C \cdot |\mathcal{H}'| \cdot w_C$ and $b_{C'} = b_C + (|\mathcal{H}'| - 1)w_C + (d_C - 1)|\mathcal{H}'| \cdot w_C = b_C - w_C + d_C \cdot |\mathcal{H}'| \cdot w_C$ for any matrix in C'_1. It follows that the difference between the Hamming weight of black and white is $d_{C'} = b_{C'} - w_{C'} = b_C - w_C = d_C$.

The set of marking matrices C'_M is construced by permuting the rows of the *marking base matrix* S'^M which is the concatenation of d_C identical copies of the elementary marking matrix. Thus, the Hamming weight of the OR of k different rows of any matrix in C'_M is at least $d_C \cdot (1 + |\mathcal{H}'| \cdot w_C) \geq b_{C'}$ if the first row (marking row) is in the set of shares and $d_C \cdot |H'| \cdot w_C = w_{C'}$ otherwise.

Theorem 6. *The above mentioned scheme $C' = (C'_0, C'_1, C'_M)$ is a traceable $(k, \frac{n}{2})$-VCS with (ε, δ)-security and parameters $m = d * l * 2^{k-1}$, $\varepsilon = 1 - \frac{1}{k}$ and $\delta = 0$.*

Proof. Since the underlying (k,k)-VCS and the basic (k,n)-VCS are visual cryptography schemes according to [1, 2] and [5], the first condition is obviously satisfied. Also the second condition is fulfilled as one can see as follows: It is sufficient to compare a single $\widetilde{SH}$ copy in an elementary white matrix with a corresponding one of an elementary marking matrix. If the set of $p < k$ rows of the marking matrix does not contain the first row (marking row) the sets are identical. If the marking row is in the set of rows chosen from the marking matrix, the difference between these two sets consists of a fragment of that particular row which is a Boolean 2^{k-1}-vector. Let q_0 (repectively q_M) denote the fragment of the white row (respectively marking row). By construction, the white Boolean row is a concatenation of rows of the K^0 matrix. Thus, the fragment q_0 corresponds to a row of the K^0 matrix. The marking row was constructed by inserting $q_M = \overline{q_0}$ where $\overline{q_0}$ is the i-th row of some matrix K_i^1 for $1 \leq i \leq k$. Since K^0 and K_i^1 are the base matrices of a (k,k)-VCS for each set of $p < k$ shares, there is a permutation of the columns corresponding to the fragment which permutes the fragment of the marking matrix into the fragment of the white matrix and thus fulfilling the second condition.

The tracing algorithm works as follows: It chooses a set P of p different shares of a white matrix (since the markings were done by modifying white pixels) and takes it as input for the Turing machine $\mathcal{A}$, published by the traitors. If the result is 1, this symbolizes that the share-holder owning the marking row is a member of the saboteur group. Otherwise, nobody can be traced.

In order to prove the correctness of the tracing algorithm we will use the following property of the scheme: Given a set P of $p < k$ rows of some matrix in $C_0' \cup C_1' \cup C_M'$ and given a set of p different indices Q then there is a permutation of the columns which applied to P results in the rows of the white base matrix with indices in Q. This follows from the properties of the (k,k)-VCS and Lemma 5 and implies that no row has any special characteristic allowing the saboteurs to recognize the marking.

According to the definition, there is a subset S' of the shares of the saboteurs such that $A(U) = \text{ThresholdDecision}(S', U)$. The saboteurs are not able to distinguish a marked share from a regular one. Thus, the marked share is in S' with probability $\varepsilon = |S'|/|S| > 1/k$. If the marking row is in S', $\text{ThresholdDecision}(S', U) = 1$ and the tracing algorithm traces the saboteur.

If no saboteur holds the marking row, $A(U) = \text{ThresholdDecision}(S', U) = 0$ since the shares of all saboteurs are shares from a white pixel regardless of S'. Thus the probability that an innocent share-holder will be found guilty is $\delta = 0$.

□

5 Traceable Sharing of a Picture

Let $0 < \Delta \leq 1$. Suppose we have a black and white picture consisting of a finite set of black and white pixels. In order to guarantee that traitors can be traced, it is necessary to insert some markings in the picture. Based on the traceable visual cryptography scheme presented in Sect. 4.3, we will only mark white pixels. If the picture contains mainly black pixels, one has to invert the picture and distribute the shares according to the inverted picture. In the following we assume that $t < k$ traitors pool their shares and publish some information $\mathcal{A}$ for each pixel. We will now focus on the construction of the shares, based on the traceable (k, n)-VCS presented in Sect. 4.3 which is $(1 - \frac{1}{k}, 0)$-secure.

For each share-holder we randomly choose at least $x = \lceil 8(k-1)\log_2(2/\delta)\rceil$ white pixels for which the share-holder will get the marking share. Then, for a white, black, respectively marked pixel we randomly choose a matrix from $\mathcal{C}_0$, $\mathcal{C}_1$, respectively $\mathcal{C}_M$. For each pixel, we also randomly choose a permutation π of $\{1, \ldots, n\}$. In case of an unmarked pixel, the j-th share-holder will get the row $\pi(j)$ of the black or white matrix, depending on the color of the pixel. In case of a marking pixel, the share-holder who will be marked at that particular pixel receives the first row (marking row) of the marking matrix. All the other share-holders will get their shares the same way as in case of a white matrix.

Assuming that $t < k$ traitors pool their shares and publish some information $\mathcal{A}$ for each pixel according to the definition of traceable (k, n)-VCS (see Definition 2), one can trace them as follows:

For each marked pixel the tracing algorithm of the traceable (k, n)-VCS is used to check whether the mark can be found. Then, for each share-holder the number of found marks is computed. A share-holder is found guilty if more than $x/2$ of his marks are found.

Theorem 7. *The probability that a traitor is not found guilty is at most Δ, and the probability that an innocent share-holder is found guilty is 0.*

Proof. At first we look at the probability that a traitor is found guilty. We can consider this situation as a random experiment with x samples and 0-1 variables $X_1, \ldots, X_x$ which have the value 1 with probability $\gamma = 1 - \varepsilon \geq 1/k$. By means of the Chernoff bound we estimate the probability p that less than $x/2 * \gamma$ marks of the traitor are found by

$$p = Pr\{\sum_{i=1}^{x} X_i < x/2 * \gamma\} \leq Pr\left\{|\sum_{i=1}^{x} \frac{X_i}{x} - \gamma| \geq \frac{\gamma}{2}\right\}$$
$$\leq 2e^{\frac{-\gamma^2 * x}{4*2\gamma(1-\gamma)}} \leq 2e^{\frac{-8(k-1)\log(2/\Delta)k}{8k(1-k)}} = \Delta \ .$$

Since the traceable visual cryptography scheme presented in Sect. 4.3 has probability zero for tracing an innocent share-holder, the probability of tracing an innocent person by tracing a whole picture is 0, too. □

6 Open Questions

Remaining open questions are e.g. whether there is a more general definition for traceable visual cryptography schemes or more efficient examples for traceable visual cryptography schemes as the one presented in Sect. 4.3, especially in respect to a higher contrast or a smaller parameter m.

Acknowledgements

The authors would like to thank Moni Naor for helping us understand the techniques for constructing visual cryptography schemes.

References

1. Ateniese, G., Blundo, C., De Santis, A., and Stinson, D.R.: *Visual Cryptography for General Access Structures.* Information and Computation, Vol. 129, No. 2, pp. 86–106, 1996 and ECCC, Electronic Colloquium on Computational Complexity (TR96-012).
2. Ateniese, G., Blundo, C., De Santis, A., and Stinson, D.R.: *Constructions and Bounds for Visual Cryptography.* Proc. 23rd International Colloquium on Automata, Languages and Programming (ICALP '96), Springer Lecture Notes in Computer Science, 1996.
3. Carter, J.L., and Wegman, M.N.: *Universal Classes of Hash Functions.* Journal of Computer and System Sciences 18, pp 143–154, 1979.
4. Droste, S.: *New Results on Visual Cryptography.* Proc. CRYPTO '96, Springer Lecture Notes in Computer Science, pp. 401–415, 1996.
5. Naor, M., and Shamir, A.: *Visual Cryptography.* Proc. EUROCRYPT '94, Springer Lecture Notes in Computer Science, pp. 1–12, 1995.
6. Shamir, A.: *How to Share a Secret.* Comm. of the ACM, Vol. 24, No. 11, pp. 118-129, 1979.
7. Wegman, M.N., and Carter, J.L.: *New Hash Functions and their Use in Authentication and Set Equality.* Journal of Computer and System Sciences 22, pp 265–279, 1981.

Remarks on the Multiple Assignment Secret Sharing Scheme

Hossein Ghodosi
Josef Pieprzyk *
Rei Safavi-Naini

Department of Computer Science
Center for Computer Security Research
University of Wollongong
Wollongong, NSW 2500
AUSTRALIA
e-mail: hossein/josef/rei@cs.uow.edu.au

Abstract. The paper analyses the multiple assignment secret sharing scheme, presented at the GLOBECOM'87 Conference, and contains three technical comments. First it is proved that the proposed multiple assignment secret sharing scheme is not perfect. In fact, the non-perfectness of the scheme is due to the non-perfectness of a certain type of Shamir secret sharing scheme defined in the paper. Next, it is shown that both the extended multiple assignment secret sharing scheme and the extended Shamir secret sharing scheme are not secure, i.e., unauthorized sets of participants can recover the secret.

1 Introduction

Secret sharing schemes allow a group of participants to share a piece of secret information in such a way that only authorized subsets of the participants can recover the secret. Any unauthorized subset is not able to determine the secret. The collection of all authorized subset is called the access structure. Secret sharing schemes have many practical applications. For instance, they can be used to control the access to a safe so only an authorized subset of bank employees can open it by pooling their shares together and reconstructing the secret combination which unlocks the safe.

A particularly interesting class of secret sharing schemes includes threshold schemes with a group of n participants. Their access structure includes all subsets of t or more participants. Such schemes are called t out of n threshold schemes or simply (t, n) schemes. Threshold schemes were independently introduced by

* Support for this project was provided in part by the Australian Research Council under the reference number A49530480 and the ATERB grant

Shamir [1], Blakley [2] and Chaum [3]. An important question of how to realize a secret sharing scheme for an *arbitrary access structure* was studied by numerous authors. Ito, Saito and Nishizeki [4], Benaloh and Leichter [5], and Simmons [6, 7, 8] suggested different solutions for constructing such schemes.

In this paper we consider a generalized secret sharing scheme for an arbitrary access structure, proposed by Ito, Saito and Nishizeki [4]. Their scheme, also called *multiple assignment scheme*, applies the Shamir threshold scheme to realize secret sharing for an arbitrary access structure. They proposed a method to extend a scheme realizing an access structure Γ_1 such that a new scheme realizes an access structure Γ_2, where $\Gamma_1 \subset \Gamma_2$. In order to achieve this goal, they have also proposed a method to extend a Shamir threshold scheme. We are going to demonstrate that:

1. the proposed multiple assignment secret sharing scheme is not perfect,
2. the extended multiple assignment secret sharing scheme is not secure.
3. the extended Shamir threshold secret sharing scheme is not secure.

2 Background

Let $\mathcal{P} = \{P_i : 1 \le i \le n\}$ be a set of n *participants*, and let $\mathcal{K}$, $\mathcal{S}$ denote a *key set* and a *share set*, respectively. Let Γ be a collection of authorized subsets of $2^{\mathcal{P}}$, called the *access structure*, where each $\mathcal{A} \in \Gamma$ is called an *access set*. A *secret sharing scheme* for an access structure Γ is a general method of sharing a secret $K \in \mathcal{K}$ among participants from $\mathcal{P}$ such that a subset $\mathcal{A} \subset \mathcal{P}$ can reconstruct the secret only if $\mathcal{A} \in \Gamma$. An access structure of (t, n) *threshold scheme* consists of all subsets whose cardinality is equal or larger than t.

A secret sharing scheme is set up by a trusted authority, called a *dealer*. The dealer chooses a secret $K \in \mathcal{K}$ and constructs shares $s_i \in \mathcal{S}$, for each participant $P_i \in \mathcal{P}$. Shares are securely delivered to the participants. The reconstruction of the secret is done by a *combiner* who collects shares, recomputes the secret and distributes the result to all collaborating participants via a secure channel.

The system is called *perfect* if

$$H(K \mid \mathcal{A}) = \begin{cases} 0 & \text{if } \mathcal{A} \in \Gamma \\ H(K) & \text{if } \mathcal{A} \notin \Gamma \end{cases}$$

that is, in a perfect secret sharing scheme any unauthorized subset cannot get any information about the secret.

2.1 The Shamir Scheme

The Shamir threshold scheme is based on polynomial interpolation over a finite field. Let $\mathcal{K} = GF(q)$ be a finite field with q elements. To construct a (t, n)

threshold scheme a dealer $\mathcal{D}$ chooses n distinct nonzero elements of $GF(q)$, denoted by $x_1, \cdots, x_n$, and sends x_i to P_i over a public channel ($i = 1, \ldots, n$). For a secret $K \in GF(q)$, $\mathcal{D}$ randomly chooses $t-1$ elements $a_1, \cdots, a_{t-1}$ from $GF(q)$ and constructs a polynomial

$$f(x) = K + \sum_{i=1}^{t-1} a_i x^i$$

The share for participant P_i is $s_i = f(x_i)$. The degree of $f(x)$ is at most $(t-1)$. It is known (see [9]) that Shamir's scheme is perfect. That is, if a group of less that t participants collaborates, their original uncertainty about K is not reduced.

2.2 The Multiple Assignment Scheme

The multiple assignment scheme [4] is a generalized secret sharing scheme that utilizes Shamir threshold scheme to realize an arbitrary access structure. The following notation is used in [4]:

- For any access set $\mathcal{A} \in \Gamma$ ($\mathcal{A} \subset \mathcal{P}$), any superset $\mathcal{A}'$ of $\mathcal{A}$ ($\mathcal{A} \subset \mathcal{A}'$) must be an access set as well. This is the well-known *monotone* property [5]. Thus we have:

$$\mathcal{A} \in \Gamma \text{ and } \mathcal{A} \subset \mathcal{A}' \subset \mathcal{P} \text{ imply that } \mathcal{A}' \in \Gamma$$

- For any access structure Γ there is a family of sets $\bar{\Gamma} = 2^{\mathcal{P}} - \Gamma$. Any set from $\bar{\Gamma}$ represents a collection of participants who are unauthorized to recover the secret. Given an unauthorized set $\mathcal{B} \in \bar{\Gamma}$, then any subset $\mathcal{B}' \subset \mathcal{B}$ must be an unauthorized set as well.
- The family of maximal sets in $\Gamma \subset 2^{\mathcal{P}}$ is denoted by $\delta^+\Gamma$. That is,

$$\delta^+\Gamma = \{\mathcal{A} \in \Gamma \ : \ \mathcal{A} \not\subset \mathcal{A}' \text{ for all } \mathcal{A}' \in \Gamma - \{\mathcal{A}\}\}$$

The multiple assignment scheme works as follows. Let $\Gamma \subset 2^{\mathcal{P}}$ be an access structure. The dealer, $\mathcal{D}$, gets $t = |\delta^+\bar{\Gamma}|$ and utilizes a Shamir (t,t) threshold scheme to generate t shares. Then, for any non-access set $\mathcal{B} \in \delta^+\bar{\Gamma}$, it assigns a distinct share to all participants in $\bar{\mathcal{B}}$ ($\bar{\mathcal{B}} = \mathcal{P} - \mathcal{B}$). For every access set, $\mathcal{A} \in \Gamma$, it is shown that the number of distinct shares given to the participants is equal to t, while for every unauthorized set, $\mathcal{B} \notin \Gamma$, the number of different shares given to its members is less than t. That is, the scheme satisfies the requirement of secret sharing scheme, since the knowledge of at least t shares enables to recover the secret. The knowledge of less than t shares, however, does not allow an unauthorized set to recover the secret (for more detail see [4]).

Example 1. Let $\mathcal{P} = \{A, B, C, D\}$ be the set of participants and let

$$\Gamma = \{\{A, B\}, \{B, C\}, \{C, D\}\}$$

be the access structure. In order to share the secret $K \in GF(q)$, the dealer gets

$$\delta^+\bar{\Gamma} = \{\{A, C\}, \{A, D\}, \{B, D\}\}.$$

Since $|\delta^+\bar{\Gamma}| = 3$, it designs a Shamir $(3, 3)$ threshold scheme and generates three shares, s_1, s_2, s_3. Then it assigns s_1 to participants B and D (which are not in $\{A, C\}$). It also assigns s_2 to participants B and C (which are not in $\{A, D\}$). Similarly, it assigns s_3 to participants A and C. In the reconstruction phase, every access set can reconstruct the secret (knowing three shares, cooperatively), while non-access sets cannot do so.

3 Multiple Assignment Scheme Made Perfect

In this section we show that the proposed multiple assignment scheme [4] is not perfect. The non-perfectness of the scheme is inherited from the underlying non-perfect Shamir scheme. In other words, we show the certain type of Shamir scheme defined in [4] is not perfect.

3.1 A Non-Perfect Shamir Scheme

A Shamir (t, n) threshold scheme, which is used in the multiple assignment secret sharing scheme, is defined as follows [4]:

1. Take a prime power q such that $q > n$ and let $\mathcal{K} = GF(q)$. Select distinct elements $x_1, \ldots, x_n \in \mathcal{K} - \{0\}$ at random.
2. Choose $a_1, \ldots, a_{t-2} \in \mathcal{K}$ and $a_{t-1} \in \mathcal{K} - \{0\}$ randomly, where $t \leq n$.
3. Let $f(x) = K + a_1 x + a_2 x^2 + \cdots + a_{t-1} x^{t-1}$.
4. Let $s_i = f(x_i)$ and assign (x_i, s_i) to P_i for each i, $1 \leq i \leq n$.

Theorem 1. *Given a Shamir (t, n) threshold scheme. If the degree of the associated polynomial $f(x)$ is equal to $t - 1$ then the scheme is not perfect.*

Proof. Let $(t - 1)$ participants collaborate with pooling their shares $s_1, \ldots, s_{t-1}$ in order to perform the Lagrange interpolation polynomial method. Certainly, they can find the unique polynomial $g(x)$ of degree at most $(t - 2)$ such that $s_i = g(x_i)$ for all $i = 1, \ldots, t - 1$, where $g(x) = b_0 + b_1 x + \ldots + b_{t-2} x^{t-2}$. At the same time from the construction of the scheme, it is possible to write $s_i = f(x_i)$ for $i = 1, \ldots, t - 1$. So we have the following system of equations

$$s_1 = g(x_1) = f(x_1)$$

$$\vdots$$

$$s_{t-1} = g(x_{t-1}) = f(x_{t-1})$$

The system can be transformed to

$$(a_0 - b_0) + (a_1 - b_1)x_1 + \ldots + (a_{t-2} - b_{t-2})x_1^{t-2} + a_{t-1}x_1^{t-1} = 0$$

$$\vdots$$

$$(a_0 - b_0) + (a_1 - b_1)x_{t-1} + \ldots + (a_{t-2} - b_{t-2})x_{t-1}^{t-2} + a_{t-1}x_{t-1}^{t-1} = 0$$

Now, we show by contradiction that $a_0 \neq b_0$. Suppose that $a_0 = b_0$. This implies that the system becomes

$$(a_1 - b_1)x_1 + \ldots + (a_{t-2} - b_{t-2})x_1^{t-2} + a_{t-1}x_1^{t-1} = 0$$

$$\vdots$$

$$(a_1 - b_1)x_{t-1} + \ldots + (a_{t-2} - b_{t-2})x_{t-1}^{t-2} + a_{t-1}x_{t-1}^{t-1} = 0$$

As the Vandermonde determinant of the system is different from zero, there is only one solution in which $a_{t-1} = 0$. This contradicts that $f(x)$ is of degree $t-1$ and proves that $a_0 \neq b_0$. Clearly, the $(t-1)$ participants has been successful in finding an integer b_0 which is not the secret, that is, the scheme is not perfect.

Corollary 2. *The multiple assignment secret sharing scheme [4] is not perfect.*

Proof. Let $\Gamma \subset 2^P$ be an arbitrary access structure, such that $|\delta^+\bar{\Gamma}| = t$. The multiple assignment secret sharing scheme generates a Shamir (t,t) threshold scheme and distributes the shares amongst participants in such a way that for access set $\mathcal{A} \in \Gamma$ the number of different shares assigned to participants in a set $\mathcal{A}$ is equal to t. Now consider an unauthorized set $\mathcal{B}$, which possess only $t-1$ shares. Although it cannot recover the secret, can reduce its uncertainty about the secret (by getting a non-secret element from the set of possible shares). That is, the multiple assignment scheme is not perfect.

In order to fix the problem, the underlying Shamir scheme has to be perfect, i.e. the selection of coefficients a_i has to be random and from all elements of $GF(q)$.

4 Extension of a Multiple Assignment Scheme

Let a multiple assignment scheme realizes an access structure $\Gamma_1 \subset 2^{\mathcal{P}_1}$ on a set $\mathcal{P}_1$. Further assume that, $\mathcal{P}_1 \subset \mathcal{P}_2$ and a multiple assignment scheme realizes an access structure $\Gamma_2 \subset 2^{\mathcal{P}_2}$.

In [4], the authors claimed that their scheme is flexible for the case in which a new member joins to the group of shareholders. They considered the following problem [4, Problem 3].

"Can a scheme realizing a access structure Γ_1 be extended so that a new scheme realizes a access structure Γ_2?"
The question was answered affirmatively provided the new access structure Γ_2 is a natural extension of Γ_1, that is, $\Gamma_1 \subset \Gamma_2$ and $\bar{\Gamma}_1 \subset \bar{\Gamma}_2$. The scheme which realizes Γ_2 is an extension of the scheme that realizes Γ_1 if:

(a) both schemes allow to recover the same secret,
(b) the collection of shares defined in Γ_1 is a subset of shares generated in Γ_2.
(c) any access set in Γ_1 is an access set in Γ_2.

It was proved in [4] that a multiple assignment secret sharing scheme is easily extendable. An initial scheme for an access structure Γ_1 over $\mathcal{P}_1$ can be extended to the scheme with an access structure Γ_2 over $\mathcal{P}_2$ ($\Gamma_1 \subset \Gamma_2$ and $\mathcal{P}_1 \subset \mathcal{P}_2$). The extension is done by generation of new shares while leaving the shares of the initial scheme intact.

Theorem 3. *The extension of a multiple assignment scheme, proposed in [4], is not secure. That is, the secret can be reconstructed by an unauthorized set of participants.*

Proof. Let $\mathcal{P}_1 = \{P_1, \ldots, P_n\}$ and $\Gamma_1 \subset 2^{\mathcal{P}_1}$ is an access structure. Let $\Gamma_1 = \{\mathcal{A}_1, \ldots, \mathcal{A}_\ell\}$ and $\delta^+\bar{\Gamma}_1 = \{\mathcal{B}_1, \ldots, \mathcal{B}_t\}$. Assume a multiple assignment scheme realizes Γ_1 and we want to extend the set of shareholders to a set $\mathcal{P}_2$, where $\mathcal{P}_2 = \{P_1, \ldots, P_n, P_{n+1}, \ldots, P_m\}$, that is, $\mathcal{P}_1 \subset \mathcal{P}_2$. Let the access structure $\Gamma_2 \subset 2^{\mathcal{P}_2}$ be as follows:

$$\Gamma_2 = \{\mathcal{A}_1, \ldots, \mathcal{A}_\ell, \{P_i, P_j\}, \text{for } i < j,\ i = 1, \ldots, m-1,\ j = n+1, \ldots, m\}. \quad (1)$$

Clearly, $\Gamma_1 \subset \Gamma_2$. On the other hand, since all subsets consist of two participants in which at least one of them is chosen from the set of new shareholders are access sets, we have,

$$\delta^+\bar{\Gamma}_2 = \{\mathcal{B}_1 \ldots, \mathcal{B}_t, \{P_{n+1}\}, \ldots, \{P_m\}\}.$$

That is, $\delta^+\bar{\Gamma}_1 \subset \delta^+\bar{\Gamma}_2$ and therefore Γ_2 is an extension of Γ_1.

Assume that the multiple assignment scheme, which realizes the access structure Γ_1 on a set $\mathcal{P}_1$, applies the set $S_1 = \{s_1, \ldots, s_{t_1}\}$ of shares. That is, s_1 is assigned to participants in set $\mathcal{P}_1 - \mathcal{B}_1$, and in general s_i is assigned to participants in set $\mathcal{P}_1 - \mathcal{B}_i$. In the new scheme, however, the share s_1 will be assigned to participants in set $\mathcal{P}_2 - \mathcal{B}_1$, and in general s_i will be assigned to shareholders in the set $\mathcal{P}_2 - \mathcal{B}_i$. So the set $\mathcal{P}_2 - \mathcal{P}_1 = \{P_{n+1}, \ldots, P_m\}$ will get the set of all shares s_i ($i = 1, \ldots, t$) from the basic scheme. Since knowing all shares of a secret sharing scheme is sufficient to recreate the secret, every new shareholder solely can recreate the secret, where none of them individually are supposed to be able to recover the secret.

Example 2. Let $\mathcal{P}_1 = \{A, B, C\}$ and let $\Gamma_1 = \{\{A, B\}, \{B, C\}\}$. Since $\delta^+\bar{\Gamma}_1 = \{\{B\}, \{A, C\}\}$, the dealer generates a Shamir (2, 2) threshold scheme and assigns s_1 to set $\mathcal{P}_1 - \{B\}$, that is, to participants A and C. Similarly, it assigns the share s_2 to set $\mathcal{P}_1 - \{A, C\}$, that is, to participant B.

Let $\mathcal{P}_2 = \{A, B, C, D, E\}$ and also let

$$\Gamma_2 = \{\{A, B\}, \{B, C\}, \{A, D\}, \{A, E\}, \{B, D\}, \{B, E\}, \{C, D\}, \{C, E\}, \{D, E\}\}.$$

Hence, $\delta^+\bar{\Gamma}_2 = \{\{B\}, \{A, C\}, \{D\}, \{E\}\}$ and the dealer generates a Shamir (4, 4) threshold scheme to generate four shares s_1, s_2, s_3, s_4. However, this set of shares contains the set of shares s_1, s_2 in which have been generated in the basic scheme. In this extended scheme, however, share s_1 will be assigned to set $\mathcal{P}_2 - \{B\} = \{A, C, D, E\}$, share s_2 will be assigned to set $\mathcal{P}_2 - \{A, C\} = \{B, D, E\}$, share s_3 will be assigned to set $\mathcal{P}_2 - \{D\} = \{A, B, C, E\}$ and finally share s_4 will be assigned to set $\mathcal{P}_2 - \{E\} = \{A, B, C, D\}$. Note that A knows the set of shares $\{s_1, s_3, s_4\}$ and B possesses the set of shares $\{s_2, s_3, s_4\}$. As $\{A, B\} \in \Gamma_1$, shares s_1 and s_2 are sufficient to recover the secret. On the other hand, D holds the set of shares $\{s_1, s_2.s_4\}$ and E possesses the set of shares $\{s_1, s_2, s_3\}$. Clearly, both D and E can individually recover the secret as both know s_1 and s_2.

5 Extension of a Shamir Scheme

Let $\mathcal{P}_1 = \{P_1, \ldots, P_n\}$ be a set of participants and let a Shamir (t, n) threshold scheme be designed on the set $\mathcal{P}_1$. Assume we want to design a Shamir (k, m) threshold scheme on a set $\mathcal{P}_2 = \{P_1, \ldots, P_n, P_{n+1}, \ldots, P_m\}$, such that the set of old shares is still acceptable in the new scheme, that is, the new scheme is an extension of the old scheme. In [4] the authors proposed a method to achieve this goal. They have shown that; if polynomials $f_1(x)$ and $f_2(x)$ which are associated with the Shamir (t, n) and (k, m) threshold schemes satisfy the condition, $k \geq t + 2$ (in their scheme $t = n$), then the extension is possible by generating a polynomial $f_2(x)$ (of degree at most $k - 1$) such that the t shares generated by polynomial $f_1(x)$ still can be generated by polynomial $f_2(x)$. In this section, we show how $(k - 1)$ participants can collaborate to recover the secret in this extended Shamir (k, m) threshold scheme.

Theorem 4. *The extension of a Shamir threshold scheme, proposed in [4], is not secure. That is, any subset of $(k - 1)$ participants can also recover the secret.*

Proof. Consider a Shamir (t, n) threshold scheme which is designed on a set $\mathcal{P}_1 = \{P_1, \ldots, P_n\}$ of n participants and let the associated polynomial be $f_1(x)$. Thus, $f_1(x)$ is a polynomial of degree at most $t - 1$. Also, let a Shamir (k, m) threshold scheme which is designed on a set $\mathcal{P}_2 = \{P_1, \ldots, P_n, P_{n+1}, \ldots, P_m\}$ be an extension of the above mentioned (t, n) scheme. That is all t shares $(t = n)$

of the scheme are also acceptable shares in the new scheme. In the extended scheme, however, the associated polynomial is of degree at most $k-1$. We assume $k \geq t+2$, that satisfies the condition given in [4]. Thus, we have,

$$f_2(x) = K + b_1x + \cdots + b_{k-1}x^{k-1}.$$

Although without knowing that $f_2(x)$ is an extended Shamir scheme, less than k participants obtain absolutely nothing about the secret, here we show, the knowledge of the fact that $f_2(x)$ is an extension of the $f_1(x)$ enables $k-1$ collaborating participants of the extended scheme exactly determine the secret.

Let $f_1(x) = K + a_1x + \cdots + a_{t-1}x^{t-1}$. Let $\mathcal{B} \subset \mathcal{P}_2$ ($|\mathcal{B}| = k-1$) is a set of collaborating participants. Since $f_2(x_i) = f_1(x_i)$, ($1 \leq i \leq n$), the collaborating participants of the set $\mathcal{B}$ know the set of following equations (corresponding to polynomial $f_1(x)$).

$$\begin{aligned} K + a_1x_1 + \ldots + a_{t-1}x_1^{t-1} &= s_1 \\ &\vdots \\ K + a_1x_t + \ldots + a_{t-1}x_t^{t-1} &= s_t \end{aligned}$$

They also know the set of following equations regarding to the set $\mathcal{P}_1$ on polynomial $f_2(x)$.

$$\begin{aligned} K + b_1x_1 + \ldots + b_{k-1}x_1^{k-1} &= s_1 \\ &\vdots \\ K + b_1x_t + \ldots + b_{k-1}x_t^{k-1} &= s_t \end{aligned}$$

Moreover, the collaborating $k-1$ participants $\{P_{n+i_1}, \ldots, P_{n+i_{k-1}}\}$ of the set $\mathcal{P}_2 - \mathcal{P}_1$ can provide the set of following $k-1$ equations:

$$\begin{aligned} K + b_1x_{t+i_1} + \ldots + b_{k-1}x_{t+i_1}^{k-1} &= s_{t+i_1} \\ &\vdots \\ K + b_1x_{t+i_{k-1}} + \ldots + b_{k-1}x_{t+i_{k-1}}^{k-1} &= s_{t+i_{k-1}} \end{aligned}$$

It is not difficult to see that the above three sets of $t+t+k-1$ linearly independent equations have $1+t-1+k-1+t$ unknowns (corresponding to k, a_is, b_js and shares, respectively). Since the number of equations are equal to the number of unknowns, the system of equations has a unique solution for K, that is $k-1$ participants exactly can recreate the secret.

Acknowledgment

The first author would like to thank the University of Tehran for financial support of his study.

References

1. A. Shamir, "How to Share a Secret," *Communications of the ACM*, vol. 22, pp. 612–613, Nov. 1979.
2. G. Blakley, "Safeguarding cryptographic keys," in *Proceedings of AFIPS 1979 National Computer Conference*, vol. 48, pp. 313–317, 1979.
3. D. Chaum, "Computer Systems Established, Maintained, and Trusted by Mutually Suspicious Groups," tech. rep., Memorandum No. UCB/ERL M/79/10, University of California, Berkeley, CA, Feb. 1979.
4. M. Ito, A. Saito, and T. Nishizeki, "Secret Sharing Scheme Realizing General Access Structure," in *Proceedings IEEE Global Telecommun. Conf., Globecom '87*, pp. 99–102, Washington: IEEE Communications Soc. Press, 1987.
5. J. Benaloh and J. Leichter, "Generalized Secret Sharing and Monotone Functions," in *Advances in Cryptology - Proceedings of CRYPTO '88* (S. Goldwasser, ed.), vol. 403 of *Lecture Notes in Computer Science*, pp. 27–35, Springer-Verlag, 1990.
6. G. Simmons, "How to (Really) Share a Secret," in *Advances in Cryptology - Proceedings of CRYPTO '88* (S. Goldwasser, ed.), vol. 403 of *Lecture Notes in Computer Science*, pp. 390–448, Springer-Verlag, 1990.
7. G. Simmons, "Robust Shared Secret Schemes or 'How to be Sure You Have the Right Answer Even Though You Don't Know the Question'," in *18th Annual Conference on Numerical mathematics and Computing*, vol. 68 of *Congressus Numerantium*, (Manitoba, Canada), pp. 215–248, Winnipeg, May 1989.
8. G. Simmons, "Prepositioned Shared Secret and/or Shared Control Schemes," in *Advances in Cryptology - Proceedings of EUROCRYPT '89* (J.-J. Quisquater and J. Vandewalle, eds.), vol. 434 of *Lecture Notes in Computer Science*, pp. 436–467, Springer-Verlag, 1990.
9. D. Stinson, "An Explication of Secret Sharing Schemes," *Designs, Codes and Cryptography*, vol. 2, pp. 357–390, 1992.

Secret Sharing in Hierarchical Groups

Chris Charnes[1], Keith Martin[2] *, Josef Pieprzyk[1] ** and Rei Safavi-Naini[1]

[1] Department of Computer Science, Centre for Computer Security Research
University of Wollongong, Wollongong, NSW 2500, AUSTRALIA
[2] Katholieke Universiteit Leuven, Dept Elektrotechniek-ESAT
Kardinaal Mercierlaan 94, B-3001, Heverlee, Belgium

Abstract. We introduce the idea of hierarchical delegation within a secret sharing scheme and consider solutions with both conditional and unconditional security.

1 Introduction

In this paper we introduce the concept of secret sharing in hierarchical groups, given a group of *participants* which is partitioned into different *levels.* We assume that there is a *secret* which is to be protected in such a way that only certain subsets of participants, after being *delegated* to do so, can jointly retrieve the secret. Each level is associated with an *access structure* which specifies the collection of sets of participants who are authorised to collectively recreate the secret (the *authorised sets*). At first only authorised sets at the highest level can retrieve the secret. However, should the need arise, it is possible for this level to delegate the ability to retrieve the secret to a lower level. Only once this has been done, authorised sets belonging to the lower level can retrieve the secret. Hence this system has a dynamic access structure which changes through the delegation process. Such systems have potential applications in any situation that a secret sharing scheme might be used [15, 18] within any organisation or trust structure that is layered hierarchically [7].

In the first part we consider *unconditionally secure* systems. There is already an extensive body of literature dealing with unconditionally secure secret sharing schemes (see for example [15, 18]). Our model for secret sharing in hierarchical groups uses delegation to dynamically change the overall access structure of the system [3, 7, 12]. Simmons [16] defined the entire lattice of access structures that can arise from other access structures through trust relationships between participants. MTA-free schemes [8, 10] approach this from a reverse angle.

Unconditionally secure secret sharing schemes can normally be used once only. If a scheme is to be reused, it is usually necessary to reissue the shares. To overcome this problem we transform an unconditionally secure linear scheme into a *conditionally secure* scheme that can be used many times.

---

* This work done while the author was at Department of Pure Mathematics, The University of Adelaide, Adelaide, SA 5005, AUSTRALIA

** Support for this project was provided in part by the Australian Research Council under the reference number A49530480 and the 1995 ATERB grant

2 Hierarchical delegation in secret sharing

Assume there is a secret $k \in \mathcal{K}$ and a group of participants $\mathcal{P}$. An access structure Γ is the collection of authorised subsets of $\mathcal{P}$. Any subset from Γ is able to collectively recover the secret. We assume that Γ is *monotone* and can be determined uniquely by its *minimal sets* [18]. If the collection Γ^- of minimal sets of Γ is $\{C_1, \ldots, C_r\}$ then we can represent Γ by the boolean expression $\Gamma = C_1 + \cdots + C_r$ (see [1, 12, 17]).

In hierarchical secret sharing, $\mathcal{P}$ is partitioned into λ levels, $\mathcal{P}_1, \ldots, \mathcal{P}_\lambda$, and each level $\mathcal{P}_i$ is associated with a corresponding access structure Γ_i for the purposes of retrieving the secret. We make the following assumptions about the process of delegation from a higher level to a lower level: (1) a higher level can only delegate the power for secret reconstruction to certain pre-specified lower levels; (2) the delegation is successful if certain specified participants at the higher level *jointly* agree to the delegation. For each level i ($i = 1, \ldots, \lambda - 1$), let $D[i] \subseteq \{i+1, \ldots, \lambda\}$ be the collection of levels to which level i can delegate. Further, for each level i ($i = 1, \ldots, \lambda - 1$) we associate a *delegation access structure* $\mathcal{B}_i$, defined on $\mathcal{P}_i$, which consists of the subsets of $\mathcal{P}_i$ that are desired to be able to jointly delegate any lower level $j \in D[i]$. A *delegation structure* is determined by the collections $\{\mathcal{B}_1, \ldots, \mathcal{B}_{\lambda-1}\}$ and $\{D[1], \ldots, D[\lambda - 1]\}$.

To effect delegation we will use the concept of *delegation tickets*. To each $p \in \mathcal{P}_i$ ($i = 1, \ldots, \lambda - 1$) there is associated a collection of tickets $\{T_{pj} \mid j \in D[i]\}$. Ticket $T_{pj} \in \mathcal{T}_{pj}$ is the piece of information that p uses when p wants to participate in delegation of level j. If participants at level j are presented with a collection of tickets $\{T_{pj} \mid p \in A \in \mathcal{B}_i\}$, then participants in $B \in \Gamma_j$ can subsequently recover the secret.

We say that a set of tickets T *directly delegates from level i to level j* ($j \in D[i]$) if there exists some $A \in \mathcal{B}_i$ such that $T_{pj} \in T$ for all $p \in A$. Further let $1 = b_0 < b_1 < b_2 < \cdots < b_a = k$. We say that a collection of tickets *delegates to level k* if (1) $b_i \in D[b_{i-1}]$ (for $i = 1, \ldots, a$) and (2) for each i ($i = 0, \ldots, a-1$), T directly delegates from level b_i to level b_{i+1}. Tickets can be transferred from level i to level j using either broadcasting or secure transfer using a confidentiality channel. We assume broadcast transfer for the rest of this paper.

Some interesting classes of delegation structure are: *complete delegation*, for which $D[i] = \{i + 1, \ldots, \lambda\}$ for each $i = 1, \ldots, \lambda - 1$; *sequential delegation*, for which $D[i] = \{i+1\}$ and $i = 1, \ldots, \lambda - 1$; *ranked delegation*, for $1 \leq i < j < k$, if $k \in D[j]$ then $k \in D[i]$ (hence $D[j] \subseteq D[i]$) (we assume that $D[1] = \{2, \ldots, \lambda\}$, since any level not in $D[1]$ is never delegated); *transitive delegation*, for $1 \leq i < j < k$, if $k \in D[j]$ and $j \in D[i]$ then $k \in D[i]$. For simplicity we restrict the rest of this paper to complete delegation.

3 Unconditionally secure approach

A *hierarchical delegation secret sharing scheme* (HD-SSS) with the access structure $(\Gamma_1, \ldots, \Gamma_\lambda)$ and delegation structure $\{\mathcal{B}_1, \ldots, \mathcal{B}_{\lambda-1}\}$, $\{D[1], \ldots, D[\lambda - 1]\}$, comprises two algorithms: dealer and combiner. The dealer generates and distributes shares and tickets (via secure channels) to all participants from $\mathcal{P}$. The combiner can be called by any collection of participants $A \subset \mathcal{P}_i$ on the level i. The combiner can recover the secret $k \in \mathcal{K}$ if the set $A \in \Gamma_i$ and level i has been delegated by some higher level (level one is initially delegated). We will use the following access structure Ω:

$$\Omega = \Gamma_1 + \mathcal{B}_1(\Gamma_2 + \mathcal{B}_2(\Gamma_2 + \cdots + (\Gamma_{\lambda-1} + \mathcal{B}_{\lambda-1}\Gamma_\lambda)\cdots)).$$

Theorem 1. *The access structure Ω can be used to distribute shares and tickets to form a HD-SSS.*

The proof of the above theorem involves treating the access structures $\mathcal{B}_i$ as being on 'dummy' participants and issuing shares corresponding to them as the tickets. During a delegation, participants at a higher level broadcast these shares as the tickets that delegate to a lower level. Note that we can write $\Omega = \Gamma_1 + \Gamma_2\Lambda_2 + \ldots \Gamma_\lambda\Lambda_\lambda$, where Λ_i represents all the different delegation chains which can delegate the i-th level. An HD-SSS is called *perfect*, if for the i-th level, $H(K|AT) = 0$ if $A \in \Gamma_i$ and T is a collection of tickets that delegates to level i, and $H(K|AT) = H(K)$ otherwise. In such an HD-SSS the following simple bounds must hold:

Lemma 2. *In a perfect HD-SSS scheme if $p \in A \in \Gamma_i$ (for some level i) then $H(p) \geq H(K)$. If $p \in B \in \mathcal{B}_i$ (for some level i) then for any $j \in D[i]$, $H(T_{pj}) \geq H(K)$, where $H(p)$ is the entropy of shares assigned to participant $p \in \mathcal{P}_i$.*

Using Lemma 2 it is possible to prove some further bounds on the storage complexity for participants in an HD-SSS.

4 Conditionally secure approach

If an HD-SSS is to be practically implemented it is highly desirable that such a scheme can be used more than once without redistribution of shares and tickets. An alternative is to issue several shares and tickets in advance, but this leads to considerable increases in storage requirements. We permit re-use by converting an unconditionally secure scheme to a conditionally secure one and note that while we have hierarchical delegation schemes in mind, such transformations can be applied to any unconditionally secure scheme. We perform the conversion by applying a one-way function to the shares of the original scheme. Following previous conditionally secure schemes [4, 5, 6, 11, 13] we use exponentiation as our one-way function.

The transformation can be applied to any secret sharing scheme that is *linear*, which includes schemes based on polynomial interpolation [14], monotone circuit constructions [1, 9], linear code constructions [2], and finite geometrical constructions [17]. Informally, a secret sharing scheme is *linear* if, for every $A \in \Gamma$, secret K and corresponding shares $s_1, \ldots, s_{|A|}$, there exists (publicly known) constants $\mu_1, \ldots, \mu_{|A|}$ such that $K = \mu_1 s_1 + \cdots + \mu_{|A|} s_{|A|}$. Thus the transformation idea is simple – participants hold their secret shares s_i and when they decide to cooperate, they send a *modified shares* g^{s_i} to the combiner who computes the secret $g^K = g^{\sum_{i=1}^{\ell} \mu_i s_i} = \prod_{i=1}^{\ell} g^{\mu_i s_i}$ (and hence for this instance of the scheme the secret is actually g^K). The exponent K and shares s_i are never revealed, and to reuse the scheme it is enough to broadcast a new generator g'. The participants new modified shares are g'^{s_i} and the new secret is g'^K. Properties of these kind of schemes were considered in [4].

Before exponentiation is used, two problems have to be solved. Firstly, some representation has to be used which maps shares s_i and coefficients μ_i into the integers. The second issue concerns the form of the modulus to be used. We consider three possibilities: (1) the modulus is an irreducible polynomial $p(x)$ which generates $GF(2^m)$ where $2^m - 1$ is prime, (2) the modulus N is prime (Diffie Hellman exponentiation), and (3) $N = pq$ where p and q are primes (RSA exponentiation).

Theorem 3. *For exponentiation in Galois fields $GF(2^m)$ (generated by an irreducible polynomial $p(x)$) and $2^m - 1$ is prime, the congruence $g^K = \prod_{i=1}^{\ell} g^{\frac{a_i}{b_i} s_i}$ (mod $p(x)$) has a unique solution and $K = \sum_{i=1}^{\ell} \frac{a_i}{b_i} s_i \pmod{2^m - 1}$.*

Theorem 4. *For exponentiation in Galois fields $GF(N)$ and $N - 1 = p_1 p_2 \ldots p_r$ where p_j are primes for $j = 1, \ldots, r$, the congruence $g^K = \prod_{i=1}^{\ell} g^{\frac{a_i}{b_i} s_i} \pmod{N}$ has r unique solutions $k_j = \sum_{i=1}^{\ell} \frac{a_i}{b_i} s_i \pmod{p_j}$ if all b_j are co-prime to $N - 1$. Combining these using Chinese Remainder Theorem, the numbers k_j ($j = 1, \ldots, r$) generate the solution $K = \sum_{i=1}^{\ell} \frac{a_i}{b_i} s_i \pmod{N - 1}$.*

Theorem 5. *For exponentiation modulo N where $N = pq$, p and q are two RSA primes ($p - 1 = 2p'$ and $q - 1 = 2q'$), the congruence $g^K = \prod_{i=1}^{\ell} g^{\frac{a_i}{b_i} s_i} \pmod{N}$ has three unique solutions $k_j = \sum_{i=1}^{\ell} \frac{a_i}{b_i} s_i \pmod{p_j}$ for $p_j \in \{2, p', q'\}$ if all b_j are co-prime to $\lambda(N) = 2p'q'$. Combining these using the Chinese Remainder Theorem, the numbers k_j ($j = 1, \ldots, r$) generate the solution $K = \sum_{i=1}^{\ell} \frac{a_i}{b_i} s_i$ (mod $\lambda(N)$).*

Example: Implementation based on linear codes

In [2] it was shown how linear block codes over $GF(q)$ could be used to implement linear secret sharing schemes for general access structures. We show how this construction can be converted.

We firstly recall the setting in [2]. Given an arbitrary access structure Ω on participant set $\mathcal{P} = \{p_1, p_2, \ldots\}$, there exists an algorithm [2] yielding a generator matrix G of a linear block code over $GF(q)$, which can be used to share a secret. (There is a choice of the parameter $GF(q)$ in their algorithm.) The shares of a secret which is to be distributed to the participants $\mathcal{P}$ are components of a certain code vector $(c_0, c_1, \ldots, c_{n-1})$, see Definition 4 of [2]. Where c_0 is the secret and is not distributed. The recovery of c_0 is by means of the parity check matrix

H of the code ($GH^T = 0$). The secret satisfies the equation $c_0 = -\sum_{i=0}^{k-1} l_i s_i$. Where the vector $(s_0, s_1, \ldots, s_{k-1})$ is defined by $s = Hv^T$. The *received* vector v is formed as follows. Suppose that participants from the set $A = \{p_i, p_j, \ldots\}$ are authorised, then the non-zero entries of v consist of the collection of shares allocated to participants from the set A. (For more details see [2].)

This scheme can be converted using exponentiation as follows. Let g be a primitive generator of $GF(q)^*$ and $GF(q)$ is chosen so that the discrete logarithm problem for this field is difficult. Corresponding to c_0 in the scheme above, we define the secret to be g^{c_0} and show how to share this secret among the participants. We will assume, for simplicity of notation, that each participant has a single share. The received vector for $\{p_i, p_j, \ldots\}$ is $v = (0, 0, a_i, a_j, \ldots)$. Which can be split up as

$$v = (0, 0, a_i, 0, \ldots) + (0, 0, 0, a_j, 0, \ldots) + \cdots. \tag{1}$$

Each vector v_i in the direct sum decomposition of v, has only one non-zero component, which is the share of the secret of participant p_i. As $s = Hv^T$, we can get that $s = H{v_1}^T + H{v_2}^T + \cdots$.

Suppose participants from A decide to recover the secret. In [2] it is shown that these participants can determine the l-vector $l = (l_0, l_1, \ldots l_{k-1})$ from the public parity check matrix H of the code, see Section 4 of [2]. Hence participants $\{p_i, p_j, \ldots\}$ know the l-vector, consequently each participant P_i knows $v_i' = H{v_i}^T$ and l. Each participant p_i can therefore compute their modified shares as $g^{(l, v_i')} = g^{(l, H{v_i}^T)}$, where (l, v_i') is the usual inner product.

Theorem 6. *The combiner can recover the secret g^{c_0} from the set of modified shares $g^{(l, v_i')}$.*

Thus, using the above scheme it is possible to use the techniques of the Section 3 to distribute shares and tickets of an HD-SSS that can be used more than once.

5 Comments

Clearly we have only briefly touched on the subject of hierarchical delegation in secret sharing schemes, and the intent of this short paper is only to introduce the topic. Further work in this area includes determining useful bounds on the necessary storage in such systems and on finding solutions that minimise such storage. Also of interest is the consideration of different types of delegation and of delegation by secret transfer rather than broadcast transfer.

References

1. J. Benaloh and J. Leichter. Generalized secret sharing and monotone functions. *Proceedings of CRYPTO'88, Lecture Notes in Computer Science, Advances in Cryptology*, 403:27–35, 1990.

2. M. Bertilsson and I. Ingemarsson. A construction of practical secret sharing schemes using linear block codes. *Advances in Cryptology - AUSCRYPT'92, Lecture Notes in Computer Science, J. Seberry and Y. Zheng (Eds)*, 718:67–79, 1993.
3. C. Blundo, A. Cresti, A. De Santis, and U. Vaccaro. Fully dynamic secret sharing schemes. *Advances in Cryptology - CRYPTO'93, Lecture Notes in Computer Science*, 773:110–125, 1994.
4. C. Charnes, J. Pieprzyk, and R. Safavi-Naini. Conditionally secure secret sharing schemes with disenrollment capability. In *Proceedings of the 2nd ACM Conference on Computer and Communication Security, November 2-4, 1994, Fairfax, Virginia*, pages 89–95, 1994.
5. Y. Desmedt. Threshold cryptography. *ETT*, 5(4):449–457, August 1994.
6. W. Diffie and M.E. Hellman. New directions in cryptography. *IEEE Trans. Inform. Theory*, IT-22(6):644–654, November 1976.
7. H. Ghodosi, J. Pieprzyk, C. Charnes, and R. Safavi-Naini. Cryptosystems for hierarchical groups. *Information Security and Privacy - ACISP'96, Lecture Notes in Computer Science*, 1172:275–286, Springer 1996
8. I. Ingemarsson and G.J. Simmons. The consequences of trust in shared secret schemes. *Advances in Cryptology - EUROCRYPT'90, Lecture Notes in Computer Science*, 473:266–282, Springer 1991.
9. M. Ito, A. Saito, and T. Nishizeki. Secret sharing schemes realizing general access structures. In *Proceedings of Global Telecommunications Conf., Globecom'87*, pages 99–102. IEEE Communications Soc. Press, 1987.
10. W.A. Jackson, K.M. Martin, and C.M. O'Keefe. Efficient secret sharing without a mutually trusted authority. *Advances in Cryptology - EUROCRYPT'95, Lecture Notes in Computer Science*, 921:183–193, 1995.
11. Hung-Yu Lin and Lein Harn. A generalized secret sharing scheme with cheater detection. In *Advances in Cryptology - ASIACRYPT'91, Lecture Notes in Computer Science (H. Imai, R.L. Rivest and T. Matsumoto (Eds))*, pages 149–158. Springer-Verlag, 1993. Fujiyoshida, Yamanashi, Japan, November 11-14, 1991.
12. K.M. Martin. Untrustworthy participants in perfect secret sharing schemes. In *Cryptography and Coding III*, pages 255–264. Oxford University Press, 1993.
13. T. P. Pedersen. Non-interactive and information-theoretic secure verifiable secret sharing. *Proceedings of CRYPTO'91, Lecture Notes in Computer Science, Advances in Cryptology*, 576:129–140, 1992.
14. A. Shamir. How to share a secret. *Communications of the ACM*, 22(11):612–613, November 1979.
15. G.J. Simmons (Ed). *Contemporary Cryptology - The Science of Information Integrity*. IEEE Press, New York, 1992.
16. G.J. Simmons. The consequences of trust in shared secret schemes. *Advances in Cryptology - EUROCRYPT'93, Lecture Notes in Computer Science*, 765:448–452, 1994.
17. G.J. Simmons, W. Jackson, and K. Martin. The geometry of shared secret schemes. *Bulletin of the ICA*, 1:71–88, 1991.
18. D.R. Stinson. An explication of secret sharing schemes. *Designs, Codes and Cryptography*, 2:357–390, 1992.

Stateless Connections

Tuomas Aura, Pekka Nikander
Helsinki University of Technology, FIN-02015 HUT, Finland
Tuomas.Aura@hut.fi, Pekka.Nikander@hut.fi

Abstract We describe a secure transformation of stateful connections or parts of them into stateless ones by attaching the state information to the messages. Secret-key cryptography is used for protection of integrity and confidentiality of the state data and the connections. The stateless protocols created in this way are more robust against denial of service resulting from high loads and resource exhausting attacks than their stateful counterparts. In particular, stateless authentication resists attacks that leave connections in a half-open state.

1 Introduction

In the open networks, malicious denial-of-service attacks and resource exhaustion by unexpectedly high demand for a service have become increasingly serious threats. In this paper we show how stateless protocols can be used to make systems more robust against denial-of-service. Our goal is to limit the number of potential attackers and to make recovery after an attack easier. This is done by saving the server state in the client rather than in the server.

The paper is structured as follows. We first discuss resource exhaustion problems typical of stateful services in Sec. 2. Sec. 3 shows how to securely make protocols stateless and compares their behavior to stateful ones under denial-of-service attacks. Sec. 4 describes partially stateless protocols. Sec. 5 contains some examples of real-world protocols and Sec. 6 concludes the paper.

The closest resemblants to our ideas in the literature are the HTTP cookie mechanism [7] and the stateless Sprite file server [9] that distributed the storage of state information in order to recover from server crashes. Perrochon [8] suggests stateless front-ends to stateful services. The cost of saving the connection state has been considered as one aspect affecting scalability of authentication protocols, for example, in [6]. Most literature concentrates on application-specific design techniques and reactive countermeasures.

2 Denial of service in stateful protocols

In Sec. 2.1 we explain how storing connection state makes stateful protocols vulnerable to resource exhaustion by overload. Sec. 2.2 discussed malicious attacks.

2.1 Running out of connection table space

In stateful protocols, there is always an upper limit on the number of clients that can connect to a server simultaneously, even if the clients might still be satisfied

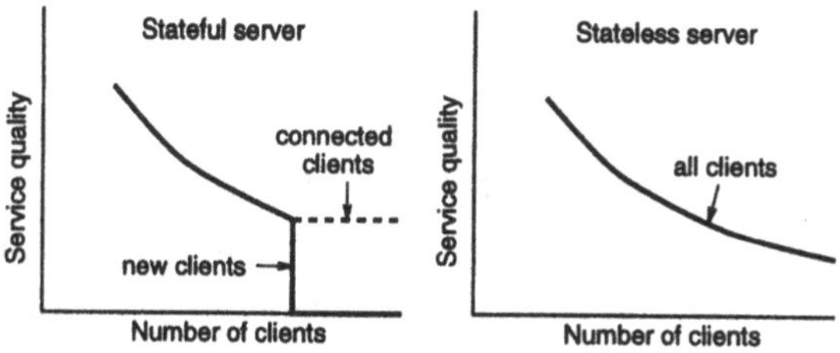

Figure1. Service quality as a function of load in stateless and stateful servers

with a thinner share of the server capacity. Eventually, the restriction is caused by the limited space that is available for storing connection state information. When more and more clients attempt to connect to the server, its storage space becomes exhausted and new connections must be refused. In the worst case, the connected clients do not consume the full capacity of the server, and the server remains partially idle while the refused clients are waiting to connect.

An unintentional mistake or a communication error may also leave a connection in an eternally open state. The client can forget to close the connection, lose its connection table data, or it may be unable to reach the server for the closing commands. In the end, the stale connections must be purged from the server table but it is difficult do this without sometimes closing valid connections.

One way to describe the problem with stateful servers is that their behavior under stress is unideal. Fig. 1 compares the quality of stateful and stateless service when the number of clients or connections to the server increases. The service quality in stateful server shows sharp fall at a certain threshold while in the stateless server it declines slowly.

2.2 Attacks exhausting connection limit

In a denial-of-service attack, an attacker can exhaust the connection limit of a server. If an open connection does not obligate the client to actively use the service, the malicious party needs to sacrifice only little of its capacity to continually block other clients' access to the service. Connection state maintenance becomes a critical resource, which it normally would not be. The attack is particularly disturbing because the attacker is not utilizing the service but merely consuming a secondary resource that is needed for the service access.

Of course, the attacker tries to keep the connections in a state where the number of simultaneous clients is as limited as possible, the effect of blocking maximally harmful, and its own efforts minimal. Furthermore, the attacker would prefer not to reveal its own identity to the server. Therefore, the danger of this type of attacks is usually greatest at the beginning of connections.

3 Making connections stateless

In Sec. 3.1, we show how to make protocols stateless by passing the state information between the protocol principals along the messages. Sec. 3.2 and 3.3 add integrity check to the state data and to the entire connection. Sec. 3.4 compares the behavior of stateful and stateless protocols under denial-of-service attacks.

3.1 Transformation from stateful into stateless

Assuming that the communication channels are reliable and flooding attacks the only concern, we can transform any stateful client/server protocol or communication protocol with initiator and responder into a stateless equivalent. This is done by sending state information from the server to the client with every message. Along the next message from the client, the state information is returned to the server. A stateful protocol and an equivalent stateless protocol:

1. $C \longrightarrow S : Msg_1$	S stores $State_{S1}$.
2. $S \longrightarrow C : Msg_2$	
3. $C \longrightarrow S : Msg_3$	S stores $State_{S2}$.
4. $S \longrightarrow C : Msg_4$	
... ...	

1. $C \longrightarrow S : Msg_1$
2. $S \longrightarrow C : Msg_2, State_{S1}$
3. $C \longrightarrow S : Msg_3, State_{S1}$
4. $S \longrightarrow C : Msg_4, State_{S2}$
... ...

Usually the server or the responding principal is the primary target of the denial-of-service threats and it is sufficient to make this principal stateless. In a symmetric protocol it is also possible to make both principals stateless by passing the states of both principals between them. Similar transformations are possible for multi-party protocols if the messages travel suitably. There one must take care that the state information is returned to the stateless principal in time.

The main reason for the stateless transformation is that it makes the system behavior more ideal. When there is no limit on the number of clients, the limit cannot be exploited by denial-of-service attacks. The ideal protocol properties also simplify quantitative analysis of system behavior under stress.

Moreover, the stateless protocol moves the responsibility of saving the state information from the server to the client. The client, who has requested the service, is better motivated to maintain the information and to recover from error conditions and data loss. The server does not have to reserve its resources for a single client for the indefinite time that may pass between protocol messages.

Another application for stateless protocols is information services that divide the server load between several identical machines. The servers can be geographically distributed or clustered in one place. When the servers are stateless, client requests can be routed to an arbitrary server without giving any consideration to where the previous messages were processed. Routing decisions and reply addresses can be changed dynamically in order to level the load on the servers and to minimize communication costs.

As a drawback, the stateless protocols require additional bandwidth for transferring the state data. If the states are large, the cost may be too high. File or document servers are therefore examples of promising applications for stateless protocols while intensely interactive sessions most often are not.

Although stateless protocols implemented in the above way resist denial-of-service attacks by server flooding, we still have to address other security issues. This is the topic of the next sections.

3.2 Integrity and confidentiality of the state data

When the state data is repeatedly transferred through insecure channels, its integrity and confidentiality become an important security concern. Since the state messages are sent and eventually received by the server itself, we can protect their integrity with message authentication codes that are relatively short and inexpensive to compute.

Also, the freshness of the state data should be checked in order to limit the number of times the data can be replayed. Timestamps can be applied liberally, because they are checked by their creator against the same clock that is used for the timestamping. Expired messages can be simply ignored. (The client is responsible for taking any corrective steps after such error conditions.) It is, however, necessary to allow long lifetimes for the states so that the data does not expire before the client wants to continue the message exchange and succeeds in sending its next request. Hence, the timestamp lifetime should be longer than the expected duration of a denial-of-service attack, usually on the order of several hours or a few days.

One consequence of timestamping is that distributed servers that accept state data packets created by each other must have synchronized clocks. Fortunately, the accuracy does not need to be very high if the timestamp lifetimes are long.

The improved transformation of a stateful service into a stateless one is illustrated by the protocol schema below. Every message leaving the server contains a timestamped state of the connection, authenticated with a key K_S^s known only by the server. The state is then returned to the server along the next message from the client.

$$
\begin{array}{l}
1.\ C \longrightarrow S : Msg_1 \\
2.\ S \longrightarrow C : Msg_2,\ T_{S1}, State_{S1},\ MAC_{K_S^s}(T_{S1}, State_{S1}) \\
3.\ C \longrightarrow S : Msg_3,\ T_{S1}, State_{S1},\ MAC_{K_S^s}(T_{S1}, State_{S1}) \\
4.\ S \longrightarrow C : Msg_4,\ T_{S2}, State_{S2},\ MAC_{K_S^s}(T_{S2}, State_{S2}) \\
\quad \ldots \qquad \ldots
\end{array}
$$

There is often redundancy in the actual message and the state information. In that case, it is not necessary to repeat the redundant data. The MAC can be computed over all state information that is explicitly or implicitly returned to the client in the next step.

Another method for checking freshness is to change signature keys periodically. A few of the newest keys should be kept in the server's memory for accepting fresh messages. State information signed with older keys is then discarded as invalid. A key identifier should be added to the messages. The period of generating new signature keys becomes thus the resolution of message expiration times. The period of validity is the period of key generation multiplied by the number of newest keys accepted. Keys with different lifetimes can be used for different purposes. The computation required for maintaining the keys is independent of the number of clients or of the amount of traffic in the system.

Also, any secret state data is easily concealed by encrypting it with a secret key K_S^e known only by the server. The mechanism is illustrated below.

$$\begin{array}{|rl|}\hline i.\ C \longrightarrow S: & Msg_i,\ T_{S,i}, E_{K_S^e}(State_{S,i}),\ MAC_{K_S^s}(T_{S,i}, State_{S,i}). \\ i+1.\ S \longrightarrow C: & Msg_{i+1},\ T_{S,i}, E_{K_S^e}(State_{S,i}),\ MAC_{K_S^s}(T_{S,i}, State_{S,i}) \\ \hline \end{array}$$

3.3 Integrity and confidentiality of the connection

So far, the described stateless protocols have not addressed the integrity or confidentiality of the actual protocol messages in any way. In this section we will describe a technique for authenticating and encrypting stateless connections. We have an additional reason for the security enhancements. Namely, the statelessness opens a new line of attack against connection integrity: replay of connection states. The stateless principals have no means for detecting replays, because they cannot remember which messages have already been received and processed. An integrity check that links the state data to the actual messages will limit the ways in which third parties can utilize recorded server states in attacks.

In the following protocol schema, the client and the server have a shared secret key K_{CS} for signing and encrypting connection data. The server passes the key to the client along with all other state data.

$$\begin{array}{|rl|}\hline i.\ C \longrightarrow S: & Msg_i',\ T_{S,i}, E_{K_S^e}(K_{CS}), State_{S,i},\ MAC_i \\ i+1.\ S \longrightarrow C: & Msg_{i+1}',\ T_{S,i}, E_{K_S^e}(K_{CS}), State_{S,i},\ MAC_i \\ \hline \end{array}$$

The message authentication code is $MAC_i = MAC_{K_S^s}(T_{S,i}, K_{CS}, State_{S,i})$ and the protected messages $Msg_k' = E_{K_{CS}}(Msg_k), MAC_{K_{CS}}(k, Msg_k, MAC_i)$ for $k = i, i+1$.

It is necessary to encrypt the messages only if their contents are secret. Authentication codes, on the other hand, should be used in in all systems where replay attacks are considered a threat, even on anonymous connections. Binding the state data together with the corresponding messages in this way effectively shields the system against third-party replay attacks that attempt to manipulate the logic of the protocol. (Replay flooding attacks will be discussed in Sec. 3.4.) Replays to the server still can result in multiple processing of requests and duplicate responses to the client. The protocol designer should ensure that the client is able to detect the duplicates or is not affected by them.

A dishonest client, on the other hand, cannot be stopped from replaying state data from earlier stages of the protocol run. By replaying old states, the client can return to any previous point in the protocol run. Consequently, the client can execute parts of the protocol several times, or go through several alternative branches of the protocol run. In some protocols, the possibility of collecting information from several alternative execution paths is catastrophic. For example, in many zero knowledge proofs, allowing two choices for the prover or verifier could result in a false proof or disclosure of secret knowledge, respectively. Also, stateless protocols cannot be used when accounting of service use is needed, for example, for billing the clients. Therefore, not all protocols can be securely made stateless. The stateless transformation is not suitable for protocols where the client is not entitled to the combined benefit from a small number of alternative protocol runs. Luckily, most communication protocols do not account resource usage and are deterministic enough so that the client will not gain any advantage by replaying the states.

3.4 Replays and denial of service

An attacker with access to the communication channel between the server and its clients might try to exhaust a stateless server by continuously sending replays of old messages. In this section, we compare the replay-flooding attack against stateless protocols to the connection-limit-exhaustion attack against stateful ones and show that the stateless protocol performs better.

We first consider the stateless server under a replay flooding attack. The best the attacker can do is to replay messages at the maximum throughput rate of the server so that no service capacity remains for honest clients. (This is the worst case scenario. In most communication systems, some legitimate messages will still get through.)

Another danger is that the legitimate connections start breaking because the time stamps on the state data expire while the attacker blocks the service. This can be avoided by making the timestamp lifetimes longer than it takes to detect the attack and to take countermeasures. After the attack, the clients can continue the connections without delay. This is the reason why the time stamps should last days rather than seconds or minutes. The timestamp lifetime, however, should not be infinite, because we want to limit the amount of replayable material in circulation.

Next, we consider the behavior of a stateful server when an attacker is creating new connections and leaving them open. If the attacker opens connections at the maximum rate C allowed by the server, no other clients can access the service. The stateful server must purge the idle connections from its memory after a certain time to make space for new ones. If the server has enough memory to save the state of M connections, each connection will remain in the memory at most time M/C. In a typical system, this time will be much shorter than the duration of an average attack. Thus, the attacker is able to break the existing connections.

Comparing the stateless and stateful protocols under their characteristic attacks, we observe that an equal rate of replays against the stateless server and connection openings against the stateful server have approximately the same effect on the service quality during the attack. Both servers can be clogged by the attacks. The stateless server, however, recovers automatically after an attack.

Another major advantage for the stateless server is that the described worst-case scenario is less likely to happen for it. We assumed that the attacker can record enough messages for the replay attacks. On a large network, most nodess never see any such messages. Hence, the number of parties that can mount the replay attack against the stateless server is very small in comparison to the group that can open false connections to the stateful server. For example, on the Internet, anyone in the world can open connections to almost any public server but very few hosts can record connections to a particular server.

We conclude that stateless protocols are, in general, more robust against denial-of-service attacks than their stateful counterparts. Stateless protocols make recovery after an attack easy and dramatically reduce the number of potential attackers.

4 Partially stateless protocols

Often the benefits of statelessness are biggest at certain specific parts of the protocol. Sections 4.1 and 4.2 discuss stateless connection opening and idle periods. In Sec. 4.3 we consider stateless layers in protocol stacks. Sec. 4.4 discusses optimizations based on state caching.

4.1 Stateless handshake

When abuse of the service is expected, the obvious solution is to authenticate the clients at the beginning of the connection. Attackers usually do not want to reveal their identity. Furthermore, authentication helps the server administration in resolving problems off-line. In this section, we show how statelessness improves robustness of the protocol at the beginning of the connection, before sufficient client authentication has taken place.

The level of authentication may vary from strong cryptographic identification to anonymous verification of access rights or electronic payment. In any case, the first few steps of the protocol before the client has been authenticated are vulnerable to the same kinds of denial-of-service attacks as the completely unauthenticated connections. The attacker may start the authentication procedure and then leave the server waiting at an intermediate state. Therefore, we suggest that authentication protocols should always remain stateless until the client authentication is complete or the client has in some other way clearly shown its commitment to honest use of the service. After the authentication, the server can change to stateful mode. Especially in pay-per-use services this is most natural because the server must save accounting information.

We now demonstrate the importance of authenticating the client before the server becomes stateful. In the three-way X.509 authentication protocol [5], an attacker can replay a large number of old copies of the first message. This could exhaust the space that the principal B has reserved for saving the state of the protocol between sending Message 2 and receiving Message 3. (Note that the three-way X.509 protocol uses nonces for verifying freshness of the messages. Hence, principal B only knows that Message 1 is fresh after receiving Message 3.)

1. $A \longrightarrow B : S_A(N_A, B, E_B(K_{AB}))$
2. $B \longrightarrow A : S_B(N_B, A, N_A, E_A(K_{BA}))$,
3. $A \longrightarrow B : S_A(N_B, B)$

In the following modification of the X.509 protocol, the responding principal B does not need to save the state of the protocol until it has positively authenticated A.

1. $A \longrightarrow B : S_A(N_A, B, E_B(K_{AB}))$
2. $B \longrightarrow A : S_B(N_B, A, N_A, E_A(K_{BA}))$,
   $T_B, E_{K_B^e}(K_{AB}, K_{BA}),\ MAC_{K_B^a}(T_B, N_B, A, K_{AB}, K_{BA})$
3. $A \longrightarrow B : S_A(N_B, B), T_B, E_{K_B^e}(K_{AB}, K_{BA}), MAC_{K_B^a}(T_B, N_B, A, K_{AB}, K_{BA})$

The above protocol is deterministic in the sense that there is only one execution path. Its runs differ only in that fresh nonces and keys are generated every time. Therefore, an attacker could not possibly collect any interesting information by replaying old states and thus causing the principals to re-execute steps. Luckily, most cryptographic protocols have a similar deterministic nature. The biggest benefit attackers can sometimes gain from repeating protocol runs or steps is a little more material for cryptanalysis, but this should not endanger the security of strong cryptographic algorithms.

Most key exchange protocols aim at producing unique keys for every session and purpose. For the stateless principals, however, there is no difference between one and many sessions with the same parameters. Furthermore, the branching of the key exchange process can lead to the generation of several alternative end results. Thus, one-to-one correspondence between session and keys may be partially lost in the stateless protocols.

In anonymous services that are free to everyone, the return address of the client may be the only available identifier, but it is often enough. If the client responds to a message from the server, the server at least knows how to reach the client, which can be construed as a level of authentication. In the following protocol schema, the server is stateless only during the first roundtrip from it to the client and back. This is sufficient to prevents attacks like the SYN-flooding against the TCP protocol [2].

1. $C \longrightarrow S$:	$Msg1$
2. $S \longrightarrow C$:	$Msg2,\ T_S, State_S,\ MAC_{K_S^s}(T_S, State_S)$
3. $C \longrightarrow S$:	$Msg3,\ T_S, State_S,\ MAC_{K_S^s}(T_S, State_S)$
	Return address valid. S moves to stateful mode.
4. $S \longrightarrow C$:	$Msg4$
...	...

4.2 Statelessness during idle periods

In long-lived protocol runs, activity often ceases and is resumed later. In a stateful protocol, the server will have to maintain connection state data such as session keys throughout the idle periods. The server can be relieved of this duty by sending the state information to the client for temporary storage. The client has the responsibility for saving the data and recovering from data losses.

Depending on the protocol, the server can send its state to the client with every message, after certain messages, or after the connection has been idle for a threshold time. The client can either automatically return the state in its next message or check first whether the server still has the state in its cache.

The session parameters are usually known to both principals. Hence, it is not necessary to send everything to the client. Often only the session key needs to be encrypted with the server's secret key and sent over the channel along with a message authentication code for the key and the rest of the session parameters. The client knows the parameters and can return them with the message authentication code in its next message. If necessary, the client can encrypt confidential session parameters with the session key.

$$S \longrightarrow C : Msg1, T_S, E_{K_S^s}(K_{SES}),\ MAC_{K_S^s}(T_S, C, \text{sespar}, K_{SES})$$

Possibly long idle period follows.

$$C \longrightarrow S : Msg2, T_S, C, \text{sespar}, E_{K_S^s}(K_{SES}), MAC_{K_S^s}(T_S, C, \text{sespar}, K_{SES})$$

When the session keys are transfered in this way, the server's key encryption keys must be treated with as much care as any master key and preferably changed periodically.

4.3 Stateless layers in protocol stacks

Communication protocols are organized in stacks where each layer uses the services of the layer below and provides services to the layer above. The stateful and stateless protocols should be viewed in this context. It seems that statelessness offers most benefits at the transport and application layers of the protocol stack. For example, in the TCP/IP protocol stack, denial-of-service attacks are usually targeted either at the TCP protocol [2] or at the application layer. In UDP based protocols, the application layer protocols are the natural target [3]. When an application layer protocol on the top of the stack is stateless, it usually makes sense to have all layers down to the network layer equally stateless.

A protocol can have more than one upper layer protocol accessing it. Then, regardless of whether the upper protocols are stateless or not, it is beneficial to make the lower layers stateless. The reason is that the alternative upper layer protocols should be able to operate independently of the resources consumed by each other and attacks against each other. To accomplish this, it is not always necessary to send the state data to the other end of the connection. The lower layer could also pass the information to the upper layers for storage.

4.4 State caching

The main disadvantage of fully stateless protocols is the hit on protocol performance. Fortunately, most performance benefits of stateful servers can be retained in stateless protocols by caching state data in the server.

The stateless server can cache state information as long as its memory capacity is not exhausted. Under normal server load, a caching server can behave just as a stateful server except for the cost of transmitting the state data. It can, for example, detect lost and duplicated messages, report the errors to the client, collect statistics on the channel characteristics, and dynamically adjust the transfer rate and packet size for optimum performance. Windowing techniques can be applied to avoid waiting for acknowledgements from the other party after each request. If server memory becomes scarce or the maintenance of the state data too burdensome, the server can immediately purge the data from its memory. In this way, the system can have nearly the performance of stateful servers while being resistant to denial-of-service. Although it may be difficult to design state caching servers with optimal performance, it should be feasible to find reasonable compromises. State caching has been succesfully applied in the stateless Sprite file server [9].

5 Application examples

In Sec. 5.1 we improve both robustness and performance of a key exchange protocol from the ISAKMP specification. Sec. 5.2 describes how the statefulness of the TCP protocol leads to the SYN flooding attack.

5.1 Stateless ISAKMP/Oakley

We show how to make stateless a version of the Oakley key exchange [4] used in the Internet Security Association and Key Management Protocol (ISAKMP).

The initiator A and the responder B start by exchanging nonces in order to to ensure that the initiator has given a correct reply address. The protocol uses public key signatures and Diffie-Hellman key exchange with reusable secret parameters and nonces for freshness.

$$
\begin{array}{|l|}
\hline
1.\ A \longrightarrow B : N_A \\
2.\ B \longrightarrow A : N_B, N_A \\
3.\ A \longrightarrow B : N_A, N_B, S_{K_A}(g^x, A, B, N'_A) \\
4.\ B \longrightarrow A : N_B, N_A, S_{K_B}(g^y, B, A, N'_B, N'_A) \\
5.\ A \longrightarrow B : N_A, N_B, S_{K_A}(g^x, A, B, N'_A, N'_B) \\
\hline
\end{array}
$$

The responder knows the initiator's key generation parameter to be fresh only after receiving Message 5. This leaves the protocol vulnerable to attacks where someone initiates a connection, executes Steps 1, 2 and 3 of the protocol, and then leaves the responder waiting forever.

We enhance the protocol by making the responder stateless until the receipt of the last message. The initial cookie exchange may be unnecessary, because the stateless responder is not as greatly affected by opening messages with forged return address. Thus, the protocol is more robust and has less messages than the original one.

$$
\begin{array}{|l|}
\hline
1.\ A \longrightarrow B : N_A, g^x, A, B, N'_A \\
2.\ B \longrightarrow A : N_B, N_A, S_{K_B}(g^y, g^x, B, A, N'_B, N'_A), \\
\qquad\qquad MAC_{K_B^s}(T_B, g^y, g^x, N_B, N_A, N'_B, N'_A),\ E_{K_B^s}(y) \\
3.\ A \longrightarrow B : N_A, N_B, S_{K_A}(g^x, g^y, A, B, N'_A, N'_B), \\
\qquad\qquad MAC_{K_B^s}(T_B, g^y, g^x, N_B, N_A, N'_B, N'_A),\ E_{K_B^s}(y) \\
\hline
\end{array}
$$

5.2 TCP resistance to SYN flooding

Recently, attention has been paid to the so called SYN flooding attack against the TCP/IP Transmission Control Protocol (TCP). In the TCP connection establishment, the parties exchange message sequence numbers in the so called SYN and SYN ACK messages. After receiving the first message, the responder creates a state called Transmission Control Block (TCB). In the SYN flooding attack, the attacker fills up the responder's TCB table by sending SYN messages to the server with forged IP return adresses.

Several people have independently suggested versions of this protocol where the server does not create a state initially. Instead, the responder can compute a message authentication code of the initiator sequence number and other session parameters. It sends the MAC to the client and receives it again in the next message.

The original and enhanced protocols are shown below. The MAC plays the dual role of functioning as ISN_S and the MAC for the server's state. ($MAC = MAC_{K_S^s}(ISN_C, parameters)$)

1. $C \longrightarrow S : ISN_C, SYN$	1. $C \longrightarrow S : ISN_C, SYN$
2. $S \longrightarrow C : ISN_S, ISN_C + 1, SYN\|ACK$	2. $S \longrightarrow C : MAC, ISN_C + 1, SYN\|ACK$
3. $C \longrightarrow S : ISN_C + 1, ISN_S + 1, ACK$	3. $C \longrightarrow S : ISN_C + 1, MAC + 1, ACK$

After these changes, the server avoids creating the TCB before it knows the initiator's IP address to be valid. Early experiments by the present authors indicate that the TCP protocol could be futher enhanced by making the responder completely stateless.

6 Conclusion

We described weaknesses in stateful protocols and showed how they they can be avoided by making the connections stateless. The stateless protocols behave more ideally under overload or denial-of-service attacks than their stateful counterparts, they recover faster from the attacks, and they effectively limit the number of potential attackers. In particular, stateless protocols resist attacks that leave connections in a half-open state.

References

1. Tuomas Aura and Pekka Nikander. Stateless connections. Technical Report A46, Helsinki University of Technology, Digital Systems laboratory, May 1997.
2. TCP SYN flooding and IP spoofing attack. CERT Advisory CA-96.21, CERT, November 1996.
3. UDP port denial-of-service attack. CERT Advisory CA-96.01, CERT, August 1996.
4. D. Harkins and D. Carrel. The resolution of ISAKMP with Oakley. Internet draft, IETF IPSEC Working Group, June 1996.
5. Recommendation x.509 (11/93) - the directory: Authentication framework. ITU, November 1993.
6. P. Janson, G. Tsudik, and M. Yung. Scalability and flexibility in authentication services: The KryptoKnight approach. In *IEEE INFOCOM'97*, Tokyo, April 1997.
7. David M. Kristol and Lou Montulli. HTTP state management mechanism. Internet draft, IETF HTTP Working group, July 1996.
8. Louis Perrochon. *Gateways in globalen Informationssystemen.* PhD thesis, ETH Zürich, 1996. Diss. ETH Nr. 11708.
9. Brent Welch, Mary Baker, Fred Douglis, John Hartman, Mendel Rosenblum, and John Ousterhout. Sprite position statement: Use distributed state for failure recovery. In *Proc. 2nd Workshop on Workstation Operating Systems WWOS-II*, pages 130–133, September 1989.

Design of a Security Platform for CORBA based Application

Rakman Choi, Jungchan Na, Kwonil Lee, Eunmi Kim, Wooyong Han

S/W Engineering Section, Electronics and Telecommunications Research Institute, Yousung P.O. Box 106, Taejeon, KOREA

Abstract. This paper proposes a security platform, SCAP(Security platform for CORBA based APplication), to cope with potential threats in a distributed object system. SCAP supports CORBA security specification announced by OMG. SCAP is composed of four functional blocks which co-work with ORB to provide security services: Authentication Block, Security Association Block, Access Control Block, and Security Information Management Block. It is designed to support Common Secure Interoperability functionality level 2 which is useful for large scale intra- or inter-enterprise network based applications. Actual security services which are dependent on supporting security technology will be provided as external security services for replaceability. Implementation issues such as how to simulate an interceptor mechanism using a commercial ORB product without source code, and how to extend Current object required for security services are also described.

1 Introduction

Recently, information technology is characterized by the distributed computing system which is composed of a set of co-operating software components running in a number of computers inter-connected by various networks. Moreover, many of the focuses have been on applying object oriented technology to distributed systems. New paradigm of information technology which is integrated with distributed computing and object-oriented technology has led to a vision of componentized software, that is a software system dynamically constructed from prefabricated components and ready for network installation.

While the distributed object system may improve the efficiency of application development and the reusability of components, it is more vulnerable to security threats than the traditional system as there are more places that the system can be attacked. Common threats in a distributed object system are access to hidden information by an authorized user, user masquerading, by-passing security controls, eavesdropping on a communication line, tampering with communication between objects, and a lack of accountability[4]. Security techniques used for centralized system are not suitable for the distributed object system. Therefore, new security technology that takes account of their inherent distributed and object- oriented natures is needed.

Some security technologies such as Kerberos[8], Security Framework for POSIX Open System Environment[1], OSF DCE(Distributed Computing Environment)[9], and SESAME (Secure European System for Applications in a Multivender Environment) [10] are surveyed to support the security of distributed computing system. The deployment and use of Kerberos in many live user environment has been proven in practice in its ability to provide the foundation for authentication and building secure distributed applications within a local network, or within closed groups of users. But Kerberos is constrained by its current limitations of supporting purely symmetric key distribution, and an identity-based authorization model. Though POSIX provides a well-defined security framework and interfaces for secure open environment, it does not provide secure object model and architecture to be implemented easily. DCE security service is common in the distributed processing environment and SESAME is recognized as one of the security mechanisms efficient for distributed and heterogeneous environments. Because they have an integrated asymmetric key distribution, and privilege-based authorization support. However, DCE and SESAME do not support the object-oriented concept and security unaware applications.

Common Object Request Broker Architecture (CORBA)[6] announced by Object Management Group (OMG) supports the construction and integration of client-server applications in heterogeneous distributed environments. The transmission of service requests and responses between clients and servers is handled by CORBA so that applications need not deal with concerns like: where clients and servers are located on the network; differences between hardware platforms, operating systems, and implementation languages; networking protocols; and others. To provide the core security facilities and interfaces required to ensure a reasonable level of security of CORBA based applications as a whole, OMG announced CORBA security specification[4][7]. It suggests an architecture which allows various security technologies to be used. It is strongly recommended to use a set of generic security interfaces such as the Generic Security Services Application Program Interface (GSS-API)[3] to insulate the implementations of the object security service from detailed knowledge of the underlying mechanism. Those security technologies may be provided by existing security components which actually implement the security services. OMG also announced Common Secure Interoperability (CSI) specification[5] to provide the standards for secure interoperability between Object Request Brokers (ORB). Recently, CORBA is generally recognized as a platform for new distributed object applications or an application integration middleware. Therefore it is expected that it will have a significant impact on the future development of distributed system and applications.

This paper proposes a security platform, SCAP, to provide security services required to provide protection against threats in a distributed object system. It is composed of four functional blocks: authentication block, security association block, access control block, and security information management block.

This paper is organized as follows. In Section 2, we present design issues and overall structure of SCAP. In Section 3, detailed design for each functional block of SCAP is described. Section 4 describes implementation issues confronted when making existing ORBs into secure ones. Finally, In Section 5, we conclude with future works.

2 Overview of SCAP

2.1 Design Issues

SCAP is a middleware and a security platform for CORBA based applications in a distributed object environment. Security services such as authentication, access control, security of communication, and an ease-to use security administration will be provided by SCAP to provide security in standalone and distributed CORBA-compliment systems.

It is also designed to support CSI functionality level 2 which is particularly useful for large scale intra- or inter-enterprise network based applications. In SCAP, actual security services which are dependent on security technology will be provided as external security services for replaceability. Through technical review on existing security technologies, Trusted Third Party(TTP) components of SESAME was selected to provide external security services in SCAP.

For some technical and strategic reason, security facilities for security aware applications, privilege delegation, secure interoperability, and security services such as security auditing and non-repudiation are not included in the current version of SCAP. Those will be supported in the next version.

2.2 Overall Structure of SCAP

As shown in Fig. 1, our security platform is based on a client-server model and it is composed of three machines: domain security server, client machine, and target machine.

The domain security server is composed of a security information management block and a external security server. The security information management block provides facilities and interfaces for a security administrator to manage the security policy domain and it also provides security information to other blocks. The external security server provides external security services such as authentication services, privilege attribute services, security association services, and cryptographic support facilities. Physically, the external security server is composed of TTP components of SESAME such as the Authentication Server(AS), the Privilege Attribute Server(PAS), the Key Distribution Server(KDS), the Local Registration Authority(LRA), the Certification Authority Agent(CAA) and the Certification Authority(CA). Their perspective roles are as follows: the AS authenticates principals acting as initiators, the PAS certifies access privileges of principals acting as initiators, the KDS provides keys to principals, acting either as initiators or targets, to protect the exchange of security information(i.e. Privilege Attribute Certificates(PAC) and related control information), the LRA is the authority from which entities may request the generation of asymmetric key pairs, the CAA receives and responds to such requests when forwarded from LRA, the CA produces Directory Certificates containing public keys so that

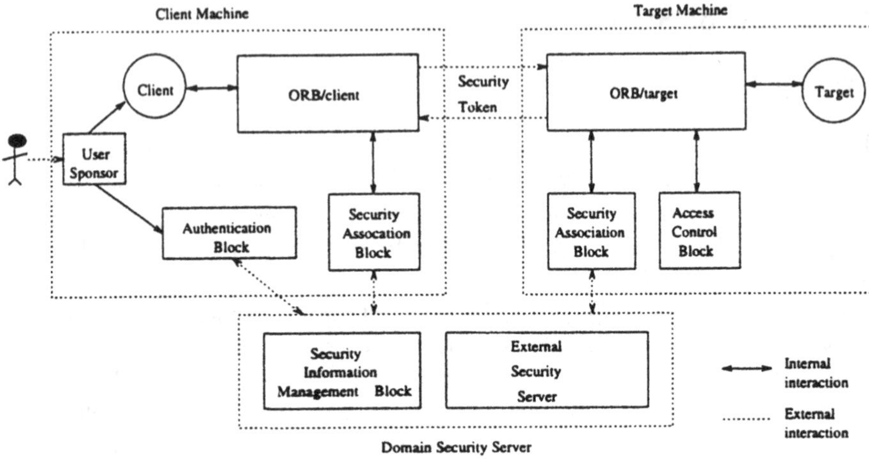

Fig. 1. Operational structure of security platform

anybody who trusts the CA can check a signature produced by an entity whose public key is known[10].

Fundamentally, the client and target machine are composed with following components; authentication block, security association block, access control block, and ORB. The authentication block provides identification and authentication of principals to verify they are who they claim to be. The security association block provides security of communication via establishing trust between the client and target. It also provides integrity protection and confidentiality protection of message in transit between objects. The access control block provides access control which decides whether a principal can access an object, by using the identity and/or other privilege attributes of the principal and the control attributes of the target object in order to access it.

To enter the system, a user should submit his identity, password, and role name to the user sponsor. Using user information, the user sponsor invokes the authentication block to authenticate the person, who specifies the name of the roll he wishes to use. Actual authentication service for the authentication block is provided by the AS and the PAS of the external security server. A new credentials object is created by the authentication block using the PAC received from the AS. After a successful authentication, a client application is invoked and it submits a service request to the ORB/client. The ORB/client calls the security association block to create a security context. The security association block creates a new security context and receives keying information from the KDS of the external security server. Thereafter the ORB/client sends it to the ORB/target in security token format to establish the security association with the target object. According to the client's request, the ORB/target also calls the target-side security association block to establish a security association with the client and sends an acknowledge message that the security context establishment was completed. After completion of security association, the ORB/client calls the security association block again to encrypt the client's request message and sends the encrypted request message to the ORB/target. The ORB/target decrypts the client's request message using the security association block and calls the access control block to decide whether access to the target is allowed. The request will be transmitted to the target application, if access is allowed.

3 Detailed Design of SCAP

3.1 Authentication Block

As shown in Fig. 2, the authentication block includes Principal Authentication object, and Credentials object.

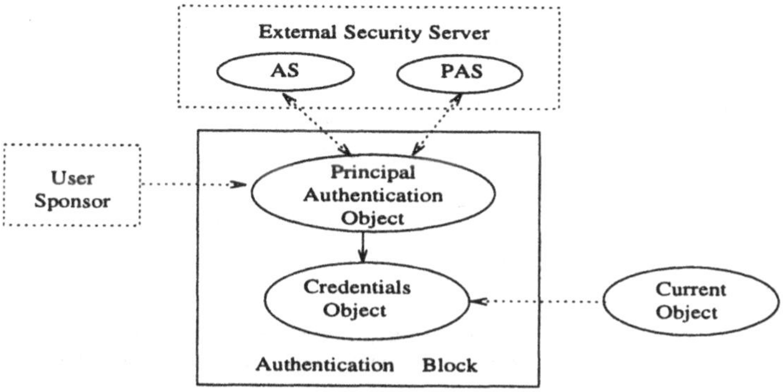

Fig. 2. Structure of Authentication Block

For user authentication, the user sponsor invokes the Principal Authentication object by calling the member function of the Principal Authentication object. The Principal Authentication object requests actual authentication from the AS of the external security server by providing a user identity and password(or secret key). As a response to the request, the AS returns a PAS ticket and a session key encrypted under the secret key of the claimed user. If the user wants privilege attributes by specifying his role name, the Principal Authentication object will request the privilege attributes of the user from the PAS of the external security server by using the PAS ticket and session key. It creates a Credentials object based on the privilege attributes returned and authenticated identities received previously.

The Credentials object holds the security attributes of a principal. These security attributes include its authenticated identities and privileges and information for establishing security associations. It provides operations to obtain and set security attributes of the principal it represents. A reference of Credentials object is retained by Current object. The Current object represents the current thread of activity at both client and target object. In a secure object system, It supports the secure current interfaces, which gives access to security information.

3.2 Security Association Block

Security association block provides initialization and management function related to security context needed to establish security association between the client and target object. As shown in Fig. 3, this block is composed of Message Interceptor, Vault object, Security Context object, and Secure Association Context Management(SACM).

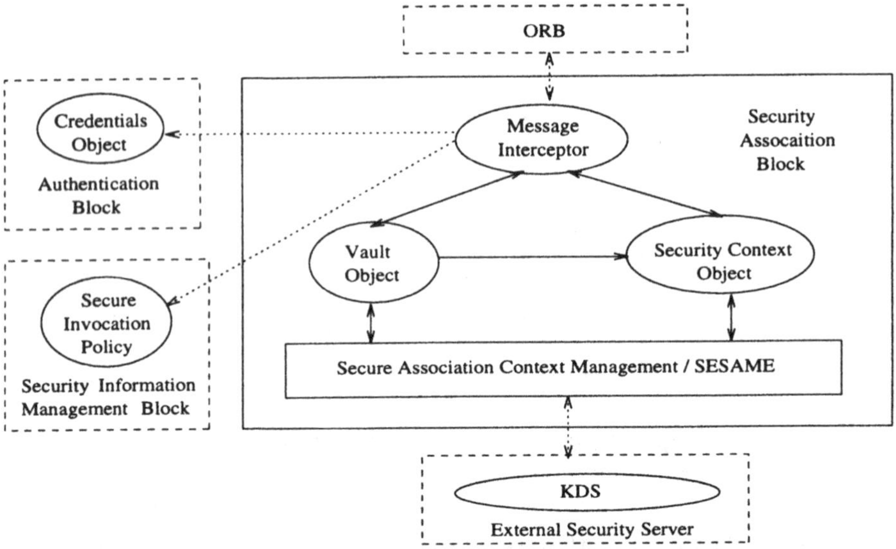

Fig. 3. Structure of Security Association Block

The Message Interceptor provides an interaction mechanism between ORB core and the object security services. At bind time, the interceptor establishes a security context which the client initiating the binding can use to securely invoke the target object designated by the object reference used in establishing the binding. At object invocation time, the interceptor is called to use the previously established security context to protect the message data transmitted from the client to the invoked target object.

It intercepts client's service requests and checks if there is already a suitable Security Context object for this client's use of this target. If a suitable context already exists, it is used. If no suitable context exists, the interceptor obtains a credentials information from the authentication block and a secure invocation policy from the security information management block and establishes a security association between the client and target object.

The client interceptor calls the Vault object to request the security features (such as quality of protection) required by the client policy and target. Through the SACM, the Vault obtains session keys and SESAME-made credentials from external security server. The SACM is a component of SESAME which provides actual interactions with the KDS of external security server. The SACM provides interfaces such as GSS-API for the object security services. Using the obtained SESAME-made credentials, the interceptor creates a new Security Context object and returns a security token to be sent to the target, and indicates whether a continuation of the exchanges is needed. It also returns a reference to the Security Context object for this client-target security association. The interceptor constructs the association establishment message (including the security token, which must be transferred to the target to permit it to establish the target side Security Context object).

On receiving an association establishment message at the target side, the target's message interceptor separates it into the security token and the request message and use the Vault of the appropriate Security Context object to process the security token.

Once the association is established, the message interceptor is used for message protection providing integrity and/or confidentiality protection of requests and responses, according to quality of protection requirements specified for this security association in the active Security Context object.

3.3 Access Control Block

In SCAP, the access control block protects unauthorized users from performing operations on target objects. Based on a credential that was produced by the authentication block, the target's control attributes, such as the domain access policy and the required rights policy, and context information related to execute operations, the block can enforce an access control mechanism based on rights scheme to check access permission.

According to the CORBA security specification, access control functions may be enforced by the ORB services on the object invocation and/or by the application itself[4]. However, SCAP is designed to support access control function automatically enforced by the ORB during the object invocation. In addition, our access control decision is made at the target side depending on the target's domain access policy and access rules such as the POSIX.6 rules[11]. As a consequence, access control for applications in SCAP is automatically made by the target side ORB on the basis of privileges of request users, the target domain's access control policy, the required rights policy, and the access rules.

As shown in Fig. 4, the access control block consists of two object classes: Request Interceptor and Access Decision object. The Request Interceptor intercepts all requests from the clients to application targets, and it invokes the access decision function in Access Decision object for validation of the requested access. Depending on the result of access decision, the Access Decision object may forward the request to the target object or may raise an exception error.

In order to decide whether the access is allowed, first of all, the interceptor must acquire the user's privilege attributes from the received credentials through the Security Context object created by the target side security association block. Thereafter, the interceptor calls the function of access decision object with the acquired privileges.

The Access Decision object takes both required rights policy and domain access policy from the security information management block. Based on user's privileges, required rights from required rights policy, and granted rights from the domain access policy, this object makes a decision whether the client's request is allowed to execute an operation of a target object or not.

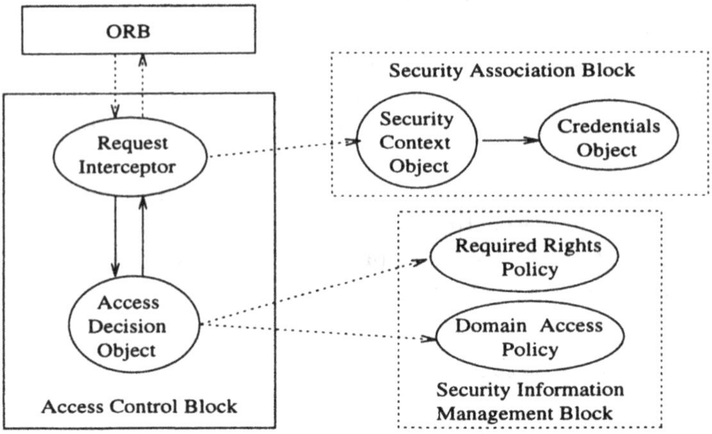

Fig. 4. Structure of Access Control Block

3.4 Security Information Management Block

The functions of security information management block can is categorized into two groups. The one is a set of functions for security administration, and the other is functions to support other three blocks of SCAP.

As shown in Fig. 5, this block is composed of three components: User Attribute Manager(UAM), Role Attribute Manager(RAM), and Security Policy Manager(SPM). Security information of each domain such as authentication information, user's attributes, roll attributes, etc. is maintained in a centralized security information base.

The UAM and the RAM provide functions required to manage authentication information, user's attributes, and roll attributes needed by the authentication block.

The SPM is composed of three sub-managers which manage required rights policy, domain access policy, and secure invocation policy. The required rights policy and the domain access policy are used by the access control block to decide whether the client's request is allowed or not. The secure invocation policy is used by the security association block to support the establishment of the security association and the message protection.

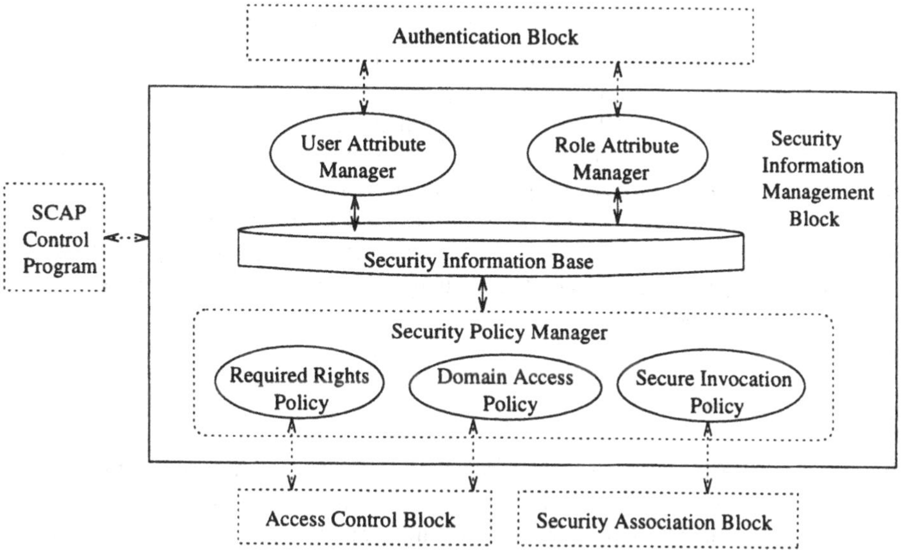

Fig. 5. Structure of Security Information Management Block

The user who wants to manage security policy domain should log-in as a security administrator. The SCAP control program provides functions needed to manage the domain itself, the members of the domain, and security information such as authentication, user's attributes, role names, and security policies. Using the functions provided by the UAM, RAM, and SPM, the security administrator may create, modify, delete contents of the security information base.

4 Implementation Issues

The SCAP design has been completed and we are now implementing each block of SCAP. For the strategic reason, we decide to adopt an outsourcing approach. According to the strategy, we are implementing the SCAP using the commercial ORB "IONA's Orbix"[2] as a building block and TTP components of SESAME to provide external security services. However, the current version of our Orbix does not support the facilities interceptor and Current object which are needed to provide security services defined in CORBA security specification. Interceptor is responsible for the execution of one or more ORB services. Logically, it is interposed in the invocation path between a client and a target object.

To implement interceptor mechanism to be embedded in the ORB core, modification of the ORB core is required. This is impossible, because the ORB source code is not available for us. As an alternative solution in current situation, we use a filter mechanism provided by the Orbix to link the ORB and the external interceptor. Both the message interceptor, for the security association block, and the request interceptor, for the access control block, may be initiated through filters. The external interceptor interacting with the ORB through the filter can be operated as if ORB-embedded interceptors does. However, this approach may cause a problem to the application transparency of security services.

According to the CORBA security specification, the Current object should provide some information associated with the execution context at both the client and the target object. In particular, it associates security information, such as credentials, with the Current object and provides means to access it. But the current version of Orbix does not support the Current object. To solve the above problem, one instance of the Current object will be created externally whenever it is needed.

5 Conclusions

This paper suggests a security platform, SCAP, for CORBA based applications. The platform is composed of four functional blocks: authentication block, access control block, security association block, and security information management block. It is designed to provide security services defined in CORBA security specification to cope with potential threats in a distributed object system, especially for large scale intra- or inter- network environment.

SCAP is designed to be implemented using currently available IONA's Orbix and TTP components of the SESAME as a building block. At the next version of SCAP, the filter and the external interceptor of current version SCAP will be replaced with a standard interceptor embedded in ORB core. The external security server of current SCAP implemented using TTP components of the SESAME may also be replaced with our own version.

For some technical and strategic reason, security facilities for security aware applications, privilege delegation, and secure interoperability and security services such as security auditing and non-repudiation are excluded from the current version of SCAP. Those will be supported in the next version.

References

1. IEEE 1003.22/D6, *Draft Guide to the POSIX Open System Environment: A Security Framework*, IEEE Doc. Number NO.13, Aug. 1995.
2. IONA Technologies Ltd., *Orbix Reference Guide*, 1997.
3. J. Linn, *Generic Security Service Application Programming Interface*, IETF RF C1508, Sep. 1993.

4. Object Management Group, *CORBA Security*, OMG Document NO 95-12-1, Dec., 1995.
5. Object Management Group, *Common Secure Interoperability Specification*, OMG Document no. orbos/96-06-20 , Jun. 1996.
6. Object Management Group, *The Common Object Request Broker: Architecture and Specification, 2.0ed.*, Jul., 1995.
7. OMG Security Working Group, *OMG White Paper on Security*, OMG Doc. No. 94-4-16, Apr., 1994.
8. R. Oppliger, "Authentication Systems for Secure Networks," Artech House, pp.29-6 2, 1996.
9. OSF, *Open Software Foundation Training Course*, OSF DCE System Administration Course , Student Guide, Vol. 1.0 Dec. 1992.
10. T. Parker and D. Pinkas, *SESAME V4 Overview*, SESAME Issue 1, Dec. 1995
11. POSIX, *Protection, Audit, and Control Interfaces*, IEEE P1003.6.1, 1995.

Secure Document Management and Distribution in an Open Network Environment

Antonio Lioy, Fabio Maino, Marco Mezzalama

Politecnico di Torino,
Dip. di Automatica e Informatica,
Torino, Italy

Abstract. This paper analyzes the problem of secure document management and distribution in an open network environment. Reader and author authentication, document integrity, origin, and privacy are addressed by a public-key based solution which exploits a combination of the PEM format with SSL-enhanced FTP and HTTP servers and clients. The solution is being implemented as part of a project to provide network security to the Italian public administration.

1 Introduction

The work presented in this paper origins from the need of the Italian public administration to securely exchange documents externally with the citizens and internally between various departments.

Within this general framework, several initiatives are being carried out; many of them share the common architecture of a central repository to store and distribute documents with different security levels. The documents must be accessible to several people but the security levels and the distribution lists must be individually controlled by the author of the document. In other words, the central repository is used only as a distribution point but the minimum level of trust compatible with its functionality is put in its operation. This is usually due to the problem that a central repository is useful for the users of the information but it is normally external to many organizations, which therefore don't trust it very much as they don't have full control over it.

The analysis of the users' needs has led to the identification of several system requirements that can be summarized by the following terms:

document integrity the document must contain code to prove that it has not been modified in any way since it was generated and sealed

document origin authentication the document must contain information to prove the (electronic) identity of its author and must be useful for non-repudiation

document privacy if required, the document must be readable only by authorized users; this must be true even in case the document is stolen from the repository that distributes it

document destination authentication the system must keep track of the identity of people who have received a copy of a document

secure remote document management it must be possible to remotely manage the repository in a secure way, including document addition/removal and authorization granting/revoking

On the implementation ground, the system has to be an open one: people that interact are not supposed to belong to the same organization and not even to the same country. It was therefore a natural choice to base security on public-key techniques and X.509 certificates. This allows each actor to be identified in his own country (with his own language and procedures).

Additionally we wanted the system to work equally well in an Intranet, Extranet, or Internet environment (i.e. LAN or private/public WAN), with non-proprietary components as much as possible. Therefore the system had to work in a TCP/IP network with standard protocols.

Before developing our own solution, we surveyed various products that could possibly satisfy the requirements. We have found that existing proposals for cooperative systems don't provide a strong solution to reader and author authentication, document integrity and privacy. As stated in [1] the reason can be twofold: first, access control models for groupware tend to be complex and require a deep analysis of relations between principals and objects involved in the cooperative system; second, the need for a running prototype normally postpones the security issues to a "future" release of the product.

To see how security is managed in cooperative systems, we will shortly analyze three very different approaches to groupware:

- BSCW *(Basic Support for Cooperative Work)* [2], a shared workspace written in the Python programming language, freely available for non-commercial use from GMD, the Germany's national Research Center for Information Technology
- DOMINO, the Lotus Development Corporation product that, integrated with the Lotus Notes suite of products, offers an integrated set of services for groupware
- a "self-made" solution for cooperative work based on the filesystem workgroup sharing of a Windows-NT WWW server.

BSCW is widely open to different platforms and operating systems: emphasis is placed on the sharing of documents and security issues have been addressed in the latest version only. The HTTP basic authentication method is used to authenticate both readers and writers, but this is a very weak kind of authentication that clearly exposes the readers and author's password. A small improvement is obtained by using an SSL protected channel, but nothing is done in order to grant document integrity or protection of reserved documents when stored on the server.

The Lotus solution acts as a front-end to the World Wide Web for documents stored in a Lotus Notes database: using private Lotus client authors can spread documents over the Internet through DOMINO. Author authentication is done using standard X.509 certificates, that allows documents to be signed or encrypted even when stored in the database. On the WWW side, the use of SSL allows channel-oriented security and reader authentication but document security is restricted only to those users equipped with Notes clients. This, of course, forbids the diffusion of the products to companies that have an heterogeneous set of applications. Moreover the restrictions on exporting cryptographic products from USA allow the use of short-key encryption only, significantly reducing the security level for cooperative systems outside the United States.

The wide diffusion of corporate networks based on Windows-NT suggests the use of its workgroup sharing capabilities, in order to allow access to the filesystem of a NT WWW server. This *naive* approach to cooperative working is widely used in small Intranets thanks to its easy implementation and management. By using the NT audit capabilities it is possible to set up a simple logging of writer's access to shared documents, but this solution doesn't seem to provide enough protection to documents inserted from different authors of the same workgroup. No kind of encryption, integrity, or authentication is available for stored documents that are unprotected from unauthorized access via the WWW server.

The analysis of these solutions shows the need for a cooperative system really open to different platforms and operating systems, based on simple WWW interfaces and able to ensure document integrity, origin, destination authentication, and privacy, additionally the system must be securely manageable from a remote site.

Since, for different reasons, none of the various products examined satisfied all our requirements we moved on to designing our own solution, which is described in the sequel of the paper.

2 System architecture

Three different roles can be identified in the system. The *author* is the one who generated a document and he is the only one that controls its security level and distribution scope. The *document master* is the system manager in charge of properly setting up and managing the server used as document repository. As this role and this server are often external to the author's organization (e.g. because they are outsourced), we want to place the minimum level of trust in their actions. Finally, the *reader* is any user wishing to get an electronic copy of a document.

There is no restriction about the basic document formats supported: any standard electronic format (Postscript, PDF, GIF, JPEG, MS-Word, ...) can be managed because the document itself is treated as an opaque field as long as the security functions are concerned.

The system is designed to handle two security levels: *open* and *reserved* documents. In both cases, the document is sealed in an envelope which carries an integrity and origin authentication code. We find useless in a formal environment to provide electronic documents without ensuring at least these two basic security properties. Additionally, a reserved document is also encrypted to prevent its disclosure to unauthorized parties.

2.1 Secure document distribution

Open documents have unlimited distribution and are immediately delivered by the document server upon request, with no need for reader's authentication or authorization checks (see Fig. 1).

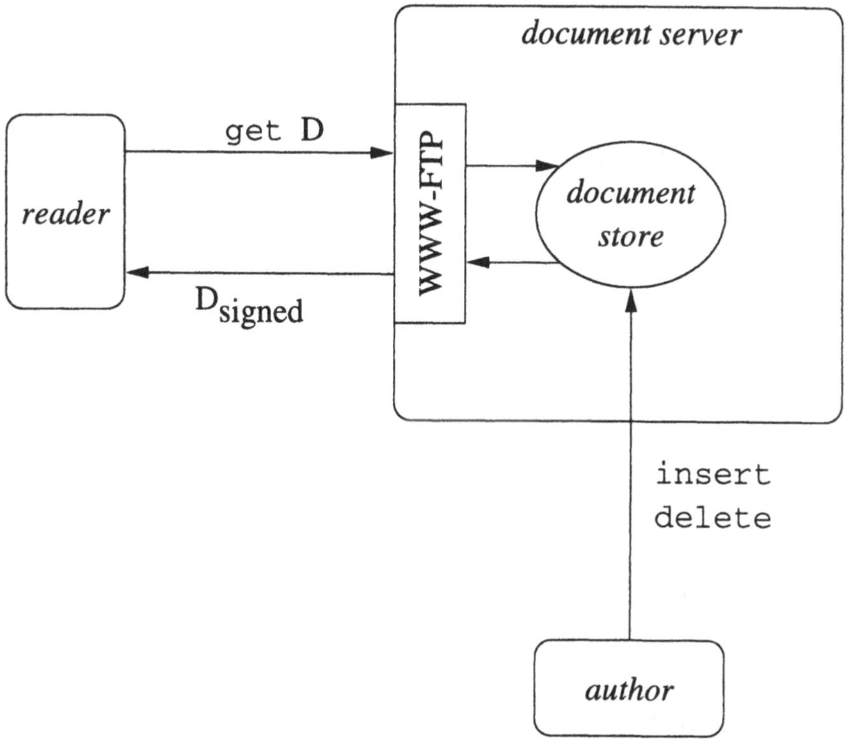

Fig. 1. System architecture for documents of the open type

On the contrary, a reserved document is not only encrypted but has limited distribution too (see Fig. 2). A pre-requisite to access a reserved document is that the reader be authenticated. We chose to use public-key authentication based on the RSA algorithm and X.509v3 public-key certificates. In turn this requires the existence of a public-key infrastructure (PKI) with certification authorities (CA) that emit the public-key certificates to formally bind public-keys to personal attributes, such as an identity. Since there is not yet a world-wide accepted PKI, our system is able to trust several roots simultaneously. However, our preference (and default root) is that of the ICE-TEL European project [3] which has set up an infrastructure which is increasingly being used in several countries of the European Union.

Limited distribution is enforced by means of an *access control list* (ACL): each document in the repository has an associated ACL which contains the list of all

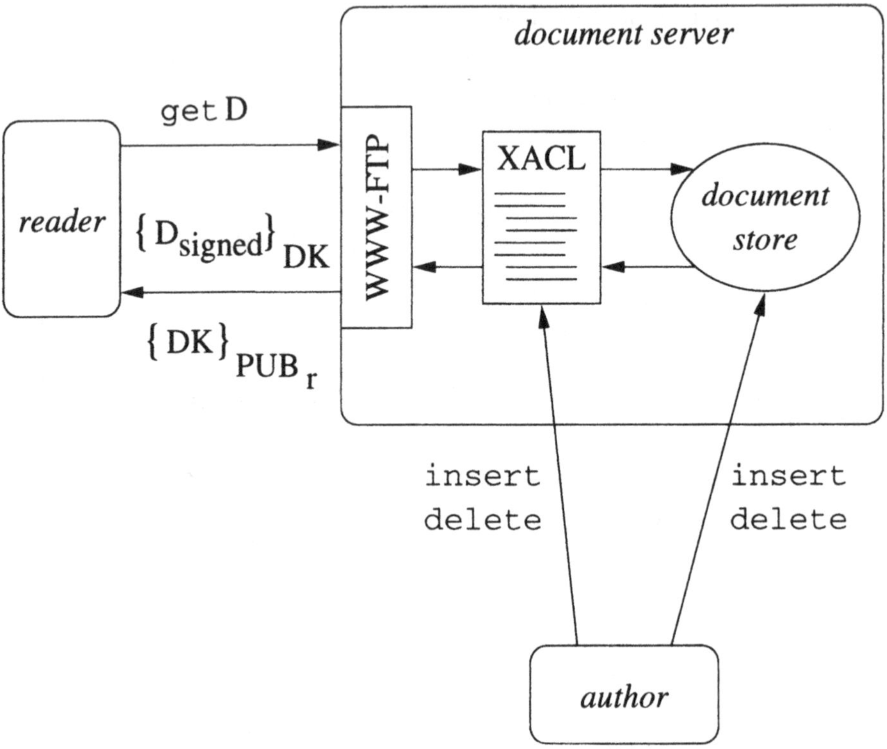

Fig. 2. System architecture for documents of the reserved type

the distinguished names (DN) authorized to retrieve a copy of the document.

Privacy is guaranteed by storing documents on the server only in encrypted form. This shields the documents from direct attacks by the document master as well as from illegal distribution. To allow the authorized recipients of a document to decrypt it we use a modified ACL, named an *extended ACL* or XACL. To describe its operation, let DK be the secret key used to encrypt the document D. The extended ACL for D is the set of pairs that associate the DN of an authorized reader to the document key encrypted with the public-key of the reader. This can be formally expressed as[1]:

$$\mathrm{XACL(D)} = \{\ \langle \mathrm{DN}_i, \{\mathrm{DK}\}\mathrm{PUB}_i \rangle : \forall i \in \{\textit{authorized users for document D}\}\ \}$$

In this way, when a generic user X requests a reserved document the following actions take place:

1. X authenticates himself via the SSL protocol, either to the WWW server or to the FTP one

[1] The notation $\{X\}K$ represents X encrypted with the key K.

2. the server extracts the user's identity and public key from the X.509 certificate received from the client and checks them against those contained in the the XACL components:

$$\text{authorized(X)}\ \mathit{iff}\ \exists i : \text{DN}_X = \text{DN}_i \ \wedge\ \langle DN_i, \{\text{DK}\}\text{PUB}_i \rangle \in \text{XACL(D)}$$

3. if in the previous step authorization is granted, then the document is delivered to the user along with the key to decrypt it:

$$\text{X} \longleftarrow \{\text{D}\}\text{DK}, \{\text{DK}\}\text{PUB}_X$$

4. when the reader receives the document, an helper application is available to check its integrity and authentication and to decrypt it to permit viewing and/or printing.

It should be noticed that on the reader's side documents are always saved in their protected form. In particular reserved documents are kept encrypted to prevent unauthorized disclosure. The helper application permits to save the document in an unprotected form too, but this is strongly discouraged: the user is warned that his action lowers the security of the system and he is legally responsible in case that the unprotected copy is illegally disclosed. (As an aside, this issue has stirred interest in investigating methods to insert into the distributed documents deliberate modifications to permit tracing documents even in their unprotected form. However, this is a completely different research.)

As long as the implementation is concerned, we chose the PEM message format [4] to seal and encrypt the document, mainly due to its simplicity and flexibility. This is in contrast with the current industrial emphasis on S/MIME as preferred format for securing e-mail messages. However, we feel that S/MIME has several disadvantages if applied in our context:

- it is more complex than PEM
- it is less flexible than PEM as long as the specification of the encryption, digest, and authentication algorithms is concerned
- its standardization status is still uncertain, as it has not yet attained RFC status and several of its components are under control of RSADSI only

For all these reasons our first implementation uses PEM with triple-DES, MD5, and 1024-bit RSA. We are already considering to move towards RC5, IDEA, and SHA. Nonetheless, there is no technical reason which prevents the use of S/MIME as an envelope format for the document and we will closely monitor the evolution of this format - especially with respect to standardization issues - ready to adopt it when it will eventually become widely adopted.

We have also paid attention to the DSIG initiative [5] of the W3C but it was found unsuitable to our context as it provides digital signatures only (i.e. integrity and authentication) and doesn't address document privacy. Moreover DSIG is strictly bound to HTML pages, which makes it difficult to be used as an encapsulation format for objects that can live outside of the WWW too.

The PEM-encapsulated documents are distributed by an SSL-enabled WWW or FTP server that has been modified to properly handle XACLs. In this context, SSL [6] is used to provide server authentication when distributing open documents, and mutual authentication (i.e. both client and server authentication) for the reserved documents.

2.2 Secure remote document management

One of the system requirements was to put the author in complete control of the privacy and distribution of the documents. This target has been achieved by developing a simple document management interface that allows an author to perform on his local workstation several tasks:

- to choose a document from the local storage to be uploaded to the repository
- to seal a document with the proper integrity and authentication codes (i.e. to digitally sign the document)
- to encrypt the document with a random key, which is maintained in a local personal security environment for future use
- to upload a document (sealed and possibly encrypted) to the repository
- to manage the XACL for a document D, by adding, deleting, or modifying entries
- to upload the XACL to the repository and to compare the local XACL with the one on the server, for verification purposes

The interface has been developed by using the Tcl/Tk language for the GUI, and our own version of SSL-ftp to provide channel security in the form of mutual authentication between the author's workstation and the repository. In particular, SSL-ftp has been chosen because the SSL protocol provides adjustable channel security (single or mutual authentication and integrity with or without encryption) while the FTP protocol allows remote management of a file system in a way very similar to that possible with a local one.

Since availability for several platforms is an issue in our project, we developed our own version of SSL-ftp by starting from the WU-ftpd package and the SSLeay crypto library [7]. The resulting package implements an SSL-ftp client for WIN32 and UNIX machines, while the server side is currently available for UNIX servers only (a Windows-NT porting is in progress). We are looking into porting the client to the Macintosh environment too.

3 Open issues and future work

The system has been designed to resist to a certain number of attacks but others are still possible. For example, the system is vulnerable to denial-of-service attacks leaded by the document manager who can refuse documents to authorized users or provide them with the wrong key. While this is annoying from the reader's point of view, it doesn't lower the security of the documents in any way: reserved data can never be exposed to unauthorized people. In the same fashion, if logging of the people who received a document is wanted this can be altered by the document manager as well.

However the biggest problem we have still to solve is related to ACLs: they work fine for a limited number of users but are impractical for large sets. A solution can be searched in using some form of group authorization. However it was an explicit design choice to avoid group authentication to enforce the principle of personal responsibility. We are therefore looking into using groups only for the DN-based part of the authorization procedure, while maintaining disclosure of the document key strictly connected to personal keys. For example, in this case a group entry in an XACL could look like this:

⟨ `(C=IT, O="Politecnico di Torino", CN=*)` , *empty* ⟩

With this syntax we would authorize all the people from the Politecnico di Torino to access a document; however, in this case the component $\{DK\}PUB_i$ cannot be pre-computed and must be generated on the fly as needed. This has the obvious disadvantage of requiring the author to be on-line when a request arrives but, at the same time, offers a solution for secure logging and for key revocation. We will continue to explore this issue and look for alternative solutions too.

We are continuing the development and the deployment of the system to gain valuable feedback from the user community, ready to improve the architecture according to the user requests.

Acknowledgment

This project is being carried out with the support of CNR (the Italian national research council) under grant "Progetto strategico Informatica nella Pubblica Amministrazione - sottoprogetto DEMOSTENE".

References

1. K. Sikkel, "A Group-based Authorization Model for Cooperative Systems", *Proceedings European Conference on Computer-Supported Cooperative Work (ECSCW'97)*, Lancaster (UK), September 1997.
2. R. Bentley, W. Appelt, U. Busbach, E. Hinrichs, D. Kerr, S. Sikkel, J. Trevor, G. Woetzel, "Basic Support for Cooperative Work on the World Wide Web", *International Journal of Human-Computer Studies: Special issue on Innovative Applications of the World Wide Web*, Spring 1997
3. "ICE-TEL: Interworking Public-Key Certification Infrastructure for Europe", project RE 1005 of the Telematics programme of the European Commission `http://www.darmstadt.gmd.de/ice-tel/`
4. J. Linn, "Privacy Enhancement for Internet Electronic Mail, Part I: Message Encryption and Authentication Procedure", RFC-1421, February 1993
5. "Digital Signature Initiative", World Wide Web Consortium `http://www.w3.org/Security/DSig/Overview.html`

6. A.O. Freier, P.L. Karlton, P.C. Kocher, "The SSL Protocol (version 3.0)", Netscape Communications Corporation
`http://home.netscape.com/eng/ssl3`
7. E.A. Young, T .Hudson, "SSLeay and SSLapps FAQ",
`http://psych.psy.uq.oz.au/~ftp/Crypto/`

A^2-code = Affine Resolvable + BIBD

Satoshi OBANA and Kaoru KUROSAWA

Department of Electrical and Electronic Engineering,
Faculty of Engineering, Tokyo Institute of Technology
2-12-1 O-okayama, Meguro-ku, Tokyo 152, Japan
E-mail {obana, kurosawa}@ss.titech.ac.jp
http://tsk-www.ss.titech.ac.jp/~kurosawa/

Abstract. We say that an A^2-code is optimum if it has the minimum cheating probabilities and the minimum sizes of keys. We first show that an optimum A^2-code implies an affine α-resolvable design. Next, we define an affine α-resolvable + BIBD design and prove that an optimum A^2-code is equivalent to an affine α-resolvable + BIBD design.

1 Introduction

Unconditionally secure authentication codes (A-codes) have been studied by many researchers. Especially, combinatorial lower bounds on cheating probabilities and the size of keys were derived by [1, 5]. We say that an A-code is optimum if it has the minimum cheating probabilities and the minimum sizes of keys. Then optimum A-codes are closely related to combinatorial designs. An optimum general A-code is equivalent to a BIBD [4, 5]. An optimum A-code without secrecy is equivalent to an orthogonal array [5]. This means that we can construct an optimum A-code by using a BIBD or an orthogonal array.

In the model of A-codes, a transmitter and a receiver are both honest and trust each other. However, it is not always the case. Simmons introduced authentication codes with arbitration (A^2-codes) to include protection against cheating by the transmitter and the receiver [2, 3]. This model uses an arbiter, who distributes partial keys to the transmitter and the receiver and decides in case of controversy between the transmitter and the receiver.

T. Johansson showed entropy based lower bounds on cheating probabilities and the size of keys of A^2-codes [6]. Kurosawa and Obana showed combinatorial lower bounds [7]. We say that an A^2-code is optimum if it has the minimum cheating probabilities and the minimum sizes of keys. However, no connection is known between combinatorial design and optimum A^2-codes. In other words, the combinatorial structure of optimum A^2-codes is not known.

In this paper, we first show that an optimum A^2-code implies an affine α-resolvable design. Next, we define an affine α-resolvable + BIBD design and prove that an optimum A^2-code is equivalent to an affine α-resolvable + BIBD design.

2 Authentication Code with Arbitration (A^2-code)

2.1 Model

In the model of authentication codes (A-codes), there are three participants, a transmitter T, a receiver R and an opponent O. O tries to cheat R, but T and R are honest and trust each other.

On the other hand, in the model of authentication codes with arbitration (A^2-codes), there is a fourth person, an arbiter [2, 3]. T and R are not necessarily honest and caution is taken against deception of T and R as well as that of O. The arbiter has access to all key information of T and R, and solves disputes between them. Let

- $S \triangleq \{s\}$ be the set of source states (plaintexts),
- $M \triangleq \{m\}$ be the set of messages,
- $E_T \triangleq \{e\}$ be the set of the transmitter's encoding keys and
- $E_R \triangleq \{f\}$ be the set of the receiver's decoding keys.

On input a source state s, T sends a message m such that $m = e(s)$ to R by using his encoding key $e \in E_T$. R accepts m iff $f(m)$ is valid, where $f \in E_R$ is his decoding key. The arbiter accepts m as authentic iff e can generate m.

We denote an A^2-code by (S, M, E_R, E_T).

2.2 Attacks

In the model of A^2-codes, there are five different kinds of attacks. In what follows, the cheating probabilities are taken over S, E_R and E_T, where S is assumed to be independent of E_R and E_T.

1. Impersonation by O. the opponent O sends a message m to R. O succeeds if m is accepted by R as authentic. The impersonation attack probability P_I is defined by

$$P_I \triangleq \max_{m \in M} \Pr[\text{R accepts } m] \tag{2.1}$$

2. Substitution by O. O observes a message m that is transmitted by T and substitutes m with another message $\hat{m}$. O succeeds if $\hat{m}$ is accepted by R as authentic. The substitution attack probability P_S is defined by

$$P_S \triangleq \sum_{m \in M} \Pr(M = m) \max_{\hat{m} \neq m} \Pr[\text{R accepts } \hat{m} | \text{R accepts } m] \tag{2.2}$$

The maximum is taken over $\hat{m}$ such that the source state conveyed by $\hat{m}$ is different from that of m.

3. Impersonation by T. T sends a message m to R and denies having sent it. T succeeds if m is accepted by R as authentic and if m is not one of the messages that T could have generated due to his encoding rule e. This cheating probability P_T is defined as follows

$$P_T \triangleq \max_{e \in E_T} \max_{m \in M} \Pr[\text{R accepts } m \text{ and } m \text{ is not generated by e} | \text{T has e}] \tag{2.3}$$

4. Impersonation by R. R claims to have received a message m from T. R succeeds if m could have been generated by T due to his encoding rule e. Remember that the arbiter accepts m as authentic iff e can generate m. This cheating probability P_{R_0} is defined by

$$P_{R_0} \triangleq \max_{f \in E_R} \max_{m \in M} \Pr[\text{T can generate } m | \text{R has } f \in E_R] \tag{2.4}$$

5. Substitution by R. R receives a message m from T but claims to have received another message $\hat{m}$. R succeeds if $\hat{m}$ could have been generated by T due to his encoding rule e. This cheating probability P_{R_1} is defined by

$$P_{R_1} \triangleq \max_{f \in E_R} \sum_{m \in M} \Pr(m)$$
$$\max_{\hat{m} \neq m} \Pr[\text{T can generate } \hat{m} | \text{R has } f \text{ and T sends } m] \tag{2.5}$$

2.3 Lower Bounds

From entropy argument, T. Johansson derived a lower bound on the size of keys such as follows [6].

Proposition 1. *[6]*

$$|E_R| \geq (P_I P_S P_T)^{-1}$$
$$|E_T| \geq (P_I P_S P_{R_0} P_{R_1})^{-1}$$

Kurosawa and Obana showed a combinatorial lower bound on the cheating probabilities such as follows [7]. In an A^2-code without secrecy, a message m is written as $m = (s, a)$, where s is source state and a is an authenticator. For $f \in E_R$ and $s \in S$, let

$$M(f, s) \triangleq \{a \mid f(s, a) = valid\}$$

$M(f, s)$ is the set of authenticators that f accepts for a source state s. We first note that $|M(f, s)| > 1$ to prevent the impersonation attack of R. In what follows, we assume that each $f \in E_R$ accepts c authenticators for each $s \in S$.

Assumption 2. *For* $\forall f \in E_R$ *and* $\forall s \in S$,

$$|M(f, s)| = c.$$

Let

$$l \triangleq |M|/|S|.$$

Proposition 3. *[7] In an A^2-code without secrecy,*

- $P_I \geq \frac{c}{l}$. *If* $P_I = \frac{c}{l}$, *then* $P_S \geq \frac{c}{l}$.
- $P_T \geq \frac{c-1}{l-1}$.
- $P_{R_0} \geq \frac{1}{c}$, $P_{R_1} \geq \frac{1}{c}$.

From Proposition 1 and Proposition 3, we can obtain a combinatorial lower bound on the size of keys such as follows.

Corollary 4. *If all the equalities of Proposition 3 are satisfied, then*

$$|E_R| \geq l^2(l-1)/c^2(c-1), \tag{2.6}$$

$$|E_T| \geq l^2. \tag{2.7}$$

Proposition 3 and Corollary 4 are tight because there exists an A^2-code which satisfies all the equalities [6, Construction I].

Definition 5. We say that an A^2-code without secrecy is optimum with respect to cheating probabilities if all the equalities of Proposition 3 are satisfied.

Definition 6. We say that an A^2-code without secrecy is optimum if it is optimum with respect to cheating probabilities and the equalities of (2.6) and (2.7) are satisfied.

3 Optimum A^2-code implies Affine Resolvable Design

3.1 Affine Resolvable design

In this section, we show that an optimum A^2-code implies an affine c-resolvable design.

A block design is a pair $(V, \mathcal{B})$, where $V = \{a_1, \cdots, a_v\}$ is a set of symbols. $\mathcal{B} = \{B_1, \cdots, B_b\}$ is a set of blocks, where $B_j \subseteq V$.

Definition 7. The incidence matrix of a block design $(V, \mathcal{B})$ is a $|V| \times |\mathcal{B}|$ binary matrix whose (i,j)-th element x_{ij} is defined as follows

$$x_{ij} \triangleq \begin{cases} 1 & \text{if } a_i \in B_j, \text{ where } a_i \in V,\ B_j \in \mathcal{B} \\ 0 & \text{otherwise} \end{cases}$$

Definition 8. A block design $(V, \mathcal{B})$ is α-resolvable if the blocks can be grouped into t classes $C_1, \ldots, C_t$ such that in each class, every symbol appears in α blocks.

(Fig. 1 shows the incidence matrix of an α-resolvable block design with $\alpha = 2$ and $t = 3$.)

Definition 9. An α-resolvable incomplete blocks design is called affine α-resolvable if any pair of blocks of the same class intersect in q_1 symbols and any pair of blocks from different classes intersect in q_2 symbols. We denote this design by $(V, \mathcal{B}, C_1, \cdots, C_t, q_1, q_2)$.

(Fig. 1 shows the incidence matrix of an 2-resolvable block design with $q_1 = 2$ and $q_2 = 3$.)

	C_1				C_2				C_3			
	B_{11}	B_{12}	B_{13}	B_{14}	B_{21}	B_{22}	B_{23}	B_{24}	B_{31}	B_{32}	B_{33}	B_{34}
a_1	1	1	0	0	1	1	0	0	1	1	0	0
a_2	1	1	0	0	0	0	1	1	0	0	1	1
a_3	0	0	1	1	1	1	0	0	0	0	1	1
a_4	0	0	1	1	0	0	1	1	1	1	0	0
a_5	0	1	0	1	1	0	1	0	1	0	1	0
a_6	0	1	0	1	0	1	0	1	0	1	0	1
a_7	1	0	1	0	1	0	1	0	0	1	0	1
a_8	1	0	1	0	0	1	0	1	1	0	1	0
a_9	0	1	1	0	1	0	0	1	1	0	0	1
a_{10}	0	1	1	0	0	1	1	0	0	1	1	0
a_{11}	1	0	0	1	1	0	0	1	0	1	1	0
a_{12}	1	0	0	1	0	1	1	0	1	0	0	1

Fig. 1. affine α-resolvable design

3.2 Optimum A^2-code implies Affine Resolvable Design

Let A denote the set of authenticators. Generally, $M \subseteq S \times A$.

Lemma 10. *In an optimum A^2-code with respect to cheating probabilities,*

$$M = S \times A \text{ and } |A| = l$$

Proof. If $P_I = c/l$, then [8]

$$\Pr[R \text{ accepts } (s, a)] = c/l > 0$$

for $\forall (s, a) \in S \times A$. Therefore, $M = S \times A$. In this case, from (2.3), $|A| = |M|/|S| = l$. □

Lemma 11. *In an optimum A^2-code with respect to cheating probabilities, for $\forall (s, a), \forall (s', a') \in S \times A$ with $s \neq s'$, there exists $e \in E_T$ such that*

$$e(s) = (s, a) \text{ and } e(s') = (s', a') \tag{3.1}$$

Proof. Fix $(s,a),(s',a') \in S \times A$ with $s \neq s'$ arbitrarily. Then [8]

$$\Pr[R \text{ accepts } (s,a) \text{ and } (s',a')] = (c/l)^2 > 0$$

if $P_I = P_S = c/l$. Therefore, there exists $f_0 \in E_R$ such that

$$f_0(s,a) = \text{valid and } f_0(s',a') = \text{valid}$$

Further, if there exists such f_0 and if $P_{R_0} = P_{R_1} = 1/c$, then [8]

$$\Pr[T \text{ can generate } (s,a) \text{ and } (s',a') \mid R \text{ has } f_0] = (1/c)^2 > 0.$$

Therefore, there exists $e \in E_T$ which satisfies (3.1). □

Lemma 12. *In an optimum A^2-code with respect to cheating probabilities, for $\forall(s,a), \forall(s',a'), \forall(s',a'') \in S \times A$ with $s \neq s'$ and $a' \neq a''$, there exists $f \in E_R$ such that*

$$f(s,a) = f(s',a') = f(s',a'') = \textit{valid} \tag{3.2}$$

Proof. Fix $(s,a),(s',a') \in S \times A$ with $s \neq s'$ arbitrarily. Then from Lemma 11, there exists $e_0 \in E_T$ such that

$$e_0(s) = (s,a), \quad e_0(s') = (s',a')$$

R accepts any message which T can send. Therefore,

$$\Pr[\text{R accepts } (s,a) \text{ and } (s',a') \mid E_T = e_0] = 1$$

On the other hand, if $P_T = (c-1)/(l-1)$, then [8]

$$\Pr[\text{R accepts } (s',a'') \mid E_T = e_0] = (c-1)/(l-1)$$

for $\forall a'' \neq a'$. Therefore,

$$\Pr[\text{R accepts } (s,a),(s',a') \text{ and } (s',a'') \mid E_T = e_0] = (c-1)/(l-1) > 0$$

This means that there exists $f \in E_R$ which satisfies (3.2). □

Definition 13. For $(s,a) \in S \times A$, let

$$E_R(s,a) \triangleq \{f \mid f \in E_R, f(s,a) = \text{valid}\}$$

Lemma 14. *In an optimum A^2-code with respect to cheating probabilities,*

$$|E_R| = l^2(l-1)/c^2(c-1)$$

if and only if

$$|E_R(s,a) \cap E_R(s',a') \cap E_R(s',a'')| = 1 \tag{3.3}$$

for $\forall(s,a), \forall(s',a'), \forall(s',a'') \in S \times A$ with $s \neq s'$, $a' \neq a''$. In this case,

$$|E_R(s,a) \cap E_R(s',a')| = (l-1)/(c-1) \tag{3.4}$$

$$|E_R(s',a') \cap E_R(s',a'')| = l/c \tag{3.5}$$

$$|E_R(s,a)| = l(l-1)/c(c-1) \tag{3.6}$$

for $\forall(s,a), \forall(s',a'), \forall(s',a'') \in S \times A$ with $s \neq s'$, $a' \neq a''$.

Proof. Lemma 12 is equivalent to say

$$|E_R(s,a) \cap E_R(s',a') \cap E_R(s',a'')| \geq 1$$

for $\forall(s,a), \forall(s',a'), \forall(s',a'') \in S \times A$ with $s \neq s'$, $a' \neq a''$. Then from Lemma 10, we have

$$\sum_{a''\neq a'} |E_R(s,a) \cap E_R(s',a') \cap E_R(s',a'')| \geq 1 \cdot (|A|-1) = l-1 \tag{3.7}$$

On the other hand, from Assumption 2, we see that

$$\sum_{a''\neq a'} |E_R(s,a) \cap E_R(s',a') \cap E_R(s',a'')| = (c-1) \cdot |E_R(s,a) \cap E_R(s',a')| \tag{3.8}$$

From (3.7) and (3.8),

$$|E_R(s,a) \cap E_R(s',a')| \geq \frac{l-1}{c-1} \tag{3.9}$$

Next,

$$\sum_{a'\in A} |E_R(s,a) \cap E_R(s',a')| \geq |A| \cdot \frac{l-1}{c-1} = \frac{l(l-1)}{c-1} \tag{3.10}$$

On the other hand, since $|M(f,s')| = c$,

$$\sum_{a'\in A} |E_R(s,a) \cap E_R(s',a')| = c \cdot |E_R(s,a)| \tag{3.11}$$

From (3.10) and (3.11),

$$|E_R(s,a)| \geq \frac{l(l-1)}{c(c-1)} \tag{3.12}$$

By repeating the same argument once more, we obtain

$$|E_R| \geq l^2(l-1)/c^2(c-1)$$

Now it is clear that the above equality is satisfied if and only if (3.3) holds.

Next, if (3.3) holds, then the equalities of (3.9) and (3.12) must hold. Therefore, (3.4) and (3.6) hold. Finally, from (3.3),

$$\sum_{a\in A} |E_R(s,a) \cap E_R(s',a') \cap E_R(s',a'')| = l$$

On the other hand, from Assumption 2,

$$\sum_{a\in A} |E_R(s_i,a) \cap E_R(s_j,a') \cap E_R(s_j,a'')| = c \cdot |E_R(s_j,a') \cap E_R(s_j,a'')|$$

Therefore, $|E_R(s',a') \cap E_R(s',a'')| = l/c$. □

Theorem 15. *For an optimum A^2-code (S, M, E_R, E_T), Let*

$$\begin{aligned} V &= E_R, \\ \mathcal{B} &= \{E_R(s,a) \mid (s,a) \in S \times A\}, \\ C_i &= \{E_R(s_i,a) \mid a \in A\} \text{ for } s_i \in S \\ q_1 &= l/c \\ q_2 &= (l-1)/(c-1) \end{aligned}$$

Then $(V, \mathcal{B}, C_1, \cdots, C_{|S|}, q_1, q_2)$. is an affine c-resolvable design.

Proof. Assumption 2 means that in each C_i, each $f \in E_R$ appears in c blocks. Therefore, this design is c-resolvable. Further, from (3.5) of Lemma 14, any pair of blocks of the same class intersect in q_1 symbols. From (3.4) of Lemma 14, any pair of blocks from different classes intersect in q_2 symbols. □

4 Optimum A^2-code = Affine Resolvable + BIBD

In this section, we define an *affine c-resolvable + BIBD* design and prove that an optimum A^2-code is equivalent to this design. (Remember that $E_R(s,a)$ is defined in Definition 13.)

4.1 Affine Resolvable + BIBD

Definition 16. A (v, k, λ)-*BIBD* is a block design $(V, \mathcal{B})$ such that $|V| = v$, $|B_j| = k$ $(< v)$ and every pair of symbols occurs in exactly λ blocks.

Now we define a new block design.

Definition 17. We say that an affine c-resolvable design $(V, \mathcal{B}, C_1, \cdots, C_t, q_1, q_2)$ is an *affine c-resolvable + BIBD* design if it satisfies the following properties. It is denoted by $(V, \mathcal{B}, C_1, \cdots, C_t; l, c, \lambda)$.

1. There exist positive integers l and c such that $|C_i| = l$ for $\forall i$ and $q_1 = l/c$, $q_2 = (l-1)/(c-1)$.
2. For any distinct classes C_i, C_j, C_k and $\forall B_i \in C_i, \forall B_j \in C_j$, there exists $B_k \in C_k$ such that $B_i \cap B_j = B_i \cap B_k$.
3. For any distinct C_i, C_j and $\forall B_i \in C_i$, define a block design $(C_j, X(C_j, B_i))$ as follows. For each $a_u \in V$, let

$$X_u \triangleq \{B_j \mid B_j \in C_j, a_u \in B_i \cap B_j\}$$

   Let

$$X(C_j, B_i) \triangleq \{X_u \mid a_u \in B_i\},$$

   Then $(C_j, X(C_j, B_i))$ is a (l, c, λ)-*BIBD*.

Fig. 1 shows an incidence matrix of an *affine 2-resolvable + BIBD* design with $t = 3, l = 4, c = 2, \lambda = 1$. It is verified as follows.

1. $|C_i| = l = 4$ for all i. Also, $q_1 = 4/2 = 2$ and $q_2 = (4-1)/(2-1) = 3$.
2. For example, for $B_{11} \in C_1$ and $B_{21} \in C_2$, there exists $B_{32} \in C_3$ and
$$B_{11} \cap B_{21} = \{a_1, a_7, a_{11}\} = B_{11} \cap B_{32}$$
3. For example, consider $(C_2, X(C_2, B_{11}))$,
where $B_{11} = \{a_1, a_2, a_7, a_8, a_{11}, a_{12}\}$ and $C_2 = \{B_{21}, \cdots, B_{24}\}$. Then
$$X_1 = \{B_{21}, B_{22}\}, \quad X_2 = \{B_{23}, B_{24}\}, \quad X_7 = \{B_{21}, B_{23}\},$$
$$X_8 = \{B_{22}, B_{24}\}, \quad X_{11} = \{B_{21}, B_{24}\}, \quad X_{12} = \{B_{22}, B_{23}\}$$
$$X(C_2, B_{11}) = \{X_1, X_2, X_7, X_8, X_{11}, X_{12}\}$$
It is easy to see that $(C_2, X(C_2, B_{11}))$ is a (4, 2, 1)-BIBD.

4.2 Optimum A^2-code implies Affine Resolvable + BIBD

Theorem 18. *For an optimum A^2-code (S, M, E_R, E_T), define a block design $(V, \mathcal{B}, C_1, \cdots, C_{|S|}, q_1, q_2)$ as shown in Theorem 15. Then it is an affine c-resolvable + BIBD design $(V, \mathcal{B}, C_1, \cdots, C_{|S|}; l, c, 1)$.*

Property (1) of Definition 17 is proved in Theorem 15. The other properties of Definition 17 will be proved in the final version.

4.3 Affine Resolvable + BIBD implies Optimum A^2-code

Lemma 19. *In an affine c-resolvable+BIBD design $(V, \mathcal{B}, C_1, \ldots, C_t; l, c, 1)$, for any (m, i_m) and (n, i_n) such that $\{m, n\} \neq \{1, 2\}$, there exists a unique (i_1, i_2) such that*
$$B_{1,i_1} \cap B_{2,i_2} = B_{m,i_m} \cap B_{n,i_n}$$
where $B_{1,i_1} \in C_1,\ B_{2,i_2} \in C_2,\ B_{m,i_m} \in C_m,\ B_{n,i_n} \in C_n$

Proof. Suppose that there exist (i_1, i_2) and (i'_1, i'_2) such that
$$B_{m,i_m} \cap B_{n,i_n} = B_{1,i_1} \cap B_{2,i_2} = B_{1,i'_1} \cap B_{2,i'_2}$$
Then,
$$(B_{m,i_m} \cap B_{n,i_n}) \supseteq (B_{1,i_1} \cap B_{2,i_2})$$
$$(B_{m,i_m} \cap B_{n,i_n}) \supseteq (B_{1,i'_1} \cap B_{2,i'_2})$$
Therefore,
$$\begin{aligned} q_2 &= |B_{m,i_m} \cap B_{n,i_n}| \\ &\geq |(B_{1,i_1} \cap B_{2,i_2}) \cup (B_{1,i'_1} \cap B_{2,i'_2})| \\ &= |(B_{1,i_1} \cap B_{2,i_2})| + |(B_{1,i'_1} \cap B_{2,i'_2})| - |(B_{1,i_1} \cap B_{2,i_2}) \cap (B_{1,i'_1} \cap B_{2,i'_2})| \\ &\geq |(B_{1,i_1} \cap B_{2,i_2})| + |(B_{1,i'_1} \cap B_{2,i'_2})| - |B_{1,i_1} \cap B_{1,i'_1}| \\ &= 2q_2 - q_1 \end{aligned}$$

Hence, $q_1 \geq q_2$. However, $q_1 = l/c < (l-1)/(c-1) = q_2$. This is a contradiction. Further,

$$|\{i_1, i_2\}| = |\{i_m, i_n\}| \ (= l^2)$$

Therefore, there exists a unique (i_1, i_2) for any (m, i_m) and (n, i_n). □

Theorem 20. *If there exists an affine c-resolvable + BIBD design* $(V, \mathcal{B}, C_1, \cdots, C_t; l, c, 1)$, *then there exists an optimum* A^2*-code* (S, M, E_R, E_T) *such that*

1. $E_R = V$, $E_R(s_i, a) = C_i$ *for any* i.
2. $|S| = t$, $|A| = |M|/|S| = l$, $|M(f, s)| = c$ *for* $\forall f \in E_R$ *and* $\forall s \in S$.

Proof. Let

$$E_R(e) \triangleq \{f \mid f \in E_R, \Pr(E_T = e, E_R = f) > 0\}$$

$$E_T(f) \triangleq \{e \mid e \in E_T, \Pr(E_T = e, E_R = f) > 0\}$$

We define $e_{(i_1,i_2)} \in E_T$ for each $B_{1,i_1} \in C_1$ and each $B_{2,i_2} \in C_2$ as follows. From (2) of Definition 17, for $\forall B_{1,i_1} \in C_1$ and $\forall B_{2,i_2} \in C_2$, there exists $B_{3,i_3} \in C_3, \ldots, B_{t,i_t} \in C_t$ such that

$$B_{1,i_1} \cap B_{2,i_2} = B_{1,i_1} \cap B_{3,i_3} = \cdots = B_{1,i_1} \cap B_{t,i_t} \tag{4.1}$$

For this $(B_{1,i_1}, B_{2,i_2}, \cdots, B_{t,i_t})$, define $e_{(i_1,i_2)} \in E_T$ as follows.

$$e_{(i_1,i_2)}(s_1) = (s_1, a_{i_1}),\ e_{(i_1,i_2)}(s_2) = (s_2, a_{i_2}),\ \cdots,\ e_{(i_1,i_2)}(s_t) = (s_t, a_{i_t}) \tag{4.2}$$

Further, let

$$E_R(e_{(i_1,i_2)}) = E_R(s_1, a_{i_1}) \cap E_R(s_2, a_{i_2}) \tag{4.3}$$

Then from (4.1),

$$E_R(e_{(i_1,i_2)}) = E_R(s_1, a_{i_1}) \cap E_R(s_2, a_{i_2}) \cap \cdots \cap E_R(s_t, a_{i_t})$$

Now, we will show that the equality of each bound is satisfied.
$|E_T|$: From Definition 17, $|\forall C_i| = l$. Therefore,

$$|E_T| = |C_1 \times C_2| = l^2.$$

$|E_R|$: Since $(C_j, X(C_j, E_R(s_i, a)))$ is $(l, c, 1)$-*BIBD*, for any two distinct $B_1, B_2 \in C_j$, there exists just one $X_u \in X(C_j, E_R(s_i, a))$ such that

$$B_1 \in X_u \text{ and } B_2 \in X_u$$

Therefore, there exists just one $a_u \in V$ such that

$$a_u \in B_i \cap B_1 \text{ and } a_u \in B_i \cap B_2$$

That is,

$$a_u \in B_i \cap B_1 \cap B_2$$

This means that

$$|B_i \cap B_1 \cap B_2| = 1$$

for $\forall C_i \neq C_j$ and $\forall B_i \in C_i$. Hence, (3.3) holds. Therefore, from Lemma 14,

$$|E_R| = (l^2(l-1))/(c^2(c-1))$$

Suppose that $E_T, E_R, E_T(f), E_R(e)$ are all uniformly distributed.

P_I, P_S: Since (3.3) holds, (3.4) and (3.6) also hold from Lemma 14. Hence,

$$P_I = \frac{|E_R(s,a)|}{|E_R|} = \frac{\frac{l(l-1)}{c(c-1)}}{\frac{l^2(l-1)}{c^2(c-1)}} = \frac{c}{l},$$

$$P_S = \frac{|E_R(s,a) \cap E_R(s',a')|}{|E_R(s,a)|} = \frac{\frac{l-1}{c-1}}{\frac{l(l-1)}{c(c-1)}} = \frac{c}{l}$$

P_T: Suppose that T has $e \in E_T$ such that

$$e(s) = (s,a) \text{ and } e(s') = (s',a'), \text{ where } s \neq s'$$

For a'' such that $e(s') \neq (s',a'')$, from (4.3), (3.3) and (3.4),

$$\begin{aligned} P_T &= |E_R(s,a) \cap E_R(s',a') \cap E_R(s',a'')|/|E_R(e)| \\ &= |E_R(s,a) \cap E_R(s',a') \cap E_R(s',a'')|/|E_R(s,a) \cap E_R(s',a')| \\ &= (c-1)/(l-1) \end{aligned}$$

P_{R_0}, P_{R_1}: Suppose that R has $f \in E_R$. Each $e_{(i_1,i_2)} \in E_T(f)$ is defined for any (s_1,i_1) and (s_2,i_2) such that

$$f(s_1,i_1) = f(s_2,i_2) = valid.$$

Therefore,

$$|E_T(f)| = |M(f,s_1)| \times |M(f,s_2)| = c^2$$

For any (s_m,i_m) and (s_n,i_n), there exists a unique key $e_{(i_1,i_2)} \in E_T$ from Lemma 19. Now for any (s_m,i_m) such that $f(s_m,i_m)$ =valid, choose $n(\neq m)$ arbitrarily. Then

$$\begin{aligned} |\{e \mid e(s_m) = (s_m,i_m)\} \cap E_T(f)| &= |\{i_n \mid f(s_m,i_m) = f(s_n,i_n) = valid\}| \\ &= |\{i_n \mid f(s_n,i_n) = valid\}| \\ &= |M(f,s_n)| = c \end{aligned}$$

Therefore,

$$\begin{aligned} P_{R_0} &= \Pr[\text{T can generate } (s_m,i_m) \mid E_R = f] \\ &= |\{e \mid e(s_m) = (s_m,i_m)\} \cap E_T(f)|/|E_T(f)| \\ &= c/c^2 = 1/c \\ P_{R_1} &= \Pr[\text{T can generate } (s_n,i_n) \mid E_R = f \text{ and T sent } (s_m,a_m)] \\ &= 1/|\{e \mid e(s_m) = (s_m,i_m)\} \cap E_T(f)| \\ &= 1/c \end{aligned}$$

□

References

1. G. J. Simmons: Message authentication: a game on hypergraphs, Congresus Numerantium, Vol. **45**, pp.161–192 (1984)
2. G. J. Simmons: Message Authentication with Arbitration of Transmitter/Receiver Disputes, Proceedings of Eurocrypt'87, Lecture Notes in Computer Science, LNCS **304**, Springer Verlag, pp.150–16 (1987)
3. G. J. Simmons: A Cartesian Product Construction for Unconditionally Secure Authentication Codes that Permit Arbitration, Journal of Cryptology, Vol. **2**, no. **2**, 1990, pp.77–104 (1990)
4. J. L. Massey: Cryptography – a selective survey, in *Digital Communications*, North Holland (pub.), pp.3–21 (1986)
5. D. R. Stinson: The combinatorics of authentication and secrecy codes, Journal of Cryptology, Vol. **2**, no. **1**, 1990, pp.23–49 (1990)
6. T. Johansson: Lower Bounds on the Probability of Deception in Authentication with Arbitration", IEEE Transaction on Information Theory, Vol. **40**, no. **5**, pp.1573–1585 (1994)
7. K. Kurosawa and S. Obana: Combinatorial bounds for authentication codes with arbitration, Proceedings of Eurocrypt'95, Lecture Notes in Computer Science, LNCS **921**, Springer Verlag, pp.289–300 (1995)
8. K. Kurosawa and S. Obana: Combinatorial bounds for authentication codes with arbitration (revised version), (1997) http://tsk-www.ss.titech.ac.jp/~kurosawa/97.html

Multisender Authentication Systems with Unconditional Security

K. M. Martin[1] *
R. Safavi-Naini[2] **

[1] Katholieke Universiteit Leuven, Dept. Elektrotechniek-ESAT
Kardinaal Mercierlaan 94, B-3001 Heverlee, Belgium
Email: keith.martin@esat.kuleuven.ac.be
[2] Department of Computer Science
University of Wollongong, Northfields Ave
Wollongong 2522, Australia
Email: rei@cs.uow.edu.au

Abstract. We consider an extension of the classical model of unconditionally secure authentication in which a single transmitter is replaced by a group of transmitters such that only certain specified subsets can generate authentic messages. We provide a model and derive sufficient conditions for systems that provide perfect protection. We give two generic constructions using secret sharing schemes and authentication codes as the underlying primitives and show that key-efficient and fast SGA-systems can be constructed by proper choice of the two primitives.

1 Introduction

Collaborative authentication is common practise for applications requiring a sufficiently high level of security that no single entity is trusted to be able to authenticate (sign) a message. In the banking community sensitive documents are usually signed by more than one person [8].

Collaborative generation of authentic messages that are intended for a designated recipient could be obtained by collaborative generation of an authenticator in a message authentication code (MAC) [8, 9]. In [8] Desmedt proposed an unconditionally secure threshold authentication system in which generation of an authentic message is only permitted by authorised groups of transmitters and verification is by an intended receiver who owns a secret key. Unconditionally secure systems are important because they maintain their security despite progress in hardware and software technologies, or algorithm development. In the case of authentication systems, the pioneering construction of Wegman and Carter [] allows for the construction of very efficient unconditionally secure systems. Numerous constructions for MACs [13, 16], with provable security and high speed,

* This work done while the author was at Department of Pure Mathematics, The University of Adelaide, Adelaide, SA 5005, AUSTRALIA

** Support for this project was partly provided by Australian Research Council under the reference number A49703076.

provide practically attractive solution for message authentication and hence the natural question is how to extend such systems to a multisender scenario.

The contribution of this paper is twofold. Firstly we develop a model for shared generation of authenticators (SGA-systems) and give a description of the system in terms of a matrix, called the ***key distribution matrix***, that can be seen as a generalized secret sharing system in which shares and secrets are vectors and focusing on a particular coordinate gives the authentication code for a sender or the receiver. This way of representing the system is especially useful in revealing the multi-faceted relationship between SGA-systems and two well-known cryptographic primitives, authentication codes (A-codes) and secret sharing schemes (SSS), which are instrumental in the construction of secure SGA-systems. We define perfect protection in this model and derive sufficient conditions on the key distribution matrix of the SGA-systems that provide perfect protection.

Secondly, we give two generic constructions for SGA-systems that combine certain classes of authentication codes and secret sharing systems in such a way that the security of the resulting system can be derived from the security of the underlying A-code and SSS. We note that by replacing a single sender with authorised groups of senders, many new points of attack are introduced into the system. That is, the enemy in this case may not only be an outsider but may also be an unauthorised group of insiders who have access to some key information. The proposed constructions effectively reduce the security analysis of SGA-systems to simpler primitives and attack models and allow us to employ the extensive body of work on secret sharing and authentication codes to construct SGA-systems with unconditional security. Moreover, specific choices of A-codes and SSS results in SGA-systems with extra properties. For example using a SSS with cheater detection [6, 14], or a SSS with disenrollment capabilities [2], gives an SGA-system that provides protection against cheating senders, or allows efficient regrouping of the senders if one sender is denounced as dishonest. Similarly, by employing various A-codes constructions, or equivalently using ϵ-almost strongly universal hash functions, we can obtain SGA-systems with high speed and small key size.

The remainder of the paper is structured as follows. Section 2 contains definitions and notations. In Section 3 we introduce SGA-systems. In doing so we show an elaborate and interesting relationship between SGA-systems, secret sharing schemes and A-codes. The security of SGA-systems is studied in Section 4. We define systems that are secure against certain attacks and give some necessary conditions and some sufficient conditions for the existence of such SGA-systems. In Section 5 we give two generic constructions of SGA-systems. In Section 6 we give a key-efficient and fast SGA-system based on one of the generic constructions and finally in Section 7 present some concluding remarks.

Due to space limitations the proofs of results are omitted, but are given in the full paper.

2 Preliminaries

Unconditionally secure authentication codes have been studied extensively since the paper of Gilbert et al [11]. We use the classical model and terminology proposed by Simmons [20]. In this model a *transmitter* wishes to send authenticated messages to a *receiver*. Both parties agree on a secret *encoding rule* that determines the *codeword* representing each plaintext message. The plaintext messages are referred to as *source states*. In this paper we only consider *systematic Cartesian* authentication codes in which there there is no secrecy involved, and a codeword is obtained by appending an *authenticator* to the source state. The authenticator is generated by the transmitter using the agreed encoding rule and the source state. The receiver verifies the authenticity of the received message by checking that the authenticator is valid for the source state under the agreed encoding rule.

A (q,k,b) *(Cartesian) authentication code* is a collection $\mathcal{E}$ of b distinct *encoding rules* defined on a set $\mathcal{S}$ of k *source states* and a set $\mathcal{T}$ of q *tags* (or *authenticators*). Each encoding rule $e \in \mathcal{E}$ is a function $e: \mathcal{S} \rightarrow \mathcal{T}$. Each pair $e \in \mathcal{E}$ and $s \in \mathcal{S}$ thus give rise to a unique *codeword* (or *message*) of the form $(s : t)$ (the concatenation of s and t), where $t = e(s)$. For economy of representation, and when the source state is known from the context, we denote this codeword by t. An *encoding matrix* of a (q,k,b) A-code is a $b \times k$ matrix $\mathcal{A}$, whose rows are labeled by the encoding rules, columns are labeled by the source states and entries given by the tags, such that $\mathcal{A}(e,s) = e(s)$.

Assessment of the performance of authentication codes is made by calculating the probabilities of success of an attacker (who does not know the secret encoding rule) constructing a fraudulent codeword that is acceptable to the receiver [18, 21, 23]. In a *spoofing attack of order* i, an attacker observes i genuine codewords being sent from transmitter to receiver and then attempts to construct a fraudulent codeword. Spoofing attacks of orders 0 and 1 are often referred to as *impersonation* and *substitution* attacks respectively. Let P_i denote the maximum probability of success of an attacker who performs a spoofing attack of order i. It can be shown [23] that for any $i \geq 0$ we have that $P_i \geq 1/q$. A (q,k,b) A-code is said to have *perfect protection against a spoofing attack of order* i if $P_i = 1/q$. The code is *r-fold secure against spoofing* if it has perfect protection against a spoofing attack of order i for each $i = 0, \ldots, r$.

In a *secret sharing schemes (SSS)* a secret is protected among a group of participants in such a way that only certain subsets of participants can reconstruct the secret. An *access structure* Γ defined on a set $\mathcal{P}$ is a collection of subsets of $\mathcal{P}$ with the property that if $B \in \Gamma$ and $B \subseteq C \subseteq \mathcal{P}$ then $C \in \Gamma$. A set $B \in \Gamma$ is *minimal* if $C \subseteq B$, $C \in \Gamma$ implies that $C = B$. Note that Γ is uniquely determined by its collection Γ^- of minimal sets. A *secret sharing scheme* for Γ is a method of sharing a *secret* $k \in \mathcal{K}$ among a set of *participants* $\mathcal{P}$ in such a way that only sets in Γ can determine k. This is done by using a trusted third party $d \notin \mathcal{P}$ to issue each participant $p \in \mathcal{P}$ with a value from a set $\mathcal{H}$, known as a *share* of k. Following Stinson [24], we define a secret sharing scheme to be a collection $\mathcal{F}$ of *distribution rules*. Each rule is a function

$f: \mathcal{P} \cup d \to \mathcal{K} \cup \mathcal{H}$ such that $f(d) \in \mathcal{K}$ and $f(p) \in \mathcal{H}$ (for each $p \in \mathcal{P}$). For $k \in \mathcal{K}$, let $\mathcal{F}_k = \{f \in \mathcal{F} \mid f(d) = k\}$. To share a secret $k \in \mathcal{K}$, d chooses (randomly) a distribution rule $f \in \mathcal{F}_k$ and issues p with share $f(p)$ (for each $p \in \mathcal{P}$). The following holds in order that a set $B \in \Gamma$ can reconstruct k:

[SS1] Let $B \in \Gamma$ and $f, g \in \mathcal{F}$. If $f(p) = g(p)$ for all $p \in B$ then $f(d) = g(d)$.

We say that the secret sharing scheme is *perfect* if $B \notin \Gamma$ can not obtain any information about the secret k. This is ensured if the following property holds:

[SS2] Let $B \notin \Gamma$ and $f \in \mathcal{F}$. Then there exists a non-negative $\lambda(f, B)$ such that for every $k \in \mathcal{K}$,

$$|\{g \in \mathcal{F}_k \mid g(p) = f(p), \text{ for all } p \in B\}| = \lambda(f, B).$$

We can represent the scheme $\mathcal{F}$ by a matrix M, whose rows are labeled by the distribution rules, columns labeled by $\mathcal{P} \cup \{d\}$ and entries from $\mathcal{H} \cup \mathcal{K}$, such that $M(f, x) = f(x)$, where if $x \in \mathcal{P}$ then $f(x) \in \mathcal{H}$ and if $x = d$ then $f(x) \in \mathcal{K}$.

We will be particularly interested in a class of secret sharing schemes which we call ***linear*** secret sharing schemes. These schemes have previously been defined in a variety of different terms; vector spaces [3, 5], error correcting codes [10] and finite geometry [22]. Let $GF(q)^r$ be the vector space of all r-tuples over $GF(q)$, where q is a prime power and $r \geq 2$. For $X \subseteq GF(q)^r$, let $\langle X \rangle$ denote the subspace of $GF(q)^r$ spanned by X. Let ϕ be a mapping from $\mathcal{P} \cup d$, where for each $p \in \mathcal{P}$, $\phi(p)$ is a subspace of $GF(q)^r$ (let the dimension of $\phi(p)$ be given by $\sigma(p)$) and $\phi(d) \in GF(q)^r$, such that for $A \subseteq \mathcal{P}$,

$$\phi(d) \in \langle \cup_{p \in A} \phi(p) \rangle \iff A \in \Gamma.$$

Such a mapping ϕ gives rise to a perfect secret sharing scheme $\mathcal{F}$ for Γ (for details of this see [24]). The resulting scheme $\mathcal{F}$ has a number of special properties, two of which are:

[L1] For each $f \in \mathcal{F}$ and each $p \in \mathcal{P}$, $f(p)$ is a $\sigma(p)$-tuple over $GF(q)$.
[L2] Write $f(p) = ([f(p)]_1, \ldots, [f(p)]_{\sigma(p)})$. For any $A \in \Gamma$ there exists non-negative integers $\lambda_{p,i}$ ($p \in A$, $1 \leq i \leq \sigma(p)$) such that for any $f \in \mathcal{F}$, $f(d) = \sum_{p \in A, 1 \leq i \leq \sigma(p)} \lambda_{p,i} [f(p)]_i$ (where multiplication is over $GF(q)$).

If $\sigma(p) = 1$ for each $p \in \mathcal{P}$ then we call $\mathcal{F}$ an ***ideal*** linear scheme. A number of examples of linear schemes are given in Section 5.

3 The model

In this paper we generalise the classical model of authentication to the case where there is a group of potential transmitters who can only compute authenticators if certain specified subsets of them (***authorised sets***) jointly agree to do so. It might seem that secret sharing schemes offer a simple solution to the problem discussed here. The group of transmitters could all be regarded as participants of a secret sharing scheme which has as its secret the encoding rule of a classical authentication code (with one transmitter). The major shortcoming of this solution is that the system requires a *combiner* to collect the shares of an authorised group and reconstruct the secret (encoding rule). However the combiner now has complete control over the system and can not only change the current source state but can also successfully authenticate any further source states without any future communication with the (other) transmitters. The combiner in an SGA-system does not have any key information and only collects the authenticators generated by the transmitters and computes the authenticator that is appended to the source state and forwarded to the receiver.

In our model, the transmitters and the receiver each have an A-code and each is given an encoding rule of their A-code by a trusted third party. To authenticate a source state, each transmitter in an authorised group uses its A-code and encoding rule to construct an authenticator which is forwarded with the source state to a combiner who constructs a codeword as mentioned before. The trusted third party only participates in the initial generation and distribution of secret encoding rules. We assume subsequent communications are all through public channels and hence the security of the system relies only on the secrecy of the transmitters' and the receiver's encoding rules.

We now discuss a model for SGA-systems. Suppose a set $\mathcal{U} = \{u_1, u_2, \cdots u_n\}$ of transmitters wish to be able to jointly authenticate source states in $\mathcal{S}$ according to access structure Γ. An *SGA-system* for Γ consists of a set $\mathcal{C}$ of *key distribution rules*. Let $\mathcal{E}_1, \ldots, \mathcal{E}_n, \mathcal{E}_R$ be finite sets of k-tuples with entries from sets $\mathcal{T}_1, \ldots, \mathcal{T}_n, \mathcal{T}_R$ respectively. Each key distribution rule is a function $f: \mathcal{U} \cup R \rightarrow (\cup_{i=1}^{n} \mathcal{E}_i) \cup \mathcal{E}_R$, such that for each $u_i \in \mathcal{U}$, $f(u_i) \in \mathcal{E}_i$ and $f(R) \in \mathcal{E}_R$. We can represent $\mathcal{C}$ as a matrix with columns labeled by $u_1, \ldots, u_n, R$, rows labeled by key distribution rules and entry $\mathcal{C}(f, x) = f(x)$ (for each $x \in \mathcal{U} \cup R$). We denote the component of $f(x)$ in position j by $[f(x)]_j$. We require an SGA-system to satisfy the following property:

- [SGA1] Let $B \in \Gamma$, $s_j \in \mathcal{S}$ and $f, g \in \mathcal{C}$. If $[f(u)]_j = [g(u)]_j$ for all $u \in B$ then $[f(R)]_j = [g(R)]_j$.

In order to understand the need for [SGA1] we explain the implementation of an SGA-system. We assume that $\mathcal{C}$ is public and that there exists a trusted third party to initialise the system. The SGA-system has three distinct phases:

(1) Initialisation Phase. The trusted third party selects a key distribution rule $f \in \mathcal{C}$ at random and then securely delivers key $f(u_i)$ to transmitter u_i $(i = 1, \ldots, n)$ and $f(R)$ to the receiver R.

(2) Transmission Phase. Suppose a set B of transmitters, such that $B \in \Gamma$, wish to authenticate a source state $s_j \in \mathcal{S}$. Each transmitter $u_i \in B$ calculates $[f(u_i)]_j$. By [SGA1] there is a unique tag t_R such that any $g \in \mathcal{C}$ with $[g(u_i)]_j = [f(u_i)]_j$ for all $u_i \in B$, has the property that $[g(R)]_j = t_R$. This tag t_R is determined from $\mathcal{C}$ and then the codeword $(s_j : t_R)$ is forwarded to the receiver.

(3) Authentication Phase. The receiver accepts the codeword $(s_j : t_R)$ as authentic if $t_R = [f(R)]_j$.

We now consider some of the structures underlying an SGA-system.

3.1 A-codes and secret sharing systems from SGA-systems

Note that for $i = 1, \ldots, n$, each key of the form $f(u_i)$ is a k-tuple and so can be thought of as a function $e^i : \mathcal{S} \to \mathcal{T}^i$, where $e^i(s_j) = [f(u_i)]_j$. Interpreting $f(R)$ in a similar way, a typical key distribution rule of $\mathcal{C}$ has the form

u_1	$\cdots$	u_n	R
$\cdots$	$\cdots$	$\cdots$	$\cdots$
$(e^1(s_1), \ldots, e^1(s_k))$	$(\cdots)$	$(e^n(s_1), \ldots, e^n(s_k))$	$(e^R(s_1), \ldots, e^R(s_k))$
$\cdots$	$\cdots$	$\cdots$	$\cdots$

Thus for each $i = 1, \ldots, n$, $\mathcal{C}$ induces an A-code $\mathcal{A}_i$, where the encoding rules are given by the set $\{e^i = f(u_i) \mid f \in \mathcal{C}\}$. It follows that $\mathcal{A}_i$ is a (q_i, k, b_i) A-code, where $q_i = |\mathcal{T}_i|$ and $b_i = |\{f(u_i) f \in \mathcal{C}\}|$. In a similar way $\mathcal{C}$ induces a (q_R, k, b_R) A-code $\mathcal{A}_R$ consisting of the encoding rules $\{e_R = f(R) \mid f \in \mathcal{C}\}$.

An SGA-system induces a number of secret sharing schemes. ¿From properties [SGA1] and [SS1] we see that for each $j = 1, \ldots k$, an SGA-system $\mathcal{C}$ for Γ induces a secret sharing scheme M_j for Γ, where the distribution rules are $M_j = \{([f(u_1)]_j, \cdots [f(u_n)]_j, [f(R)]_j) \mid f \in \mathcal{C}\}$. Note also that $\mathcal{C}$ also induces a secret sharing scheme M for Γ where M is formed from $\mathcal{C}$ by independently labeling the columns of M by identifying the k-tuple $\mathcal{C}(f, x)$ with an element from a finite set of size $|\{g(x) \mid g \in \mathcal{C}\}|$. We call this secret sharing scheme the *underlying* secret sharing scheme of $\mathcal{C}$.

3.2 SGA-systems from secret sharing schemes and A-codes

Conversely, suppose we have k secret sharing schemes $M_1, \ldots, M_k$ for Γ. By choosing k distribution rules $f^1 \in M_1, \ldots, f^k \in M_k$, we can construct a key distribution rule of an SGA-system for Γ of the form,

u_1	$\cdots$	u_n	R
$\cdots$	$\cdots$	$\cdots$	$\cdots$
$(f^1(u_1),\ldots,f^k(u_1))$	$(\cdots)$	$(f^1(u_n),\ldots,f^k(u_n))$	$(f^1(R),\ldots,f^k(R))$
$\cdots$	$\cdots$	$\cdots$	$\cdots$

By continuing to choose tuples $(f^1,\ldots,f^k)$ of distribution rules, we can construct many different SGA-systems for Γ. These systems will satisfy [SGA1] because each of the schemes M_i satisfies [SS1].

Construction of an SGA-system can also be attempted by starting with a collection $\mathcal{A}_1,\ldots,\mathcal{A}_n,\mathcal{A}_R$ of A-codes and an underlying secret sharing scheme M for Γ. The encoding rules of the A-code can then be used to replace the entries of the secret sharing scheme M, resulting in a matrix $\mathcal{C}$ with entries corresponding to encoding rules of $\mathcal{A}_1,\ldots,\mathcal{A}_n,\mathcal{A}_R$. In general there is no guarantee that $\mathcal{C}$ will satisfy [SGA1] and thus be an SGA-system for Γ. However in Section 5 we will describe a construction which allows us to use this technique to find examples of optimally secure SGA-systems.

4 Security analysis

A group of attackers of an SGA-system can either be outsiders (not any of the transmitters) or include a set B of transmitters such that $B \notin \Gamma$. The latter is clearly advantaged over the former, since they hold extra information about the system. The goal of a group of attackers is to persuade the receiver to accept a codeword that was not sent by an authorised set of transmitters. This is achieved by constructing a codeword that is valid under the receiver's encoding rule and that has not previously been sent by an authorised set of transmitters. We assume that the information available to a group of attackers is:

[I1] a full description of the SGA-system $\mathcal{C}$;
[I2] the secret keys (encoding rules) held by any of the attackers;
[I3] knowledge of any previous communications sent using the SGA-system. This knowledge consists of the previous tags sent from the combiner to the receiver ***and*** the corresponding previous tags sent from the transmitters to the combiner.

A ***spoofing attack of order*** i is an attack based on knowledge of [I1],[I2] and [I3](2) relating to i previous communications over the system. The aim of attackers in a spoofing attack is to construct a codeword acceptable to the receiver. A spoofing attack of order 0 is called an ***impersonation attack*** and a spoofing attack of order 1 is called a ***substitution attack***. For simplicity, in the following analysis we assume that the key distribution rule that initialises an SGA-system is chosen uniformly.

4.1 Impersonation attacks

Let $\mathcal{C}$ be an SGA-system for Γ. Suppose that the system has been initialized by selecting key distribution rule f. Let $B \notin \Gamma$ be an unauthorized set of transmitters who attempt to conduct an impersonation attack. As the attackers already have some information about which key distribution rule has been chosen, they are only interested in analyzing $\mathcal{C}(f, B)$, obtained from $\mathcal{C}$ by restricting $\mathcal{C}$ to key distribution rules g that have $g(u_i) = f(u_i)$ for all $u_i \in B$. Since B knows that f is one of the rows of $\mathcal{C}(f, B)$, the best strategy available to B for selection of a fraudulent codeword is to choose a codeword $(s : t)$ that arises maximally from the encoding rules in column R of $\mathcal{C}(f, B)$. More precisely, let $b(f, B)$ be the number of rows of $\mathcal{C}(f, B)$ and let $E(s_i, t)$ be the number of rows of $\mathcal{C}(f, B)$ with $[g(R)]_i = t$. Then by using the above strategy, the maximum probability $P_0(f, B)$ of B succeeding in an impersonation attack is

$$P_0(f, B) = \max_{(s_i, t) \in \mathcal{S} \times \mathcal{T}_R} \frac{E(s_i, t)}{b(f, B)}.$$

Finally, we define the maximum probability P_0 of success for an impersonation attack to be

$$P_0 = \max_{f \in \mathcal{F}} \max_{B \notin \Gamma} P_0(f, B).$$

Recall that the tags of codewords forwarded to the receiver come from a set $\mathcal{T}_R$ such that $|\mathcal{T}_R| = q_R$. A secure SGA-system is one when an attacker can do no better than choose a tag uniformly from $\mathcal{T}_R$. Hence we say that an SGA-system has *perfect protection against impersonation* if $P_0 = 1/q_R$.

Theorem 1. *Let $\mathcal{C}$ be an SGA-system for Γ that has perfect protection against impersonation. Then for all $f \in \mathcal{C}$ and $B \notin \Gamma$,*

1. *the secret sharing schemes $M_1, \ldots, M_k$ for Γ that are induced by $\mathcal{C}(f, B)$ are perfect;*
2. *the A-code $\mathcal{A}_R$ induced by $\mathcal{C}(f, B)$ has perfect protection against impersonation.*

We now give some sufficient conditions for the existence of an SGA-system for Γ that has perfect protection against impersonation.

Theorem 2. *An SGA-system for Γ has perfect protection against impersonation if the following conditions hold.*

1. *the secret sharing scheme underlying $\mathcal{C}$ is a perfect secret sharing scheme for Γ;*
2. *the A-code $\mathcal{A}_R$ induced by $\mathcal{C}$ has perfect protection against impersonation.*

4.2 Substitution Attacks

Again, suppose that an SGA-system for Γ has been initialised by selecting a key distribution rule f. Let $B \notin \Gamma$ be an unauthorised set of transmitters who wish to try to conduct a substitution attack. In this case the attackers have all the information that would be available for conducting an impersonation attack *plus* knowledge of a codeword $(s_j : t)$ that was previously sent by the combiner to the receiver and the tags that were sent from an authorised group A to the combiner. In this case the attackers B are interested in $\mathcal{C}(f, B, s_j, A)$, obtained from $\mathcal{C}$ by restricting $\mathcal{C}$ to key distribution rules g satisfying $g(u) = f(u)$, $u \in B$ and $[g(u)]_j = [f(u)]_j$, $u \in A$. Note that although the attackers also see $[f(R)]_j$, this is not any additional use to them because of [SGA1]. Thus the probability of succeeding in a substitution attack is

$$P_1(f, B, s_j, A) = \max_{Y \in A} \max_{(s_i:t') \in S \times T_R, i \neq j} \frac{E(s_i, t')}{b(f, B, s_j, A)},$$

where $E(s_i, t')$ is the number of rows g of $\mathcal{C}(f, B, s_j, A)$ with $[g(R)]_i = t'$ and $b(f, B, s_j, A)$ is the number of rows of $\mathcal{C}(f, B, s_j, A)$. The maximum probability P_1 of succeeding in a substitution attack is then

$$P_1 = \max_{(s:t) \in S \times T^R} \max_{f \in \mathcal{C}} \max_{B \notin \Gamma} \max_{A \in \Gamma} P_1(f, B, s_j, A).$$

An SGA-system has *perfect protection against substitution* if $P_1 = 1/q_R$. An SGA-system is *1-fold secure against spoofing* if it has perfect protection against impersonation and substitution.

Theorem 3. *Let $\mathcal{C}$ be an SGA-system for Γ that has perfect protection against substitution. Then for all $f \in \mathcal{C}$, $s_j \in S$, $B \notin \Gamma$ and $A \in \Gamma$,*

1. *the secret sharing schemes $M_1, \ldots, M_k$ for Γ that are induced by $\mathcal{C}(f, B)$ are perfect;*
2. *the A-code $\mathcal{A}_R$ induced by $\mathcal{C}(f, B, s_j, A)$ has perfect protection against substitution.*

Theorem 4. *An SGA-system for access structure Γ is 1-fold secure against spoofing if the following two conditions hold:*

1. *the secret sharing scheme underlying $\mathcal{C}$ is an ideal linear secret sharing scheme for Γ;*
2. *the A-codes $\mathcal{A}_i$ $(i = 1, \ldots, n)$ and $\mathcal{A}_R$ induced by $\mathcal{C}$ are 1-fold secure against spoofing.*

We note that higher order attacks, where an opponent sees more than one message before attempting to conduct a spoofing attack, are natural generalisations of the attacks in Sections 4.2 and 4.3. Definitions and Theorems 3 and 4 can all be suitably generalised (details can be found in the full paper).

5 Construction of SGA-Systems

Desmedt [8] proposed a solution to the problem of shared generation of authenticators. The Desmedt Construction starts with a special type of secret sharing scheme and produces an SGA-system whose induced A-codes have a very special structure. We provide generic constructions which allow more general type of A-code and SSS to be utilised in an SGA-system.

We describe two constructions for SGA-systems that are t-fold secure against spoofing. In the first construction the system is built around an arbitrary A-code that is linear while in the second one we construct the key distribution matrix of the SGA-system by combining an arbitrary SSS and a linear A-code.

Let $GF(q)$ denote the finite field of q elements. A (q,k,b) A-code is called *linear* if the q-ary linear combination of encoding rules gives another encoding rule. That is $\sum_i a_i e_i = e \in \mathcal{E}, a_i \in GF(q)$.

5.1 The Benaloh and Leichter Construction

In [1] an algorithm was described that constructs perfect secret sharing schemes for any monotone access structure Γ. This algorithm considers Γ as a monotone formula. We will use the description of the algorithm given in [25] and assume $\mathcal{K}=\mathcal{H}=GF(q)$. By taking the direct product of secret space and share space we can extend the algorithm to the case that secrets and shares are n-tuples over $\mathcal{K}$ and $\mathcal{H}$ respectively such that the ith components of the share vectors are the shares of the ith component of the secret vector.

Now suppose $\mathcal{A}$ is a linear A-code and let $\Gamma^- = \{ab, ac\}$. We can represent Γ by the monotone formula $\Gamma = a(b+c)$. For an arbitrary encoding rule assigned to the receiver, trusted party uses the algorithm of [25] to determine the encoding rules that must be used by the transmitters. This results in a $\mathcal{C}$ such that if the secret of $f \in \mathcal{C}$ is encoding rule e_i then under f, a and b are given encoding rules e_j and e_k respectively, where $e_j + e_k = e_i$, and c is also given rule e_k.

In this construction transmitters and the receiver are using the same A-code and the role of trusted third party is to determine the specific encoding rules that must be used in a particular communication.

Theorem 5 extends the security properties of the underlying primitives to the security of the SGA-system.

Theorem 5. *Let $b = q^m$ be a prime power such that there exists a linear (q,k,b) A-code $\mathcal{A}$ that is t-fold secure against spoofing. Applying a Benaloh and Leichter Construction for Γ to the encoding rules of $\mathcal{A}$ results in an SGA-system for Γ that is t-fold secure against spoofing.*

5.2 Using linear schemes

Benaloh and Leichter Construction is interesting because it can be applied directly to encoding rules of a linear A-code. A second method is to start with a 'nice' underlying secret sharing scheme and build up an SGA-system based on it. Theorems 2 and 4 suggest that the underlying scheme should be perfect. We choose our underlying secret sharing scheme to be linear and map encoding rules of $\mathcal{A}$ onto the shares of $\mathcal{F}$ in order to construct an SGA-system $\mathcal{C}$. Property [L2] ensures that the components of the entries also have a linear relationship between authorised sets of shares and the secret, and hence because [SS1] holds for the underlying scheme, [SGA1] holds for the resulting SGA-system.

Let $b = q^m$ be a prime power. Let $\mathcal{F}$ be an ideal linear secret sharing scheme for Γ defined over $GF(q) = \{0, \ldots, q-1\}$ with secret denoted R. Let $\mathcal{A}$ be a linear (q, k, b) A-code. Form an SGA-system $\mathcal{C}$ as follows:

1. Since the rows of the encoding matrix of $\mathcal{A}$ necessarily form an m-dimensional subspace of $[GF(q)]^k$, pick a basis $b_1, \ldots, b_m$ of $\mathcal{A}$ (m linearly independent encoding rules) and label the encoding rule of $\mathcal{A}$ in terms of this basis by $(0, \ldots, 0), \ldots, (q-1, \ldots, q-1)$.
2. Construct the ideal linear secret sharing scheme $\mathcal{F}^m$ for Γ as follows. For every m distribution rules $f_1, \ldots, f_m \in \mathcal{F}$ form a distribution rule $f_{1,\ldots,m} \in \mathcal{F}^m$ such that for $x \in \mathcal{P} \cup R$ $f_{1,\ldots,m}(x) = (f_1(x), \ldots, f_m(x))$. Note that $\mathcal{F}^m$ is a perfect secret sharing scheme for Γ [7, 15] and further, that $\mathcal{F}^m$ is ideal linear with [L2] holding *for the same constants* $\lambda_{p,i} \in GF(q)$ as for $\mathcal{F}$.
3. For each $f \in \mathcal{F}^m$ and $x \in \mathcal{P} \cup R$ define a key distribution rule $g \in \mathcal{C}$ by replacing $f(x) = (x_1, \ldots, x_m)$ in $\mathcal{F}^m$ by $g(x) = e_{f(x)}$ in $\mathcal{C}$, where $e_{f(x)}$ is the encoding rule of $\mathcal{A}$ labeled $(x_1, \ldots, x_m)$ with respect to basis $b_1, \ldots, b_m$.

Theorem 6. *Let $b = q^m$ be a prime power such that there exists an ideal linear secret sharing scheme $\mathcal{F}$ for Γ defined over $GF(q)$ and a linear (q, k, b) A-code $\mathcal{A}$ that is t-fold secure against spoofing. Then [1]-[3] above result in an SGA-system for Γ that is t-fold secure against spoofing.*

Theorem 6 is particularly significant because it is well known [19] that for $q \geq n$ there exist ideal linear (l, n)-*threshold schemes*, that is, secret sharing schemes whose access structure consists of all subsets of at least l out of n participants. The procedure can be generalised to deal with the case that $\mathcal{F}$ is linear but not necessarily ideal.

In the above constructions A-codes with t-fold security can be replaced by A-codes that are *t-fold ϵ-secure*. That is probability of success in spoofing of order $i, 0 \leq i \leq t$, is at most ϵ. Extension of Theorem 5 and 6 results in SGA-systems that are *t-fold ϵ-secure*. More details will be provided in the full paper.

6 A key-efficient SGA-system

In this section we describe an example of an implementation of the Benaloh-Leichter construction and show that the resulting system requires a very small number of keys and achieves very high speed.

The constructions of Sections 5.1 and 5.2 are mainly of theoretical interest. This is because in both cases we need to generate, store and distribute encoding rules that have $|S|$ components and hence for authenticating a source of size 2^{20} a vector of more than 1000000 components must be dealt with.

In the following we will show that by putting an extra condition on the A-code in the construction given in Section 5.1 it is possible to operate on the encoding rule indices, rather than the encoding rules themselves.

We say an A-code has an *indexing function* if there exists a function I that for every element of $GF(q)^\ell$ gives an encoding rule. That is, $I : GF(q)^\ell \rightarrow \mathcal{E}$. An indexing function for a linear A-code is called *homomorphic* if $I(a+b) = I(a)+I(b)$ where '+' on the left hand side is addition over $GF(q)$ and on the right hand side is vector addition of the encoding rules. An A-code with homomorphic indexing is called a *homomorphic A-code*.

Using A-codes with homomorphic indexing enables us to use the Benaloh-Leichter algorithm on indices rather than on the whole encoding rule, and so for an A-code with b encoding rules we only need to distribute vectors of size $\log_2 b$ bits. This is independent of the size of the source.

In the following we give an example of such A-codes. The A-code used for the construction is based on a modified version of the den Boer code [4]. The important property of the den Boer A-code is the small number of key bits, thus providing highly controllable security for long messages. However to use the code for an SGA-system we use the modified version proposed in [17]. The modified version retains its perfect 1-fold security for messages of length equal to the key length, but for longer messages the probability of success in a substitution attack increases with the length of the message.

Consider the finite field $F = GF(2^h)$. A source state is a string of m bits, where m is a multiple of h, and $m \leq h^2$. We can write $s = M_0.M_1.\cdots M_\ell$, $\ell \leq h-1$, where each M_i, $0 \leq i \leq \ell$, is a block of h bits. A key is a pair (μ, λ), $\mu, \lambda \in F$. For a key (μ, λ), the tag for a source state s is,

$$A(e, s) = \mu + \sum_{i=0}^{\ell} M_i \lambda^{2^i}.$$

For this code, if the size of the source is the same as the size of the key then $P_0 = P_1 = 2^{-h}$. It is easy to check that the code is homomorphic and hence instead of distributing an encoding rule it suffices to distribute the index of an encoding rule given by (μ, λ).

A recent implementation of the den Boer code [12] uses the technique of *bucket hashing* and results in a very fast implementation.

7 Concluding remarks

The work of this paper can be extended in two directions. On the theoretical side, finding bounds on the performance of the system and key requirement are the natural questions to be asked. As noted in Section 5, the definitions and constructions of this paper can be extended to ϵ-secure systems. Another extension of the model is for multiple authentication using the Wegman and

Carter construction. However in this case efficient implementation requires an efficient way of generating shared pseudorandom sequences.

On the practical side, finding efficient constructions such as those given in Section 6, or finding constructions with extra properties such as cheater detection and disenrollment, are interesting open problems that require further study.

References

1. J. Benaloh and J. Leichter, *Generalised secret sharing schemes and monotone functions*, Adv. in Cryptology - CRYPTO '88, Lecture Notes in Compt. Sci., **403**, (1990), 27–35.
2. B. Blakley and G.R. Blakley and A.H. Chan and J.L. Massey, *Threshold Schemes With Disenrollment*, Adv. in Cryptology - CRYPTO '92, Lecture Notes in Comput. Sci., **740**, (1993), 540–548.
3. G.R. Blakley and G.A. Kabatianski, *Linear algebra approach to secret sharing schemes* , Adv. in Cryptology - CRYPTO '95, Lecture Notes in Comput. Sci., **963**, (1995), 367–371.
4. B. den Boer, *A simple and key-economical unconditional authentication scheme*, J. of Computer Security, **2**, (1993), 65–71.
5. E.F. Brickell, *Some ideal secret sharing schemes* , J. Combin. Math Combin. Comput., **6**, (1989), 105–113.
6. E.F. Brickell and D.R. Stinson, *The detection of cheaters in threshold schemes* SIAM Journal of Discrete Mathematics, **4**, (1991), 502–510.
7. E. F. Brickell and D. R. Stinson, *Some improved bounds on the information rate of perfect secret sharing schemes*, J. Cryptology, **2**, (1992), 153–166.
8. Y. Desmedt, *Threshold cryptosystems*, Adv. in Cryptology - AUSCRYPT '92, Lecture Notes in Comput. Sci., **718**, (1993), 3–14.
9. Y. Desmedt, Y. Frankel and M. Yung, *Multi-receiver/multisender network security: efficient authenticated multicast/feedback*, IEEE INFOCOM '92-11th Annual Joint Conf of the IEEE Computer and Communications Societies, 1992, pp 2045-2054.
10. M. van Dijk, *A linear construction of perfect secret sharing schemes* , Adv. in Cryptology - EUROCRYPT '94, Lecture Notes in Comput. Sci., **950**, (1995), 23–34.
11. E.N. Gilbert, F.J. MacWilliams and N.J. Sloane, *Codes which detect deception*, Bell System Technical Journal, **53**, no. 3, (1974), 405–424.
12. T. Johansson, *Bucket hashing with a small key size*, Adv. in Cryptology - *EUROCRYPT '97*, Lecture Notes in Comput. Sci., **1233**, (1997), 149–162.
13. T. Johansson, B. Smeets, G. Kabatianski, *On the relation between A-codes and codes correcting independent errors*, Adv. in Cryptology - *EUROCRYPT '93*, Lecture Notes in Computer Science, **765**, (1994), 1–11.
14. H.Y. Lin and L. Harn, *A generalized secret sharing scheme with cheater detection*, Advances in Cryptology ASIACRYPT '91, Lecture Notes in Comput. Sci., **739**, (1994), 149–158.
15. K. M. Martin, *New secret sharing schemes from old* , J. Combin. Math Combin. Comput. **14**, (1993), 65–77.
16. P. Rogaway, *Bucket Hashing and its Application to Fast Message Authentication*, Adv. in Cryptology- *CRYPTO '95*, Lecture Notes in Comput. Sci., **963**, (1995), 29–42.

17. R. Safavi-Naini, *Three systems for shared generation of authenticators*, Proceedings of COCOON'96, Lecture Notes in Comput. Sci., **1090**, (1996), 401–411.
18. R. Safavi-Naini and L. Tombak, *Combinatorial characterization of A-codes with r-fold security*, Adv. in Cryptology - ASIACRYPT '94, Lecture Notes in Comput. Sci., **917**, (1995), 211–223.
19. A. Shamir, *How to share a secret*, Comm. ACM, 22(11), (1979), 612–613.
20. G.J. Simmons, *A game theory model of digital message authentication*, Congressus Numerantium **34**, (1982), 413–424.
21. G.J. Simmons, *A survey of information authentication*, in Contemporary Cryptology, The Science of Information Integrity, G.J. Simmons, ed., IEEE Press, (1992), 379–419.
22. G. J. Simmons, W.-A. Jackson and K. Martin, *The geometry of shared secret schemes* , Bull. Inst. Combin. Appl., **1**, (1991), 71–88.
23. D.R. Stinson, *The combinatorics of authentication and secrecy codes* , J. Cryptology **2**, (1990), 23–49.
24. D. R. Stinson, *An explication of secret sharing schemes* Des. Codes Cryptogr. **2**, (1992), 357–390.
25. D.R. Stinson, *Cryptography: Theory and Practice*, CRC Press, 1995.

Proposal of User Identification Scheme Using Mouse

Kenichi HAYASHI, Eiji OKAMOTO, Masahiro MAMBO

School of Information Science, Japan Advanced Institute of Science and Technology
1-1 Asahidai Tatsunokuchi Nomi Ishikawa, 923-12 Japan
Email:okamoto@jaist.ac.jp

Abstract. User identification is very effective for protecting information from illegal access. There are password and handwriting authentication system, but the password authentication system has been frequently cracked and the handwriting authentication system needs a special device. Thus we examine whether it is possible or not to identify users by a standard device, mouse. We implement a user identification system using a mouse and experiment with it. The experimental results indicate mouse can be used for user identification.

1 Introduction

Computer security is usually attained by combination of many different security mechanisms. In most situations, prevention of illegal access is one of the most fundamental mechanisms. It is very important to distinguish illegal users from legal users. Today computers are connected each other, and a world-wide network is established. Many people can get access to information through that network. Thus user identification schemes are very important from security point of view. There are many user identification schemes. But most of schemes need exclusive devices.

Password is the simplest and the most popular user identification scheme. It does not need exclusive devices. However, we generally think that the degree of its security is low because the user identification using knowledge provides lower degree of security than using motion like handwriting or handwritten signature. It is very easy to imitate knowledge like password. Once user identification using knowledge like password is cracked, we cannot distinguish illegal users from legal users. On the other hand, it is difficult to crack and imitate action like handwriting. User identification scheme based on key stroke latencies can be regarded as an improved password scheme [2, 3], and it is more secure than a scheme using password only [3].

Nowadays mouse is a standard device of computers, but a user identification scheme using a mouse is not popular at all. Hence, we propose a user identification scheme using the standard device, mouse. We examine whether it is possible or not to identify users by the mouse and verify its usefulness.

Fig. 1. Samples of circles drawn by two users

2 Proposed scheme

How do we identify users using a mouse? Can we apply fundamental skills of the handwriting on a tablet or a digitizer with a pen to the user identification scheme using a mouse? Obviously, writing exact characters by a mouse is not easy compared with writing precise characters on a tablet or a digitizer with a pen. Therefore, we examine whether it is possible or not to identify users with a simple figure. As a user identification scheme, individual characteristics must appear in drawn figures. Candidates of simple figures are straight line, triangle, four-sided figure and so on. In our research a user draws a circle in experiment 1 and any figure in experiment 2. In experiment 1 user draws a circle between concentric circles appeared on a screen. As long as circles lay between concentric circles, users are allowed to draw various shapes of circles (Fig. 1). The reasons why we show concentric circles are as follows; (1) There is a psychological test to examine personality by writing a circle between concentric circles. Users may be distinguished by the drawn circle. (2) Concentric circles shown on a screen will help users to draw a circle as a guide. In experiment 2 a user draws any figure on a template with concentric circles used in experiment 1.

2.1 Implemented scheme

We have implemented a system with an X-window system on workstations. Our identification scheme has two stages.

- Registration
  1. Drawing figures several times for registration
  2. Pre-processing drawn data
  3. Making a database DB from that pre-processed data
- Identification
  1. Drawing figures for identification several times (fewer times than in registration)
  2. Comparing input data with data in the DB
  3. Accepting or rejecting a user (Verifying a user)

Process of our system is constructed from three parts:

- **Showing the template on screen and outputting data to a file for registration or identification.**
  A template is shown on screen for registration or identification. As a mouse moves, (X, Y) coordinates of a mouse on screen is stored to a file as input data. The (X, Y) coordinates of a mouse on screen is recorded by Event precess of Xlib. Moreover, we record time to get the elapsed time since a mouse begins to move. There are differences of speed of mouse movement in the same shape of figures. Time is counted by a millisecond with use of the gettimeofday Unix system function. The data stored in the file is mouse's (X, Y) coordinate and elapsed time from pressing mouse's button until releasing the button.

- **Pre-processing.**
  The data stored in a file is pre-processed and a database DB is made. The length from the coordinate to the center of circle is added to the data in our scheme.

- **Comparing and verifying.**
  The DB data and input data are compared and verified. The input data is not pre-processed. The procedure for comparison and verification is described as follows.

  I Start points and end points in the DB are compared with start points and end points of the input data, and the error is checked. If the error distances of start points and end points are within a parameter, we set S=1.0. If only one point is within the parameter, set S=0.5. If both points are out of the parameter, set S=0.

  II The following process is applied to all input data:
    (a) Line data is read from the input data.
    (b) The time of the data is checked, and the nearest time data is searched in the DB.
    (c) Using the input data and data searched from the DB, an error distance of coordinates, and an error distance from the center of circle are checked.
    (d) TRUE is output if the error distance are within error distance parameters P and L. P is an error distance parameter of coordinate and L is an error distance parameter from the center of circle. Otherwise, FALSE is output.
    (e) Counting the percentage of TRUE. Let T be this percentage.

  III. The importance rate of process result is evaluated for step I and step II described above. (Importance of step I) : (Importance of step II) $= \alpha : \beta$

  IV. Match-rate MR is calculated by $MR = S \times \frac{\alpha}{\alpha+\beta} + T \times \frac{\beta}{\alpha+\beta}$ from all results of step I, II and III.

  V. A user is identified as a right user if the match-rate is greater than a parameter M.

Table 1. Results of experiment 1 and 2

Experiment	1	2
Right-comp(%)	85	87
Miss-reject(%)	15	13
Miss-comp(%)	7	0

For example, let us consider a case such that the importance ratio between step I and step II is 1 to 4 (=0.2:0.8) and the parameter M for judging a right user is 0.6. If the result of step I is S=1.0, and the result of step II is T=0.7, then the match-rate is calculated by

$$MR = (1.0 \times 0.2) + (0.7 \times 0.8) = 0.76 > M = 0.6.$$

This user is judged as a right user.

3 Experiments

In user identification, we can consider four situations: (i) a right user is identified as a right user, (ii) a right user is identified as a wrong user, (iii) a wrong user is identified as a right user, (iv) a wrong user is identified as a wrong user. The situations (i) and (iv) are called right-comparison, (ii), (iii) are called miss-reject and miss-comparison, respectively. Naturally, it is important to keep right-comparison-rate high and miss-comparison-rate low. The right-comparison-rate and miss-comparison-rate are usually set to be over 80% and below 5%, respectively, in the handwriting identification schemes and the identification schemes based on key stroke latencies [2, 3]. We aim to make a scheme satisfying these percentages. All experiments are conducted in our laboratory.

3.1 Experiment 1

In the first experiment a user draws a circle N(=5) times to make a DB and twice for comparison and verification. We have conducted a preparatory experiment in order to determine parameter L. In the preparatory experiment the maximum error distance from the center of circle is at most 50 for all users. So we set L=50. Concerning P, we presume P=L in this experiment. The result for the case of M=0.6 is shown in Table 1.

3.2 Averaging data

In experiment 1, for each input data, data comparison is executed with values in each of N databases. This makes verification slow. We seek to reduce database size without deteriorating identification performance. We examine the following method: (1) sorting the written data of N drawings by time, then repeat the following two steps: (2) popping N data from the top of sorted data and (3) averaging popped data and storing its result into a new database.

As a result of the experiment, the time required for comparison is shortened and the right-comparison-rate and the miss-rejection-rate keep the similar value as in experiment 1. Miss-comparison-rate is reduced to 4%. This result shows the proposed technique of averaging of data has enough efficacy.

3.3 Experiment 2

Although the miss-rejection-rate is not high in experiment 1, in rare cases some of users are rejected in all trials. For these users circles are more difficult to draw than other figures. Therefore, we have conducted an experiment allowing users to draw not only circle but also any other figures. The degree of security increases by writing various figures because this method combines knowledge of figure shapes, motion, and habits. We have utilized the averaging method described above. In this experiment we need to set new parameters because we allow users to draw various figures. By estimating from the result of experiment 1, if we set P=100, any right user is identified as a right user, and it is suitable to set L=20. The importance ratio between step 1 and 2 is set to be 1 to 4. We do not change other conditions, drawing time etc. The result of experiment 2 is also shown in Table 1. Since arbitrary figure is difficult to guess, the miss-comparison-rate has decreased. Moreover, no user is rejected for all trials in experiment 2.

4 Conclusion

We have proposed a user identification scheme using mouse. We implemented the system, and experimented with it. The implemented system achieves high enough right-comparison-rate and low miss-comparison-rate. These results imply the mouse is useful for user identification. We need to continue to conduct experiments in order to make our scheme more practical.

References

1. M. Yoshimura, I. Yoshimura, "Recent Trends in Writer Recognition Technology ", The Journal of the Institute of Electronics, Information and Communication Engineers, Vol.72, No.7, pp.788–791, 1989
2. M. Kawasaki, H. Kakuda, Y. Mori, "An Identity Authentication Method Based on Arpeggio Keystrokes", Transactions of Information Processing Society of Japan, Vol.34, No.5, pp.1198–1205, 1993
3. M. Kasukawa, Y. Mori, K. Komatsu, H. Akaike, H. Kakuda, "An Evaluation and Improvement of User Authentication System Based on Keystroke Timing Data", Transactions of Information Processing Society of Japan, Vol.33, No.5, 1992

An Effective Genetic Algorithm for Finding Highly Nonlinear Boolean Functions

William Millan, Andrew Clark, and Ed Dawson

Information Security Research Centre
Queensland University of Technology
GPO Box 2434 Brisbane 4001 Australia
Facsimile: +61 7 3221 2384
Email: {millan,aclark,dawson}@fit.qut.edu.au

Abstract. We report on the results of the first known use of Genetic Algorithms (GAs) to find highly nonlinear Boolean functions. The basic method, using a new breeding procedure, is shown to be several orders of magnitude faster than random search in locating Boolean functions with very high nonlinearity. When a directed hill climbing method is employed, the results are even better. The performance of random searches is used as a bench mark to assess the effectiveness of a basic GA, a directed hill climbing method, and a GA with hill climbing. The selection of GA parameters and convergence issues are discussed. Finally some future directions of this research are given.

1 Introduction

Modern encryption systems use Boolean functions that should satisfy certain properties in order for the cipher to resist cryptanalytic attack. Perhaps the most important cryptographic property is nonlinearity: the Hamming distance between a Boolean function and the set of all affine functions.

Definition 1. Let $f(x)$ be the binary truth table of an n variable Boolean function with Walsh-Hadamard Transform (WHT) given by

$$\hat{F}(\omega) = \sum_x (-1)^{f(x) \oplus L_\omega(x)},$$

where $L_\omega(x) = \omega_1 x_1 \oplus \omega_2 x_2 \oplus \cdots \oplus \omega_n x_n$ denotes the linear function selected by ω. The value of $\hat{F}(\omega)$ is directly proportional to the correlation that $f(x)$ has with $L_\omega(x)$:

$$c(f, L_\omega) = \frac{\hat{F}(\omega)}{2^n}.$$

Let WH_{max} denote the maximum absolute value of $\hat{F}(\omega)$, so that the nonlinearity of $f(x)$ is given by $N_f = \frac{1}{2}(2^n - WH_{max})$. □

It is known that, for n even, the maximum possible nonlinearity is given by [6] $N_{max} = 2^{n-1} - 2^{\frac{n}{2}-1}$. However, when n is odd and $n \geq 9$ the upper bound of nonlinearity is not known [7], and remains an open problem.

It is important for cryptography to have means of generating highly nonlinear Boolean functions in a random manner, in order to avoid low complexity problems associated with simple deterministic constructions. To this end we present a genetic algorithm for finding highly nonlinear Boolean functions. Also provided is a method for directed hill climbing: that is choosing a truth table position to change so that the nonlinearity will increase. We present results for both of these methods, and using them together, as compared with the results from purely random searching. Our results consistently show that the GA is more than 1000 times faster than random searching at finding examples of Boolean functions with high nonlinearity. When the hill climbing is included, the quality of the obtained functions is increased.

A genetic algorithm is a heuristic approach to combinatorial optimisation in a complex fitness space. They have been previously used to find binary sequences with certain properties for communications applications [2], and also the real-valued Walsh transform has been studied in relation to GAs in [4]. However, neither of these papers mentioned cryptographic applications. Genetic algorithms have been successfully used to cryptanalyse simple substitution and transposition ciphers [5,8].

The idea behind a GA is to start with a pool of candidate solutions (the gene pool, or parent pool) and combine pairs of them in some manner (a *breeding scheme*) to produce new solutions (children). Breeding schemes may include parent selection, and mutation of the resulting child. A new pool is selected from among the parents and children (*culling*), on the basis of a fitness function, and the process repeated until a suitably strong solution is found, or the process converges. In our application, the solutions are the truth tables of Boolean functions of n variables, the breeding scheme is a "merge" operation we have designed specifically for this purpose, and the fitness function is the nonlinearity. Using the fast Walsh-Hadamard transform, the nonlinearity of a Boolean function can be determined in $O(n2^n)$ operations.

In the next section we describe the GA, and briefly discuss the selection of its operational parameters. The merge operation is defined, and the advantage of using it over some other simple breeding schemes is described. Convergence of the basic algorithm is discussed and means to avoid it described. The issues of pool size and mutation rate are also related to the performance of the algorithm.

In Section 3 we present a fast method for hill climbing Boolean functions to increase the nonlinearity. Naive hill climbing is computationally intensive, so these results are significant. A fast way of obtaining the Walsh-Hadamard of a slightly altered Boolean function is also provided, since this improves the efficiency of the algorithm.

The results of our experiments are shown in Section 4. Firstly we provide the results of random searching, which are a bench mark for the performance evaluation of the other algorithms. Then we give the results obtained for random search with hill climbing, the basic GA, and a GA with hill climbing. It is clear that even very simple algorithms are much better than random search alone, and that well chosen algorithms are more effective. The algorithms are compared in two ways: (i) computational effort required to achieve bench mark results, and (ii) the best results achieved with bounded computation.

2 The Genetic Algorithm

The GA was designed to mimic the natural evolutionary process by operating on a gene pool (list of solutions to the problem). The genetic processes of selection, mating and mutation are combined in order to "breed" a superior race of genes (solutions). In the classical GA the solution is represented as a binary string. We utilise the same representation here for the boolean function where the binary string is just the truth table of the function in 1,0 format. Given a solution representation, there are three other requirements in order to implement a GA, namely a solution evaluation technique, a mating function and a mutation operator. The mating function allows the combining of solutions, hopefully, in a meaningful manner. Mutation is performed on a single solution in order to introduce a degree of randomness to the solution pool. These three GA requirements are now discussed individually.

The GA requires a method of assessing and comparing solutions. Typically this measure is referred to as the "fitness". Not all fitness functions are suitable for use in genetic algorithms, as shown in [1]. The fitness we use here is simply the nonlinearity (N_f) of the Boolean function $f(x)$. This is suitable for a GA, since $|N_f - N_g| \leq dist(f(x), g(x))$. In other words nonlinearity is a locally smooth fitness function. With Boolean functions there are also numerous other fitness functions possible (for example, we could minimise the maximum value taken by the auto-correlation function).

The GA requires a method for combining two (possibly more) solutions in order to obtain offspring. The usual mating process utilised in classical GAs is often referred to as "crossover". The crossover operation involves selecting two "parents" from the current solution pool, picking a random point in the binary string representing each of the parents and swapping the values beyond that point between the two parents. This process results in two "children" with some characteristics of each of the parents. Here we use a slightly different breeding process, namely "merging", which is described below.

The mutation operation simply introduces randomness to the solution pool. Mutation is generally applied to the children which result from the breeding process. In some cases the breeding step is ignored and mutation is applied to the selected parents in order to produce children. Such an option is not considered here. It is usual to mutate a child by complementing a random subset of the binary string representing the child. The number of values complemented is a parameter of the algorithm - sometimes referred to as the "mutation factor".

Combining all of the GA operations which have been described above the overall algorithm is obtained. Generally the initial solution pool is generated randomly or using some "smart" technique specific to the type of optimisation problem being tackled. (In our case a random initial pool is suitable since very few randomly generated functions have low nonlinearity. The problem with random generation is that very highly nonlinear functions are difficult to find.) The algorithm then updates the solution pool over a number of iterations (or *generations*). The maximum number of iterations is fixed, although addition stopping criteria may be specified. In each iteration a number of steps are involved: 1. selection of parents from the current solution pool, 2. mating of parents to produce offspring, 3. mutation of the offspring, and 4. selection from the mutated offspring and the current solution pool to determine the solution pool for the next iteration.

Much of the implementation detail of the GA is specific to the problem being solved. In our case we combine mating and mutation in the merge operation, which allows two good parents that are close together in Hamming distance to produce a child close to both parents. It follows that the child is expected to have a good fitness.

In our algorithm all possible combinations of parents undergo the breeding process. The number of such pairings is dependent upon the pool size P - there are $\frac{P(P-1)}{2}$ such pairings. After initial experiments we chose to use a pool of 10, as a compromise between efficiency and convergence. Another parameter of the GA is the "breed factor": the number of children produced by each parent pair. Our results show that for small, fast GAs, the most efficient breed factor is 1, when efficiency is taken to mean the fitness of the best functions found as compared with the number of functions evaluated. However using larger values may be useful in gaining quick access to the entire search space. The following algorithm describes the genetic algorithm as used in experiments for this paper. The parameter MaxIter defines the maximum number of iterations that the algorithm should do.

- **GeneticAlgorithm(MaxIter,P,BF,HC)**
  1. Generate a random pool of P Boolean functions and calculate their Walsh-Hadamard transforms.
  2. For i in 1 ... MaxIter, do
     (a) For each possible pairing of the functions in the solution pool do
        i. Perform the merge operation on the two parents to produce a number of offspring equal to the "breed factor" (BF).
        ii. For each child, do
           A. If hill climbing is desired (ie. if HC=1) call the **HillClimb** function with the Start parameter set to 1.
           B. Provided the resulting offspring is not already in the list of children, add the new child to the list. (this prevents convergence)
     (b) Select the best solutions from the list of children and the current pool of solutions. In the case where a child has equal fitness (nonlinearity) to a solution in the current solution pool, preference is given the to the child.
     (c) Check that a maximum number of functions have been considered - if so skip to Step 3.
  3. Report the best solution(s) from the current solution pool.

The merging operation is defined as follows:

Definition 2. Given the binary truth tables of two Boolean functions $f_1(x)$, $f_2(x)$ of n variables at Hamming distance d, we define the merge operation as:

If $d \leq 2^{n-1}$ then $merge_{f_1,f_2}(x) = f_1(x)$ for those x such that $f_1(x) = f_2(x)$, and a random bit otherwise;

else $merge_{f_1,f_2}(x) = f_1(x)$ for those x such that $f_1(x) \neq f_2(x)$, and a random bit otherwise. □

We note that merging is partly deterministic and partly probabilistic, and that it takes the fact that complementation does not change nonlinearity into account. The number of distinct children that can be the result of a merge is given by $2^{dist(f_1,f_2)}$, and we observe that all are equally probable. Thus the use of merge as a breeding scheme includes implicit mutation. Since random mutation of a highly nonlinear function is likely to reduce the nonlinearity, we avoid additional random mutations and rely on the merge to direct the pool into new areas of the search space. The motivation for this operation is that two functions that are highly nonlinear and close to each other will be close to some local maximum, and the merging operation produces a function also in the same region, hopefully close to that maximum. Also when applied to uncorrelated functions, the merge operation produces children spread over a large area, thus allowing the GA to search the space more fully. At the start of the GA, the children are scattered widely, then as the pool begins to consist of good functions, the merging assists convergence to local maxima. Our experiments have shown that other simple combining methods, such as XOR and crossover, do not assist convergence to good solutions. It is the use of merging that allows the GA to be effective.

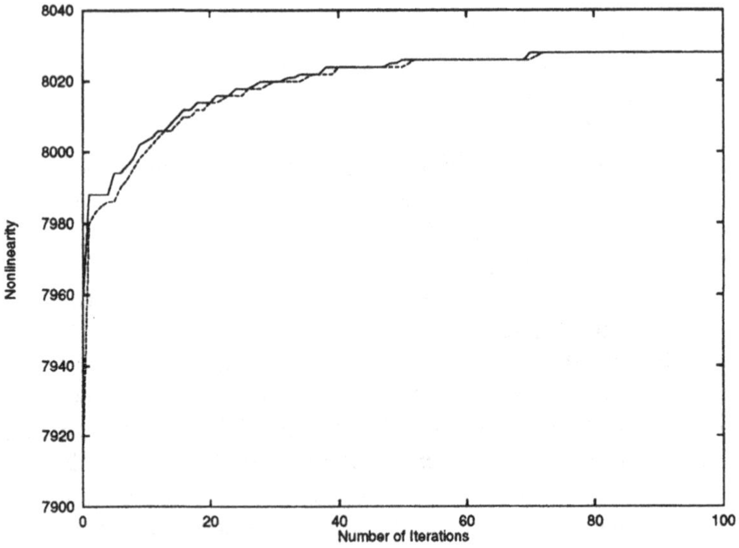

Fig. 1. Performance of a typical GA, n=14, Pool Size = 10.

Our initial experiments have shown that GAs with a pool size of less than 10 tend to converge too quickly, since a single good solution comes to dominate the pool: it is the parent of many children that survive into subsequent generations. When both parent and offspring survive and breed, their progeny tend to survive also, and that "family" soon dominates the gene pool, producing convergence. A policy precluding "incest" counters this effect. This policy may take the form of forbidding a parent pair from breeding if their Hamming distance is too small, as set by some threshold parameter.

However, when this parameter is set too large, it prevents breeding by the best pairs, thus undermining the motivation for the merge operation. Further experiments will be required to determine optimum sets of GA parameters. It is clear that to find very highly nonlinear functions, larger pools are useful, since they converge less rapidly.

As an example of how the genetic algorithm performs, we present Figure 1, which represents a search for nonlinear Boolean functions with 14 input variables. Starting with a random pool, the GA quickly achieves highly nonlinear functions, then the rate of increase drops as convergence occurs. The solid line shows the best fitness value in the current pool, and the dotted line shows the minimum fitness in the pool.

3 Hill Climbing

Any Boolean function that is not a local maximum may be made more nonlinear by complementing the output for some single inputs. We call the set of these inputs the Improvement Set of the given function. Theorem 4 gives conditions for an input to be a member of the Improvement Set. First we need to make some definitions.

Definition 3. Let $f(x)$ be the binary truth table of a Boolean function with Walsh-Hadamard Transform $\hat{F}(\omega)$ as previously defined. Let us define the following sets:

$$W_1^+ = \{\omega : \hat{F}(\omega) = WH_{max}\}$$

and

$$W_1^- = \{\omega : \hat{F}(\omega) = -WH_{max}\}.$$

We also need to define sets of ω for which the WHT magnitude is close to the maximum:

$$W_2^+ = \{\omega : \hat{F}(\omega) = WH_{max} - 2\},$$

$$W_2^- = \{\omega : \hat{F}(\omega) = -(WH_{max} - 2)\}.\square$$

When a truth table is changed in exactly one place, all WHT values are changed by either +2 or -2. It follows that in order to increase the nonlinearity we need to make the WHT values in set W_1^+ change by -2, the WHT values in set W_1^- change by +2, and also make the WHT values in set W_2^+ change by -2 and the WHT values in set W_2^- change by +2. The first two conditions are obvious, and the second two conditions are required so that all other $|\hat{F}(\omega)|$ remain less than WH_{max}. These conditions can be translated into simple tests.

Theorem 4. *Given a Boolean function $f(x)$ with WHT $\hat{F}(\omega)$, we define sets $W^+ = W_1^+ \cup W_2^+$ and $W^- = W_1^- \cup W_2^-$. For an input x to be an element of the Improvement Set, all of the following conditions must be satisfied.*

(i) $f(x) = L_\omega(x)$ for all $\omega \in W^+$

and

(ii) $f(x) \neq L_\omega(x)$ for all $\omega \in W^-$.

If the function $f(x)$ is not balanced, and we wish to reduce the Imbalance, we impose the additional restriction that

(iii) when $\hat{F}(0) > 0$, $f(x) = 0$, else $f(x) = 1$. □

Proof. We start by considering the conditions to make WHT values change by a desired amount. When $\hat{F}(\omega)$ is positive, there are more 1s than -1s in the polarity truth table, and more 0s than 1s in the binary truth table of $f(x) \oplus L_\omega(x)$. It follows that to make $\Delta\hat{F}(\omega) = -2$, we must change any single 0 to 1 in the truth table of $f(x) \oplus L_\omega(x)$. This means that we select an x to change such that $f(x) = L_\omega(x)$. We desire a -2 change for all WHT values with $\omega \in W^+$, so this proves condition (i). A similar argument proves condition (ii). A function is balanced when $\hat{F}(0) = 0$, so to reduce the imbalance we must select x according to condition (iii). □

We note that Forre [3] has previously proposed an algorithm for incremental improvement of Boolean functions. However, the algorithm used is naive. Firstly the truth table bits to change are selected randomly. This is a poor strategy since a function near a local maximum will be made less nonlinear by most changes. Secondly, a complete WHT of the new function is performed after each bit change, with complexity $O(n2^n)$. Below we give a theorem showing how to quickly modify the existing WHT to account for the single bit truth table change. This has complexity $O(2^n)$, which is n times faster. Finally, once a good function has been obtained, Forre suggests modifying it to become balanced. Our directed hill climbing algorithm is clearly superior, since it can produce balanced functions directly, no bits will be changed that reduce nonlinearity, and the new WHT can be efficiently obtained. Together, Theorems 4 and 5 provide a way for fast, directed hill climbing, which may be used to improve the nonlinearity of most Boolean functions.

Theorem 5. *Let $g(x)$ be obtained from $f(x)$ by complementing the output for a single input, x_1. Then each component of the WHT of $g(x)$, $\hat{G}(\omega) = \hat{F}(\omega) + \Delta(\omega)$, can be obtained as follows: If $f(x_1) = L_\omega(x_1)$, then $\Delta(\omega) = -2$, else $\Delta(\omega) = +2$*

Proof. When $f(x_1) = L_\omega(x)$, we have $(-1)^{f(x_1) \oplus L_\omega(x_1)} = 1$, which contributes to the sum in $\hat{F}(x_1)$. Changing the value of $f(x_1)$ changes this contribution to -1, so $\Delta\hat{F}(\omega) = -2$. Similarly when $f(x_1) \neq L_\omega(x)$, $\Delta\hat{F}(\omega) = +2$

This hill climbing technique was implemented for use in the genetic algorithm. The following function - **HillClimb(BF, WHT, Start)** - can be used to sequentially complement bits in the truth table of BF, each time improving the nonlinearity by one. The parameter "Start" is used to prevent the algorithm from complementing the same bit twice - it should initially be equal to 1 (one). Note that the function **HillClimb** is recursive as defined below. The recursion ends when no input i satisfies the conditions in Step 3b.

- **HillClimb(BF, WHT, Start)**
  1. Determine maximum value of the Walsh-Hadamard transform WH_{max}.
  2. By parsing the WHT find the values of ω which belong to the sets W_1^+, W_1^-, W_2^+ and W_2^-. At the completion of this step there should be two lists: $W^+ = W_1^+ \cup W_2^+$ and $W^- = W_1^- \cup W_2^-$. NB. Either (but not both) of W^+ and W^- may be empty.
  3. For i in Start $\ldots 2^n$, do
     (a) Let b_i denote the ith bit in the truth table of BF.

(b) Parse the sets W^+ and W^- ensuring that $L_\omega(i) = b_i$ for each ω in W^+ and $L_\omega(i) \neq b_i$ for each ω in W^-. If not skip to Step 3e.
(c) We have a candidate for improvement. Complement b_i in the truth table of BF (denote the resulting boolean function BF'), update the WHT (becoming WHT') by using Theorem 5, and call **HillClimb(BF', WHT',** $i + 1$**)**.
(d) Skip to Step 4.
(e) Increment i ($i = i + 1$).
4. BF represents a locally maximum - terminate processing.

4 Results

In this section we present typical results obtained for the four algorithms under consideration: uniformly random generation, random generation with hill climbing (R HC) to a local maximum, a plain GA, and a GA in which each new child is hill climbed to a local maximum (GA HC). Random generation of the truth tables is used as a benchmark for assessing the effectiveness of the other algorithms.

Sample Size	8	9	10	11	12	13	14	15	16
1000	110	228	468	958	1947	3946	7966	16048	32273
10000	111	229	469	959	1949	3950	7971	16054	32280
100000	112	230	470	961	1952	3952	7975	16058	32288
1000000	112	230	472	962	1955	3954	7978	16065	n/a

Table 1. Best Nonlinearity Achieved by Random Searching, Typical Results.

Table 1 shows typical results for the best nonlinearity obtained over searches with different sample sizes ranging from 1000 to 1000000 functions. The table clearly shows that increasing the sample size ten times only marginally increases the nonlinearity we might expect to obtain. This results from the shape of the probability distribution of nonlinearity of Boolean functions: most functions do not have low nonlinearity, but very highly nonlinear functions are extremely rare. This is illustrated in Figure 2, which shows a typical nonlinearity distribution from a selection of 1 million randomly generated Boolean functions of eight input variables. It should be noted that for $n = 8$, the maximum nonlinearity of Boolean functions is 120 [6]. The graph shows that functions approaching this nonlinearity are very rare.

Table 2 shows typical values for the number of functions that need to be tested before an example with nonlinearity equal to or exceeding the benchmark is obtained. The benchmark we use is the highest nonlinearity found in a random sample of 1,000,000 functions. In most cases a simple GA needs less than 1,000 functions to get a benchmark result, indicating that the GA is far more efficient than random search in finding highly nonlinear Boolean functions even when the overhead involved with the GA is taken into account. The results for hill climbing algorithms are even better than the GA alone, indicating that hill climbing is a very effective technique for finding strong functions

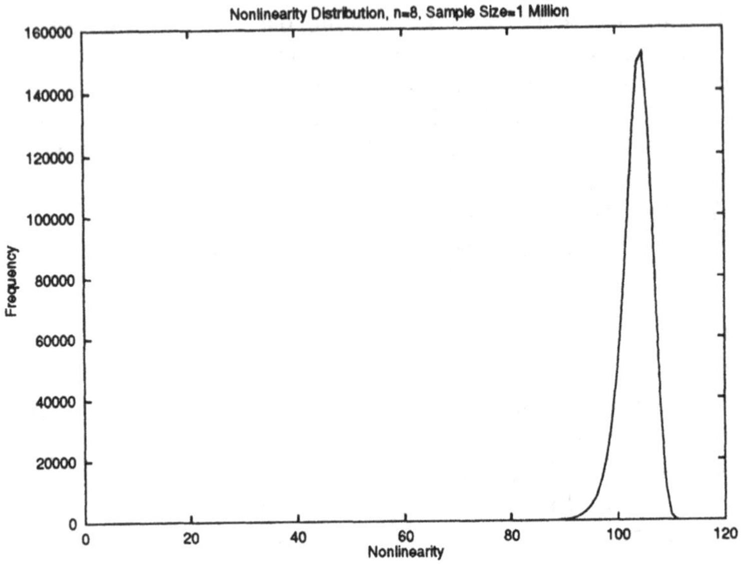

Fig. 2. Distribution of Nonlinearity, n=8

quickly. Note that the benchmark used for $n = 16$ is the highest obtained in 100,000 random generations, since we have not completed a search of a million functions for this number of inputs, as yet.

n	8	9	10	11	12	13	14	15	16
Benchmark	112	230	472	962	1955	3954	7978	16065	32288*
R HC	4	3	2	8	3	2	2	2	1
GA	591	422	767	588	639	721	722	1108	588
GA HC	2	4	4	5	9	3	2	2	1

Table 2. Number of Functions to Achieve Benchmark Results, Typical Results.

In Tables 3 and 4 we show the best results achieved by the algorithms when they are forced to terminate after a specific number of functions have been tested. A direct comparison between random generation with hill climbing, and a simple GA without hill climbing shows that these algorithms are about equally effective for 1000 or 10000 function tests. Other experiments have suggested that as the computation bound is increased, the performance of the GA will eventually exceed that of hill climbing. It is interesting to note that the best algorithm is clearly a GA with hill climbing. This hybrid algorithm is able to quickly obtain functions far better than the benchmarks.

Method	8	9	10	11	12	13	14	15	16
Random	110	228	468	958	1947	3946	7966	16048	32276
R HC	112	232	474	966	1960	3962	7991	16080	32319
GA	111	232	473	964	1955	3962	7982	16076	32289
GA HC	114	236	478	974	1972	3978	8014	16114	32366

Table 3. Best Nonlinearity Achieved After Testing 1000 Functions, Typical Results.

Method	8	9	10	11	12	13	14	15	16
Random	111	229	469	959	1949	3950	7971	16054	32280
R HC	114	232	476	968	1961	3964	7995	16090	32332
GA	113	232	475	968	1964	3968	7996	16085	32329
GA HC	114	236	482	980	1980	3994	8036	16144	32405

Table 4. Best Nonlinearity Achieved After Testing 10000 Functions, Typical Results.

5 Conclusion

New methods for combinatorial optimisation of Boolean functions have been presented. These methods are clearly useful in the design of Boolean functions for cryptography. They may be easily adapted to find Boolean functions that are strong with respect to different, and combined, criteria. Their extension to S-box design is the subject of ongoing research.

References

1. A. Clark, E. Dawson, and H. Bergen. Combinatorial Optimisation and the Knapsack Cipher. *Cryptologia*, XX(1):85–93, January 1996.
2. W. Crompton and N.M. Stephens. Using Genetic Algorithms to Search for Binary Sequences with Large Merit Factor. In *Proceedings of the Third IMA Conference on Cryptography and Coding*, pages 83–96. Clarendon Press, Oxford, December 1991.
3. R. Forre. Methods and Instruments for Designing S-Boxes. *Journal of Cryptology*, 2(3):115–130, 1990.
4. D.E. Goldberg. Genetic Algorithms and Walsh Functions: Part I, A Gentle Introduction. *Complex Systems*, 3:129–152, 1989.
5. Robert A. J. Matthews. The use of genetic algorithms in cryptanalysis. *Cryptologia*, 17(2):187–201, April 1993.
6. W. Meier and O. Staffelbach. Nonlinearity Criteria for Cryptographic Functions. In *Advances in Cryptology - Eurocrypt '89, Proceedings, LNCS*, volume 434, pages 549–562. Springer-Verlag, 1990.
7. N.J. Patterson and D.H. Wiedemann. The Covering Radius of the $(2^{15}, 16)$ Reed-Muller Code is at least 16276. *IEEE Transactions on Information Theory*, 29(3):354–356, May 1983.
8. R. Spillman, M. Janssen, B. Nelson, and M. Kepner. Use of a Genetic Algorithm in the Cryptanalysis of Simple Substitution Ciphers. *Cryptologia*, 17(1):31–44, January 1993.

Duality of Boolean Functions and Its Cryptographic Significance

Xian-Mo Zhang[1] and Yuliang Zheng[2] and Hideki Imai[3]

[1] The University of Wollongong, Wollongong, NSW 2522, Australia
xianmo@cs.uow.edu.au
[2] Monash University, Frankston, Melbourne, VIC 3199, Australia
yzheng@fcit.monash.edu.au, http://www-pscit.fcit.monash.edu.au/~yuliang/
[3] The University of Tokyo, 7-22-1 Roppongi, Minato-ku, Tokyo 106, JAPAN
imai@iis.u-tokyo.ac.jp

Abstract. Recent advances in interpolation and high order differential cryptanalysis have highlighted the cryptographic significance of Boolean functions with a high algebraic degree. However, compared with other nonlinearity criteria such propagation, resiliency, differential and linear characteristics, apparently little progress has been made in relation to algebraic degree in the context of cryptography. The aim of this work is to research into relationships between algebraic degree and other nonlinearity criteria. Making use of duality properties of Boolean functions, we have obtained several results that are related to lower bounds on nonlinearity, as well as on the number of terms, of Boolean functions. We hope that these results would stimulate the research community's interest in further exploring this important area.

1 Introduction

The algebraic degree has long been believed by many designers of block ciphers and one-way hash functions to be an important nonlinearity indicator for the cryptographic strength of Boolean functions. Recent progress in interpolation cryptanalysis [1] and high order differential cryptanalysis [5] can be viewed as a proof for the correctness of the belief. Of particular interest is the work of [5] in which the authors showed how to break in less than 20 milli-seconds a block cipher that employs low algebraic degree (quadratic) Boolean functions as its S-boxes and is provably secure against linear and (the first order) differential attacks.

Investigation into the algebraic degree of Boolean functions has been a difficult topic. This is supported by the fact that, while the past few years have seen much progress in relation to other nonlinearity criteria such as propagation, differential profile, nonlinear profile, resiliency, correlation-immunity, local and global avalanche characteristics, little progress has been made in designing Boolean functions that have a high algebraic degree and also satisfy other important nonlinearity criteria.

In this paper we tackle algebraic degree, together with nonlinearity, propagation characteristics, correlation immunity and the number of terms in a Boolean function by exploring the duality property of a Boolean function. Main contributions of this work are to show (1) two lower bounds, one on the nonlinearity and the other on the number of terms of a Boolean functions, and (2) a connection between the algebraic degree of a Boolean function and its Walsh-Hadamard transform.

2 Basic Definitions

We consider functions from V_n to $GF(2)$ (or simply functions on V_n), V_n is the vector space of n tuples of elements from $GF(2)$. The *truth table* of a function f on V_n is a (0,1)-sequence defined by $(f(\alpha_0), f(\alpha_1), \ldots, f(\alpha_{2^n-1}))$, and the *sequence of* f is a (1,−1)-sequence defined by $((-1)^{f(\alpha_0)}, (-1)^{f(\alpha_1)}, \ldots, (-1)^{f(\alpha_{2^n-1})})$, where $\alpha_0 = (0,\ldots,0,0)$, $\alpha_1 = (0,\ldots,0,1)$, ..., $\alpha_{2^{n-1}-1} = (1,\ldots,1,1)$. The *matrix* of f is a (1,−1)-matrix of order 2^n defined by $M = ((-1)^{f(\alpha_i \oplus \alpha_j)})$ where $\oplus$ denotes the addition in $GF(2)$. f is said to be *balanced* if its truth table contains an equal number of ones and zeros.

Given two sequences $\tilde{a} = (a_1, \cdots, a_m)$ and $\tilde{b} = (b_1, \cdots, b_m)$, their *component-wise product* is defined by $\tilde{a} * \tilde{b} = (a_1 b_1, \cdots, a_m b_m)$. In particular, if $m = 2^n$ and $\tilde{a}$, $\tilde{b}$ are the sequences of functions on V_n respectively, then $\tilde{a} * \tilde{b}$ is the sequence of $f \oplus g$.

Let $\tilde{a} = (a_1, \cdots, a_m)$ and $\tilde{b} = (b_1, \cdots, b_m)$ be two vectors (or sequences), the *scalar product* of $\tilde{a}$ and $\tilde{b}$, denoted by $\langle \tilde{a}, \tilde{b} \rangle$, is defined as the sum of the component-wise multiplications. In particular, when $\tilde{a}$ and $\tilde{b}$ are from V_m, $\langle \tilde{a}, \tilde{b} \rangle = a_1 b_1 \oplus \cdots \oplus a_m b_m$, where the addition and multiplication are over $GF(2)$, and when $\tilde{a}$ and $\tilde{b}$ are (1,−1)-sequences, $\langle \tilde{a}, \tilde{b} \rangle = \sum_{i=1}^{m} a_i b_i$, where the addition and multiplication are over the reals.

An *affine* function f on V_n is a function that takes the form of $f(x_1, \ldots, x_n) = a_1 x_1 \oplus \cdots \oplus a_n x_n \oplus c$, where $a_j, c \in GF(2)$, $j = 1, 2, \ldots, n$. Furthermore f is called a *linear* function if $c = 0$.

Definition 1. The *Hamming weight* of a (0,1)-sequence ξ is the number of ones in the sequence. Given two functions f and g on V_n, the *Hamming distance* $d(f,g)$ between them is defined as the Hamming weight of the truth table of $f(x) \oplus g(x)$, where $x = (x_1, \ldots, x_n)$. The *nonlinearity* of f, denoted by N_f, is the minimal Hamming distance between f and all affine functions on V_n, i.e., $N_f = \min_{i=1,2,\ldots,2^{n+1}} d(f, \varphi_i)$ where $\varphi_1, \varphi_2, \ldots, \varphi_{2^{n+1}}$ are all the affine functions on V_n.

A (1,−1)-matrix H of order m is called a *Hadamard* matrix if $HH^t = mI_m$, where H^t is the transpose of H and I_m is the identity matrix of order m. A Sylvester-Hadamard matrix of order 2^n, denoted by H_n, is generated by the following recursive relation

$$H_0 = 1,\ H_n = \begin{bmatrix} H_{n-1} & H_{n-1} \\ H_{n-1} & -H_{n-1} \end{bmatrix},\ n = 1, 2, \ldots.$$

Let ℓ_i, $0 \le i \le 2^n-1$, be the i row of H_n. By Lemma 2 of [4], ℓ_i is the sequence of a linear function $\varphi_i(x)$ defined by the scalar product $\varphi_i(x) = \langle \alpha_i, x \rangle$, where α_i is the ith vector in V_n according to the ascending alphabetical order.

Definition 2. *Let f be a function on V_n. For a vector $\alpha \in V_n$, denote by $\xi(\alpha)$ the sequence of $f(x \oplus \alpha)$. Thus $\xi(0)$ is the sequence of f itself and $\xi(0) * \xi(\alpha)$ is the sequence of $f(x) \oplus f(x \oplus \alpha)$. Set*

$$\Delta(\alpha) = \langle \xi(0), \xi(\alpha) \rangle,$$

the scalar product of $\xi(0)$ and $\xi(\alpha)$. $\Delta(\alpha)$ is also called the auto-correlation *of f with a shift α.*

Obviously, $\Delta(\alpha) = 0$ if and only if $f(x) \oplus f(x \oplus \alpha)$ is balanced, i.e., f satisfies the propagation criterion with respect to α. On the other hand, if $|\Delta(\alpha)| = 2^n$, then $f(x) \oplus f(x \oplus \alpha)$ is a constant and hence α is a linear structure of f.

A function f on $GF(2)$ can be uniquely represented by a polynomial on $GF(2)$ whose degree is at most n. Namely,

$$f(x_1, \ldots, x_n) = \bigoplus_{\alpha \in V_n} g(a_1, \ldots, a_n) x_1^{a_1} \cdots x_n^{a_n} \tag{1}$$

where $\alpha = (a_1, \ldots, a_n)$, and g is also a function on V_n. Each $x_1^{a_1} \cdots x_n^{a_n}$ is called a term (in the polynomial representation) of f.

The algebraic degree, or simply degree, of f, denoted by $deg(f)$, is defined as the number of variables in the longest term of f, i.e.,

$$deg(f) = \max\{W(a_1, \ldots, a_n) \mid g(a_1, \ldots, a_n) = 1\}.$$

Definition 3. *Let f be a function on V_n and U be s-dimensional subspace of V_n. The* restriction *of f to U, denoted by f_U, is a function on U, defined by the following rule*

$$f_U(\alpha) = f(\alpha) \text{ for every } \alpha \in U.$$

Notation 1 Let W be a subspace of V_n. Denote the dimension of W by $dim(W)$.

Notation 2 $(b_1, \ldots, b_n) \preceq (a_1, \ldots, a_n)$ means that $(b_1, \ldots, b_n)$ is covered by $(a_1, \ldots, a_n)$, namely if $b_j = 1$ then $a_j = 1$. In addition, $(b_1, \ldots, b_n) \prec (a_1, \ldots, a_n)$ means that $(b_1, \ldots, b_n)$ is properly covered by $(a_1, \ldots, a_n)$, namely $(b_1, \ldots, b_n) \preceq (a_1, \ldots, a_n)$ and $(b_1, \ldots, b_n) \neq (a_1, \ldots, a_n)$.

3 Duality of Boolean Functions

The dual of a Boolean function f is a function g that is uniquely determined by the coefficients of the terms of f. The main purpose of this section is to provide the minimum amount of knowledge on duality that is required in the rest part of this paper. A proof for the following result is provided, as we feel that understanding the proof would be helpful in studying other issues that are more directly related to cryptography.

Theorem 4. *Let f be a function on V_n. Let $\alpha, \beta \in V_n$ $\alpha = (1, \ldots, 1, 0, \ldots, 0)$ where only the first s components are one, and $\beta = (0, \ldots, 0, 1, \ldots, 1, 0, \ldots, 0)$ where only the $(s+1)$th, ..., the $(s+t)$th components are one. Then the number of the terms among $x_1 \cdots x_s$, $x_1 \cdots x_s x_{s+1}$, ..., $x_1 \cdots x_s x_{s+1} \cdots x_{s+t}$ that appear in the polynomial representation of f, is even if $\bigoplus_{\gamma \preceq \alpha} f(\gamma \oplus \beta) = 0$, and this number is odd if $\bigoplus_{\gamma \preceq \alpha} f(\gamma \oplus \beta) = 1$.*

Proof. Consider a term

$$\chi(x) = x_{j_1} \cdots x_{j_{s'}} x_{i_1} \cdots x_{i_{t'}} \tag{2}$$

in f, where $x = (x_1, \ldots, x_n)$, $1 \leq j_1 \leq \cdots \leq j_{s'} \leq s$ and $s + 1 \leq i_1 \leq \cdots \leq i_{s'+t'} \leq s + t$. For $s' < s$, there are an even number of vectors γ in V_n such that $\gamma \prec \alpha$ and $\chi(\gamma \oplus \beta) = 1$. Hence

$$\bigoplus_{\gamma \preceq \alpha} \chi(\gamma \oplus \beta) = 0. \tag{3}$$

For $s' = s$, there is only one vector in V_n, $\gamma = \alpha$, such that $\chi(\gamma \oplus \beta) = 1$. Hence

$$\bigoplus_{\gamma \preceq \alpha} \chi(\gamma \oplus \beta) = 1. \tag{4}$$

Now consider a term

$$\omega(x) = x_{j_1} \cdots x_{j_k} \tag{5}$$

in f, where $x = (x_1, \ldots, x_n)$, $1 \leq j_1 \leq \cdots \leq j_k$, and $j_k > s + t$. From (5) with $j_k > s + t$, and the structures of α and β,

$$\omega(\gamma \oplus \beta) = 0 \tag{6}$$

for each $\gamma \preceq \alpha$. Denote the set of terms given in (2) by Γ_1 if $s' < s$, and by Γ_2 if $s' = s$. And denote the set of terms given in (5) by Ω. Then we can write f as

$$f = \bigoplus_{\chi \in \Gamma_1} \chi \oplus \bigoplus_{\chi \in \Gamma_2} \chi \oplus \bigoplus_{\omega \in \Omega} \omega.$$

From (3), (4) and (6),

$$\bigoplus_{\gamma \preceq \alpha} f(\gamma \oplus \beta) = \bigoplus_{\gamma \preceq \alpha} \bigoplus_{\chi \in \Gamma_2} \chi(\gamma \oplus \beta). \tag{7}$$

$\bigoplus_{\gamma \preceq \alpha} f(\gamma \oplus \beta) = 0$ implies that $|\Gamma_2|$ is even, while $\bigoplus_{\gamma \preceq \alpha} f(\gamma \oplus \beta) = 1$ implies that $|\Gamma_2|$ is odd. This completes the proof. □

Set $\beta = 0$ in Theorem 4 and reorder the variables, we obtain a result well known to coding theorists (see p.372 of [3]):

Corollary 5. *Let f be a function on V_n and $\alpha = (a_1, \ldots, a_n)$, a vector in V_n. Then the term $x_1^{a_1} \cdots x_n^{a_n}$ appears in f if and only if $\bigoplus_{\gamma \preceq \alpha} f(\gamma) = 1$.*

With the above two results, it is not hard to verify the correctness of the following theorem:

Theorem 6. *Let f and g be function on V_n. Then the following four statements are equivalent*

(i) $f(\alpha) = \bigoplus_{\beta \preceq \alpha} g(\beta)$ *for every vector* $\alpha \in V_n$.
(ii) $g(\alpha) = \bigoplus_{\beta \preceq \alpha} f(\beta)$ *for every vector* $\alpha \in V_n$.
(iii) $f(x_1, \ldots, x_n) = \bigoplus_{\alpha \in V_n} g(a_1, \ldots, a_n) x_1^{a_1} \cdots x_n^{a_n}$ *where* $\alpha = (a_1, \ldots, a_n)$.
(iv) $g(x_1, \ldots, x_n) = \bigoplus_{\alpha \in V_n} f(a_1, \ldots, a_n) x_1^{a_1} \cdots x_n^{a_n}$ *where* $\alpha = (a_1, \ldots, a_n)$.

4 Polynomial Representation and Nonlinearity

4.1 Restriction to Cosets

Let f be a function on V_n and U be an s-dimensional subspace of V_n. Then V_n is the union of 2^{n-s} disjoint 2^s-subsets

$$V_n = \Pi_0 \cup \Pi_1 \cup \cdots \cup \Pi_{2^{n-s}-1} \tag{8}$$

where

(i) $\Pi_0 = U$,
(ii) for any $\alpha, \beta \in V_n$, α, β belong to the same class, say Π_j, if and only if $\alpha \oplus \beta \in \Pi_0 = U$. From (i) and (ii), it follows that
(iii) $\Pi_j \cap \Pi_i = \phi$ for $j \neq i$, where ϕ denotes the empty set.

As each Π_j can be expressed as $\Pi_j = \beta_j \oplus U$ for a $\beta_j \in V_n$, where $\beta_j \oplus U = \{\beta_j \oplus \alpha | \alpha \in U\}$, the definition of restriction (Definition 3) can be extended to each coset Π_j.

Definition 7. *Let f be a function on V_n and U be an s-dimensional subspace of V_n. The* restriction *of f to a coset* $\Pi_j = \beta_j \oplus U$, $j = 0, 1, \ldots, 2^{n-s}-1$, *denoted by* f_{Π_j}, *is a function on U, and it is defined by* $f_{\Pi_j}(\alpha) = f(\beta_j \oplus \alpha)$ *for every* $\alpha \in U$.

4.2 Maximal Odd Weighting Subspaces

Definition 8. *Let f be a function on V_n. A subspace U of V_n is called a* maximal odd weighting subspace *of f if the Hamming weight of f_U is odd and the Hamming weight of $f_{U'}$, where U' is any subspace with $U' \supset U$ (i.e. U is a proper subset of U'), is even.*

A maximal odd weighting subspace of a function is not necessarily a subspace with the maximum dimension, even if the Hamming weight of the restrictions of f to the subspace is odd. This is best explained with the following example.

Example 1. let $f(x_1, x_2, x_3, x_4) = x_1x_2x_3 \oplus x_1x_2x_4 \oplus x_3x_4 \oplus x_3$ be a function on V_4, whose truth table is 001000100010000. The eight vectors (0000), (0001), (0100), (0101), (1000), (1001), (1100) and (1101) form a 3-dimensional subspace, say W, such that the Hamming weight of f_W is one (odd). By a direct verification, 3 is the maximum dimension of the subspaces, the Hamming weight of the restrictions of f to these subspaces is odd. However, the four vectors (0000), (0001), (0010) and (0011) form a 2-dimensional subspace, say U, such that the Hamming weight of f_W is one (odd). There are four 3-dimensional subspaces containing U:

$$U' = \{(0000), (0001), (0010), (0011), (0100), (0101), (0110), (0111)\}$$
$$U'' = \{(0000), (0001), (0010), (0011), (1000), (1001), (1010), (1011)\}$$
$$U''' = \{(0000), (0001), (0010), (0011), (1000), (1001), (1010), (1011)\}$$

We note that the Hamming weights of $f_{U'}$, $f_{U''}$ and $f_{U'''}$ are all two (even). We also note that the 4-dimensional subspace containing U is V_4 itself and the Hamming weight of f is four (even). Hence both W and U are a maximal odd weighting subspace of f.

As will be shown in the forthcoming sections, the concept of maximal odd weighting subspace of a function plays an important role, primarily due to the fact that the dimension of a subspace is relevant to the structure of the function. In particular, we will show in the next section a connection between the dimension of a maximal odd weighting subspace of a function and the lower bound on nonlinearity of a function.

4.3 A Lower Bound on Nonlinearity

Definition 9. *Let f be a function on V_n, $x_{j_1} \cdots x_{j_t}$ and $x_{i_1} \cdots x_{i_s}$ be two terms in the polynomial representation of function f. $x_{j_1} \cdots x_{j_t}$ is said to be* covered *by $x_{i_1} \cdots x_{i_s}$ if $\{j_1, \ldots, j_t\}$ is a subset of $\{i_1, \ldots i_s\}$, and $x_{j_1} \cdots x_{j_t}$ is said to be* properly covered *by $x_{i_1} \cdots x_{i_s}$ if $\{j_1, \ldots, j_t\}$ is a proper subset of $\{i_1, \ldots i_s\}$.*

Theorem 10. *Let f be a function on V_n and U be a maximal odd weighting subspace of f. If $dim(U) = s$ then the Hamming weight of f is at least 2^{n-s}.*

Proof. Let U be a subspace defined in (8). And let $N_j = |\{\alpha|\alpha \in \Pi_j,\ f(\alpha) = 1\}|$, where Π_j is defined in (8), $j = 0, 1, \ldots, 2^{s-1}$. Since $\Pi_0 = U$, N_0 is odd. Note that $\Pi_0 \cup \Pi_j$ is a $(s+1)$-dimensional subspace of V_n, $j = 1, \ldots, 2^{n-s} - 1$.

Since $\Pi_0 = U$ is a maximal odd weighting subspace of f, Hamming weight of the restriction of f to $\Pi_0 \cup \Pi_j$ is even. In other words, $N_0 + N_j$ is even. This proves that each N_j is odd, $j = 1, \ldots, 2^{n-s} - 1$. Hence $N_0 + N_1 + \cdots + N_{2^{n-s}-1} \geq 2^{n-s}$, namely, the Hamming weight of f is at least 2^{n-s}. □

Theorem 11. *Let f be a function on V_n and U be a maximal odd weighting subspace of f. Let $dim(U) = s$ $(s \geq 2)$ then the nonlinearity of f, N_f, satisfies $N_f \geq 2^{n-s}$.*

Proof. Let φ be an affine function on V_n. Since $s \geq 2$ the Hamming weight of φ_U must be even. Hence the Hamming weight of φ_U must be even. Hence the Hamming weight of $(f \oplus \varphi)_U$ must be odd. According to Lemma 10, the Hamming weight of $f \oplus \varphi$ is at least 2^{n-s}. As the Hamming weight of $f \oplus \varphi$ determines $d(f, \varphi)$, the theorem is proved. □

Theorem 12. *Let $t \geq 2$. If $x_{j_1} \cdots x_{j_t}$ is a term in a function f on V_n and it is not properly covered (see Definition 9) by any other terms in the same function, then the nonlinearity of f, N_f, satisfies $N_f \geq 2^{n-t}$.*

Proof. Write $\alpha = (a_1, \ldots, a_n)$ where $a_j = 1$ for $j \in \{j_1, \ldots, j_t\}$ and $a_j = 0$ for $j \notin \{j_1, \ldots, j_t\}$. Set

$$U = \{\gamma | \gamma \preceq \alpha\}.$$

Obviously U is a t-dimensional subspace of V_n. Since $x_{j_1} \cdots x_{j_t}$ is a term in f on V_n, by using Corollary 5, $\bigoplus_{\gamma \preceq \alpha} f(\gamma) = 1$ or $\bigoplus_{\gamma \in U} f(\gamma) = 1$ i.e. the Hamming weight of f_U is odd.

We now prove that U is a maximal odd weighting subspace of f. Suppose U is not a maximal odd weighting subspace of f. Hence there is a s-dimensional subspace of V_n, say W, such that U is a proper subset of W i.e, $s > t$ and the Hamming weight of f_W is odd i.e. $\bigoplus_{\gamma \in W} f(\gamma) = 1$. Since U is a proper subspace of W, by using linear algebra, W can be expressed as a union of 2^{s-t} disjoint 2^t-subsets

$$W = U \cup (\beta_1 \oplus U) \cup \cdots \cup (\beta_{2^{s-t}-1} \oplus U) \tag{9}$$

where each $\beta \preceq \overline{\alpha}$, where $\overline{\alpha} \oplus \alpha = (1, \ldots, 1)$. Since both the Hamming weights of f_U and f_W are odd, there is a coset, say $\beta_k \oplus U$, $1 \leq k \leq 2^{s-t} - 1$, such that the Hamming weight of $f_{\beta_k \oplus U}$ is even or $\bigoplus_{\gamma \in U} f(\beta_k \oplus \gamma) = 0$ i.e.

$$\bigoplus_{\gamma \preceq \alpha} f(\beta_k \oplus \gamma) = 0. \tag{10}$$

Applying Theorem 4 to (10), there are even number of terms covering $x_{j_1} \cdots x_{j_t}$. Since the term $x_{j_1} \cdots x_{j_t}$ itself appears in f, there is another term properly covering $x_{j_1} \cdots x_{j_t}$. This contradicts the condition in the theorems, that the term $x_{j_1} \cdots x_{j_t}$ is not properly covered by any other terms in f. The contradiction proves that U is a maximal odd weighting subspace of f. By using Theorem 11, the proof is completed. □

Example 2. Let

$$f(x_1, \ldots, x_{10}) = x_1x_2x_3x_4x_5x_6x_7 \oplus x_3x_4x_5x_6x_7x_8x_9 \oplus x_7x_8x_9x_{10} \oplus \\ x_4x_6x_8x_{10} \oplus x_1x_5x_9 \oplus x_2x_4 \oplus x_6$$

be a function on V_{10}. term $x_1x_5x_9$ is not properly covered by any other terms in f. By using Corollary 12, the nonlinearity of f, N_f, satisfies $N_f \geq 2^{10-3} = 2^7$.

Example 3. Let

$$f(x_1,\ldots,x_{10}) = x_1x_2x_3x_4x_5x_6x_7 \oplus x_3x_4x_5x_6x_7x_8x_9 \oplus x_7x_8x_9x_{10} \oplus x_4x_6x_8x_{10} \oplus x_1x_3x_5 \oplus x_2x_8 \oplus x_1 \oplus x_2$$

be a function on V_{10}. The term x_2x_8 is not properly covered by any other terms in f. By using Corollary 12, the nonlinearity of f, N_f, satisfies $N_f \geq 2^{10-2} = 2^8$.

We note that the lower bound in Theorem 11 is tight:

Corollary 13. *For any n and any s, $2 \leq s \leq n$, there are a function on V_n, say f, and a s-dimensional subspace, say U, U be a maximal odd weighting subspace of f and the nonlinearity of f, N_f, satisfies $N_f = 2^{n-s}$.*

Proof. We prove the corollary by an example. Let g be a function on V_s, defined as $g(\beta) = 1$ if and only if $\beta = 0$. Set $f(z,y) = g(y)$, a function on V_n, where $z \in V_{n-s}$ and $y \in V_s$. Since the Hamming weight of f is 2^{n-s} $(s \geq 2)$, $d(f,h) \geq 2^{n-s}$ where h is any affine function on V_n and the equality holds if h is the zero function on V_n. Hence the nonlinearity of f, N_f, satisfies $N_f = 2^{n-s}$. On the other hand, set

$$U = \{(0,\ldots,0,b_1,\ldots,b_s)|b_j \in GF(2)\}$$

where the number of zeros is $n-s$. It is easy to verify that s-dimensional subspace U is a maximal odd weighting subspace of f. □

Finally we note that for $s = 2$, the value of 2^{n-s} in Theorem 11 is very close to $2^{n-1} - 2^{\frac{1}{2}n-1}$, the upper bound on the nonlinearity of functions on V_n [4]. However Theorem 11 cannot be further improved by extending s to $s = 1$, as the condition of $s \geq 2$ in the proof of the theorem cannot be removed. For example, let f be a function on V_n, whose truth table is given as follows

0110011010011001.

It is easy to verify that (0000), (0001) form a maximal 1-dimensional subspace, denoted by U. Theorem 11 is not applicable due to the fact that $dim(U) = 1$. In fact, f is a linear function, hence its nonlinearity is 0.

Nevertheless, Theorem 10 can be applied, which gives us $\geq 2^{4-1} = 8$ as the Hamming weight of f.

4.4 A Lower Bound on the Number of Terms

Theorem 14. *Let f be a function on V_n such that $f(\alpha) = 1$ for a vector $\alpha \in V_n$, and $f(\beta) = 0$ for every vector β with $\alpha \prec \beta$ where $\prec$ is defined as in Notation 2. Then f has at least 2^{n-t} terms where t denotes the Hamming weight of α.*

Proof. We first give Theorem 10 an equivalent statement, that we call Theorem 10', as follows

Theorem 10' Let f be a function on V_n and g be defined in (1). Let $g(\alpha) = 1$ for a vector $\alpha \in V_n$, and $g(\beta) = 0$ for every vector β with $\alpha \prec \beta$ where $\prec$ is defined as in Notation 2. Then the Hamming weight of f is at least 2^{n-t}.

The equivalence between (iii) and (iv) in Theorem 6 allows us to interchange f and g in Theorem 10'. Thus we have

Theorem 10" Let f be a function on V_n and g be defined in (1). Let $f(\alpha) = 1$ for a vector $\alpha \in V_n$, and $f(\beta) = 0$ for every vector β with $\alpha \prec \beta$ where $\prec$ is defined as in Notation 2. Then the Hamming weight of g is at least 2^{n-t}.

This completes the proof. □

Corollary 15. *Let f be a function on V_n such that $f(\alpha) = 0$ for a vector $\alpha \in V_n$, and $f(\beta) = 1$ for every vector β with $\alpha \prec \beta$ where $\prec$ is defined as in Notation 2. then f has at least*

(i) $2^{n-s} - 1$ *terms if* $f(0) = 0$,
(ii) $2^{n-s} + 1$ *terms if* $f(0) = 1$,

where s denotes the Hamming weight of α.

Proof. Set $f' = 1 \oplus f$. Hence $f'(\alpha) = 1$ and $f'(\beta) = 0$ for every $\beta \in V_n$. By using Theorem 14, f' has at least 2^{n-s} terms and hence f has at least $2^{n-s} - 1$ terms. This proves (i) of the corollary.

In the above the proof, we have already proved that f' has at least 2^{n-s} terms. Suppose $f(0) = 1$. Note that $f'(0) = 0$. Hence f has at least $2^{n-s} + 1$ terms. □

Example 4. Let f be a function on V_6, whose truth table is given as follows

1000110111110010001101001100100001111100011001101001011010001010

Note that the value of $f(001011)$ is one, while the values of $f(001111)$, $f(011011)$, $f(011111)$, $f(101011)$, $f(101111)$, $f(111011)$ and $f(111111)$ are all zero. Applying Theorem 14 to the vector (001011), we conclude that f has at least $2^{6-3} = 8$ terms.

Example 5. Let f be a function on V_6, whose truth table is given as follows

1000110111110011001101011101100101111101011101111001011110011010

Note that $f(000011)$ assumes the value zero, while $f(000111)$, $f(001011)$, $f(001111)$, $f(010011)$, $f(010111)$, $f(011011)$, $f(011111)$, $f(100011)$, $f(100111)$, $f(101011)$, $f(101111)$, $f(110011)$, $f(110111)$, $f(111011)$ and $f(111111)$ all assume the value one. Applying (ii) of Corollary 15 to the vector (000011), one can see that f has at least $2^{6-2} + 1 = 17$ terms.

The lower bounds on the number of terms given by Theorem 14 and Corollary 15 are tight, due to Corollary 13 and Theorem 6.

5 Relating Algebraic Degree to Other Criteria

Note that the algebraic degree of any function, say f, on V_n is invariant under a non-singular linear transformation on the variables, and for any vector $\alpha \in V_n$, the subset $W = \{\beta | \beta \preceq \alpha\}$ is a s-dimensional subspace, where s denotes the Hamming weight of α. Using Theorem 6 it is not difficult to prove

Theorem 16. *Let f be a function on V_n $(n \geq 2)$. Then*

$$deg(f) = \max\{dim(U) \mid U \text{ is a subspaces and Hamming weight of } f_U \text{ is odd}\}.$$

The following lemma is called "Poisson Summation" whose proof can be found in [2].

Lemma 17. *Let real valued sequences $a_0, \ldots, a_{2^n-1}$ and $b_0, \ldots, b_{2^n-1}$ satisfy*

$$(a_0, \ldots, a_{2^n-1})H_n = (b_0, \ldots, b_{2^n-1}).$$

Then for any p-dimensional subspace $1 \leq p \leq n-1$, say W,

$$\sum_{\alpha \in W} a_\alpha = 2^{p-n} \sum_{\alpha \in W^\perp} b_\alpha$$

where $W^\perp = \{\beta | \beta \in V_n, \langle \beta, \alpha \rangle = 0, \text{ for each } \alpha \in W\}$.

The next theorem shows a relationship between algebraic degree and Wlash-Hadamard transforms of a function.

Theorem 18. *Let f be a function on V_n $(n \geq 2)$, ξ be the sequence of f, and p is an integer, $2 \leq p \leq n$. If $\langle \xi, \ell_j \rangle \equiv 0 \pmod{2^{n-p+2}}$, where ℓ_j is the jth row (column) of H_n, $j = 0, 1, \ldots, 2^n - 1$, then $deg(f) \leq p - 1$.*

Proof. Let $\xi = (a_0, a_1, \ldots, a_{2^n-1})$. Note that

$$(a_0, a_1, \ldots, a_{2^n-1})H_n = (\langle \xi, \ell_0 \rangle, \langle \xi, \ell_1 \rangle, \ldots, \langle \xi, \ell_{2^n-1} \rangle).$$

Then from Lemma 17

$$\sum_{\alpha \in W} a_\alpha = 2^{p-n} \sum_{\alpha \in W^\perp} \langle \xi, \ell_\alpha \rangle \tag{11}$$

holds for each p-dimensional subspace W of V_n, where $W^\perp = \{\alpha | \alpha \in V_n, \langle \alpha, \beta \rangle = 0, \text{ for each } \beta \in W\}$ and $a_\alpha = a_j$ if α is the binary representation of integer j. From (11) and the condition that $\langle \xi, \ell_j \rangle \equiv 0 \pmod{2^{n-p+2}}$, $j = 0, 1, \ldots, 2^n - 1$, we have $\sum_{\alpha \in W} a_\alpha \equiv 0 \pmod 4$. Note that $\xi = (a_0, a_1, \ldots, a_{2^n-1})$ is the sequence of f. It is easy to verify that $\sum_{\alpha \in W} a_\alpha \equiv 0 \pmod 4$ if and only if the Hamming weight of f_W is even.

Since W is an arbitrary p-dimensional subspace, using Theorem 16, the Hamming weight of the restriction of f to any q-dimensional subspace is even, $q = p,\ p+1,\ \ldots,\ n$. So from Theorem 16, we have $deg(f) \leq p - 1$. □

Corollary 19. *Let f be a function on V_n $(n \geq 2)$ and ξ be the sequence of f, and p is an integer, $2 \leq p \leq n$. If $\Delta(\alpha) \equiv 0 \pmod{2^p}$, for each $\alpha \in V_n$, then $deg(f) \leq n + 1 - \frac{1}{2}p$ for p even, and $deg(f) \leq n + 1 - \frac{1}{2}(p+1)$ for p odd.*

Proof. From [6]

$$(\Delta(\alpha_0), \Delta(\alpha_1), \ldots, \Delta(\alpha_{2^n-1}))H_n = (\langle\xi, \ell_0\rangle^2, \langle\xi, \ell_1\rangle^2, \ldots, \langle\xi, \ell_{2^n-1}\rangle^2)$$

where ℓ_j is the jth row (column) of H_n. Since $\Delta(\alpha) \equiv 0 \pmod{2^p}$ for each $\alpha \in V_n$, we have $\langle\xi, \ell_j\rangle^2 \equiv 0 \pmod{2^p}$ for $j = 0, 1, \ldots, 2^n - 1$. Hence $\langle\xi, \ell_j\rangle \equiv 0$ $\pmod{2^{\frac{1}{2}p}}$ if p is even, and $\langle\xi, \ell_j\rangle \equiv 0 \pmod{2^{\frac{1}{2}(p+1)}}$ if p is odd. Now the corollary follows from Theorem 18. □

We note that in Theorem 18, $\langle\xi, \ell_j\rangle$ is closely related to nonlinearity [4], and in Corollary 19, $\Delta(\alpha)$ is related to propagation characteristics [6].

References

1. T. Jakobsen and L. Knudsen. The interpolation attack on block ciphers. In *Fast Software Encryption*, Lecture Notes in Computer Science, Berlin, New York, Tokyo, 1997. Springer-Verlag.
2. R. J. Lechner. *Harmonic Analysis of Switching Functions.* in Recent Developments in Switching Theory, eited by Amar Mukhopadhyay. Academic Press, New York, 1971.
3. F. J. MacWilliams and N. J. A. Sloane. *The Theory of Error-Correcting Codes.* North-Holland, Amsterdam, New York, Oxford, 1978.
4. J. Seberry, X. M. Zhang, and Y. Zheng. Nonlinearity and propagation characteristics of balanced boolean functions. *Information and Computation*, 119(1):1–13, 1995.
5. T. Shimoyama, S. Moriai, and T. Kaneko. Cryptanalysis of the cipher KN, May 1997. (presented at the rump session of Eurocrypt'97).
6. X. M. Zhang and Y. Zheng. Characterizing the structures of cryptographic functions satisfying the propagation criterion for almost all vectors. *Design, Codes and Cryptography*, 7(1/2):111–134, 1996. special issue dedicated to Gus Simmons.

Construction of Correlation Immune Boolean Functions

Ed Dawson and Chuan-Kun Wu

Information Security Research Centre
Queensland University of Technology
GPO Box 2434 Brisbane 4001 Australia

Abstract. It is shown in this paper that every correlation immune Boolean function of n variables can be written as $f(x) = g(xG^T)$, where g is an algebraic non-degenerate Boolean function of k ($k \leq n$) variables and G is a generating matrix of an $[n, k, d]$ linear code. It is known that the correlation immunity of $f(x)$ is at least $d - 1$. In this paper we further prove when the correlation immunity exceeds this lower bound. A method which can theoretically search exhaustively all possible correlation immune functions is proposed, while constructions of higher order correlation immune functions as well as algebraic non-degenerate correlation immune functions are discussed in particular. It is also shown that many cryptographic properties of g can be inherited by the correlation immune function $f(x) = g(xG^T)$ which is an important property for choosing useful correlation immune functions.

1 Introduction

Correlation immune functions are proposed by Siegenthaler [9] in order to protect some shift register based stream ciphers against correlation attacks. Their cryptographic significance is stressed further in recently published papers (see for example [1,3,4]). It is obvious that constructions of such functions are important, especially in the case where the constructed functions can be controlled to have other cryptographic properties. Several alternative ways have been presented to construct correlation immune functions [2,8,9,11,12], each has its own merits. However, the correlation immunity of the constructed function from all methods known so far is studied by lower bounds, i.e., the constructed functions should have the correlation immunity at least as large as the designed value rather than being equivalent to a concrete value. In this paper we investigate the inherent structure of correlation immune functions in terms of algebraic degeneration and subsequently constructions of functions with concrete correlation immunity are investigated. Additionally, it is shown that other cryptographic properties of the constructed functions can easily be controlled.

Denote by $F_2 = \{0, 1\}$ the binary field. A function $f : F_2^n \longrightarrow F_2$ is called a *Boolean function* of n variables. We write it as $f(x) = f(x_1, ..., x_n)$. The *truth table* of $f(x)$ is a binary vector of length 2^n generated by $f(x)$. The *Hamming weight* of $f(x)$, denoted by $W_H(f)$, is the number of 1s in its truth table. A function $f(x)$ is called *balanced* if $W_H(f) = 2^{n-1}$. The function $f(x)$ is called an *affine* function if there exist $a_0, a_1, ..., a_n \in F_2$ such that $f(x) = a_0 \oplus a_1x_1 \oplus \cdots \oplus a_nx_n$. In particular, if $a_0 = 0$, $f(x)$ is also called a *linear* function. We will denote by $\mathcal{F}_n$, the set of all Boolean functions of n variables and by $\mathcal{L}_n$, the set of affine ones.

For x and y in F_2^n, we will denote by $< x, y >= x_1y_1 \oplus x_2y_2 \oplus \cdots \oplus x_ny_n$ the inner product of x and y. It is noticed that when one of them is a constant and the other is a vector of n variables, the inner product then yields a new variable. The inner product can also be written as $x \cdot y^T$, where y^T is the transposed vector from y. Some concepts from the theory of error-correcting codes [6] are included here which will be used in the forthcoming discussion. An $[n, k, d]$ linear code C is a subspace of F_2^n of dimension k and with minimum distance d. A generating matrix G of C is a $k \times n$ matrix of which the row vectors form a basis of C.

2 Algebraic degeneration

Algebraic degeneration is an important criterion for measuring the security of cryptographic Boolean functions. For example an effective attack on nonlinear filtered generators can be found in [10] where the nonlinear filtered function was assumed to be algebraic degenerate. In order to study the algebraic degeneration of normal and CI Boolean functions we first introduce the Walsh transform of Boolean functions. Let $f(x) \in \mathcal{F}_n$. Then the Walsh transform of $f(x)$ is expressed as

$$S_f(\omega) = \sum_x f(x)(-1)^{<\omega,x>}. \tag{1}$$

Accordingly, the inverse transformation is expressed as

$$f(x) = 2^{-n} \sum_\omega S_f(\omega)(-1)^{<\omega,x>}. \tag{2}$$

Note that the summations in (1) and in (2) are over the real number field, and the Walsh transform of a Boolean function then is a real function. It should be noted that the value of $< \omega, x >$ could be treated as a real value when executing the operations.

It is easy to deduce that

Lemma 1. *Let $f(x) \in \mathcal{F}_n$, D be an $n \times n$ invertible matrix over F_2. Let $g(x) = f(xD)$. Then*

$$S_g(\omega) = S_f(\omega(D^{-1})^T), \tag{3}$$

where $(D^{-1})^T$ is the transposed matrix from D^{-1}.

Let $f(x) \in \mathcal{F}_n$. If there is a Boolean function $g(y) \in \mathcal{F}_k$ and an $n \times k$ $(k < n)$ matrix D over F_2 such that

$$f(x) = g(xD) = g(y)$$

holds for every x, then $f(x)$ is called *algebraic degenerate* and $g(y)$ is called an *algebraic degenerated function* of $f(x)$, or simply a degenerated function of $f(x)$. The largest possible value of $n - k$ is called the *algebraic degeneration* of $f(x)$ and is denoted by $AD(f)$. Here matrix D is assumed to be of rank k, because there would exist another degenerated function of lesser variables otherwise. If $AD(f) = 0$, $f(x)$ is called an *algebraic non-degenerate Boolean function* or simply a *non-degenerate function*. A useful result can be found in [5] which describes the algebraic degeneration of Boolean functions precisely.

Lemma 2. *Let $f(x) \in \mathcal{F}_n$. Denote by $V = \prec \{\omega : S_f(\omega) \neq 0\} \succ$ the vector space generated by the vectors on which the Walsh transform takes nonzero values, or the linear span of $S(f) = \{\omega : S_f(\omega) \neq 0\}$. Suppose $dim(V) = k$, and let $h_1, ..., h_k$ be a basis of V. Write $H = [h_1^T, h_2^T, ..., h_k^T]$, where h_i^T is the transposed vector of h_i. Then there must exist a Boolean function $g(y) \in \mathcal{F}_k$ such that*

$$f(x) = g(xH) = g(y). \tag{4}$$

It can also be shown that the dimension of the vector space V is the least number k that f has a degenerated function in $\mathcal{F}_k$.

Corollary 3. *Let $f(x) \in \mathcal{F}_n$, A be an $n \times n$ nonsingular matrix, and denote by $g(x) = f(xA)$. Then $AD(f) = AD(g)$.*

Corollary 4. *Let $f(x) \in \mathcal{F}_n$. If $deg(f) = n$ then f is non-degenerate.*

3 Correlation immunity of Boolean functions

Let $f(x) \in \mathcal{F}_n$. The function $f(x) \in \mathcal{F}_n$ is called *correlation immune with respect to the subset* $T \subset \{1, 2, ..., n\}$ if the probability for f to take any value from $\{0, 1\}$ is not changed given that the value of $\{x_i, i \in T\}$ are fixed and other variables are chosen independently at random. The function $f(x)$ is called *correlation immune (CI) of order* t if for every T of cardinality at most t, f is correlation immune with respect to T. It is noticed that $f(x)$ is CI of order t implies that it is CI of any order less than t as well. The largest possible value of t is called the *exact correlation immunity* of f. Let $z = \bigoplus_{i=1}^{n} c_i x_i$ be another (nonzero) variable, where $c_i \in \{0, 1\}$. Then the function $f(x)$ is said to be *correlation immune in z* if the probability for f to take any value from $\{0, 1\}$ is not changed given that z is assigned any fixed value in advance.

Lemma 5. *Let $f(x) \in \mathcal{F}_n$. Then $f(x)$ is CI of order t if and only if for every $\gamma \in F_2^n$ with $W_H(\gamma) \leq t$, $f(x)$ is CI in $z = < \gamma, x >$.*

It should be noted that $f(x)$ is CI in z_1 and z_2 individually does not imply that it is CI in $z_1 \oplus z_2$. For example, although $f(x_1, x_2, x_3) = x_3 \oplus x_1x_2 \oplus x_1x_3 \oplus x_2x_3$ is a 1-st order CI function, it is easy to verify that it is not CI in $x_1 \oplus x_2$.

Let $f(x) \in \mathcal{F}_n$, $g(y) \in \mathcal{F}_k$, $D = (d_1^T, d_2^T, ..., d_k^T)$ be an $n \times k$ binary matrix with $rank(D) = k$, where $d_i \in F_2^n$ and we have

$$f(x) = g(xD) = g(y).$$

It is known that each y_i is the linear combination of x_i's with coefficients the components of d_i, i.e., $y_i = < x, d_i > = x \cdot d_i^T$. Let $z = \bigoplus_{i=1}^{n} c_i x_i$ be another variable. Then it is obvious that $f(x)$ is CI in z if and only if $g(y)$ is such. Denote by $\gamma = (c_1, c_2, ..., c_n)$. Then we have

Lemma 6. *If $rank[D; \gamma^T] = k+1$, where $[A; B]$ means the concatenation of matrices A and B, then for any Boolean function $g(y) \in \mathcal{F}_k$, $g(xD)$ is independent of $z = < \gamma, x >$ and hence is CI in z.*

Proof: Let $y = (y_1, y_2, ..., y_k) = xD$. It is noticed that $rank[D; \gamma^T] = k + 1$ if and only if variables $y_1, y_2, ..., y_k$ together with z are all independent, and consequently $g(xD)$ is independent of z. So we have

$$Prob(g(xD) = 1|z = 1) = Prob(g(y) = 1|z = 1) = Prob(g(y) = 1).$$

This means that $g(xD)$ is CI in z. □

The following lemma has been proved in [12].

Lemma 7. *If G is a generating matrix of an* $[n, k, d]$ *linear code, then for any* $g(y) \in \mathcal{F}_k$*, the correlation immunity of* $f(x) = g(xG^T)$ *is at least* $d - 1$.

In order for the function f to have correlation immunity of order larger than $d - 1$, by the definition of correlation immunity and lemma 6, we need to make $g(y)$, or equivalently $f(x) = g(xG^T)$, to be CI in every $z = < x, \gamma >$ with $W_H(\gamma) = d$. It is obvious that $rank[G^T, \gamma^T] = k$ if and only if γ is a codeword of C_G, the linear code generated by G. By lemma 6 we know that for those γ with Hamming weight d which are not codewords of C_G, the function f is already CI in $z = < x, \gamma >$. So we have

Lemma 8. *Let G be a generating matrix of an* $[n, k, d]$ *linear code, and* $f(x) = g(xG^T)$. *Then f is CI of order* $\geq d$ *if and only if for every* $\alpha \in F_2^k$ *with* $W_H(\alpha G) = d$, $g(y)$ *is CI in* $z = < \alpha, y >$.

By generalising lemma 8 we have

Theorem 9. *Let G be a generating matrix of an* $[n, k, d]$ *linear code, and* $f(x) = g(xG^T)$. *Then a necessary and sufficient condition for the function f to be CI of order m is that for every* $\alpha \in F_2^k$ *with* $d \leq W_H(\alpha G) \leq m$, $g(y)$ *is CI in* $z = < \alpha, y >$.

Corollary 10. *If the i-th row vector of G is a codeword with nonzero minimum Hamming weight d and the function* $g(y)$ *is not CI in* y_i, *then the correlation immunity of* $f(x) = g(xG^T)$ *is exactly* $(d - 1)$.

Now we consider the inverse question for general CI functions. Given an m-th order CI function $f \in \mathcal{F}_n$, can it be written as $f(x) = g(xD)$, where $g \in \mathcal{F}_k$ is non-degenerate and D^T is a generating matrix of an $[n, k, d]$ linear code with $k \leq n$ and $d \geq 1$? The answer is yes according to lemma 2. Furthermore it can be shown that the code generated by D^T is unique.

Lemma 11. *Let* $f(x) \in \mathcal{F}_n$. *Then it can be written as* $f(x) = g(xD)$, *where* $g \in \mathcal{F}_k$ *is non-degenerate and* D^T *is a generating matrix of an* $[n, k, d]$ *linear code with* $k \leq n$ *and* $d \geq 1$. *Moreover, the linear code is unique given that* $f(x)$ *is fixed.*

Proof: From the discussion above, what we need to show is the uniqueness. On the contrary suppose we have $f(x) = g_1(xD_1) = g_2(xD_2)$, where $C_{D_1^T} \neq C_{D_2^T}$. Then there must exist a column α of D_1 which is linearly independent of the column vectors of D_2. Without loss of generality let α be the first column of D_1. Then by lemma 6 we know that $f(x)$ is independent of $< \alpha, x >$, and then $g_1(y)$ must be independent of y_1. This is a contradiction to the premise of the lemma. So the conclusion is true. □

By lemma 11 we know that theorem 9 gives a necessary and sufficient condition for a general Boolean function to be CI. Since theorem 9 applies to every CI function, it can be used for exhaustive constructions of CI functions.

4 Exhaustive construction of CI functions

Theoretically by using lemma 7 and theorem 9 the complete set of CI functions can be constructed. By applying theorem 9 we are able to see when the correlation immunity is larger than or equal to the minimum distance of the code. In order to do this, we need to construct Boolean functions which are CI in some of their variables and/or their linear combinations. Let $\hat{x_i} = (x_1, ..., x_{i-1}, x_{i+1}, ..., x_n)$. Then we have

Lemma 12. *Let $f(x) = x_i f_1(\hat{x_i}) \oplus f_2(\hat{x_i})$. Then $f(x)$ is CI in x_i if and only if*

$$W_H(f_1 \oplus f_2) = W_H(f_2). \tag{5}$$

Proof: By writing $f(x) = x_i(f_1 \oplus f_2) \oplus (1 \oplus x_i)f_2$ it can be seen that $f(x)$ is CI in x_i if and only if $W_H(f_1 \oplus f_2) = W_H(f_2) = \frac{1}{2}W_H(f)$. □

Lemma 13. *Let $f(x) \in \mathcal{F}_n$. Then $deg(f) < n$ if and only if $2|W_H(f)$, i.e., the Hamming weight of $f(x)$ is an even number.*

In [9] it was shown that if $f(x) \in \mathcal{F}_n$ is CI (of order ≥ 1), then $deg(f) \leq n - 1$. We further prove that

Lemma 14. *Let $f(x) \in \mathcal{F}_n$. If $deg(f) = n$ then $f(x)$ is not CI in any linear combination of its variables.*

Proof: Assume the contrary, $f(x)$ is CI in $< \alpha, x >$, and without loss of generality the first coordinate of α is assumed to be not zero. Denote by e_i the vector in F_2^n with i consecutive ones followed by zeros. Let $D = [\alpha, e_2, ..., e_n]$. Then $g(x) = f(xD^{-1})$ is CI in x_1 and hence can be written as $g(x) = x_1 g_1(\hat{x_1}) \oplus g_2(\hat{x_1})$. By lemma 12 we know that

$$\begin{aligned} W_H(g_1) &= W_H((g_1 \oplus g_2) \oplus g_2) \\ &= W_H(g_1 \oplus g_2) + W_H(g_2) - 2W_H((g_1 \oplus g_2) \cdot g_2) \\ &= 2W_H(g_2) - 2W_H((g_1 \oplus g_2) \cdot g_2) \end{aligned}$$

is an even number and by lemma 13 we have $deg(f) = deg(g) = deg(g_1) + 1 < (n-1) + 1 = n$. This is a contradiction. So the conclusion of lemma 14 follows. □

Let $f(x) = g(xG^T)$ be a Boolean function of $\mathcal{F}_n$, where g is non-degenerate, and G is a generating matrix of an $[n, k, d]$ linear code. It is known that by a linear transformation on the rows of G, we can always make the row vectors of G satisfy

$$W_H(g_1) \leq W_H(g_2) \leq \cdots \leq W_H(g_k),$$

and there does not exist another basis $e_1, e_2, ..., e_k$ of C_G with $W_H(e_1) \leq W_H(e_2) \leq \cdots \leq W_H(e_k)$ such that $W_H(e_i) < W_H(g_i)$ for some $1 \leq i \leq k$. Constructions can always be based on this assumption. Such a matrix will be called a *minimum weight generating matrix*.

It is noticed that under a permutation of the variables of a Boolean function, the correlation immunity of the function is an invariant. To simplify the problem we will treat two CI functions as equivalent if they are equivalent by a variable permutation. For the function $f(x) = g(xG^T)$ the permutation of variable x is equivalent to the same permutation of the column vectors of G. Complements of CI functions can be left out in the first steps and then added at last. So the exhaustive construction can be outlined as follows:

1. For all integers $k \in \{1, 2, ..., n\}$ perform the following steps:
2. Search the minimum weight generating matrices G_i, $i \in I$, of $[n, k]$ codes such that they are not column-equivalent.
3. List all nontrivial Boolean functions $g(y) \in \mathcal{F}_k$ such that $g(\mathbf{0}) = 0$.
4. Match each $g(y)$ with every G_i to see if $f_i(x) = g(xG_i^T)$ is CI of any order according to theorem 9.
5. For those $f_i(x)$ with certain order of CI, permute their variables to get an equivalent class of CI functions.
6. Complement every CI function obtained above.

Note that this exhaustive construction method can be used to construct all CI functions for small n. It is not efficient for large n due to the large number of CI functions. However, this technique can be modified for random construction.

5 Construction of high order CI functions

From above, every CI function can be written as $g(xD)$, where g is a non-degenerate function while D^T is a minimum weight generating matrix of an $[n, k, d]$ linear code. In this section we will concentrate mainly on the construction of those functions whose correlation immunity is no less than d.

For any Boolean function $f(x) \in \mathcal{F}_n$, set

$$\Delta_f = \{\delta \in F_2^n,\ \ f(x) \text{ is CI in } <\delta, x>\}.$$

Then by theorem 9 we directly have

Theorem 15. *Let $g(y) \in \mathcal{F}_k$ and G be a generating matrix of an $[n, k, d]$ linear code. Set $f(x) = g(yG^T)$. Then the correlation immunity of $f(x)$ is*

$$\min_{\alpha \notin \Delta_g} W_H(\alpha G) - 1. \tag{6}$$

Moreover we have

$$AD(f) = n - k + AD(g). \tag{7}$$

Proof: The former part comes directly from theorem 9. So we need only to prove the latter part. Assume $AD(g) = t$, i.e., there exist a non-degenerate $g_1 \in \mathcal{F}_{k-t}$ and a $k \times (k - t)$ matrix D such that $g(y) = g_1(yD)$. So $f(x) = g_1(xG^T D)$, and $AD(f) \geq n - (k - t) = n - k + AD(g)$. On the other hand, since $rank(G) = k$, we can assume, without loss of generality, the first k columns of G are linearly independent and we write $G = [G_1; G_2]$. Then $g(y) = f(yG_1^{-1}, 0, \cdots, 0)$. This means that if f can be degenerated to a function of r variables then g can be degenerated to a functions of no more than r variables, i.e., $k - AD(g) \leq n - AD(f)$ or $AD(f) \leq n - k + AD(g)$. So the conclusion follows. □

It is seen that $g(y)$ can always be chosen as non-degenerate. When we use theorem 15 to construct CI functions, it is noticed that an $[n, k, d]$ linear code normally has several code words of Hamming weight d. So it is hard to find a Boolean function which can match a generating matrix of this linear code to generate CI functions of order $\geq d$. However it is easy to find Boolean functions which are CI in part of their variables and their linear combinations as shown in the following.

Corollary 16. *Let $g(y) \in \mathcal{F}_k$ be CI in its first t variables and their nonzero linear combinations. Let G be a generating matrix of an $[n-t, k-t, d]$ linear code. Then the correlation immunity of function $f(x) = g(x\hat{G}^T)$ is at least $d-1$, where*

$$\hat{G} = \begin{bmatrix} D & 0 \\ 0 & G \end{bmatrix},$$

and D is an arbitrary nonsingular binary matrix of order $t \times t$.

We note that when corollary 16 is used in construction, the size of D is normally very small.

6 Associated cryptographic properties of CI functions

In practice a CI function is required to satisfy some other cryptographic properties as well. From the discussions above we know that every CI function can be written as $f(x) = g(xG^T)$, where g is a non-degenerate Boolean function of k variables and G is a generating matrix of an $[n, k, d]$ linear code. We will show that some cryptographic properties of g can be inherited by the CI function f. It is easy to see that

Lemma 17. *Let $f(x) = g(xD)$, where g is a non-degenerate Boolean function of k variables and D^T is a generating matrix of an $[n, k, d]$ linear code. Then*

$$Prob(f(x) = 1) = Prob(g(y) = 1).$$

Particularly, $f(x)$ is balanced if and only if $g(y)$ is such.

Proof: Denote by $KerD = \{x : xD = 0\}$. For any $y \in F_2^k$, since $rank(D) = k$, there must exist an $x \in F_2^n$ such that $y = xD$. So $x + KerD$ is the set of all solutions of equation $xD = y$. This means that when x is chosen uniformally at random, $y = xD$ will appear uniformally at random as well. Consequently we have the conclusion. □

The algebraic degeneration of f regarding that of g is given by theorem 15. Besides the balancedness and the algebraic degeneration of a Boolean function, some other cryptographic properties of Boolean functions which are commonly studied include:

- **Algebraic degree:** The algebraic degree or simply degree of a Boolean function is defined as the largest number of variables in one product term of its polynomial expression and denoted by $deg(f)$.

- **Nonlinearity:** The nonlinearity of a Boolean function $f(x) \in \mathcal{F}_n$, denoted by N_f, is the minimum distance of f from all affine functions in $\mathcal{L}_n$.
- **Linear structure:** A boolean function $f(x) \in \mathcal{F}_n$ is said to have a linear structure $\alpha \in F_2^n$ if $f(x) \oplus f(x \oplus \alpha) \equiv c$, where $c \in \{0, 1\}$ is a constant. In particular α is called an *invariant linear structure* if $c = 0$ and a *complement linear structure* if $c = 1$.
- **Propagation criterion:** A Boolean function $f(x) \in \mathcal{F}_n$ is said to satisfy the propagation criterion with respect to a non-zero vector α if $f(x) \oplus f(x \oplus \alpha)$ is balanced. A Boolean function f(x) is said to satisfy the propagation criterion of degree k if it satisfies the propagation criterion with respect to all α with $1 \leq W_H(\alpha) \leq k$.
  Note: *Strict Avalanche Criterion* (SAC) is equivalent to the propagation criterion of degree 1 and *perfect nonlinearity* defined in [7] is equivalent to the propagation criterion of degree n.

We will study these associate properties of CI functions individually upon the assumption that $f(x) = g(xG^T)$, where g is a non-degenerate Boolean function of k variables and G is a generating matrix of an $[n, k, d]$ linear code.

6.1 Algebraic degree

Algebraic degree is one criterion to measure the nonlinearity of Boolean functions. It can be shown that the degree of f is the same as that of g.

Lemma 18. : *Let $f(x) \in \mathcal{F}_n$ and A be an $n \times n$ invertible binary matrix. Then $deg(f(xA)) = deg(f(x))$.*

Proof: Denote $f_1(x) = f(xA)$. It is obvious that the expansion of $f(xA)$ does not generate a term with degree $> deg(f(x))$, so we have $deg(f_1(x)) \leq deg(f(x))$. On the other hand, from the invertibility of A we have $f(x) = f_1(xA^{-1})$ and hence $deg(f(x)) \leq deg(f_1(x))$. Therefore, $deg(f_1(x)) = deg(f(x))$. □

Lemma 19. *Let D be an $n \times k$ $(k \leq n)$ binary matrix and let $f(x) = g(xD)$, where $g \in \mathcal{F}_k$. Then $deg(f) = deg(g)$ holds for any g if and only if $rank(D) = k$.*

Proof: By row-transformation, matrix D can be written as

$$D = A\begin{pmatrix} I_r & 0 \\ 0 & 0 \end{pmatrix}P$$

where A is an $n \times n$ invertible matrix and I_r is an $r \times r$ $(r \leq k)$ identity matrix, P is a $k \times k$ permutation matrix. Then

$$f(x) = g(xD) = g(xA\begin{pmatrix} I_r & 0 \\ 0 & 0 \end{pmatrix}P)$$

Denote $f_1(x) = f(xA^{-1})$, $g_1(y) = g(yP)$, where $x \in F_2^n$ and $y \in GF^k(2)$. Then

$$f_1(x) = f(xA^{-1}) = g(xA^{-1}D) = g(x\begin{pmatrix} I_r & 0 \\ 0 & 0 \end{pmatrix}P)$$
$$= g_1(x\begin{pmatrix} I_r & 0 \\ 0 & 0 \end{pmatrix}) = g_1(x_1, \ldots, x_r, 0, \ldots, 0).$$

From the equation above we see that, $deg(f_1) = deg(g_1(x_1, \ldots, x_r, 0, \ldots, 0)) = deg(g_1(y))$ holds for any $g_1(y) \in \mathcal{F}_k$ if and only if $r = k$, i.e., if and only if rank(D)=k. Notice that by lemma 18, $deg(g_1) = deg(g)$ and $deg(f_1) = deg(f)$. So we have $deg(f) = deg(g)$ holds for any $g(y) \in \mathcal{F}_k$ if and only if $rank(D)$=k. □

6.2 Nonlinearity

Nonlinearity of Boolean functions is a measurement of the distance of Boolean functions to the nearest affine one. If the nonlinearity of a Boolean function is very small, then it can be approximated by an affine Boolean function with high coincidence and hence it is cryptographically insecure. By using the Walsh spectral techniques it is easy to deduce that

Lemma 20.

$$N(f) = min\{W_H(f),\ 2^n - W_H(f),\ 2^{n-1} - \max_{\omega \neq 0} |S_f(\omega)|\}. \tag{8}$$

Lemma 21.

$$N_f \leq 2^{n-k} N_g.$$

Proof: By the definition there exists an affine function $l(y)$ of k variables such that $W_H(g(y) \oplus l(y)) = N_g$. Hence we have $W_H(g(xG^T) \oplus l(xG^T)) = 2^{n-k} N_g$ and again by definition we have $N_f \leq 2^{n-k} N_g$. □

Furthermore we can prove

Theorem 22. *Let D be an $n \times k$ $(k \leq n)$ binary matrix. Then $rank(D) = k$ if and only if for any Boolean function $g(y) \in \mathcal{F}_k$ and $f(x) = g(xD)$ we have*

$$N_f = 2^{n-k} N_g. \tag{9}$$

6.3 Linear structures

It is known that the more linear structures a Boolean function has, the closer it is related to an affine function. In the extreme case when every vector is a linear structure of a Boolean function, it must be an affine one. From a cryptographic point of view, a Boolean function is required to have as few linear structures as possible. When a Boolean function can be written as $f(x) = g(xD)$, it definitely has linear structures if $k < n$. The relationship of linear structures of f and that of g can be described as follows.

Theorem 23. *Let $f(x) = g(xD)$, where D is an $n \times k$ ($k \leq n$) matrix with $rank(D) = k$. Then α is an invariant (a complement) linear structure of f if and only if αD is an invariant (a complement) linear structure of g.*

Proof: The sufficiency is obvious. So we only need to present the proof of the necessity. Assume the contrary, i.e., there exists a vector $\alpha \in F_2^n$ such that $f(x) \oplus f(x \oplus \alpha) \equiv c$ and $g(y) \oplus g(y \oplus \alpha D) \not\equiv c$. Let $g(y') \oplus g(y' \oplus \alpha D) \neq c$. Since $rank(D) = k$, there must exist an $x' \in F_2^n$ such that $y' = x'D$. So we have

$$f(x') \oplus f(x' \oplus \alpha) = g(x'D) \oplus g((x' \oplus \alpha)D) = g(y') \oplus g(y' \oplus \alpha D) \neq c.$$

This is a contradiction of the assumption. So the conclusion is true. □

Corollary 24. *Denote by V_f and V_g the set of linear structures of f and g respectively. Then $dim(V_f) = (n-k)dim(V_g)$, where $dim(.)$ means the dimension of a vector space.*

It can be seen from corollary 24 that even if g has no nonzero linear structures, f may have because the all-zero vector is an invariant linear structure of every function. It also implies that a Boolean function may have many invariant linear structures but no complement ones.

6.4 Propagation criterion

Unlike other properties, the propagation property is not inheritable, i.e., g satisfies propagation criterion does not guarantee that f does. For example, let

$$D = \begin{bmatrix} 1\,0\,0 \\ 0\,1\,0 \\ 0\,0\,1 \\ 1\,1\,1 \end{bmatrix}.$$

Although $g(y_1, y_2, y_3) = y_1 \oplus y_1y_2 \oplus y_1y_3 \oplus y_2y_3$ satisfies the propagation criterion of order 2, $f(x_1, x_2, x_3, x_4) = g(xD) = x_1 \oplus x_1x_2 \oplus x_1x_3 \oplus x_2x_3$ does not satisfies the propagation criterion of order 1. However when g satisfies the propagation criterion of order k we have

Theorem 25. *Let $g \in \mathcal{F}_k$ satisfies the propagation criterion of order k, or g is perfect nonlinear, and let D be and $n \times k$ matrix. Let d be the largest number of rows such that they are linearly independent. Then $f(x) = g(xD)$ satisfies the propagation criterion of order d.*

Proof: For any $\alpha \in F_2^n$ with $1 \leq W_H(\alpha) \leq d$, by the condition of the theorem we have $1 \leq W_H(\alpha D) \leq k$. Since g satisfies the propagation criterion of order k, by lemma 17 we know that

$$f(x) \oplus f(x \oplus \alpha) = g(xD) \oplus g(xD \oplus \alpha D)$$

is a balanced function. So $f(x)$ satisfies the propagation criterion of order d. On the other hand, since D has $d+1$ rows which are not linearly independent, we suppose $\beta D = 0$, where $\beta \in F_2^n$ with $W_H(\beta) = d+1$. Then $f(x) \oplus f(x \oplus \beta) = 0$ and hence $f(x)$ does not satisfy the propagation criterion of order $d+1$. □

7 Conclusion

In this paper we have revealed the inherent structure of CI functions and described constructions for such functions. Since any CI function can be written as $f(x) = g(xG^T)$, where g is a non-degenerate function of k variables and G is a generating matrix of an $[n, k, d]$ linear code, it is also shown that most other cryptographic properties of g, such as *balancedness, nonlinearity, etc.*, can be inherited by the CI function f.

References

1. R.J.Anderson, Searching for the optimum correlation attacks, *Proc. of K.U.Leuven workshop on Cryptographic Algorithms*, Leuven, Belgium, 1994, pp.56-62.
2. P.Camion, et al., On correlation-immune functions, *Advances in Cryptology, Proc. CRYPTO'91*, Springer-Verlag 1992, pp.86-100.
3. J.Dj.Golic, On the security of shift register based keystream generators, *Fast Software Encryption (Cambridge'93)*, Springer-Verlag 1994, pp.90-100.
4. J.Dj.Golic, Correlation properties of a general binary combiner with memory, *Journal of Cryptology*, Vol.9, No.2, 1996, pp.111-126.
5. R.L.Lechner, Harmonic Analysis of Switching Functions, in *Recent Developments in switching Theory*, Edited by A.Mukhopadhyay, Academic Press, 1971.
6. F.J.MacWilliams and N.J.A.Sloane, *The Theory of Error-Correcting Codes*, North-Holland 1977.
7. W.Meier and O.Staffelbach, Nonlinearity criteria for cryptographic functions, *Advances in Cryptology, Proc. of Eurocrypt'89*, Springer-Verlag 1990, pp.549-562.
8. J.Seberry, X.M.Zhang, and Y.Zheng, On constructions and nonlinearity of correlation immune functions (extended abstract), *Advances in Cryptology, Proc. of Eurocrypt'93*, Springer-Verlag 1993, pp.181-197.
9. T.Siegenthaler, Correlation-immunity of nonlinear combining functions for cryptographic applications, *IEEE Trans. on Infor. Theory*, Vol. IT-30, No.5, 1984, pp.776-780.
10. T. Siegenthaler, Cryptanalysts' representation of nonlinearly filtered m-sequences, *Advances in Cryptology, Proc. of Eurocrypt'85*, Springer-Verlag 1986, pp.103-110.
11. T. Siegenthaler, *Methoden fur den Entwurf von Stream Cipher Systemen*, Diss. ETH Nr. 8185, 1986
12. C.K.Wu, X.M.Wang, and E.Dawson, Construction of correlation immune functions based on the theory of error-correcting codes, *Proc. ISITA96*, Canada September 1996, pp.167-170.
13. G.Z.Xiao and J.L.Massey, A spectral characterization of correlation-immune combining functions, *IEEE Trans. Inform. Theory*, Vol. IT-34, 1988, pp.569-571.

An Improved Key Stream Generator Based on the Programmable Cellular Automata

Miodrag J. Mihaljević *

Institute of Applied Mathematics and Electronics
Institute of Mathematics, Academy of Science and Arts
Belgrade, Yugoslavia
E-mail: emihalje@ubbg.etf.bg.ac.yu

Abstract. An improved programmable cellular automata (PCA) based key stream generator is proposed which originates from a recently proposed scheme for key stream generators based on the PCA and a read only memory. Cryptographic security examination of the proposed key stream generator is realized through the following two steps. As the first, an equivalent model of the generator is given, and it is shown that the generator is resistant on the known attacks. Than, a novel method for the cryptanalysis, is developed, and it is shown that the generator is not vulnerable on this approach assuming that PCA length is sufficiently large.

Key words: stream ciphers, cellular automata, key stream generators, cryptanalysis.

1 Introduction

In recent years, significant attention has been paid to the theory and applications of the Cellular Automata (CA) including cryptographic applications (see [1], for example). CA appears to be a promising building block for key stream generators. In [2] and [3], the CA is introduced for pseudorandom sequence generation with possible applications for stream ciphers. Additive CA could be considered as an alternative to Linear Feedback Shift Register (LFSR) for a basic building block for key stream generators. As pointed out in [4], CA differ from cryptographic mechanisms such as LFSR in that, even if they are invertable, it is not possible to calculate the predecessor of an arbitrary state by simply reversing the rule for finding the successor. In [5], CA with dynamically alterable rule vector called Programmable Cellular Automata (PCA), is introduced as a building block for key stream generators. Two PCA based key stream generators are proposed in [5], called PCA with ROM (Read Only Memory) and Two stage PCA, together with results of theirs security examination. Also, the cryptanalysis of certain CA/PCA based key stream generators have been published. A method for reconstruction of a CA initial state based on the sequence of bits generated by a CA cell is given in [6]. Security examinations of the Two stage PCA, and the PCA with ROM have been reported in [7] and [8], respectively.

* This research was supported by the Science Fund, Grant. No. 04M02.

In this paper, starting from the PCA with ROM, [5], and the results of its cryptanalysis given in [8], a cryptographically improved key stream generator scheme based on the PCA with ROM is proposed and its security examination is presented. The scheme is designed in such a way that it is resistant on the attacks known so for. Also, a novel cryptanalytic approach is developed for cryptographic security examination of the proposed scheme. The approach employs, for reconstruction of the PCA initial state, the divide and conquer principle with standard correlation attack based on a novel result related to generation of a substring of an arbitrary PCA cell output sequence. The background is presented in Section 2. A novel PCA based key stream sequence generator is proposed in Section 3, and its cryptographic security examination is given in Section 4. Conclusions are summarized in Section 5.

2 Background

CA is an array of cells where each cell is in any one of the permissible states. At each discrete time step (clock cycle) the evolution of cell depends on some rule (the combinatorial logic) which is a function of the present state of k of its neighbors for a k-neighborhood CA. For 3-neighborhood CA, the evolution of the i-th cell can be represented as a function of the present states of $(i-1)$th, (i)th, and $(i+1)$th cells as: $x_i(t+1) = f\{x_{i-1}(t), x_i(t), x_{i+1}(t)\}$, where f represents the combinatorial logic. The CA characterized by a rule known as rule 90 specifies an evolution from neighborhood configuration to the next state according to the following combinatorial logic $x_i(t+1) = x_{i-1}(t) \oplus x_{i+1}(t)$, that is, the next state of ith cell depends on the present states of its left and right neighbors. Similarly, the combinatorial logic for rule called 150 is given by $x_i(t+1) = x_{i-1}(t) \oplus x_i(t) \oplus x_{i+1}(t)$, that is, the next state of ith cell depends on the present states of its left and right neighbors and on its own present state. A CA characterized by *exor* and/or *exnor* dependencies only is called an additive CA.

Characterization of additive CA based on matrix algebraic tools has been reported in [9] and [10]. If a characteristic polynomial of a CA is primitive, then it is referred to as a maximal length CA. Existence of primitive charactersitic polynomial for an L cell CA confirms that the CA will run through the maximum length of $2^L - 1$ distinct nonzero states, [11]. Such maximum length CA generates high quality pseudorandom patterns significantly better than that of the LFSR (see [12], for example). In [13], an algorithm to design a 90/150 CA having a primitive characteristic polynomial is given.

Positional representations of *rule* 90 and *rule* 150 show that their neighborhood dependence differ in only one position, viz., on the cell itself. Therefore, by allowing a single control line per cell, one can apply both *rule* 90 and *rule* 150 on the same cell at different time steps. Thereby, an L cell CA structure can be used for implementing 2^L CA configurations. Realizing different CA configurations on the same structure can be achieved using a control logic to control the appropriate switches and a control program, stored in a ROM, can be employed to active the control. The 1(0) state of the ith bit of a ROM word closes (opens) the switch that controls the ith cell. Such a structure is referred to as a Programmable CA (PCA). The L bit control word for an L cell PCA has 1(0) on ith cell if the rule 150(90) is applied to the ith cell.

In [8] a novel cryptological security examination of the PCA with ROM key stream generator proposed in [5] is given. A vulnerability of this scheme on certain cryptanalytic attacks is demonstrated. It is shown that, the effective secret key size is significantly smaller than its formal length. The proposed cryptanalytic attacks (basic one and fast one) are based on the divide and conquer principle, standard correlation attack [14], elements of its generalization employing the same underlying ideas as in [15] - [16], and the fast iterative error-correction approach considered in [17], - [19] developed for the shift registers based generators.

3 A Novel PCA Based Key Stream Generator

In this section, as an improvement of the PCA with ROM generator [5], a novel PCA based key stream generator is proposed. The main parts of the here proposed key stream generator are: an L-cell PCA and, a ROM which contains the configuration rules for the PCA, an L^*-length binary buffer, and an L^*-dimensional time varying permutation. A logic scheme of the generator is displayed in Fig. 1.

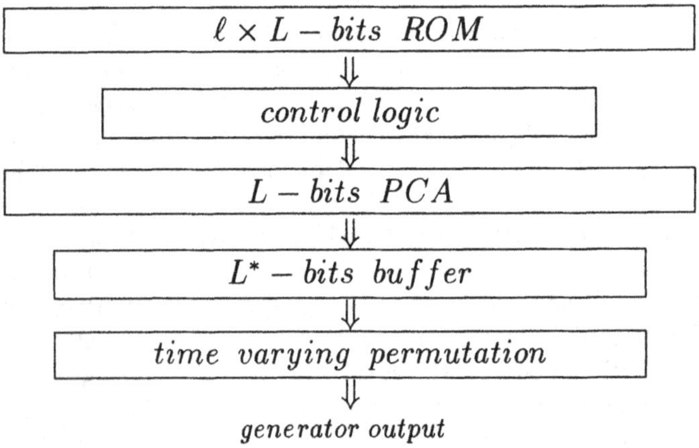

Fig. 1: A logic scheme of the proposed PCA based key stream generator.

Assume that ℓ maximal length CA's are chosen out of all possible maximal length CA's with rule 90 and 150. These rules are noted as $\{R_0, R_1, R_2, ..., R_{\ell-1}\}$. The rule configuration control word corresponding to rule R_i is stored in a ROM word. Initially the PCA is configured with rule R_0 and loaded with a non zero seed. With this configuration the PCA runs one clock cycle. Then it is reconfigured with the next rule (i.e., R_1) and runs another cycle. The rule configuration of PCA changes after every run, i.e., if in the ith run rule configuration is $R_{(i)modl}$, then in the next run, rule is $R_{(i+1)modl}$.

After each clock cycle, the content of a middle cell of the PCA is taken as an output and stored in the L^*-length binary buffer.

After each L^* clock cycles, the buffer content is permutated according the time varying permutation. Time varying permutation is controlled by the current PCA state.

Concatenation of the permutation output patterns give the binary key stream sequence.

For simplicity we assume that the cryptographic secret key determines the PCA seed only, so that the key length is L bits.

Note that the output bits generation rate of the proposed generator is L times slower than the rate of the PCA with ROM generator.

4 Security Examination of the Proposed Generator

In this section, cryptographic security examination of the proposed key stream generator is realized through the following two steps. As the first, an equivalent model of the generator is given, which yields that the generator is resistant on the published attacks. Than, a novel method for the cryptanalysis, the best one known so far, is developed, and it is shown that the generator is not vulnerable on this approach assuming that PCA length is sufficiently large.

4.1 Preliminaries and an Equivalent Model

Characterization of additive CA based on matrix algebraic tools has been reported in [9] and [10]. Note that the elementary configuration rules 90 and 150 are linear so that the whole PCA configuration rule which transforms the state $\mathbf{X}_{t-1}$ into the state $\mathbf{X}_t$ can be described by certain matrix. Accordingly, denote by A_t an $L \times L$-dimensional binary matrix such that the following is valid $\mathbf{X}_t = A_t\mathbf{X}_{t-1}$, $t = 1, 2, \dots$. Note that non-zero elements of any matrix A_t are on main diagonal and its two adjacent diagonals only. When the PCA configuration rules are known yielding that the sequence of matrix A_t, $t = 1, 2, \dots$, is known, the t-th PCA state is given by

$$\mathbf{X}_t = (\prod_{i=1}^{t} A_i)\mathbf{X}_0 \ , \quad t = 1, 2, \dots \ , \tag{1}$$

where $\mathbf{X}_0$ is the PCA initial state.

On the other hand, assuming unknown permutation rule, equal probability of all possible permutations and that inputs into the permutator are independent identically distributed (i.i.d.) random binary variables which take 0 or 1 value with probability equal to $\frac{1}{2}$. It can be directly shown that the key stream sequence can be considered as the sequence of bits generated by certain PCA cell after transmission through binary symmetric channel (b.s.c.) with error-probability equal $p = \frac{1}{2} - \frac{1}{L^{\bullet}}$. So, a problem of the secret key reconstruction can be considered as the following one. Let L-dimensional binary vector $\mathbf{X}_t = [x_t(i)]_{i=1}^{L}$ be the t-th PCA state, $t = 1, 2, ...$, and denote by $\{y_n\}$ the binary sequence which is input into the buffer so that, for selected i the following is valid:

$$y_n = x_t(i) \; , \quad n = t, \;\; t = 1, 2, ... \tag{2}$$

Let $\{e_n\}$ be a binary noise sequence. In a statistical model, it is assumed that $\{e_n\}$ is a realization of a sequence of i.i.d. binary variables $\{E_n\}$ such that $Pr(E_n = 1) = p$, $n = 1, 2, ..., N$. Let a binary sequence $\{z_n\}$ be a noisy version of the sequence $\{y_n\}$ defined by

$$z_n = y_n \oplus e_n \; , \quad n = 1, 2, ... \; , \tag{3}$$

where $\oplus$ denotes the modulo 2 addition in GF(2).

Finally, the security examination problem is equivalent to the problem of a PCA initial state $\mathbf{X}_0$ reconstruction based on the sequence $\{z_n\}$ assuming that the PCA configuration rules and the noise parameter p are known.

Accordingly note that the algorithm [6] can not be applied for cryptanalysis of the proposed generator because only noisy version of the sequence generated by certain CA cell is available, and the approach [6] works only when the error-free sequence is available. Also, it can be directly shown that the attack given in [8] can not work because it is based on the assumption that the sequence of successive noisy PCA states is available, and in the proposed scheme this assumption is not valid, and more precisely, only one noisy bit of each PCA state is available.

4.2 Main Idea for Cryptanalysis

This section presents the main idea for a cryptanalytic attack on the proposed generator. The idea assumes independent reconstruction of two nonoverlapping parts of the whole PCA initial content.

Suppose that $\{x_m\}$ is the output sequence from certain PCA cell when the PCA initial content is a binary vector X_0. Denote by $\{s_i\}$ a subsequence of $\{x_m\}$, and recall that $\{z_n\}$ is the generator output sequence which is a noisy version of $\{x_m\}$.

For further consideration we assume the following:

- A procedure exists for generating the subsequence $\{s_i\}$ using an appropriate part only of the whole PCA initial content.
- A criterion exists which ensures using the sequence $\{z_n\}$ the decision making whether or not $\{s_i\}$ is a subsequence of $\{x_m\}$.

Let the criterion be a distance measure d between the binary sequences $\{s_i\}$ and $\{z_n\}$, and suppose that d is a realization of random variable D with two possible probability distributions $\{Pr(D \mid H_0)\}$ and $\{Pr(D \mid H_1\}$ where H_0 and H_1 are the following hypothesis:

- H_0: the sequences $\{s_i\}$ and $\{z_n\}$ are realizations of two independent truly random binary sources.
- H_1: the sequence $\{s_i\}$ generated by the assumed part of the whole PCA initial content *is* a subsequence of the sequence $\{x_m\}$ which yields $\{z_n\}$.

Note that H_0 is a good model for the situation when the sequence $\{s_i\}$ generated by the assumed part of the PCA initial content *is not* a subsequence of $\{x_m\}$ which yields $\{z_n\}$.

Suppose that the probability distributions $\{Pr(D \mid H_0)\}$ and $\{Pr(D \mid H_1)\}$ are known. Then, employing the standard procedure, the decision threshold t can be determined such that achieve the given probabilities of "the missing event" P_m and "the false alarm" P_f.

Accordingly, a basic form of the procedure for cryptanalysis consists of the following steps:

- *Step 1*: Split the PCA initial content into two successive nonoverlapping parts.
- *For each possible content of the first part do the Steps 2-4.*
  Step 2: Using the assumed initial content generate the subsequence $\{s_i\}$.
  Step 3: Calculate the distance d between $\{s_i\}$ and $\{z_n\}$.
  Step 4: According to d and the threshold t accept H_0 or H_1. If H_1 was accepted memorize the considered initial content part.
- *Step 5*: Determine the whole PCA initial content by examination of all the combinations of candidates memorized in the Step 4 and all the possible candidates for the second part of the initial content.

The procedure output is the set of most probable ccndidates for the true solution or the conclusion that the true solution is missed because of non zero P_m. Obviously, the proposed approach for the initial state reconstruction by separate reconstruction of its two parts yields significant speed-up of the cryptanalysis in comparison with the exhaustive search.

4.3 Subsequence of a CA / PCA Sequence

As the first, we investigate a binary sequence generated by a maximum length CA with a primitive characteristic polynomial. The underlying problem which is under consideration is to generate an arbitrary long CA output subsequence employing a part of CA initial content only.

Denote by $\{x_m\}_{m=1}^{M}$ an output segment of an L length CA. Denote by $\mathbf{X}_q = [x_{q+l}]_{l=1}^{L}$ an L-dimensional binary vector defining the q-th state of the CA,

$$\mathbf{X}_q = A^q \mathbf{X}_0, \quad q = 0, 1, 2, \dots, \tag{4}$$

where $\mathbf{X}_0$ is the initial state, and A^q is the q-th power, in $GF(2)$, of the $L \times L$-dimensional binary state-transition matrix determined by the CA's characteristic polynomial. Note that $A^j = A^{2^L-1+j}$, $j < 0$. Our final goal is to estimate the number of the CA output sequence elements that depend of certain w elements of the initial content, $w = 1, 2, ..., L$.

Theorem 1. Let $a_q(i,j)$ be the (i,j)-th element of the matrix A^q, and let Ω be a subset of $\{1, 2, ..., L\}$ with cardinality $L - w$. The number, $\eta(w)$, of different values of q, $1 \leq q \leq 2^L - 1$, such that $\sum_{j \in \Omega} a_q(i,j) = 0$ is given by

$$\eta(w) = 2^w \quad , \quad 1 < w < L, \tag{5}$$

for each $1 < i < L$, assuming that the sum denotes the integer addition.

Sketch of the proof. Let

$$A^q = [\mathbf{A}_q(1)\mathbf{A}_q(2)...\mathbf{A}_q(L)] \ , \tag{6}$$

$$I = [\mathbf{I}(1)\mathbf{I}(2)...\mathbf{I}(L)] \ , \tag{7}$$

where $A_q(j)$ is the j-th column of A^q and $I(j)$ is the j-th column of the identity matrix I , $j = 1, 2, ..., L$. According to $A^q = A^q I$ we have:

$$\mathbf{A}_q(j) = A^q \mathbf{I}(j), \ j = 1, 2, ..., L, \ q = 1, 2, ..., 2^L - 1 \ . \tag{8}$$

In view of the basic CA matrix equation (4) it follows that $\mathbf{A}_q(j)$ can be considered as the q-th state of the PCA with the initial state $\mathbf{I}(j)$, $q = 1, 2, ..., 2^L - 1$, $j = 1, 2, ..., L$. Consequently, involving a vector linear transformation and employing the basic characteristics of the maximal length CA sequence yield the theorem statement.

Corollary 1. The expected number of output elements, from certain CA cell, which depends of arbitrary w initial content elements only, after N clock cycles is equal to $2^{w-L}N$ and is independent of the characteristic polynomial, assuming that it is a primitive one.

Starting from the previous result related to the CA, the following conjecture can be stated for a PCA when the involved configuration rules correspond to the primitive characteristic polynomial.

Conjecture 1. The expected number of output elements, from certain PCA cell, which depends of arbitrary w initial content elements only, after N clocks is equal to $2^{w-L}N$ and is independent of the dynamically alterable configuration rules, assuming that all of the applied rules correspond to the primitive characteristic polynomials.

Massive systematic computer simulations imply that the conjecture is true. Accordingly, the following is the procedure for generating a subsequence of the PCA output sequence, which elements depends only of no more than certain w elements of the initial state.

Algorithm for the PCA Subsequence Generating

- *Input:* $L \times L$ dimensional state transition matrices, A_k, $k = 1, 2, ..., \ell$, of the considered PCA.

- *Initialization:* Assume certain w elements of the PCA initial content and denote by Ω a subset of $L-w$ elements of the set $\{1, 2, ..., L\}$ which contain indexes of the initial content elements which were *not* assumed. Set $i = 1$.
- *For each* $m = 1, 2, ..., M$ *do the following two steps.*
  Step 1: Calculate $A^m = \prod_{i=1}^{m} A_i$.
  Step 2: If $\sum_{j\in\Omega} a_m(i,j) = 0$ than $s_i = x_m(i)$, and set: $m(i) = m$, $I = i$, $i \rightarrow i+1$.
- *Output:* The PCA subsequence $\{s_i\}_{i=1}^{I}$ and the sequence $\{m(i)\}_{i=1}^{I}$ such that $x_{m(i)} = s_i$, $i = 1, 2, ..., I$.

4.4 Distance Measure

According to the nature of the sequences $\{s_i\}_{i=1}^{I}$ and $\{z_n\}_{n=1}^{N}$, and according to [14], the appropriate distance measure d^* is:

$$d^* = I - 2\sum_{i=1}^{I} s_i \oplus z_{m(i)} \ , \tag{9}$$

where the sequence $\{m(i)\}_{i=1}^{I}$ is the output of the *Algorithm for the PCA Subsequence Generating.*

Assuming that d^* is a realization of an integer random variable D^*, it can be directly shown that the expected value $E\{D^*\}$ and variance $Var\{D^*\}$, when H_0 and H_1 are valid, are given by the following:

$$E_{H_0}\{D^*\} = 0 \ , \quad Var_{H_0}\{D^*\} = I \ , \tag{10}$$

$$E_{H_1}\{D^*\} = I(1-2p) \ , \quad Var_{H_1}\{D^*\} = 4Ip(1-p) \ . \tag{11}$$

For large I the random variable D^* can be assumed to be normally distributed with the expected value and variance given by the previous formulas. Accordingly a clear distinction exists between the probability distributions relevant for H_0/H_1 decision making.

The acceptance of H_0 or H_1 is based on certain threshold t^* and the calculated value d^* according to the following: for $d^* < t^*$ accept H_0, and for $d^* \geq t^*$ accept H_1. Denote by $p_{D^*|H_0}(x)$ and $p_{D^*|H_1}(x)$ the normally distributed probability density functions of D^* when H_0 and H_1 are valid, respectively. Note that, when $p = 0.5$ the two probability density functions $p_{D^*|H_0}(x)$ and $p_{D^*|H_1}(x)$ are identical, and therefore no decision can be made. The threshold t^* can be calculated based on the approach from [14], and estimation of required $\{s_i\}$ length I can be derived employing the method, also presented in [14].

4.5 Algorithm for Cryptanalysis

In this section, according to the previously presented results, the complete algorithm for the PCA initial reconstruction is proposed.

Algorithm for the PCA Initial State Reconstruction

- *Input*: $\{y_n\}_{n=1}^N$, and set of the applied configuration rules - characteristic polynomials.
- *Initialization:*
  - Split the PCA initial content into two parts of L_1 and $L - L_1$ elements such that $2^{L_1-L}N > I$ so that the subsequence $\{s_i\}$ length I is at least equal to I_{min} which ensures small number of false alarms after testing all the hypothesis.
  - For given P_m and I employing the standard procedure (see [14], for example), determine the threshold t^* such that P_f is minimized.
- *Phase I:*
  For each possible content of the first PCA part do the Steps 1-3.
  Step 1: Using the assumed initial content generate the subsequence $\{s_i\}$ employing the *Algorithm for the PCA Subsequence Generating.*
  Step 2: Calculate the distance $d^* = I - 2\sum_{i=1}^{I} s_i \oplus z_{m(i)}$ between $\{s_i\}$ and $\{z_n\}$ assuming that the sequence $\{m(i)\}_{i=1}^{I}$ is an output of the *Algorithm for the PCA Subsequence Generating.*
  Step 3: According to d^* and the threshold t^* accept H_0 or H_1. If H_1 was accepted memorize the considered initial content part.
- *Phase II:*
  Step 4: Determine the whole PCA initial content by examination of all the combinations of candidates memorized in the Step 3 and all the possible candidates for the second part of the PCA initial content employing standard correlation approach, [14].
- *Output*: The PCA initial content solution or the decision that the true solution was missed because of non-zero value of P_m.

4.6 Discussion

Note that testing of any hypothesis in the algorithm Phase I employs approximately the same number of operations as testing a hypothesis during the exhaustive search, but the total number of hypothesis that have to be checked through Phase I and Phase II is significantly smaller than the total number of hypothesis which has to checked for realization of the exhaustive search for the whole initial content.

The Phase I and Phase II employ testing of 2^{L_1-1} and 2^{L-L_1-1} hypothesis in average, respectively, so that the complexity of the algorithm is proportional to $2^{L_1-1}+2^{L-L_1-1}$, which is typically significantly smaller than 2^{L-1} the expected number of testing employed for the direct exhaustive search.

Suppose that I is greater than I_{min} which yields small number of false alarms after all 2^w testing of H_0/H_1 hypothesis. Than, assuming that $w > L/2$, it can be directly shown that the gain g obtained by the proposed algorithm in comparison with direct exhaustive search through the PCA initial states is lower bounded by the following:

$$g > \frac{2^L}{2\,2^w} = \frac{N}{2I} . \tag{12}$$

Note that, assuming that $w = L_1 = L/2$ and $I_{min} = 10^4$ (which corresponds to $p = 0.4$), the required length of the generator output is equal to $N = 2^{L/2}10^4$. Accordingly, when $p = 0.4$, it can be directly shown that the scheme is cryptographically secure, assuming that the PCA length L is sufficiently large, $L > 120$, and this is the upper bound when $p > 0.4$.

5 Conclusions

A novel PCA based key stream generator, originating from the PCA with ROM key stream generator, is proposed, and its cryptographic security examination is presented.

Roughly speaking, the key stream sequence is time varying permutation of bits inside the segments of binary sequence generated by certain PCA cell. The time varying permutation is controlled by the sampled PCA states, and the PCA initial state is controlled by the secret key.

An equivalent model of the proposed generator is pointed out, and it is shown that the proposed scheme is resistant on the relevant, so far, reported cryptanalytic attacks. Also, a novel method for cryptanalysis of the proposed scheme is developed. The main underlying idea for cryptanalysis is based on independent reconstruction of two nonoverlapping parts of the whole PCA initial content i.e. secret key. The algorithm for cryptanalysis include a novel method for generating the subsequence, of a PCA cell output sequence, which depends of a part only of the whole PCA initial state, and the hypothesis testing based on the appropriate distance measure.

The analysis implies that the scheme is cryptographically secure, assuming that the PCA length L, equal to the secret key length, is sufficiently large, $L > 120$.

Finally note that the upgrading of the cryptographic security of the proposed scheme in comparison with the PCA with ROM scheme is paid by decreasing the speed of key stream generation for a factor equal to the PCA length L.

References

1. P.P. Chaudhuri, D.P. Chaudhuri, S. Nandi, and S. Chattopadhyay, *Additive Cellular Automata, Theory and Applications vol. 1.* New-York: IEEE Press,1997.
2. S. Wolfram, "Random sequence generation by cellular automata", *Advances in Applied Mathematics* 7, pp. 123-169, 1986.
3. S. Wolfram, "Cryptography with cellular automata", *Lecture Notes in Computer Science*, vol. 218, pp. 429-432, 1986.
4. W. Diffie, "The first ten years of public-key cryptography", *Proc. IEEE*, pp. 560-577, 1988.

5. S. Nandi, B.K. Kar, and P. Pal Chaudhuri, "Theory and applications of cellular automata in cryptography", *IEEE Trans. Computers*, vol. 43, no. 12, pp. 1346-1357, Dec. 1994.
6. W. Meier and O. Staffelbach, "Analysis of pseudo random sequences generated by cellular automata", Advances in Cryptology - EUROCRYPT '91, *Lecture Notes in Computer Science*, vol. 547, pp. 186-199, 1992.
7. M. Mihaljević, "Security examination of certain cellular automata based key stream generator", *ISITA '96 - 1996 IEEE International Symposium on Information Theory and Its Applications*, Canada, Victoria, B.C., September 1996, Proceedings, pp. 246-249.
8. M. Mihaljević, "Security examination of a cellular automata based pseudorandom bit generator using an algebraic replica approach", Applied Algebra, Algebraic Algorithms and Error-Correcting Codes, AAECC - 12, *Lecture Notes in Computer Science*, vol. 1255, pp. 250-262, 1997.
9. A.K. Das, A. Gonguly, A. Dasgupta, S. Bhawmik, and P. Pal Chaudhuri, "Efficient characterization of cellular automata", *IEE Proc., Pt. E*, vol. 137, no. 1, pp. 81-87, Jan. 1990.
10. A.K. Das and P. Pal Chaudhuri, "Vector space theoretic analysis of additive cellular automata and its applications for pseudo-exhaustive test pattern generation", *IEEE Trans. Computers*, vol. 42, no. 3, pp. 340-352, Mar. 1993.
11. P.D. Hortensius, R.D. McLeod, and H.C. Card, "Parallel pseudorandom number generation for VLSI systems using cellular automata", *IEEE Trans. Computers*, vol. 38, no. 10, pp. 1466-1473, Oct. 1989.
12. S. Nandi and P. Pal Chaudhuri, "Analysis of periodic and intermediate boundary 90/150 cellular automata", *IEEE Trans. Computers*, vol. 45, no. 1, pp. 1-11, Jan. 1996.
13. M. Serra, T. Slater, J.C. Muzio, and D.M. Miller, "The analysis of one-dimensional linear cellular automata and their aliasing properties", *IEEE Trans. Computer-Aided Design*, vol. 9, no. 7, pp. 767-778, July 1990.
14. T. Siegenthaler, "Decrypting a class of stream ciphers using ciphertext only", *IEEE Trans. Computers*, vol. C-34, no. 1, pp. 81-85, Jan. 1985.
15. M. Mihaljević, "A correlation attack on the binary sequence generators with time-varying output function", Advances in Cryptology - ASIACRYPT '94, *Lecture Notes in Computer Science*, vol. 917, pp. 67-79, 1995.
16. J. Golić and M. Mihaljević, "A generalized correlation attack on a class of stream ciphers based on the Levenshtein distance", *Journal of Cryptology*, vol. 3, pp. 201-212, 1991.
17. M. Mihaljević and J. Golić, "A fast iterative algorithm for a shift register initial state reconstruction given the noisy output sequence", Advances in Cryptology - AUSCRYPT '90, *Lecture Notes in Computer Science*, vol. 453, pp. 165-175, 1990.
18. M. Mihaljević and J. Golić, "A comparison of cryptanalytic principles based on iterative error-correction", Advances in Cryptology - EUROCRYPT '91, *Lecture Notes in Computer Science*, vol. 547, pp. 527-531, 1992.
19. M. Mihaljević and J. Golić, "Convergence of a Bayesian iterative error-correction procedure on a noisy shift register sequence", Advances in Cryptology - EUROCRYPT '92, *Lecture Notes in Computer Science*, vol. 658, pp. 124-137, 1993.

A Trust Policy Framework

Audun Jøsang *

Department of Telematics, NTNU, 7034 Trondheim, Norway
Information Security Research Centre, QUT, Brisbane Qld 4001, Australia **
Email: ajos@item.ntnu.no

Abstract. This paper describes a formal framework for implementing trust policies in security systems. It consists of a model for representing opinions about trust and a set of operations to manipulate these opinions and to make decisions.

1 Introduction

A trust policy can be described as a a set of rules about trust which is used for guiding action which includes risk. Many security schemes contain trust policies, as for example schemes for secure distribution of public keys where it can be specified that a key can only be trusted when it is certified by a sufficiently trustworthy agent, or alternatively by a certain number of trustworthy agents.

In this paper we present a flexible framework for defining and implementing trust policies in security systems. It is based on a model for representing belief in general and belief about trust in particular. The model is more complete than previously proposed trust models (see e.g. [RS97, Jøs97] for an analysis of some models). The framework consists of a set of operations for manipulating these opinions and for making decisions based on opinions. An example will illustrate how the framework can be applied.

2 The Trust Model

In order to better understand the philosophy behind the model, we will start with a short phenomenologic account for the concept belief, and show how this leads to a concise model for expressing trust.

We assume the world to be in a particular state at any given time. Our knowledge about the world is never perfect so we can never determine the state exactly. For the purpose of believing a proposition about the world, we assume that it is either true or false, and not something in between. Because of our imperfect knowledge, it is impossible to know with certainty whether it is true or false, so that we can only have an *opinion* about it, which translates into degrees of belief or disbelief.

* This research was supported by Norwegian Research Council Grant No.116417/410.
** This research was carried out while the author was visiting the ISRC at QUT.

In addition it is necessary to take into consideration degrees of ignorance, which can be described as a vacuous belief which fills the void in the absence of both belief and disbelief. For a single opinion about a proposition, we assume that

$$b + d + i = 1, \quad \{b, d, i\} \in [0, 1] \tag{1}$$

where b, d and i designate belief, disbelief and ignorance respectively. Eq.1 describes a triangle as illustrated in Fig.1, and an opinion about a proposition can now be uniquely described as a point (b, d, i) in the triangle.

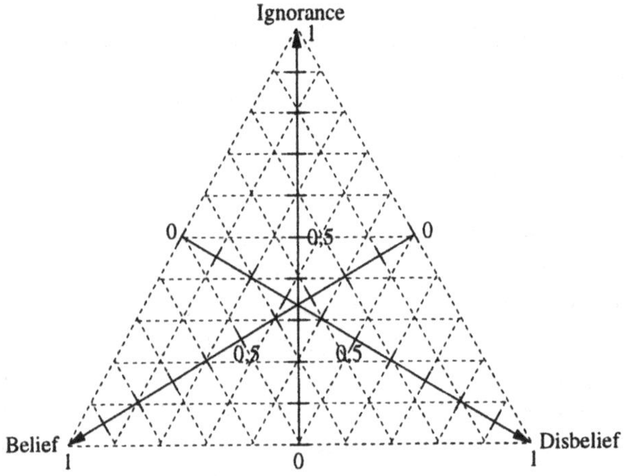

Fig. 1. Opinion Triangle

Definition 1. *Let $\pi = (b, d, i)$ be a triplet which satisfies Eq.1, where the first, second and third element correspond to belief, disbelief and ignorance respectively. Then π is called an opinion.*

The bottom line between belief and disbelief in Fig.1 represents situations with zero ignorance and is equivalent with a traditional probability model. The degree of ignorance can be interpreted as the lack of evidence to support either belief or disbelief. The justification for Eq.1 and for the opinion triangle is that the human mind maintains belief and disbelief simultaneously as quasi-complementary, and that when both beliefs become weaker, due to insufficient evidence, ignorance increases.

For example, the opinion (0, 0, 1) which represents total ignorance can be interpreted as the belief that it is absolutely certain that the proposition is either true or false, but there is no evidence to indicate that one is more likely than the other. The difference between (0, 0, 1) and (0.5, 0.5, 0.0) is that in the first case there is no reason to believe that the proposition is true or false, whereas in the second case there are equally strong reasons to believe both. This way of modelling beliefs and uncertainty corresponds to the Dempster-Shafer Theory of Evidential Reasoning.

The mathematical theory of evidence was first set forth by Dempster in the 1960s and subsequently extended by Shafer who in 1976 published *A Mathematical Theory of Evidence*[Sha76]. A more concise presentation can be found in [LVDG91] from which Def.2 is taken.

The first step in applying evidential reasoning is to define a set of possible situations, called the *frame of discernment*, which delimits the set of possible states of the world, exactly one of which is assumed to be true at any one time.

Definition 2. *Let Θ be a frame of discernment. If with each subset $x \subseteq \Theta$ a number $m_\Theta(x)$ is associated such that:*

$$\begin{aligned}&(1)\ m_\Theta(x) \geq 0\\&(2)\ m_\Theta(\emptyset) = 0\\&(3)\ \textstyle\sum_{x \subseteq \Theta} m_\Theta(x) = 1\end{aligned}$$

then m_Θ is called a belief mass distribution *on Θ. For each subset $x \subseteq \Theta$, the number $m_\Theta(x)$ is called the* belief mass *of x.*

For the purpose of trusting an agent for a particular task, we assume that it will either cooperate or defect. Trust and distrust is then the belief or disbelief that an agent will cooperate. We then have

$$\begin{aligned}&\Theta = \{\text{cooperate, defect}\} \quad \text{and}\\&2^\Theta = \{\{\text{cooperate, defect}\},\ \text{cooperate, defect},\ \emptyset\}.\end{aligned}$$

Belief in {cooperate}, {defect} and {cooperate, defect} can be translated into trust, distrust and ignorance designated by b, d and i respectively, . We then get $m_\Theta(\text{cooperate}) = b$, $m_\Theta(\text{defect}) = d$ and $m_\Theta(\{\text{cooperate, defect}\}) = i$.

In a similar way, trust in the authenticity of a cryptographic key can be expressed by defining

$$\Theta = \{\text{authentic, not authentic}\}.$$

3 The Trust Policy Framework

Opinions, as defined in Sec.2, are considered subjective, and will therefore have an ownership assigned whenever relevant. In our notation, superscripts indicate ownership, and subscripts indicate the proposition to which the opinion apply. For example

$$\pi_p^A$$

is an opinion held by agent A about the truth of proposition p.

The framework presented here, which consists of a subset of a general framework for artificial reasoning currently under development by the author, contains the following operations:

(1) conjunction
(2) conditional implication
(3) parallel combination of opinions
(4) serial combination of opinions
(5) opinion thresholds

Operation 1) is equivalent with the conjunction defined in [Bal86]. Operation 3) is the same as Dempster's rule defined in [Sha76]. The definitions of the operations 2), 4), and 5) have as far as we know not been proposed before.

3.1 Conjunction

A conjunction of two opinions about propositions consists of determining from the two opinions a new opinion reflecting the conjunctive truth of both propositions. This corresponds to the logical binary "AND" operation in standard logic.

Definition 3. *Let $\pi_p = (b_p, d_p, i_p)$ and $\pi_q = (b_q, d_q, i_q)$ be an agent's opinions about two distinct propositions p and q. Let $\pi_{p \wedge q} = (b_{p \wedge q}, d_{p \wedge q}, i_{p \wedge q})$ be the opinion such that*

$$(1)\ b_{p\wedge q} = b_p b_q$$
$$(2)\ d_{p\wedge q} = d_p + d_q - d_p d_q$$
$$(3)\ i_{p\wedge q} = b_p i_q + i_p b_q + i_p i_q$$

Then $\pi_{p\wedge q}$ is called the conjunction of π_p and π_q, representing the agents opinion about both p and q being true. By using the symbol "$\wedge$" to designate this operation, we get $\pi_{p\wedge q} = \pi_p \wedge \pi_q$.

It is easy to prove that conjunction of opinions is both commutative and associative. It must be assumed that the opinion arguments in a conjunction are independent. This means for example that the conjunction of an opinion with itself will be meaningless, because that would necessarily imply that the propositions are identical, whereas the conjunction rule will see them as if they were distinct propositions.

3.2 Conditional Implication

The goal of the conditional implication is to determine an opinion about the conclusion of the conditional, based on the opinion about the validity of the implication and the condition of the implication. This corresponds to Modus Ponens axiom of inference in standard logic defined by:

$$p \rightarrow q\ ;\ \ p\ ;\ \ \text{then deduce } q,$$

meaning that if it is true that p implies q and that p is true, then q must be true.

Definition 4. *Let* $\pi_{p \to q} = (b_{p \to q}, d_{p \to q}, i_{p \to q})$ *be an agent's opinion about the validity of* $p \to q$ *as a true conditional implication, and let* $\pi_p = (b_p, d_p, i_p)$ *be the same agent's opinion about the truth of proposition* p. *The opinion* $\pi_{q/p}$ *defined by*

$$\pi_{q/p} = \pi_{p \to q} \wedge \pi_p$$

is then called the conditional opinion about q *by* p, *representing the agent's opinion about the truth of* q *as a function of her opinion about* $p \to q$ *and about* p.

The justification for the conditional opinion is that both the conditional implication and the argument must be true for the conclusion to be true, and this is the same as logical conjunction of $\pi_{p \to q}$ and π_p. This definition is consistent with Prade's extension of conditional implication[Pra85] in which truth values are replaced by probabilities, so that Modus Ponens becomes:

$$\text{Prob}(q/p) \geq a \; ; \;\; \text{Prob}(p) \geq b \; ; \;\; \text{then deduce } \; \text{Prob}(q) \geq ab$$

Although our definition is more general, it reduces to Prade's definition by collapsing the ignorance and belief components into probability.

3.3 Parallel Combination of Opinions

A parallel combination of opinions consists of combining two or more opinions about the same proposition into a single opinion. One way of doing this is to combine two belief mass distributions according to Dempster's Rule[Sha76].

Definition 5. *Let* $\pi_p^A = (b_p^A, d_p^A, i_p^A)$ *and* $\pi_p^B = (b_p^B, d_p^B, i_p^B)$ *be opinions respectively held by agents* A *and* B *about the same proposition* p. *Let* $\pi_p^{A,B} = (b_p^{A,B}, d_p^{A,B}, i_p^{A,B})$ *be the opinion such that*

$$\begin{aligned}
&(1)\; b_p^{A,B} = (i_p^A b_p^B + b_p^A i_p^B + b_p^A b_p^B)/(1-\kappa) \\
&(2)\; d_p^{A,B} = (i_p^A d_p^B + d_p^A i_p^B + d_p^A d_p^B)/(1-\kappa) \\
&(3)\; i_p^{A,B} = (i_p^A i_p^B)/(1-\kappa)
\end{aligned}$$

where $\kappa = b_p^A d_p^B + d_p^A b_p^B$ *such that* $\kappa \neq 1$. *Then* $\pi_p^{A,B}$ *is called a parallel combination of* π_p^A *and* π_p^B, *representing an imaginary agent* $[A, B]$*'s opinion about* p, *as if she represented both* A *and* B. *By using the symbol* $\oplus$ *to designate this operation, we get* $\pi_p^{A,B} = \pi_p^A \oplus \pi_p^B$.

It is easy to prove that $\oplus$ is both commutative and associative which means that the order in which opinions are combined has no importance. Opinion independence must be assumed, which obviously translates into not allowing an entity's opinion to be counted more than once

Without going into detail, it can be mentioned that Dempster's rule has been criticised, especially for the way it deals with highly conflicting opinions(see e.g. [Zad84]).

3.4 Serial Combination of Opinions

Assume two agents A and B where A has an opinion about B, and B has an opinion about a proposition p. A serial combination of these two opinions consists of combining A's opinion about B with B's opinion about p in order for A to get an opinion about p. This can be described as a recommendation of opinions. Def.6 describes one possible way of doing this.

Definition 6. *Let A, B and be two agents where $\pi_B^A = (b_B^A, d_B^A, i_B^A)$ is A's opinion about B's recommendations, and let p be a proposition where $\pi_p^B = (b_p^B, d_p^B, i_p^B)$ is B's opinion about p expressed in a recommendation to A. Let $\pi_p^{AB} = (b_p^{AB}, d_p^{AB}, i_p^{AB})$ be the opinion such that*

$$
\begin{aligned}
(1)\ b_p^{AB} &= b_B^A b_p^B, \\
(2)\ d_p^{AB} &= d_B^A + b_B^A d_p^B \\
(3)\ i_p^{AB} &= i_B^A + b_B^A i_p^B
\end{aligned}
$$

then π_p^{AB} is called a serial combination of π_B^A and π_p^B expressing A's opinion about p as a result of the recommendation from B. By using the symbol $\otimes$ to designate this operation, we get $\pi_p^{AB} = \pi_B^A \otimes \pi_p^B$.

This operation is associative but not commutative, and opinion independence must be assumed in a chain with more than one recommending entity.

B's recommendation must be interpreted as what B actually recommends to A, and *not* as B's real opinion. It is obvious that these can be totally different if B for example defects.

It must be assumed that the trust relationships are transitive. More precisely it must be assumed that the entities in the chain do not change their behaviour (i.e. cooperate or defect) as a function of which entities they interact with. It has however been pointed out [BFL96, Jøs96] that trust is not necessarily transitive, as defection can be motivated for example by antagonism between certain entities. The serial combination rule must therefore be used with care, and can only be applied in environments where behaviour invariance can be assumed.

3.5 Opinion Thresholds

An opinion threshold is simply a condition on an opinion for it to be acceptable in a particular situation. For example, it can be specified that a proposition can only be accepted if the opinion about it is such that b, d and i satisfy certain conditions. A threshold can for example be defined as

$$\pi : [(b \geq c_b) \wedge (d < c_d)]$$

where c_b and c_d are constants from the interval [0,1].

An opinion threshold will directly reflect the opinion policy in a particular application, and can be set individually by each agent.

3.6 Mixing Parallel and Serial Combinations of Opinions

Fig.2 illustrates an example of mixed parallel and serial combination of opinions.

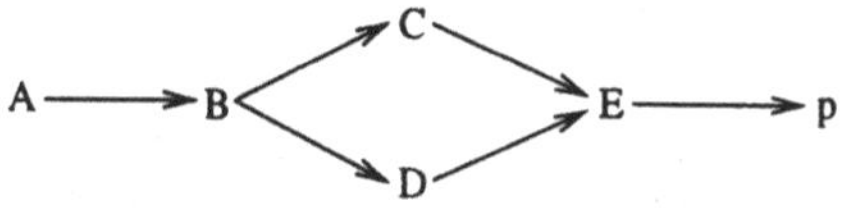

Fig. 2. Mixing parallel and serial combinations of opinions.

The serial combination rule is not distributive relative to the parallel combination rule, because that would violate the requirement of opinion independence.

Let π_B^A, π_C^B, π_D^B, π_E^C, π_E^D and π_p^E represent the trust relationships in Fig.2. We then have

$$\pi_B^A \otimes ((\pi_C^B \otimes \pi_E^C) \oplus (\pi_D^B \otimes \pi_E^D)) \otimes \pi_p^E \neq (\pi_B^A \otimes \pi_C^B \otimes \pi_E^C \otimes \pi_p^E) \oplus (\pi_B^A \otimes \pi_D^B \otimes \pi_E^D \otimes \pi_p^E) \quad (2)$$

which according to the notation in Defs.5 and 6 can be written as

$$\pi_p^{A(BC,BD)E} \neq \pi_p^{ABCE,ABDE} \quad (3)$$

The inequality may seem counterintuitive at first, but the right sides of Eqs.2 and 3 violate the requirement of independent opinions because they include two cases of dependent opinions, represented by π_B^A and π_p^E. Only the left sides of Eqs.2 and 3 thus represent correct mixing of parallel and serial combination of opinions.

4 Example: Key Authentication in PGP

4.1 Brief Introduction to PGP

PGP (Pretty Good Privacy) [Zim95] is a software tool which provides confidentiality and authentication service that can be used for electronic mail and file storage applications. A crucial component of PGP is the usage of public key cryptography. We will not go into detail on how PGP works, but concentrate on the problem of key distribution and authentication. According to the PGP documentation, "*protecting public keys from tampering is the single most difficult problem in practical public key applications. It is the "Achilles heel" of public key cryptography, and a lot of software complexity is tied up in solving this one problem*" [Zim95]. We will show how our framework can be applied to enhance the efficiency and flexibility of public key distribution.

Basically, keys can be exchanged manually or electronically. For manual distribution, person A can for example meet person B physically and give him a diskette containing her public key k_A, and B can give his public key k_B to her in return. The keys can then be considered authenticated through the persons' physical recognition of each other. These keys can then be trusted and used for confidential message exchange, or for certification of other keys, as will be explained below.

For electronic key distribution, keys need to be certified by someone whom the recipient trusts for certifying keys, and who's authenticated public key the recipient possesses. For example if A possesses B's public key k_B and B possesses C's public key k_C, then B can send C's public key to A, signed by his private key k_B^{-1}. Upon reception, A will verify B's signature, and if correct, will know that the received public key of C is authentic, and can then communicate confidentially with C.

The key authentication scheme of PGP can be referred to as *anarchic*, because there exists no hierarchy of agents and no defined roles. The trust policy is based on mutual trust between pairs of agents and can be summarised as follows:

- Each agent decides individually which other agents she will trust to produce certificates. The trust level can be specified as *undefined trust*, *distrusted*, *marginally trusted*, or *completely trusted.*
- For a correct ownership of a public key to be trusted, it can be specified how many certificates of *marginally trusted* and *completely trusted* agents are needed. Alternatively, the agent can chose to certify the public key herself, in which case no other certificates are needed.

4.2 Implementing the Trust Policy

The trust policy and key distribution scheme of PGP can be made more general and more flexible by using the framework presented here. Trust can be specified as an opinion instead of simply *marginally trusted* or *completely trusted.* Instead of specifying how many certificates of each type is needed, the different opinions can be combined using the parallel combination rule, and an opinion threshold can be used to decide whether the certified key can be accepted.

Fig.3 illustrates a possible structure of public keys and their certificates possessed by A. This structure can be maintained as a relational database.

In addition to this structure, A maintains a table of opinions about the authenticity of the public keys and opinions about their owners' trustworthiness regarding certification and recommendation, like for example Tab.1 below. Although it is not shown, a one-to-many binding between an agent and her different keys can perfectly well be accommodated within this structure.

It can be assumed that A knows B, C, D, E and F personally and therefore has first-hand evidence about their trustworthiness regarding certification and recommendation, so that the corresponding opinions have been determined on an intuitive basis by A. It can also be assumed that A has physically exchanged public keys with them, except for B who's key A might have received electronically but without certificate. This scenario is reflected by the respective opinions about the keys' authenticity in the third column of Tab.1.

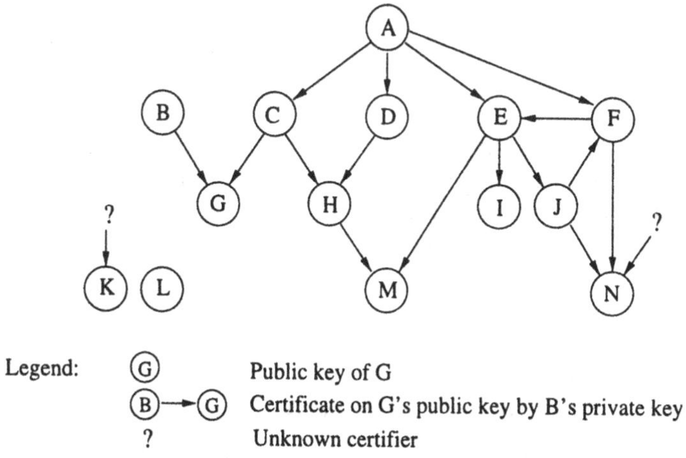

Fig. 3. Structure of public keys and their certificates.

Key k_X	Key owner X	Key Authenticity $\pi^A_{\mathrm{KA}(k_X)}$	Certification Trustworthiness $\pi^A_{\mathrm{CT}(X)}$	Recommendation Trustworthiness $\pi^A_{\mathrm{RT}(X)}$
k_A	A	(1.00, 0.00, 0.00)	(1.00, 0.00, 0.00)	(1.00, 0.00, 0.00)
k_B	B	(0.00, 0.00, 1,00)	(0.80, 0.00, 0.20)	(0.50, 0.00, 0.50)
k_C	C	(0.99, 0.00, 0.01)	(0.80, 0.00, 0.20)	(0.00, 0.70, 0.30)
k_D	D	(0.99, 0.00, 0.01)	(0.80, 0.00, 0.20)	(0.90, 0.00, 0.10)
k_E	E	(0.99, 0.00, 0.01)	(0.60, 0.00, 0.40)	(0.60, 0.20, 0.20)
k_F	F	(0.99, 0.00, 0.01)	(0.90, 0.00, 0.10)	(0.95, 0.00, 0.05)

Table 1. Table of A's opinions about public keys and their owners

A key which is received electronically, can be considered authentic if it has been certified by someone who is considered trustworthy, and who's public key is considered authentic. There are of course other conditions, such as e.g. that the cryptographic algorithm can not be broken, but it will be assumed that these conditions are met. Let us define the propositions:

$\mathrm{KA}(k_X)$: *"My copy of the agent X's public key is authentic"*
$\mathrm{CT}(X)$: *"I trust agent X to certify other public keys"*
$\mathrm{KR}(\{k_Y\}_{k_X^{-1}})$: *"I have received Y's public key certified by X's private key* k_X^{-1}*"*
$\mathrm{KA}(K_Y)$: *"My copy of agent Y's public key is authentic"*

and finally the proposition

$$q \equiv (\mathrm{KA}(k_X) \wedge \mathrm{CT}(X) \wedge \mathrm{KR}(\{k_Y\}_{k_X^{-1}})) \rightarrow \mathrm{KA}(k_Y)$$

which in normal language translates into q : "*If I trust agent X to certify other public keys, and my copy of agent X 's public key is authentic, and I have received Y 's public key certified by X 's private key, then my copy of Y 's public key is authentic*".

Anyone who believes in the conditional implication q will have an opinion π_q about its validity. Whenever a certified key of an agent Y is received, and the recipient has an opinion about $\mathrm{KA}(k_X)$ and $\mathrm{CT}(X)$ regarding the certifier X, the recipient can form an opinion about $\mathrm{KA}(k_Y)$ by using conditional implication as defined in Sec.3.2.

$$\pi^A_{\mathrm{KA}(k_Y)/(\mathrm{KA}(k_X)\wedge \mathrm{CT}(X)\wedge \mathrm{KR}(\{k_Y\}_{k_X^{-1}}))} = (\pi^A_{\mathrm{KA}(k_X)} \wedge \pi^A_{\mathrm{CT}(X)} \wedge \pi^A_{\mathrm{KR}(\{k_Y\}_{k_X^{-1}})}) \wedge \pi^A_q$$

4.3 Deriving Opinions about Key Authenticity

Let agent A's opinions $\pi^A_{\mathrm{KR}(\{k_Y\}_{k_X^{-1}})} = \pi^A_q = (1.00, 0.00, 0.00)$, which means that they can be omitted in the calculation. Using the opinion values from Tab.1, A can for example determine an opinion about the authenticity of G's public key

$$\pi^A_{\mathrm{KA}(k_G)} = \pi^A_{\mathrm{KA}(k_C)} \wedge \pi^A_{\mathrm{CT}(C)} = (0.79, 0.00, 0.21)$$

which can then be used to make a new entry for G in the table. Let A's requirement threshold for accepting a key to be used for enciphering messages be $b^A_{\mathrm{KA}(k_X)} \geq 0.80$. In this case, G's public key can not be accepted, and B's certificate gives no additional support because B's public key has not been certified. However, H's public key will be accepted, because it is certified by both C and D. The resulting opinion about the authenticity of H's public key is:

$$\pi^A_{\mathrm{KA}(k_H)} = (\pi^A_{\mathrm{KA}(k_C)} \wedge \pi^A_{\mathrm{CT}(C)}) \oplus (\pi^A_{\mathrm{KA}(k_D)} \wedge \pi^A_{\mathrm{CT}(D)}) = (0.96, 0.00, 0.04)$$

4.4 Deriving Opinions from Recommendations

At least two types of recommendation can be imagined.

1. Recommending other agents for key certification
2. Recommending other agents for further recommendation

Although different opinions about an agents trustworthiness regarding the two types of recommendation can be imagined, we have assumed them to be equal so that the opinions in column 5 of Tab.1 encompass both types.

According to Tab.1, A does not trust C to recommend other agents, but she trusts D. If D recommends trusting H for certification by sending to A the signed opinion $\pi^D_{\mathrm{CT}(H)} = (0.90, 0.00, 0.10)$, then A can take H's certificate into account together with E's certificate in order to form an opinion about the authenticity of M's public key. First A calculates $\pi^A_{\mathrm{CT}(H)}$ and subsequently $\pi^A_{\mathrm{KA}(k_M)}$:

$$\begin{aligned}\pi^A_{\mathrm{CT}(H)} &= (\pi^A_{\mathrm{KA}(k_D)} \wedge \pi^A_{\mathrm{RT}(D)}) \otimes \pi^D_{\mathrm{CT}(H)} = (0.80, 0.00, 0.20)\\ \pi^A_{\mathrm{KA}(k_M)} &= (\pi^A_{\mathrm{KA}(k_H)} \wedge \pi^A_{\mathrm{CT}(H)}) \oplus (\pi^A_{\mathrm{KA}(k_E)} \wedge \pi^A_{\mathrm{CT}(E)}) = (0.91, 0.00, 0.09)\end{aligned}$$

The new entries for the derived opinions are shown in Tab.2 below.

Key k_X	Key owner X	Key Authenticity $\pi^A_{\mathrm{KA}(k_X)}$	Certification Trustworthiness $\pi^A_{\mathrm{CT}(X)}$	Recommendation Trustworthiness $\pi^A_{\mathrm{RT}(X)}$
k_G	G	(0.79, 0.00, 0.21)	(0.00, 0.00, 1.00)	(0.00, 0.00, 1.00)
k_H	H	(0.96, 0.00, 0.04)	(0.80, 0.00, 0.20)	(0.00, 0.00, 1.00)
k_M	M	(0.91, 0.00, 0,09)	(0.00, 0.00, 1.00)	(0.00, 0.00, 1.00)

Table 2. New entries for derived opinions

5 Conclusion

The trust model presented in this paper is more complete than previously proposed models, and we believe that it is sufficient to capture the aspects of trust necessary to make a trust policy framework general and flexible. Presently the framework contains the most basic operations for manipulating opinions, and we believe that it can be directly applied to implement trust policies in practical security applications.

References

[Bal86] J.F. Baldwin. Support logic programming. In A.I Jones et al., editors, *Fuzzy Sets: Theory and Applications*, Dordrecht, 1986. Reidel.

[BFL96] Matt Blaze, Joan Feigenbaum, and Jack Lacy. Decentralized trust management. In *Proceedings of the 1996 IEEE Conference on Security and Privacy*, Oakland, CA, 1996.

[Jøs96] A. Jøsang. The right type of trust for distributed systems. In C. Meadows, editor, *Proc. of the 1996 New Security Paradigms Workshop*. ACM, 1996.

[Jøs97] A. Jøsang. Prospectives for modelling trust in information security. In Vijay Varadharajan, editor, *Proceedings of the 1997 Australasian Conference on Informatiojn Security and Privacy*. Springer-Verlag, 1997.

[LVDG91] P. Lucas and L. Van Der Gaag. *Principles of Expert Systems*. Addison-Wesley Publishing Company, 1991.

[Pra85] Henry Prade. A combinational approach to approximate and plausible reasoning with applications to expert systems. *IEEE Trans. on PAMI*, 7(3):260–283, 1985.

[RS97] Michael K. Reiter and Stuart G Stubblebin. Toward acceptable metrics of authentication. In *Proceedings of the 1997 IEEE Conference on Security and Privacy*, Oakland, CA, 1997.

[Sha76] G. Shafer. *A Mathematical Theory of Evidence*. Princeton University Press, 1976.

[Zad84] L.A. Zadeh. Review of Shafer's A mathematical theory of evidence. *AI Magazine*, 5:81–83, 1984.

[Zim95] P.R. Zimmermann. *The Official PGP User's Guide*. MIT Press, 1995.

Critical Analysis of Security in Voice Hiding Techniques*

LiWu Chang and Ira S. Moskowitz

Information Technology Division, Mail Code 5540
Center for High Assurance Computer Systems
Naval Research Laboratory
Washington, DC 20375 USA

Abstract. This paper provides a comparative assessment of detection in certain voice hiding techniques. The assessment is based on the complexity of breaking the stego key, the robustness of the hiding techniques, and the stego transmission rate. Unlike cryptography, to break the stego key in voice communication requires not only an extensive search but also estimation techniques for determining the values of parameters used in data embedding. We also consider disturbing embedded data in case resource constraints are imposed.

1 Introduction

A new challenge for secure voice communication comes from the rapidly growing techniques for hiding data inside a host message. The host message with added hidden data, which causes little or no perceivable degradation in the host message, can be transmitted via a variety of electronic media (e.g., telephone line, e-mail, network packages) as usual. Such techniques are named *information hiding* and have been an intensive subject of study (e.g., [1], [13]). More formally, information hiding refers to a collection of techniques that use an array of methods to communicate messages, embedded inside other signals (referred to as the cover signals), in such a way that these hidden messages do not cause significant perceivable change to the cover signal, may or may not be easily removed and are recoverable by using a designated code (i.e, *stego* key).

Hiding techniques have been found to be valuable in several applications (e.g. [5], [10], [11]). Information hiding complements traditional cryptography (e.g., [15] in the sense that it protects decrypted materials, enhances portability and offers real-time identification. In terms of integrated services (e.g., teleconferencing), information hiding can transmit useful data, such as visual or audio aids, in addition to the original signal. In this case, information hiding increases the utility of transmission media (e.g., telephone line). Nevertheless, those hiding techniques can also be used as instruments by intentional attackers to covertly communicate a confidential message. Conventional cryptography does not help

* Research supported by the Office of Naval Research.

to detect the existence of hidden messages. In the absence of proper security scrutiny, a sender can transmit messages by using various hiding techniques.

Our goal in this paper is to analyze the security issue in terms of techniques for hiding data inside voice signals (referred to here as voice hiding). Our main concern will be with *detection* rather than hiding. The voice signal is not necessarily transmitted via pre-recorded or real-time person to person conversation. With the advance of digitized media, a voice signal can be transmitted by digitized tape, audio files via the internet, and e-mail. Our goal is to begin laying a formal foundation for multimedia information security by analyzing voice communication transmitted over telephone lines.

In voice communication, the sender and the receiver may collude to exchange secret information by voice hiding. Even if the sender and the receiver do not knowingly collude, some information can still leak through by a Trojan horse. Using the traditional Alice-Bob scenario, it is Alice who sends a hidden message to Bob via communication channels monitored by Eve. Eve attempts to determine if there is illicit information transfer between Alice and Bob.

Detection differs from copyright verification in that, in the former, the original signal (i.e., the voice file) may be known only to the sender (i.e., Alice) but not to other agents, while in the latter, the agent at the receiver end (i.e., Bob) usually has a clean copy of the original host signal. In the absence of prior knowledge about the host signal, detection needs to be carried out by decrypting the stego key used for hiding. In this respect, the analysis of complexity of extracting hidden data with different hiding methods is essential. In case the extraction complexity is too high and time constraints are imposed, the strategy for Eve may be to disturb the hidden data, instead of extracting it. It is also important to know the amount of hidden data that may be compromised through the channels. This quantitative measure may indicate how difficult it can be to detect the hidden data in a statistical sense. We shall show that in voice communication, to determine the stego key, the critical factor may not only be the length of the key, but also the criteria (referred to as *discriminating functions*) used in searching for the desirable set of parameter values.

This paper provides a comparative assessment of voice hiding techniques. The assessment is based on the complexity of extraction, the robustness of the hiding techniques and the transmission rate of hidden data. Notations used in this paper are given in the appendix.

2 BACKGROUND

In Figure 1, we show one possible realization of voice hiding via communication channels. We refer [13] to the original input voice file as the cover voice signal, the data to be embedded as hidden data, and the output of the hiding process as the stego signal. Since the hidden data can be either text or voice and both may be digitized, we assume, without loss of generality, that the hidden data is in the form of a binary string. The (stego) *key* refers to the set of parameters

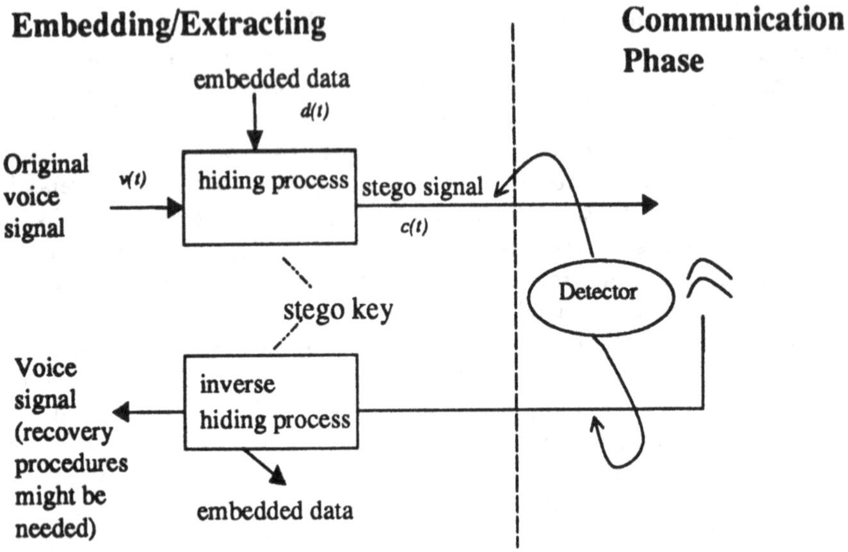

Figure 1: Diagram for hiding data in voice signal

used for embedding and extraction. In this scheme, a shared secret key protocol is in use.

In our current implementation, communication modes include either ~3 kHz bandwidth analog transmission, 64 kbits/s ISDN high data rate transmission, 16 kbits/s medium data rate, or 2.4kbits/s low data rate digital transmission via High-Frequency channels or telephone line. High data rate communication can transmit either sample-by-sample, un-compressed, or frame-by-frame [10] compressed digital data.

3 METHODS FOR HIDING DATA IN VOICE SIGNAL

We consider low bit coding, phase coding, spread spectrum-based coding and echo hiding as the initial set of voice hiding techniques. Implementation details are given in [4] (also see [2]).

3.1 Low-bit coding

Bits in a discretized sample are not equally significant from a perceptual perspective. This fact implies that low significant bits can be used for hiding (e.g., [9]). Basically, the embedding in low-bit coding done by replacing the value of the low bits of the cover signal by the value of bits of the binary hidden data. In a 16-bits per sample representation, the last four bits can be used for hiding. The retrieval of the hidden data in low-bit coding is done by reading out the value from the low bit(s).[2] The stego key is the position of altered bit(s). Low-bit coding can be applied in all ranges of transmission rates with digital communication modes.

3.2 Phase coding

Early studies (e.g., [8], [9]) indicate that listeners might not hear any difference caused by a smooth phase shift, even though the signal pattern may change dramatically.

In phase coding, a hidden datum is represented by a particular phase or phase change in the phase spectral. If we split the voice signal (e.g., vowel sounds) into segments (e.g., pitch segments),[3] data is usually hidden only in the first segment under two conditions. First, the phase difference between two consecutive segments (i.e, the relative phase) needs to be preserved. $\Phi_i(f)$ is the phase spectral of the ith segment and f is defined over the whole bandwidth. The second condition states that the final phase spectral with embedded data needs to be smoothed; otherwise, an abrupt phase change causes hearing awareness (e.g., flutter). Let the minimum spacing of imperceptible phase shifts be denoted as ps (in unit of frequency). Once the embedding procedure is finished, the last step is to update the phase spectral of each of the remaining segments by adding back the relative phase. Consequently, the stego signal, $c(t)$, can be constructed from this set of new phase spectral.

For the extraction process, the hidden data can be obtained by detecting the phase values from the phase spectral of the first segment. The stego key in this implementation includes the phase shift sequence and the size of one segment. Phase coding can be used in both analog and digital modes.

3.3 Spread spectrum-based coding

Because the hidden data is usually not expected to be destroyed by operations such as compressing and cropping, broadband spread spectrum-based techniques, which make small modifications to a large number of bits for each hidden datum, are expected to be robust against the above operations [5]. Among several variations, direct sequence spread spectrum [6] is currently considered. In general, spreading is accomplished by modulating the original signal with a sequence of random binary pulses $w(t)$ (referred to as *chip*) with values 1 and -1. The chip rate is an integer multiple of the data rate. The bandwidth expansion is typically of the order of 100 and higher [14].

For the embedding process, the stego signal $c(t)$ is given by the addition of spread hidden data $\alpha \times d(t) \times w(t)$ with the cover voice signal $v(t)$,

$$c(t) = \alpha \times d(t) \times w(t) + v(t) \tag{1}$$

where α is an attenuation factor used to reduce the noise level induced by spreading the hidden data. For extraction the same binary pseudo-random pulses applied for the embedding will be synchronously (in phase) multiplied with the stego signal

[2] In our setup with speech, hidden data is encoded in coefficients of the Linear Predictive Coding (LPC) filter.

[3] In this paper, each segment corresponds to a pitch period.

$$c(t) \times w(t) = \alpha \times d(t) \times w(t) \times w(t) + v(t) \times w(t) \quad (2)$$

(note: $w(t) \times w(t)$ is the unity function). Since $v(t) \times w(t)$ is insignificant with respect to the attenuated $d(t)$, we can recover $d(t)$. Spread spectrum-based coding requires a high bit per second rate. The stego-key is just the pseudo-random binary pulses.

3.4 Echo Hiding

Echo hiding embeds data by introducing an echo [7]. The value of a hidden datum corresponds to the time delay of the echo, with, for example, Δd for 1 and $\Delta d'$ for 0. The echo delays are selected to be less than the detectable hearing limit [12].

For the embedding process, the voice signal is divided into segments and one echo is planted in each segment. In a simple case, the stego signal can, for example, be expressed as follows:

$$c(t) = v(t) + v(t - \Delta d) = v(t) \otimes (\delta(t) + \delta(t - \Delta d)) \quad (3)$$

The stego key is the two echo time delays, of Δd and $\Delta d'$. For extraction, we need to recover the value of Δd and $\Delta d'$. The extraction is based on the autocorrelation of the ceptrum (i.e., $\log F(c(t))$) [3] of the stego signal [7]. The result in the time domain is $F^{-1}(\log(F(c(t))^2)$. The decision of a Δd or a $\Delta d'$ delay can be made by examining the position of a spike that appears in the autocorrelation diagram.

4 COMPARATIVE ANALYSIS

The comparison invokes an information-theoretical criteria, which includes the transmission rate of hidden data, the vulnerability of the stego key, and the robustness of the voice hiding technique. The analysis of the information transmission rate, R (bits/sec), can be used to provide an understanding of how serious the threat of the hiding technique may be. We analyze the degree of complexity required to crack the key.

4.1 Information-Transmission Rate Analysis

Let the amount of hidden data transmitted in a second (i.e, the information transmission rate) be R. In the noiseless case, the maximal measure of R is evaluated by the measure of channel capacity, C_{nf}. Suppose the amplitude of a sample is described by M discrete levels and M is represented by L ($= \log_2 M$) bits. With a signal bandwidth W and a sampling rate of $2W$, $C_{nf} = 2WL$. C_{nf} gives an upper band for the transmission rate when there is noise. It is a useful worst-case metric. Thus, we will discuss the transmission rate of hidden messages computed in the noise-free case. In particular, we compute the idealized maximal number of bits of hidden data transmitted given one use of the communication channel (referred to here as *channel utility*). A signal can be transmitted via different communication modes and hence, the data transmission rate of each mode in terms of bits per second varies accordingly. Channel utility is basically concerned only with the coding technique and thus is independent of transmission modes in use.

Low-bit coding In low-bit coding any sample in the cover voice signal can be modified, so that the rate of low-bit coding should be in proportion to the data sampling rate of the cover signal. Suppose the cover voice signal has bandwidth W_v (Hz). Then, for sampling at the rate of $2W_v$ samples per second, the maximal information rate is $2LW_v$. If the last b significant bits are used for embedding, then the throughput of the hidden data is $2LW_v \times (b/L) = 2bW_v$. Thus, for each use of channel, there are $2bW_v/2LW_v = b/L$ bits of hidden message sent out. In case b=1 (bit) and L=8 (bits), the channel utility is 0.125 (this is the common eight bit coding).

Phase coding The constraint on phase coding is the spacing, ps (Hz), required between two consecutive phase shifts. Suppose only the first segment is used for an embedding. For an input voice signal with bandwidth W_v, the maximum number of phase changes is W_v/ps, assuming that the band-width of the first segment is close to W_v.

Suppose B samples are taken from one segment and on the average the data hiding occurs every S segments. Then, for the amount of $B \times S$ samples transmitted, there are W_v/ps hiding bits sent out. Thus, for every channel use (i.e., one bit), there are $(W_v/ps)/(S \times B \times L)$ hidden bits transmitted. In the case of embedding data in the first pitch period of a vowel of a male voice, the channel utility with $W_v/ps = 16$ (hidden bits), L=8 (bits), S=100 (segments) and B=128 (samples), is 0.00015625. The choice of B=128 samples is close to what human ears can hear at a time.

Spread spectrum coding Let T_d denote the time duration of hidden binary string and T_w the duration of the spreading code. Also, let the bandwidth of the hidden signal and the bandwidth of the spreading code be estimated by $2/T_d$ and $2/T_w$, respectively. In this case, the bandwidth of the spread signal is approximately $2/T_w+2/T_d$. Since T_w is much smaller than T_d (e.g., $T_d/T_w= 30$ for 64 kbits/s transmission rate), the value of $2/T_w$ dominates the bandwidth of the spread signal. The transmission of the same voice signal now takes $(2 \times (2/T_w))/(2W_v)$ more samples, with W_v the bandwidth of the voice. As a result, it takes $(2 \times (2/T_w))/(2W_v)$ more times of channel usage to transmit the same voice signal. Hence, the channel utility drops by the factor of $(2\times(2/T_w))/(2W_v)$. Suppose the hidden data is also a voice signal. Thus, we may assume $W_d \approx W_v$. In this case, the value of $(2 \times (2/T_w))/(2W_v)$ is given by the ratio T_d/T_w. For $T_d/T_w = 30$, the channel utility is around 0.033.

Echo In the case of echo hiding, one echo is introduced to one segment and hence, the rate of transmission is in proportion to the length of the segment. Suppose one segment contains B samples. Then the utility of the channel is $1/B$. For B=128 samples, $1/B$ yields 0.0078125. To avoid perceivable new sounds, the occurrence of an echo might be much slower than this figure. The amount of 128 samples is about the limit that human ears can handle each time.

4.2 Complexity Analysis

In cryptography it is basically a "yes" or "no" proposition to decide whether a given set of codes matches the secret key. In voice communication, the key refers to parameters (e.g., temporal, spatial) involved in embedding. Hidden data may remain "clear", rather than be transformed. For detection, various values of parameters will be tested against the stego key. However, due to the uncertain nature of signal operations, the judgment of acceptance of a given set of parameters is no longer a binary decision; it can be multiple-valued. The true parameter values often can only be approximated through estimation. In this aspect, in addition to the length of the key, extraction also relies on selected discriminating functions.

We shall analyze the complexity of extracting embedded messages based on different discriminating functions for different coding techniques in the context of a brute force search. It is reasonable to assume that monitoring takes place at any point of the channel. We assume that the modulation schemes are known a priori. This assumption makes sense since communication media are likely to be standardized. We consider detection with stego signal only.

Low-bit coding For extraction one simply copies the least few bits to get the hidden message out. However, the exact low bits for hiding need to be determined. Procedures to determine what low-bits are used may proceed as follows:

I. Use the stego signal as a reference. Starting with the most significant bit, we replace this bit for all samples in the reference with a 0 and then a 1.

II. Evaluate the quality degradation of the modified signal and restore the value of the replaced bit.

III. Repeat replacement with the next significant bits until no degradation can be detected. Bits that are below the cutoff bit are considered the hidden bits.

The discriminating function, $D(.)$, here is just the quality of the sound. The evaluation of quality degradation is given by using the basic dimensions of sound. The replacement procedure needs to be repeated L times, with L being the number of bits in a sample (or symbol). This approach is able to reduce the search space. However, we are still not able to tell whether selected bits are all the hidden bits. The situation is worse in frame representation. Not all low-bits in frame representation are used for data, some are reserved for protection, synchronization or error-correction. It is unclear whether there are effective methods to differentiate data and designated bits.

Phase coding Recall that in the proposed approach only the phase spectral of the first pitch segment carries hidden messages. Phase spectrals of the remaining segments are obtained from adding back differential phases between two consecutive segments. For extraction, the phase spectral needs to be reconstructed. Without prior knowledge, it is necessary to examine every possible size of a segment varying from 0 to a maximum of K bits. Suppose the number of frequency elements in the DFT is equal to the number of samples in a segment (i.e., the size of a segment). For a given size of segment $F (F \leq K)$, we want to measure how likely this F is the true size used by Alice.

Let $\delta\Phi_{l,l-1}(f)$ denote the relative phase between the lth and the $(l-1)$th segments. Also, let c_l^F be the integration of the absolute value of relative phases, $\int |\delta\Phi_{l,l-1}(f)| df$, where the range of the integration is given by the frequency spectral of the voice signal. The discriminating function $D(F)$ is defined to be the summation of all c_l^Fs,

$$D(F) = \sum_{i=1}^{L_F} c_i^F \tag{4}$$

where L_F is the number of segments under a F. The desired F is the one that minimizes the $D(F)$. When values of $D(F)$ obtained form different Fs' are equal, the smallest F will be selected. With $D(F)$, we can evaluate and compare the result of the brute force search. Detection proceeds as follows:

I. For $F \leq K$ and every size of DFT, s, $1 \leq s \leq F$, compute DFT to obtain phase spectral.
II. Compute $D(F)$.
III. Store the minimum value of $D(F)$.

The search space is spanned by the number of samples taken, the dimension of DFT and the total number of segments, L_F, i.e., $K \times K \times L_F$. Let k be 128 as before. We choose L_F to be 100 segments. The result will be in the range of 2^{21}.

Spread spectrum coding In direct sequence spread spectrum-based coding, critical components of the stego key are the binary pseudo-random pulses (i.e., the spreading code) and the starting phase of the pulse sequence. For extraction, it is necessary to use a regenerated binary pseudo-random sequence (i.e., chip) as close to the one used for embedding as possible.

If two pseudo-random sequences are out of synchronization by as little as one chip (i.e, one pulse), reliable data detection will not be reached. Conventionally, synchronization involves an initial code acquisition phase and the code tracking phase. Phase tracking can be accomplished by using phase lock loop. We consider the complexity of the initial code acquisition under a brute force search. If the sender (i.e., Alice) sends out N chips for clock synchronization, then we evaluate the similarity between the re-generated pseudo-random sequences and the original one based on the magnitude of their correlation as in existing communication systems. However, if no chip is sent for synchronization, code acquisition proceeds as follows. We search the time uncertainty interval in the T_w/k time step. For each trial, the re-generated sequence, $w'(t)$, is multiplied to the stego for the sake of despreading as follows:

$$w'(t) \times c(t) = w'(t) \times w(t) \times d(t) + w'(t) \times v(t) \tag{5}$$

Let α=1 for simplicity. The discriminating function, $D(f)$, is defined in terms of the power spectral density of $S_{w \times c}(f)$, the Fourier transform of the autocorrelation of $w'(t) \times c(t)$. When $w'(t)$ and $w(t)$ are identical and synchronized, $D(f)$ will reach its peak value. For an incorrect or inappropriate chip $w'(t)$, one will receive much smaller value of $D(f)$. Thus, we search for $w'(t)$ that maximizes $D(f)$ and the maximum occurs when $w'(t) = w(t)$. Detection involves the following steps:

I. Re-generate the pseudo-random sequence for the first N chips.
II. Apply phase tracking once the initial phase acquisition is completed.
III. Search for the $w'(t)$ that maximizes $D(f)$.

Suppose the time interval of the initial uncertainty is T_u. The complexity will be determined in terms of the number of steps during T_u, the bandwidth expansion ratio of T_d/T_w and the number of initial N chips, i.e.,

$$\frac{T_u}{\frac{1}{k}T_w} \times \frac{T_d}{T_w} \times 2^N \tag{6}$$

Suppose k=2, T_d/T_w=30, N=32 and $T_u = T_w$. The result is about 2^{38}. The window size N is critical to the accuracy as well as to the complexity.

This discriminating functions makes correct decisions when the chip period $w(t)$ is close to maximal length (i.e., long period). But, if chip period is too short, we do not get good results. However, for a short chip period, the hidden data cannot be reliably retrieved at the receiver end and should not be used by the sender.

Echo Detection in echo coding is deterministic in that the decision about the position where a hidden datum is placed can be made by taking the autocorrelation of the *ceptrum* of the encoded signal as shown in the previous section. A "1" or a "0" occurs at the peak value of the magnitude of auto-correlation. There is no key search involved.

4.3 Perturbation Strategy

To prevent data from being covertly transmitted, it may be more effective just to disturb the hidden message rather than to break the key, if a time constraint is imposed. Ideally, perturbation should affect only the hidden data. In this section, we examine three types of disturbance including noise, filtering and re-sampling. Re-sampling adjusts the sampling rate either slower or faster. It is used only if the Nyquist criterion, with respect to the cover voice signal, is not violated. For detection, filtering might be useful to separate the original voice signal from hidden data. Noise in the current scenario refers to a relatively broader band random signal. Disturbance is assumed to be added to the stego signal. We discuss the robustness of these coding techniques. We consider the disturbance added prior to the transmission. Perturbation is, at present, an open issue.

Low-bit coding •*noise* – Because the magnitude of noise may well exceed that of the hidden data, low-bit coding is susceptible to noise. This point can be shown by examining the power density of the hidden data. Let the input signal be quantized into M levels, each spaced by an increment with quantization level distributed at $(2m-1)/2$, where $m \leq M/2$. Assume that all levels are equiprobable. The mean-square (power) voice signal, P_v, can be computed as

$$P_v = \int_{-\infty}^{\infty} v^2 p(v)dv = \frac{2}{M}\left[\sum_{m=1}^{M/2}\left(\frac{2m-1}{2}\right)^2 a^2\right] = \frac{M_2-1}{12}a^2 \tag{7}$$

By the same token, the mean-square of hidden data, P_d, can be calculated to be 2.5, by taking the first two terms of (EQ 7). The ratio between the two mean-square in dB, $10\log(P_v/P_d)$, is around 34 dB, with an 8-bit representation or M=256. Therefore, to perturb the hidden data, noise (e.g., white) with amplitude as low as P_v/N (dB) = 30 will do. This noise level is below the

PCM signal-to-noise transmission requirement, which is 20 dB, and hence, has no impact on the original voice signal.

The impact of noise becomes less significant if it is introduced during transmission. As matter of a fact, voice encoded sample by sample (for example, PCM) is tolerant to bit errors because they affect only one sample time period (125 micro-sec if the sample frequency is 8 kHz) which is very brief. If speech is encoded frame by frame, an error in any parameter affects for one frame period (around 20 millisecond) which is rather long. Many low-bit-rate encoders use error protection for certain perceptually sensitive parameters. Thus, encoded voice is intelligible even under bit error rates as much as 1% to 2%.

•*filtering*— We use a (low-pass) filter to show the effect of filtering on low bits. One such filter is the averaging operator which averages values of nearby samples. Suppose averaging takes two terms, i.e., $y(n) = [c(n) + c(n-1)]/2$. Although filtering alters values of the input signal, the change may or may not affect the lower bits. This is also true for other types of filters. Therefore, in general filtering can perturb low bits, but its effectiveness is not known.

•*re-sampling* — In case the hidden data is embedded in samples, re-sampling (e.g., sampling at lower rate) affects the low-bit coding.

Phase coding •*noise* — In the noisy environment, since the shape of the signal changes, the phase spectral will vary accordingly. The phase spectral of $c(t)+n(t)$ is expressed as

$$\Phi[F(c(t)+n(t))] \cong \arctan \frac{Im(C(f)) + Im(F(\overline{n(t)}))}{Re(C(f)) + Re(F(\overline{n(t)}))} \quad (8)$$

where $\overline{n(t)}$ is the mean value of $n(t)$, i.e. $\frac{1}{T}\int_{-\frac{T}{2}}^{\frac{T}{2}} n(t)dt$. T is the observation time and is assumed to coincide with the sampling interval in the current context.

The sensitivity of the arc-tan function depends on the signal to noise ratio and the phase with which hidden data is encoded. However, at $\pm\pi/2$ even the power density of the noise is much smaller than that of the stego signal, there will still be a large distortion of the phase spectral of the stego, and hence, hidden data could be destroyed. In general, noise (e.g., Gaussian white, shot) affects the phase spectral of the stego.

•*filtering*— Both linear or non-linear filters introduces new phase terms. Take the averaging operation shown in the low-bit coding as an example. We compute the phase spectral. Applying the Z-transform to the averaging operation and substituting Z with $\exp(j)$, the frequency response of this filter, $H(f) = Y(f)/C(f)$, can be expressed as follows:

$$|H(f)| = \frac{1}{2}[1 + \cos(2\pi fT)], \quad \Phi[H(f)] = -2\pi fT/2 \quad (9)$$

where T is the sample interval. The phase spectral is a linear function of f. Adding this linear phase term to the phase spectral of the stego, phase values of higher frequency might either be altered or have its sign inverted. Therefore, filtering has effects upon phase coding.

•*re-sampling* — As long as the Nyquist criterion is preserved, there will be no deterioration to the phase spectral of the cover voice signal.

Spread spectrum coding •*noise* — Spreading reduces the magnitude of the frequency spectral of hidden data. Hence, the spread data becomes more sensitive to ambient noise. The reduction of the magnitude of hidden data in frequency is in proportion to the ratio of chip vs. data rates (T_d/T_w); the larger this ratio, the worse the signal-to-noise ratio of hidden data will be. For ratio $(T_d/T_w) = 30$, the power spectral density of hidden data drops $10\log(30) \approx 14.7$ dB. Thus, noise may affect hidden data but cause little degradation to the cover signal. The impact of noise perturbation will become even more significant if the magnitude of the hidden data is attenuated by a factor α.

•*filtering*— A bandpass filter can be used to disturb hidden data in the spread spectrum coding. The bandpass filter can be designed to have bandwidth compatible to the bandwidth of the voice signal. It passes frequency components of the cover signal, but eliminates portions of the frequency components of the spread code that lie outside the range of the bandwidth. As a consequence, hidden messages will be disturbed.

• *re-sampling* — Since lowering the sampling rate might not affect the cover signal, but may affect the spread code, re-sampling will have an impact upon the hidden data.

Echo •*noise* — Recall that a filter with a number of equally spaced impulses is employed in echo generation. The result is that only the first impulse gets reinforced. The remaining impulses and the noise both decay. Hence, echo coding is relatively resistant to the ambient noise during embedding. The noise also does not have a large effect on extracted hidden data. This result lies in the use of autocorrelation of *ceptrum* in the extraction of hidden data. Autocorrelation can help to suppress the impact of added noise.

•*filtering*— In the frequency domain it can be shown that echo generates a new term $F[v(t)]\exp^{-j\Delta d 2\pi f}$ to the Fourier transform of the echoed (i.e., stego) signal. A filter that has a system function given by $1/(1-z^{-\Delta d})$ can be used to cancel echoes. Thus, filtering affects hidden data in echo hiding.

•*re-sampling* — Since the frequency spectrum with properly chosen time delays does not change after embedding echoes, re-sampling will have no effect on echo hiding.

5 DISCUSSION

We summarize our results in Table 1.The first row shows the maximal measure of the transmission rate of hidden bits, which is the utility times the data (cover) transmission rate (to facilitate the comparison, the transmission rates are calculated in terms of the high data rate transmission of 64 kbits, since this rate is good for all hiding techniques). For example, in low-bit coding the utility is 0.125 and the data rate is 64 kbits/sec. Low-bit coding is ranked the best. In terms of phase coding, the rate of hidden transmission depends on the number of phase changes that the original signal can tolerate. Since it utilizes only the first segment, the transmission rate is low. A continuous voice communication is assumed.

The key security in the second row describes how difficult it is to extract embedded messages. As our analysis indicates, the length of the key is not the only factor, discriminating functions are also crucial. The analysis is based on the simplified conditions that detection is operated with a stego signal and parameters associated with the communication channel (e.g., the frame structure, modulation, transmission mode) are known a priori. In low-bit coding, although only a limited number of bits are suitable for embedding, there is no effective way to tell their exact positions. Our analysis shows that even in the case of phase coding or spread spectrum coding, the complexity of exhaustive search is not very high. But this does not imply that it is easy for the person in the middle to extract covert messages embedded with these two coding techniques. Estimation is needed to figure out the correct values of parameters used in embedding. The fact is that the uncertainty associated with estimation makes the decision problem difficult. The complexity of spread spectrum coding was derived for initial phase acquisition. No key search is needed for echo coding.

Perceivable noise listed in line 3 shows whether embedding will cause perceptual differences. As we mentioned earlier, low-bit coding generates imperceptible disturbances. Phase coding is undetectable by human ears. Echo coding generates new sound. However, retarding the transmission rate (e.g., 64 bits/s) can make echo coding produce no objectionable distortion to the original voice signal [7]. Spread spectrum-based coding will introduce noise to the cover signal, where the noise is just the spread code of hidden data. To alleviate the noise problem in spread spectrum-based coding, one needs to either attenuate or split the spread code of the hidden data.

criteria	*style*			
	low-bit	phase	spread	echo
R kbits/s	8.0	0.01	2.0	0.5
key security	partial	yes	yes	no
perceivable noise	no	no	yes	mod
robustness -:				
(a) noise (white)	low	low	partial	ok
(b) filtering	partial	low	partial	low
(c) re-sampling	low	ok	partial	ok

Robustness in line 4 indicates the degree that a hiding technique can tolerate different types of disturbance. In the presence of (Gaussian white) noise, low-bit coding, spread spectrum coding and phase coding are all affected to different degrees. With proper selection of bandwidth gain ratio, spread spectrum coding is more resistent to noise than low-bit or phase coding. Echo coding is relatively robust in this category. Basically, all hiding techniques are affected by one or more types of filtering. In the case of low-bit coding, the impact of filtering is inconclusive. Phase coding is affected by filtering. Bandpass filtering has impact on spread spectrum coding. Echoes can be cancelled by using a adaptive recursive filter. As for re-sampling, both phase coding and echo hiding are robust to this operation.

The pros and cons of each technique should be considered from respective aspects. For a sender, if background noise is tolerable, spread spectrum coding and echo coding can be chosen; otherwise, phase coding is preferred. For an analog transmission mode, phase coding is the choice. One constraint of using spread spectrum is the need of high data rate transmission channels (e.g., 64 kbits/s or above).

Echo coding is robust to noise and does not require a high data transmission channel. The shortcoming of echo coding is its unsafe stego key. For applications with less security concerns (e.g., graphic aid), echo coding can be useful. In our models, we are considering making the time delay an interval-valued random variable. As a result, the stego key in echo hiding can be safer.

In terms of low-bit coding, a datum does not have to be embedded inside a low bit of a sample of the voice signal. It is possible to embed data inside other devices. For example, we have embedded data inside coefficients of LPC. Those coefficients and other information will then be transmitted over to the receiver end and used to synthesize the original input voice. Hidden data can be extracted from received coefficients. Filter coefficients are dynamically determined by the input voice signal. In this case, low-bit coding is more efficient and secure.

6 SUMMARY

In this paper, we analyzed the detection issue in voice hiding. The security of stego key in hiding differs from that in cryptography in that not only the length of the key matters, but also the parameters used and discriminating functions selected all have impact on the decision. We mentioned that various estimation techniques (e.g., synchronization) are required for recovering the stego key. In many cases, even if there is no noise, we can only obtain an approximation of the true values of the underlying parameters (e.g., phase) of the stego key under certain assumptions (e.g., slow phase variation). We also investigated the degree of complexity of detection at various spots during transmission.

Acknowlegement The authors would like to thank George Kang and Daniel Gruhl for their advice and helpful discussions.

References

1. Anderson, R. (1996) "Stretching the Limits of Steganography". In Proceedings of Information Hiding Workshop, University of Cambridge, pp. 39-48.
2. Bender, W., Gruhl, D., Morimoto, N. & Lu, A. (1996) "Techniques for data hiding". IBM Systems Journal, Vol 35, NOS 3&4, pp. 313-336.
3. Bogert, B., Healy, M. & Tukey, J. (1963) "The Quefrency Analysis of Time Series for Echoes". Proc. Symp. on Time Series Analysis, (ed. Rosenblatt, M.), pp. 209-243, Wiley
4. Chang, L. (1997) "Analysis of Hiding Data in Speech Signal". in preparation.
5. Cox, I., Kilian, J., Leighton, T. & Shamoon, T. (1995) "Secure Spread Spectrum Watermarking for Multimedia". NEC Research Institute, TR 95-10.
6. Dixon, R. (1994) *Spread Spectrum Systems with Commercial Applications.* John Wiley & Sons.

7. Gruhl, D., Lu, A. & Bender, W. (1996) "Echo Hiding". In Proceedings of Information Hiding Workshop, University of Cambridge, pp. 295-315.
8. Hall, D (1972) Journal of the Acoustical Society, 1972, Vol 51, pp. 1863-1871 & 1872-1881.
9. Flanagan, D. (1972) *Speech Analysis.* Springer-Verlag.
10. Kang, G. (1985) "Narrowband Integrated Voice Data System Based on the 2400b/s LPC". NRL TR 8942.
11. Meadows, C. and Moskowitz, I. (1996) "Covert Channels—A Context Based View". In Proceedings of Information Hiding Workshop, University of Cambridge, pp. 73-94.
12. Pierce, H. (1995) *The Science of Musical Sound.* Freeman.
13. Pfitzmann, B. (1996) "Information Hiding Terminology". In Proceedings of Information Hiding Workshop, University of Cambridge, pp. 347-350.
14. Proakis, J. (1996) *Digital Communication.* McGraw-Hill.
15. Schneier, B. (1996) *Applied Cryptography.* Wiley.

Appendix (Notations)

We adopt the following notations in this paper. Let
the input voice file be denoted $v(t)$,
the embedded data be $d(t)$,
the composite signal (i.e., stego) be $c(t)$, and
the noise be $n(t)$.

Let $F(x(t)) \equiv X(f)(= ReX(f) + jImX(f))$ be the Fourier or the Discrete Fourier transform of a signal $x(t)$. Let $|X(f)|$ denote the magnitude spectral and $\Phi[X(f)] = \arctan(ImX(f)/ReX(f))$ denote the phase spectral of $X(f)$. The symbol $\otimes$ denotes convolution. The Fourier transform of the auto-correlation function of $x(t)$,

$$R_x(s) = \int \tilde{x}(t)x(t+s)ds, \tag{10}$$

is called the power spectral density $S_x(f)$, where $\tilde{x}(t)$ is the complex conjugate of $x(t)$. The power of signal $x(t)$ is the integration of the power spectral density, i.e., $P_x = \int S_x(2\pi f)df$, $-\infty < f < \infty$.

For a linear filter with input signal, $x(t)$, and the system function (or the transient response of a system), $h(t)$, the output $y(t)$ is equal to the convolution of $x(t)$ and $h(t)$, i.e., $y(t) = x(t) \otimes h(t)$. The frequency response of $y(t)$ (i.e., $Y(f)$), is described by its magnitude relation $Y(f) = X(f)H(f)$ and its phase relation $\Phi[Y(f)] = \Phi[X(f)] + \Phi[H(f)]$.

Two Efficient RSA Multisignature Schemes

Sangjoon Park[1], Sangwoo Park[1], Kwangjo Kim[1], Dongho Won[2]

[1] #0710, ETRI, Yusong P.O.BOX 106, Taejon, 305-600, Korea
E-mail : {sjpark, psw, kkj}@dingo.etri.re.kr

[2] Dept. of Information Engineering, Sung-Kyun-Kwan Univ., 300 Chunchun-dong, Suwon, Kyunggi-do, 440-746, Korea
E-mail : dhwon@simsan.skku.ac.kr

Abstract. In this paper, we propose two efficient RSA multisignature schemes, one is an improved version of Okamoto's scheme [6] and the other is that of Kiesler-Harn's scheme [3]. The first one causes bit expansion in block size of a multisignature, but the bit length of the expansion is no more greater than the number of signers regardless of their RSA modulus. The second one has no bit expansion, in which all signers have a RSA modulus with the same bit size and the same most significant l bits pattern. An average number of the required exponentiations to obtain a multisignature is about $(1+\frac{1}{2^{l-1}})m$, where m denotes the number of signers. Futhermore, our schemes have no restriction in signing order and are claimed to be more efficient than Okamoto's scheme and Kiesler-Harn's scheme respectively.

1 Introduction

In 1978, Rivest, Shamir and Adleman proposed new type of public-key cryptosystem, so called "RSA cryptosystem", whose security is based on the difficulty of factoring a large integer [8]. The practical implementation of RSA cryptosystem for multiple operations of a given message causes bit expansion problem inherently. As early works to solve this problem, there are Kohnfelder's reblocking method[4] and Levine-Brewley's repeated exponentiation method[5].

Itakura and Nakamura first suggested a new notion of a multisignature scheme [2] in which multiple signers generate a digital signature for a given document. To solve the difficulty of bit expansion in a RSA multisignature, they allowed a signer to have a RSA modulus with a different bit size according to his position in a hierachical structure. Thus, the signing order is restricted.

On the other hand, Okamoto proposed a multisignature scheme with no restriction of the signing order [6]. In his scheme, if the length of intermediate signature exceeds a pre-determined threshold value, then the extra bits exceeding the threshold value are appended to a message. So, the length of expanded message depends on the number of signers and the bit size of each signer's RSA modulus.

Harn and Kiesler proposed two multisignature schemes with no bit expansion[1, 3]. In one of their schemes, based on Kohnfelder's method, the signing order is chosen according to the size of signers' public keys. The other scheme is based on

Levine and Brawley's re-encryption method. Even though their multisignature schemes have no bit expansion problem and the signing order is not restricted, all signers must have a modulus with the same size and the computational complexity of obtaining a multisignature is increased.

In this paper, we propose two efficient RSA multisignature schemes, one is an improved version of Okamoto's scheme [6] and the other is that of Kiesler-Harn's scheme [3]. The first one causes bit expansion in block size of a multisignature, but the bit length of the expansion is no more greater than the number of signers regardless of their RSA modulus. The second one has no bit expansion, in which all signers have a RSA modulus with the same bit size, and the same most significant l bits pattern. In this scheme, an average number of the required exponentiations to obtain a multisignature is about $(1+\frac{1}{2^{l-1}})m$, where m denotes the number of signers.

This paper is organized as follows : In Section 2, we propose new RSA multisignature schemes. In Section 3, we discuss the security of our proposed schemes. Finally, we state concluding remarks in Section 4.

2 Multisignature Schemes

In this section, we propose two efficient RSA multisignature schemes. The following notations are used in this section.

- U_i : one of m signers, $U_1, \ldots, U_m$.
- n_i : RSA modulus of U_i.
- (e_i, n_i) : public key of U_i, (d_i, n_i) : secret key of U_i $(e_i \cdot d_i = 1 \pmod{\phi(n_i)})$.
- $|n_i|$: bit size of n_i.
- $A||B$: concatenation of A and B
- $h(\cdot)$: a secure hash function

<u>Scheme 1</u>

First, we introduce a new reblocking method in which the size of an enciphering block varies with the size of a message block. Let n be a RSA modulus and e a public key with $\gcd(e, \phi(n)) = 1$. Assume an odd M with $0 < M < 2^l n$. Then, $\phi(2^l n) = 2^{l-1}\phi(n)$ and $\gcd(e, 2^{l-1}\phi(n)) = 1$. If $e \cdot d = 1 \pmod{2^{l-1}\phi(n)}$, then $M^{e \cdot d} = M \pmod{2^l n}$. So, l varies with the size of a message M and d varies with l. If $C = M^e \pmod{2^l n}$ and $e \cdot d_1 = 1 \pmod{2^{l-1}}$, then $C \pmod{2^l} = M^e \pmod{2^l}$ and $M \pmod{2^l} = C^{d_1} \pmod{2^l}$. Thus, the proposed reblocking method can't be directly used for enciphering M with large block size.

Now, we show that this new reblocking method can be applied to a multisignature scheme. First, each user computes l_i from n_i as followings.

$$l_i = \begin{cases} 1 & \text{if } i = 1 \text{ or } 2^{l_i - 1} n_{i-1} < 2n_i \\ 2^{l_i - 1} n_i < 2^{l_i - 1} n_{i-1} < 2^{l_i} n_i & \text{otherwise.} \end{cases}$$

The generation and verification of a multisignature is done as follows :

- Signing by $U_1 : S_1 = (2h(M)+1)^{d_1} \pmod{2n_1}$ and he sends a message M and S_1 to the next signer U_2.
- Signing by U_i $(i = 2,\ldots,m)$: $S_i = S_{i-1}^{d_i} \pmod{2^{l_i}n_i}$, where $e_i \cdot d_i = \pmod{2^{l_i-1} \cdot \phi(n_i)}$ and he sends M and S_i to the next signer.

Now, a receiver verifies S_m to be a multisignature of M by signers $U_1,\ldots,U_m$.

$$\begin{cases} S_{j-1} = S_j^{e_j} \pmod{2^{l_j}n_j} \ (j = m, m-1,\ldots,2) \\ 2h(M)+1 = S_1^{e_1} \pmod{2n_1} \end{cases}$$

If $e_{i+1} \cdot d' = 1 \pmod{2^{l_{i+1}-1}}$ and $e_i \cdot d'' = 1 \pmod{2^{l_i-1}}$, then $S_i = C_i^{d'} \pmod{2^{l_{i+1}}}$ and $S_i = S_{i-1}^{d''} \pmod{2^{l_i}}$. However, we can't obtain the most significant $|n_{i+1}|$ bits of S_i from C_i and the most significant $|n_i|$ bits of S_i from S_{i-1}.

If $L = \max(|n_1|, |n_2|,\ldots,|n_m|)$, then the bit length of the multisignature S_m is less than or equal to $L+m$. So, the length expanded by the proposed scheme is not greater than the number of signers. For example, if $|n_1| = |n_3| = |n_5| = 768$ and $|n_2| = |n_4| = |n_6| = 512$, then $|S_m| \leq 774$. So, the expanded bit length is 6. But, in this case, Okamoto'scheme has an expansion of 509 bits.

Scheme 2

Now, we propose another RSA multisignature scheme, which is a generalized version of Kiesler-Harn's scheme[3]. All users must choose a RSA modulus of the same number of bits - say m bits and the same most significant l bits pattern of all users' modulus must be the same. Let C be the l bits pattern which is pre-determined. Then the modulus of an user i can be represented as follows :

$$n_i = C \cdot 2^{k-l} + R_i (0 \leq R_i < 2^{k-l}). \tag{1}$$

Let $C \cdot 2^{k-l}$ be a threshold value u, and e_i and d_i be the RSA public key and secret key of user i, respectively. A multisignature by m signers is generated as follows :

- signer $U_1 : U_1$ generates a signature $S_1 = h(M)^{d_1} \pmod{n_1}$ for the original message M. If $S_1 \geq u$, he applies the repeated exponentiation technique to S_1 until $S_1 < u$ and sends M and S_1 to the second signer.
- signers U_i $(i = 2,\ldots,m)$: U_i computes a signature $S_i = S_{i-1}^{d_i} \pmod{n_i}$. If $S_i \geq u$ then he computes $S_i = S_i^{d_i} \pmod{n_i}$, repeatedly, until $S_i < u$. He sends M and S_i to the next signer.

The final signature S_m is the multisignature of M by the signers $U_1,\ldots,U_m$. Note that the signing order is independent of signers' public keys. To verify that S_m is the multisignature of M, the receiver also applies repeated exponentiation technique : For $i = m, m-1,\ldots,2$, he computes $S_{i-1} = S_i^{e_i} \pmod{n_i}$ and if $S_{i-1} \geq u$, then he repeates exponentiations $S_{i-1} = S_{i-1}^{e_i} \pmod{n_i}$ until $S_{i-1} < u$. Finally, the receiver confirms $h(M) \stackrel{?}{=} S_1^{e_1} \pmod{n_1}$.

Since each signer's modulus n_i is of the form as equation (1), the probability that a random number $x(0 \leq x < n_i)$ is less than $h{=}C{\cdot}2^{k-l}$ is greater than $1{-}2^{-l+1}$,

$$Pr[0{\leq}x{<}u|0{\leq}x{<}n_i] = \frac{C \cdot 2^{k-l}}{n_i} = 1{-}\frac{R_i}{n_i} > 1{-}\frac{2^{k-l}}{2^{k-1}} = 1{-}2^{-l+1}.$$

So, if l is sufficiently large, then the average number of exponentiations required for obtaining a multisignature is close to m. For example, if $l = 32$ and $m = 10$, the average number of exponentiations of Kiesler-Harn's scheme is $1.5{\times}10{=}15$, but that of our scheme is $(1 + 2^{-31}){\times}10 \approx 10$. Thus, our scheme is more efficient than Kiesler-Harn's scheme.

Now, to make our multisignature scheme practical, we propose a method for generating a RSA modulus [7] which is required for our multisignature scheme. First of all, the key management center opens the bit length of the modulus, k, and some(fixed) pattern of l bits, C, to all users. For the sake of convenience, we suppose k is even. Each user's RSA modulus n must be k bits long and its most significant l bits pattern must be C. And, we expect that n becomes the product of two primes p and q, where $p - 1$ and $q - 1$ have large prime factors. A RSA modulus for the multisignature is generated as follows :

Step 1 Generate a random number R of $k{-}l$ bits, and compute $N{=}C{\cdot}2^{k-l}{+}R$.

Step 2 Generate a random number P of $\frac{k}{2}$ bits, and two prime numbers p' and q' of $\frac{k}{2}{-}l{-}t$ bits. And, compute $s{=}\lfloor\frac{P}{2\cdot p'}\rfloor$.

Step 3 If $p{=}2{\cdot}p'{\cdot}s{+}1$ is not a prime, then $s{=}s{+}1$ and repeat step 3, until p becomes a prime.

Step 4 Compute $Q{=}\lfloor\frac{N}{p}\rfloor$ and $s{=}\lfloor\frac{Q}{2\cdot q'}\rfloor$.

Step 5 If $q{=}2{\cdot}q'{\cdot}s{+}1$ is not a prime, then $s{=}s{+}1$ and repeat step 5, until q becomes a prime.

Step 6 Compute $n{=}p{\cdot}q$, and if $\lfloor\frac{n}{2^{k-l}}\rfloor$ equals to C, then, n is a RSA modulus which is required. Otherwise, return to step 1.

In our method, the most significant l bits of n is always C, but p and q are random. By the variable t in step 2, the most significant l bits of n in step 6 is not changed, even though s is incremented in step 3 and step 5. To generate efficiently n, we choose $t = 16$. By the proposed algorithm, $p - 1$ and $q - 1$ have large prime factors p' and q', respectively.

3 Security

First, we will discuss the security of the Scheme 1 which we proposed in Section 2.

Theorem 1 *If we can compute the secret key d with* $e \cdot d = 1 \pmod{2^{l-1} \cdot \phi(n))}$*, then a RSA signature of arbitrary message M can be obtained.*
(proof) If $d' = d \pmod{\phi(n))}$*, then* $C = M^d = M^{d'} \pmod{n}$ *and* $e \cdot d' = e \cdot d = 1 \pmod{\phi(n))}$*. So, C is a RSA signature of M.*

Theorem 2 *If, for any odd M* $(0 < M < 2^l \cdot n)$*, we can compute C with* $C = M^e \pmod{2^l \cdot n}$*, then the RSA signature of M can be computed.*
(proof) Let $C' = C \pmod{n}$*. Then,* $C' = M^e \pmod{n}$*. So, C′ is a RSA signature.*

By Theorems 1 and 2, the security of the Scheme 1, based on the new reblocking method, depends on the security of a RSA signature scheme.

Now, we will discuss the security of the Scheme 2. Let n be a RSA modulus for the Scheme 2. Since n have large prime factors p and q, we can not factor it by any integer factoring algorithm. Moreover, $p-1$ and $q-1$ have large prime factors p' and q', respectively. Even if all users have n_i's, the most significant l bits of which are of the same value, the prime factors p_i and q_i of n_i are random. So, U_i can not guess the prime factors p_j and q_j of other user U_j.

4 Concluding Remarks

We have proposed two RSA multisignature schemes. First, we have suggested a new reblocking method in which the size of an enciphering block varies with the size of a message block and have applied the new reblocking method to a multisignature scheme. Each signer is allowed to have a RSA modulus with different bit size. It causes bit expansion which depends only on the number of signers regardless of the bit length of RSA modulus. The length of the expansion is less than or equal to the number of signers. If each signer has a RSA modulus with the same size, then our scheme and Okamoto's one have the same expansion. But, ours has smaller bit expansion than Okamoto's one.

The second multisignature scheme does not cause any bit expansion. All users must have a RSA modulus of a fixed length, k bits, the most significant l bits of which are the same. To obtain a multisignature, Kiesler-Harn's scheme requires an average exponentiation of $1.5m$, but our scheme requires about $(1 + \frac{1}{2^{l-1}})m$. So, our scheme is said to be more efficient than Kiesler-Harn's one.

References

1. Harn, L. and Kiesler, T., "New scheme for digital signatures", *Electronics Letters*, 1989, 25, (22), pp.1527-1528
2. Itakura, K., and Kakamura, K., "A public-key cryptosystem suitable for digital signatures", NEC J. Res. Dev. 71 (Oct. 1983).
3. Kiesler, T. and Harn, L., "RSA blocking and multisignature schemes with no bit expansion", *Electronics Letters*, 1990, 26, (18), pp.1490-1491

4. Kohnfelder, L. M., "On the signature reblocking problem in public-key cryptography", *Commun. ACM*, 1978, 21, (2) pp.179
5. Levine, J. and Brawley, J. V., "Some cryptographic applications of permutation polynomials", *Cryptologia*, 1977, 1, pp. 76-92
6. Okamoto, T., "A digital multisignature scheme using bijective public-key cryptosystems", *ACM Trans. Computer Systems*, 1988, 6, (8), pp.432-441
7. Rivest, R. L., "Remarks on a proposed cryptanalytic attack on the M.I.T. public-key cryptosystem", *Cryptologia*, 1978, Vol.2, No. 1,pp. 62-65
8. Rivest, R. L., Shamir, A., and Adleman, L., "A method for obtaining digital signatures and public-key cryptosystem", *Commun. ACM*, 1978, 21, (2), pp. 120-126

Proxy Signatures, Revisited

Seungjoo Kim[1] Sangjoon Park[2] and Dongho Won[1]

[1] Dept. of Information Engineering, Sung-Kyun-Kwan Univ.,
300 Chunchun-dong, Suwon, Kyunggi-do, 440-746, Korea
E-mail : {sjkim, dhwon}@simsan.skku.ac.kr
[2] #0710, ETRI, Yusong P.O.BOX 106, Taejon, 305-600, Korea
E-mail : sjpark@dingo.etri.re.kr

Abstract. Proxy signatures, introduced by Mambo, Usuda and Okamoto allow a designated person to sign on behalf of an original signer. This paper first presents two new types of digital proxy signatures called partial delegation with warrant and threshold delegation. Proxy signatures for partial delegation with warrant combines the benefit of Mambo's partial delegation and Neuman's delegation by warrant, and then in threshold delegation the proxy signer's power to sign messages is shared. Moreover, we also propose straightforward and concrete proxy signature schemes satisfying our conditions.

1 Introduction

Proxy signatures as embodied in [1] allow a designated person, called a proxy signer, to sign on behalf of an original signer in such a way that

- (*unforgeability*) Besides an original signer, a designated signer can create a valid proxy signature for the original signer. But third parties who are not designated as a proxy signer cannot create a valid proxy signature of the proxy signer.
- (*verifiability*) From proxy signatures a verifier can be convinced of the original signer's agreement on the signed message.

Such a signature scheme can for example be used in delegation of the power to sign messages without relying on any physical device. An employee in a company needs to go on a business trip to someplace which has no computer network access. During the trip he will receive e-mail, and expect to respond to some messages quickly.

Before going on a trip, he forwards his e-mail to his secretary, and instructs his secretary to respond to the e-mail in place of the employee according to a prearranged plan. Then the secretary responds to the e-mail using the proxy signature for the employee.

1.1 Related Work

So far, there has been three types of delegation, *full delegation*, *partial delegation*, and *delegation by warrant*.

Throughout this paper, It is assumed that an original signer Alice gives a proxy to a designated signer, called a proxy signer, Bob, in order to carry out signing instead of her.

Definition 1. *(full delegation) In full delegation, Bob is given the same secret s that Alice has, so that a proxy signature created by Bob is indistinguishable from the signature created by Alice.*

Definition 2. *(partial delegation) In partial delegation, a new secret σ is computed from a secret s of Alice, and σ is given to Bob in a secure way. From security requirement s should not be computed from σ. There are two types of signature schemes for this approach [1][2][3].*

- *(proxy-unprotected proxy signature) Besides Alice, Bob can create a valid proxy signature instead of Alice. But the third parties who are not designated as a proxy signer cannot create a valid proxy signature of Bob.*
- *(proxy-protected proxy signature) Only Bob can create a valid proxy signature for Alice. But Alice and the third parties cannot create a valid proxy signature of Bob.*

Definition 3. *(delegation by warrant) There are two types of schemes for this approach [4][5].*

- *(delegate proxy) In delegate proxy, Alice signs a document, declaring Bob is designated as a proxy signer, under her secret key by an ordinary signature scheme. The created warrant is given to Bob.*
- *(bearer proxy) In bearer proxy, a warrant is composed of a message part and an original signer's signature for newly generated public key. The secret key for a newly generated public key is given to Bob in a secure way.*

Proxy signature schemes can be constructed for each of these delegation types. The partial delegation, and the delegation by warrant, are more secure than the full delegation. The advantage of the partial delegation is the processing speed. The proxy signature for partial delegation has a computational advantage over the proxy signature by a warrant.

The delegation by warrant can be implemented by ordinary signature schemes without any modification, and it is appropriate for restricting documents to be signed, e.g. a warrant states its valid period. But the partial delegation does not have such a property. To solve this problem partial delegation requires an additional proxy revocation protocol, by which the original signer can revoke a created signature or the signing capability of the proxy signer. So while delegation by warrant can countermove before the fact, partial delegation considers a counterplan after the fact.

2 New Results of This Paper

This paper presents two new types of delegation, called *partial delegation with warrant* and *threshold delegation*.

Definition 4. *(partial delegation with warrant) In partial delegation with warrant, a new secret σ is computed from a secret s of Alice and a warrant, and σ is given to Bob in a secure way. From security requirement s should not be computed from σ and a warrant.*

Partial delegation with warrant combines the benefit of the partial delegation and the delegation by warrant. So this delegation has fast processing speed and is appropriate for the restricting documents to be signed. Furthermore, since a proxy for partial delegation with warrant can specify its valid period, our scheme doesn't need an additional proxy revocation protocol.

Well, now, in a group–oriented society it is often desired that the proxy signer's power to sign messages is shared. An employee Alice, for instance, has instructed her secretary Bob to respond in place of her accordingly. But suppose a secretary Bob doesn't follow prearranged instructions given by Alice. He does not sign a document which needs to be responded quickly, or he signs, at his will, what Alice has instructed to hold back. So, for security reasons, it may be a company's policy that documents be signed by k proxy signers rather than one person.

Definition 5. *(threshold delegation) In threshold delegation, n proxy signers are given shares such that $t \leq n$ are needed to generate a proxy signature instead of an original signer but less than t can not. This is called (t, n)-threshold delegation. i.e., in a (t, n)-threshold proxy signature scheme,*

1. *t out of n proxy signers must cooperate to issue a proxy signature.*
2. *Any $t - 1$ dishonest proxy signers cannot forge a signature.*

We first present two new definitions of delegation (Sec. 2) and propose efficient and concrete proxy signature schemes for partial delegation with warrant – one is a proxy–unprotected version and the other is proxy–protected version – (Sec. 4.1). The proposed schemes are widely applicable to schemes based on the discrete logarithm problem. Moreover, since our proxy–protected scheme requires only a linear combination of two shared secrets, our scheme can be extended to the threshold proxy signature schemes easily (Sec. 4.2). Finally our paper is concluded (Sec. 5.).

3 Mambo's Scheme for Partial Delegation

It is assumed that a signer Alice asks a proxy signer Bob to carry out signing for her, and a verifier Carol checks the validity of created signatures. Throughout this paper, p is a large prime with $2^{511} < p < 2^{512}$, and g is a generator for Z_p^*. Denote by v a public key such that $v = g^s \bmod p$, where $s \in_R Z_{p-1} \backslash \{0\}$

Basic Protocol :

1. (Proxy generation) An original signer Alice generates

$$\begin{aligned} k &\in_R Z_{p-1}\backslash\{0\}, \\ K &= g^k \bmod p, \\ \sigma &= s_{Alice} + kK \bmod p-1. \end{aligned}$$

2. (Proxy delivery) Alice gives (σ, K) to a proxy signer, Bob, in a secure manner.
3. (Proxy verification) Bob checks

$$g^{\sigma} \stackrel{?}{=} v_{Alice} K^K \bmod p.$$

4. (Signing by the proxy signer) When Bob signs a document m_p for the sake of Alice, he uses the σ as an alternative to s_{Alice}, and executes the ordinary signing operation. Then, the created proxy signature by Bob on m_p is,

$$(m_p, Sign_{\sigma}(m_p), K)$$

5. (Verification of the proxy signature) The computed value $v' = v_{Alice} K^K \bmod p$ is dealt with as a new public value, and the verification of the proxy signature is carried out by the same checking operation as in the original signature scheme.

4 Proposed Schemes Widely Applicable to Schemes Based on the Discrete Logarithm Problem

4.1 Proxy Signatures for Partial Delegation with Warrant

Basic Protocol (Proxy–Unprotected) :

1. (Proxy generation) An original signer Alice generates a random number $k \in_R Z_{p-1}\backslash\{0\}$, and computes $K = g^k \bmod p$. She concatenates m_w, and K, and hashes the result : $e = h(m_w, K)$ (where, the information on the delegation should be described in a warrant m_w e.g., its valid period).
After that, Alice computes $\sigma = es_{Alice} + k \bmod p-1$.
2. (Proxy delivery) Alice gives (m_w, σ, K) to a proxy signer, Bob, in a secure manner.
3. (Proxy verification) Bob confirms

$$\begin{aligned} e &= h(m_w, K), \\ g^{\sigma} &\stackrel{?}{=} v_{Alice}^e K \bmod p. \end{aligned} \tag{1}$$

4. (Signing by the proxy signer) For signing a document m_p, Bob uses the σ as an alternative to s_{Alice}, and executes the ordinary signing operation. Then, the proxy signature on m_p is m_p, $Sign_{\sigma}(m_p)$, K, and m_w (where, $Sign_{\sigma}(m_p)$ refers to signing a message m_p with private key σ).

5. (Verification of the proxy signature) The verification of the proxy signature is carried out by the same checking operation as in the original signature scheme except for the extra computation $e = h(m_w, K)$ and $v' = v_{Alice}^e K \bmod p$. The computed value v' is dealt with as a new public value explicitly showing the involvement of Alice.

Note that the scheme in a context [3] has the following form similar to the congruence (1).

$$g^\sigma = v^{h(K)} K \bmod p \tag{2}$$

This is a kind of partial delegation, since the congruence (2) does not have any warrant.

In the first version the reliability of an original signer is assumed, but what if an original signer forges the signature of the proxy signer. By replacing Step 3.– 5. with the following steps, we can easily extend the basic protocol into a proxy-protected proxy signature scheme. In the second version the reliability of an original signer is not assumed in terms of signature forgery.

Proxy-Protected Proxy Signature :

3'. (Verification and alteration of the proxy) After confirming the validity of (m_w, σ, K), where m_w should be composed of original signer's ID, proxy signer's ID, and other information on the delegation from the security requirement, the proxy signer Bob calculates an alternative proxy (σ_p, K) :

$$\sigma_p = \sigma + s_{Bob} \cdot h(m_w, K) \bmod p - 1. \tag{3}$$

4' (Signing by the proxy signer) For signing a document m_p, Bob uses the σ_p as an alternative to s_{Alice}, and executes the ordinary signing operation. Then, the proxy signature on m_p is m_p, $Sign_{\sigma_p}(m_p)$, K, and m_w.

5' (Verification of the proxy signature) The verifier Carol carries out the same checking operation as in the original signature scheme except for the extra computation

$$e = h(m_w, K)$$
$$v_p' = (v_{Alice} \cdot v_{Bob})^e K \bmod p.$$

The computed value v_p' is dealt with as a new public value explicitly showing the involvement of Alice.

Performance : If we choose an ElGamal signature scheme[6] both for creating a proxy signature and for verifying that, then the amount of computational work is smaller in the proxy signature for partial delegation with warrant than that by a warrant. As a total, the delegation by warrant requires $2956 + 2WI(512)$ of computational work, while our partial delegation with warrant needs $2158 + WI(512) + 2WH(|m_w|)$ $(2160 + WI(512) + 2WH(|m_w|)$ – computational work with and without parentheses mean a value for a proxy–unprotected signature scheme and that for a proxy–protected proxy signature scheme, respectively – (where, a method used by Kaliski[7] is adopted to assess the amount of computational work. Numbers means the amount of work to perform modular multiplication in 512 bits modulus, $WI(b)$ means the amount of work to perform b-bit modular inversion, and $WH(b)$ means the amount of work to compute a hash function with a b-bit long input. See [1] for detail).

In comparison with partial delegation, our partial delegation with warrant needs $641 + WH(|m_w|)$ $(642 + WH(|m_w|))$ of computational work in the proxy creation stage, $642 + WI(512)$ $(642 + WI(512))$ in the signature creation stage, $875 + WH(|m_w|)$ $(876 + WH(|m_w|))$ in the signature verification stage. However partial delegation requires 641(642) of computational work in the proxy creation phase, $642 + WI(512)$ $(642 + WI(512))$ in the signature creation phase, 875(906) in the signature verification phase, and the additional proxy revocation protocol (this may require 1282 of additional work). Note that as in [1], the following congruence can be used as an alternative to the congruence (3).

$$\sigma_p = \sigma + s_{Bob} v_{Bob} \bmod p - 1$$

But this needs $906 + WH(|m_w|))$ of computational work in the signature verification stage.

In the sequel, from the point of computational advantage, the partial delegation with warrant reduce the amount of computational work over the delegation by warrant and, from the point of organization, our scheme requires no supplementary protocol such as proxy revocation in the partial delegation (a proxy signature for the Schnorr scheme[8] or the Okamoto scheme[9] is evaluated in a similar way).

4.2 Threshold Proxy Signatures

Suppose that an original signer Alice wants to delegate the power to sign messages in such a way that the proxy signature can be created by any set of t or more proxy signers from a designated group PG of n proxy signers but that any subset with $t-1$ or less proxy signers cannot. In this section, we show an efficient (t, n)-threshold proxy signature scheme for the Schnorr scheme, by using the scheme which we developed in Sec. 4.1 and Ceredo's Schnorr type threshold digital signature scheme [10]. For convenience, we assume that $PG = \{P_i | 1 \le i \le n\}$. The public parameters are the same as those of Sect. 3, except that $p-1$ has large(160 bits) prime factor q and selects $g \in Z_p$ with the order of q.

Protocol for Generating Random Number : [11][12]

Suppose that a dealer with a random secret R chooses a random polynomial such that $f(x) = R + a_1 x + \cdots + a_{t-1} x^{t-1}$, sends $s_i = f(i)$ to P_i secretly for $i = 1, \cdots, n$, and broadcasts $y = g^R \bmod p$ and $g^{a_1}, \cdots, g^{a_{t-1}} \bmod p$. This procedure is simulated by the following protocol without the dealer.

1. Each proxy signer P_i picks $r_i \in_R Z_q$ at random and broadcasts $y_i = g^{r_i} \bmod p$ to all other proxy signers.

2. To distribute r_i, each P_i randomly selects a polynomial f_i of degree $t-1$ in Z_q such that $f_i(0) = r_i$, i.e.,

$$f_i(x) = r_i + a_{i,1}x + a_{i,2}x^2 \cdots + a_{i,t-1}x^{t-1}$$

with $a_{i,1}, \cdots, a_{i,t-1} \in_R Z_q$, and sends $f_i(j) \bmod q$ to P_j in a secure manner $(\forall j \neq i)$. P_i also broadcasts the value

$$g^{a_{i,1}}, \cdots, g^{a_{i,t-1}} \bmod p.$$

3. From distributed $f_j(i)$ $(\forall j \neq i)$, P_i checks whether, for each j $(j \neq i)$,

$$g^{f_j(i)} \stackrel{?}{=} y_j \cdot (g^{a_{j,1}})^{i^1} \cdots (g^{a_{j,t-1}})^{i^{t-1}} \bmod p.$$

4. Let $H \triangleq \{P_j | P_j$ is not detected to be cheating at step 3$\}$. Every P_i computes the share

$$s_i = \sum_{j \in H} f_j(i)$$

secretly, and computes

$$y = \prod_{j \in H} y_j,\ g^{a_1} = \prod_{j \in H} g^{a_{j,1}},\ \cdots,\ g^{a_{t-1}} = \prod_{j \in H} g^{a_{j,t-1}}$$

Proxy Sharing (Proxy–Unprotected) :

1. (Proxy generation) An original signer Alice picks at random $k \in_R Z_q \backslash \{0\}$, and computes $K = g^k \bmod p$. She concatenates a warrant m_w and K, and computes $e = h(m_w, K)$.
After that, Alice computes $\sigma = es_{Alice} + k \bmod q$.
2. (Proxy sharing) To share a proxy σ in a threshold-scheme with threshold t, Alice randomly selects elements $b_j \in_R Z_q$, $j = 1, \cdots, t-1$, and publishes the values $B_j = g^{b_j}$, $j = 1, \cdots, t-1$. Then she computes the proxy share σ_i,

$$f'(x) = \sigma + b_1 x + b_2 x^2 + \cdots + b_{t-1}x^{t-1},$$
$$\sigma_i = f'(i).$$

3. (Proxy–Share delivery) Alice sends to each proxy signer P_i (for $i = 1, \cdots, n$) the proxy share σ_i in a secure manner, and broadcasts (m_w, K).
4. (Proxy–Share verification) To verify a proxy share σ_i, the proxy signer P_i can compute $e = h(m_w, K)$, and check whether

$$g^{\sigma_i} \stackrel{?}{=} (v_{Alice}^e K) \cdot \prod_{j=1}^{t-1} B_j^{(i^j)} \bmod p.$$

Proxy Signature Issuing without Revealing Shares :

Let m_p be a message, and $H \subseteq PG$ issue a proxy signature. For convenience, we assume that $H = \{P_i | 1 \le i \le t\}$.

1. H execute *Protocol for Generating Random Number*, and obtain the public output,

$$y(= g^r \bmod p),\ g^{a_1}, \cdots, g^{a_{t-1}} \bmod p,$$

and secret output of P_i, s_i (where, $s_i = f(i) = r + a_1 i + \cdots + a_{t-1} i^{t-1}$).
2. Each proxy signer P_i computes

$$e = h(y, m_p),$$
$$\gamma_i = s_i + \sigma_i e \bmod q,$$

and reveals γ_i.
3. Every P_i verifies that

$$g^{\gamma_l} = (y \prod_{j=1}^{t-1} (g^{a_j})^{l^j}) \cdot ((v_{Alice}^e K) \prod_{j=1}^{t-1} (g^{b_j})^{l^j})^{h(y,m_p)} \bmod p \text{ for } \forall l.$$

4. Each $P_i \in H$ computes t satisfying

$$t = r + \sigma e = f(0) + f'(0)e \bmod q$$

by applying Lagrange formula to $\{\gamma_i\}$. The proxy signature is (m_p, t, e, K, m_w).
5. The validity of the signature (m_p, t, e, K, m_w) is verified by

$$y' = g^t \cdot (v_{Alice}^{h(m_w,K)} K)^{-e} \bmod p,$$
$$e \stackrel{?}{=} h(y', m_p).$$

Furthermore, by replacing the above *Proxy Sharing* with the following *Proxy–Protected Proxy Sharing*, we can easily extend the basic threshold proxy signature scheme into a proxy–protected threshold proxy signature scheme.

Proxy–Protected Proxy Sharing :

1. (Group key generation) PG execute *Protocol for Generating Random Number*, and obtain the public output of the group PG,

$$v_{PG}(= g^{s_{PG}} \bmod p),\ g^{c_1}, \cdots, g^{c_{t-1}} \bmod p,$$

and secret output of P_i, $s_{PG,i}$ (where, $s_{PG,i} = f''(i) = s_{PG} + c_1 i + \cdots + c_{t-1} i^{t-1}$).
2. (Proxy generation) Alice picks $k \in_R Z_q \backslash \{0\}$, and computes $K = g^k \bmod p$. She concatenates a warrant m_w and K, and computes $e = h(m_w, K)$, where m_w contains original signer's ID, proxy signer's ID, etc. After that, Alice computes $\sigma = e s_{Alice} + k \bmod q$.
3. (Proxy sharing) To share a proxy σ in a threshold-scheme with threshold t, Alice selects elements $b_j \in_R Z_q$, $j = 1, \cdots, t-1$, and publishes the values $B_j = g^{b_j}$, $j = 1, \cdots, t-1$. Then she computes the proxy share σ_i (where, $\sigma_i = f'(i) = \sigma + b_1 i + \cdots + b_{t-1} i^{t-1}$).
4. (Proxy–Share delivery) Alice sends to each P_i (for $i = 1, \cdots, n$) σ_i in a secure manner, and broadcasts (m_w, K).

5. (Verification and alteration of the proxy) After confirming the validity of (m_w, σ_i, K), P_i calculates an alternative proxy $(\sigma_{p,i}, K)$:

$$\sigma_{p,i} = \sigma_i + s_{PG,i} \cdot h(m_w, K) \bmod q,$$

and uses $\sigma_{p,i}$ as an alternative to σ_i.

5 Conclusion

We proposed two new types of proxy signatures. One is the partial delegation with warrant, proxy signature for which has a computational advantage over the proxy signature by warrant and has a structure advantage over the proxy signature for partial delegation.

The other is the proxy signature for threshold delegation. In the up-coming highly group–oriented society, it is desirable to delegate the power to sign messages to a group of n proxy signer. We described necessary conditions related to our new ideas, and showed a proxy signature scheme for partial delegation with warrant and a (t, n)-threshold proxy signature scheme without revealing proxy shares.

References

1. M. Mambo, K. Usuda, and E. Okamoto, "Proxy signatures: Delegation of the power to sign messages," IEICE Trans. Fundamentals, vol.E79-A, no.9, 1996, pp.1338-1354
2. M. Mambo, K. Usuda, and E. Okamoto, "Proxy signatures for delegating signing operation," Proc. Third ACM Conf. on Computer and Communications Security, 1996, pp.48-57
3. K. Usuda, M. Mambo, T. Uyematsu, and E. Okamoto, "Proposal of an automatic signature scheme using a compiler," IEICE Trans. Fundamentals, vol.E79-A, no.1, 1996, pp.94-101
4. V. Varadharajan, P. Allen, and S. Black, "An analysis of the proxy problem in distributed systems," Proc. 1991 IEEE Computer Society Symposium on Research in Security and Privacy, 1991, pp.255-275
5. B.C. Neuman, "Proxy-based authorization and accounting for distributed systems," Proc. 13th International Conference on Distributed Computing Systems, 1993, pp.283-291
6. T. ElGamal, "A public-key cryptosystem and a signature scheme based on discrete logarithms," IEEE Trans. Inf. Theory, vol.IT-31, no.4, 1985, pp.469-472
7. B.S. Kaliski, "A response to DSS," Nov. 1991.
8. C.P. Schnorr, "Efficient signature generation by smart cards," Journal of Cryptology, vol.4, no.3, 1991, pp.161-174
9. T. Okamoto, "Provably secure and practical identification schemes and corresponding signature schemes," Proc. Crypto'92, Lecture Notes in Computer Science, LNCS 740, Springer–Verlag, 1993, pp.31-53
10. M. Cerecedo, T. Matsumoto, and H. Imai, "Efficient and secure multiparty generation of digital signatures based on discrete logarithms," IEICE Trans. Fundamentals, vol.E76-A, no.4, 1993, p.532-545.

11. T.P. Pedersen, "A threshold cryptosystem without a trusted party," Proc. Eurocrypt'91, Lecture Notes in Computer Science, LNCS 547, Springer–Verlag, 1991, pp.522-526
12. T.P. Pedersen, "Distributed provers with applications to undeniable signatures," Proc. Eurocrypt'91, Lecture Notes in Computer Science, LNCS 547, Springer–Verlag, 1991, pp.221-238
13. A. Shamir, "How to share a secret," Commun. ACM, vol.22, no.11, 1979, pp.612-613
14. Y. Desmedt and Y. Frankel, "Shared generation of authenticators and signatures"' Proc. Crypto'91, Lecture Notes in Computer Science, LNCS 576, Springer–Verlag, 1991, pp.457-469
15. C. Park and K. Kurosawa, "New ElGamal type threshold digital signature scheme," IEICE Trans. Fundamentals, vol.E79-A, no.1, 1996, pp.86-93.

Related-Key Cryptanalysis of 3-WAY, Biham-DES,CAST, DES-X, NewDES, RC2, and TEA

John Kelsey Bruce Schneier
Counterpane Systems
{kelsey,schneier}@counterpane.com

David Wagner
U.C. Berkeley
daw@cs.berkeley.edu

Abstract. We present new related-key attacks on the block ciphers 3-WAY, Biham-DES, CAST, DES-X, NewDES, RC2, and TEA. Differential related-key attacks allow both keys and plaintexts to be chosen with specific differences [KSW96]. Our attacks build on the original work, showing how to adapt the general attack to deal with the difficulties of the individual algorithms. We also give specific design principles to protect against these attacks.

1 Introduction

Related-key cryptanalysis assumes that the attacker learns the encryption of certain plaintexts not only under the original (unknown) key K, but also under some derived keys $K' = f(K)$. In a chosen-related-key attack, the attacker specifies how the key is to be changed; known-related-key attacks are those where the key difference is known, but cannot be chosen by the attacker. We emphasize that the attacker knows or chooses the relationship between keys, not the actual key values. These techniques have been developed in [Knu93b, Bih94, KSW96].

Related-key cryptanalysis is a practical attack on key-exchange protocols that do not guarantee key-integrity—an attacker may be able to flip bits in the key without knowing the key—and key-update protocols that update keys using a known function: e.g., K, $K+1$, $K+2$, etc. Related-key attacks were also used against rotor machines: operators sometimes set rotors incorrectly. If the operator then corrected the rotor positions and retransmitted the same plaintext, an adversary would have a single plaintext encrypted in two related keys [DH79]. Hash functions built from block ciphers can also be vulnerable to a related-key attack against the block cipher [Win84, RIPE92].

In [KSW96] we gave a summary of key-schedule attacks against block ciphers, showed practical protocols that allow related-key attacks to be mounted, and presented related-key attacks against GOST [GOST89], IDEA [LMM91] with a reduced number of rounds, SAFER K-64 [Mas94], DES with independent subkeys, G-DES [PA90a, PA90b], and three-key triple-DES. This paper continues the research undertaken in that work.

2 New Differential Related-Key Attacks

2.1 3-WAY

3-WAY is an 11-round cipher on 96-bit blocks [Dae94]. Ignoring trivialities such as the input and output transformations, the 3-WAY round function $F(x)$ has an equivalent representation as:

$$y = N(x), \qquad z = L(y), \qquad F(x) = z \oplus K \oplus C_i$$

where N is a fixed nonlinear layer built out of 32 parallel 3-bit permutation S-boxes, L is a fixed linear function, K is the 96-bit master key, and C_i is a fixed, round-dependent public constant.

3-WAY is vulnerable to a simple related-key differential attack. It is trivial to find a differential characteristic for one S-box with probability 1/4, so we can construct a characteristic $\Delta x \to \Delta y$ with probability 1/4 for the non-linear layer N by using only one active S-box. By linearity we see that $\Delta y \to \Delta z = L(\Delta y)$ with probability 1 under the linear layer L. If we pick $\Delta K = \Delta x \oplus \Delta z$, then $\Delta x \to \Delta x$ by F with probability 1/4, which is a one-round iterative differential characteristic. In this way we can derive a 9-round characteristic with probability 2^{-18} to cover rounds 1–9, and apply a 2R analysis to the last two rounds. This breaks 3-WAY with one related-key query and about 2^{22} chosen plaintexts.

2.2 DES-X

DES-X is a DES variant proposed by Rivest [Riv95] to strengthen DES against exhaustive attacks. The DES-X encryption of P with key (K_1, K_2, K_3) is simply

$$C = K_1 \oplus \mathrm{DES}_{K_2}(K_3 \oplus P)$$

where K_3 is the pre-whitening key and K_1 is the post-whitening key. DES-X has many complementation properties. Furthermore, every DES-X key (K_1, K_2, K_3) has another equivalent key $(\overline{K_1}, \overline{K_2}, \overline{K_3})$. Therefore, DES-X cannot be used in a Davies-Meyer-like hash function construction.

This complementation property leads to an attack which requires roughly $2^{56+64-n}$ trial encryptions when 2^n chosen plaintexts are available [Dae91]. Note that Kilian and Rogaway [KR96] have proven that this attack is theoretically approximately optimal when DES is viewed as a black box, so any better (non-related-key) attack would have to take advantage of the internal structure of DES. However, their proof doesn't deal with related-key attacks. We give a related-key differential attack on DES-X, using key differences modulo 2^{64} and plaintext differences modulo 2. The attack requires 64 chosen key relations to recover the key, with one plaintext encrypted under each new key.

We start with a simple intuition. Suppose we have some unknown number Z. We are allowed to add any number we like modulo 2^{64}, and then XOR it with another number of our choosing. We are told whether or not the result of our calculation is equal to Z. Thus, we choose T and U, and test whether

$$(Z + T \bmod 2^{64}) \oplus U = Z$$

It is clear that we can learn the value of Z with enough queries. This is essentially the position we are in with DES-X. We can add T to K_1, and XOR U into our plaintext block or visa versa. If the resulting ciphertext block is the same as the ciphertext that results from encrypting the unaltered plaintext block under the unaltered DES-X key, then we can restrict the list of possible values for K_1. With enough such restrictions, we recover all of K_1 except for its high-order bit. This then allows attacks against the remainder of DES-X.

The simplest version of this attack uses T and U values each with the same single bit on. For each bit except the high-order bit, we try a T, U pair with the same bit on. If this results in the same ciphertext as resulted when $T = U = 0$, then we learn that that bit in K_1 was a zero. If it results in a different ciphertext, then we learn that that bit in K_1 was a one.

Some have suggested [KR96] using a DES-X variant which replaces the XOR pre- and post-whitening steps by addition modulo 2^{64}:

$$C = K_1 + \mathrm{DES}_{K_2}(K_3 + P).$$

From the discussion above, it should be clear that this would be vulnerable to a related-key attack very similar to the one that works against regular DES-X. [KR96] recommends a method of deriving DES-X keys from a single starting key, using SHA-1. This method seems to defend against related-key attacks.

2.3 CAST

CAST is a Feistel cipher whose key schedule uses nonlinear S-boxes [Ada94].[1] The key schedule for 8 round CAST with a 64 bit master key is as follows:

$$(k_1, k_2, \ldots, k_8) = \text{Master Key}$$

$$\begin{aligned}(k'_1, k'_2, k'_3, k'_4) &= (k_1, k_2, k_3, k_4) \oplus S5[k_5] \oplus S6[k_7] \\ (k'_5, k'_6, k'_7, k'_8) &= (k_5, k_6, k_7, k_8) \oplus S5[k'_2] \oplus S6[k'_4]\end{aligned}$$

$$\begin{array}{llll} K_1 = (k_1, k_2) & K_2 = (k_3, k_4) & K_3 = (k_5, k_6) & K_4 = (k_7, k_8) \\ K_5 = (k'_4, k'_3) & K_6 = (k'_2, k'_1) & K_7 = (k'_8, k'_7) & K_8 = (k'_6, k'_5) \end{array}$$

$$(K_{r,1}, K_{r,2}) = K_r \qquad r = 1, \ldots, 8$$

$$sk_r = S5[K_{r,1}] \oplus S6[K_{r,2}] \qquad r = 1, \ldots, 8.$$

where $S5$ and $S6$ are different 8-bit to 32-bit S-boxes. The r-th round subkey, sk_r, is XORed into the input of the F function as is conventional for Feistel ciphers.

[1] The variant of CAST analyzed here is an older version of CAST, not the CAST-128 that is used in Entrust products and described in Internet RFC 2144 [Ada97].

CAST is an interesting example of a cipher designed to resist Biham's rotational related-key cryptanalysis, but not differential related-key cryptanalysis. We apply a key-difference to the master key which changes only the byte k_1; this will lead to a difference only in round subkeys sk_1 and sk_6. When Δk_1 is known, there are only 256 possible differences for Δsk_1; by encrypting 2^{16} chosen plaintexts under each key, we can ensure that the first round is bypassed for some pair. Cover rounds 2–5 with the trivial differential characteristic of probability 1, and use a 2R attack. Note that sk_7 and sk_8 have only 32 bits of entropy in total, so we can try all 2^{32} possibilities for them, decrypt the last two rounds, and recognize correct guesses by 32 zero bits in the block difference. We recover the rest of the key with 2^{16} offline guesses by auxiliary techniques. In the end, we can recover the entire CAST master key with a total of about 2^{17} chosen plaintexts, one related-key query, and 2^{48} offline computations.

2.4 Biham-DES

Biham and Biryukov have suggested strengthening DES against exhaustive attacks by using extra key bits to modify the F-function slightly [BB94]. One of their modifications uses 5 key bits to select from 32 possible reorderings of the 8 DES S-boxes. We consider related keys which differ only in those 5 bits, and we apply related-key differential cryptanalysis. Specifically, suppose one key uses ordering 15642738 and another uses ordering 75642138 (both are from the 32 suggested reorderings listed in [BB94]). The only difference between the two F-functions is that S-boxes 1 and 7 have been swapped. Observe that:

$$\Pr_x\left(S1[x] \oplus S7[x \oplus 2] = 0\right) \;=\; 14/64.$$

The input differential 2 appears only in the middle input bits of the S-box, and will not spread to neighboring S-boxes. Hence, we can construct a one-round characteristic with probability $(\frac{14}{64})^2$.

This leads to a 13-round iterative characteristic with probability $(\frac{14}{64})^{12} = 2^{-26}$. The differential techniques of Biham and Shamir [BS93] will break Biham-DES with 2^{27} chosen plaintexts when this special related-key pair is available.

If two related keys allow the above attack (i.e. differ only in the key orderings as defined above), we call them partners. There is a $\frac{1}{16}$ chance that a randomly chosen key will have a partner; if it does, this can be detected with one related-key probe. Furthermore, we can always obtain one useful pair of related-key partners from any starting key after 32 related-key queries. Therefore, when using Biham-DES with the 32 recommended DES S-box reorderings, we have a $\frac{1}{16}$ probability of success when 2^{27} chosen plaintexts and one related-key query are available; success is nearly guaranteed with 2^{31} chosen plaintexts and 32 related-key queries.

Biham and Biryukov also mention the possibility of using 2^{15} reorderings of the s^3-DES S-boxes [KPL93]. They don't present the recommended reorderings, so it is impossible to present any specific results. Still, in general, increasing the number of reorderings gives the cryptanalyst more degrees of freedom to find more efficient attacks. Therefore, using this variant is not expected to increase security against our attack.

2.5 RC2

RC2 is a block cipher designed by Ron Rivest [Riv97]. The RC2 key schedule takes an arbitrary length master key and expands it to 128 bytes with the help of a public non-linear 8-bit permutation ρ; the result is converted to 64 16-bit round subkeys.[2] We have analyzed RC2, and found single-bit differential characteristics which pass through most rounds with probability $\frac{1}{2}$.

Consider a 64-byte master key $K = (x_0, x_1 ..., x_{63})$; its related-key partner will be $K^* = (x_0^*, x_1, ..., x_{63}^*)$. In other words, K and K^* differ only in their first and last bytes. We choose x_0, x_{63}, x_0^*, and x_{63}^* so that $\rho[x_0 + x_{63}] = \rho[x_0^* + x_{63}^*]$. This is easy—we just subtract t from x_0 and add it to x_{63} to obtain K^*, where t is a byte quantity to be carefully chosen below. The RC2 key schedule expands K to the 128-byte expanded key $xk_{0..127}$ as follows:

$$xk_{0..63} = x_{0..63} \qquad\qquad xk_i = \rho[xk_{i-1} + xk_{i-64}] \quad \forall i \geq 64.$$

We observe that $xk_{0..127}$ and $xk^*_{0..127}$ differ only in positions 0, 63, and 127.

Next, note that we know the difference t between xk_0 and xk_0*. This makes it very easy to bypass the subkey difference entering round 0 in a chosen plaintext attack by using a suitable plaintext pair P, P^*. P^* is just P with t added to its high byte. Let P_i be P after i rounds, where each round is $\frac{1}{4}$ of a cycle. We have

$$\begin{aligned} P_0^* &= P_0 + 2^{56}t \\ P_i^* &= P_i \qquad i = 1, \ldots, 31 \\ P_{32}^* &= P_{32} + t \end{aligned}$$

If we choose a difference t with only one bit set, then we've just dropped a one-bit difference into the middle of the cipher. Note that there is an iterative four-round (one-cycle) differential characteristic with this one-bit difference as input and probability 2^{-4}. This leads to a 28-round characteristic with probability 2^{-28}, which can be used in a 4R attack.

The probability of the characteristic is slightly decreased by two different cycles in the middle of encryption processing. There are eight such rounds; each has a 2^{-5} chance of hitting one of the two changed key words and destroying the propagation of a right pair. The chance of successfully missing all of these of $(1-2^{-5})^8 \approx 0.775$. Furthermore, one of those variant rounds adds a quantity with difference 0 to a quantity with a one-bit difference, which halves the probability of our characteristic. Finally, a subsequent variant takes the low 6 bits of a

[2] There is also an optional key-weakening stage, intended for export control use. For our purposes, we will assume it is not used.

quantity with a one-bit difference as input; a careful choice of t can ensure that the one-bit difference falls in the high 2 bits, so that the characteristic is not disrupted. We have to multiply the earlier estimate by $0.775 \cdot 0.5$, obtaining a total probability of $2^{-29.4}$ for our characteristic. With this technique, RC2 can be broken with one related-key query and about 2^{34} chosen plaintexts.

2.6 NewDES

NewDES [Sco85] is a 17-round 64-bit block cipher with a 120-bit key. The key schedule is simple: each cycle (which consists of 2 rounds) uses 56 bits from the key and then shifts the key by 56 bits. NewDES succumbs to standard rotation related-key techniques: it can be broken with 2^{32} known plaintexts, one related key, and about 2^{56} offline trial encryptions.

When informed of this attack, Scott modified the NewDES key schedule to resist rotational related-key cryptanalysis [Sco96]. NewDES-1996 in turn falls to differential related-key cryptanalysis.

The NewDES-1996 key schedule expands 15 bytes $K0 \ldots K14$ of the master key K into 60 round subkey bytes $SK0 \ldots SK59$ according to the following pattern:

$$\begin{array}{lllll} K0 & K1 & K2 & \ldots & K14 \\ K0 \oplus K7 & K1 \oplus K7 & K2 \oplus K7 & \ldots & K14 \oplus K7 \\ K0 \oplus K8 & K1 \oplus K8 & K2 \oplus K8 & \ldots & K14 \oplus K8 \\ K0 \oplus K9 & K1 \oplus K9 & K2 \oplus K9 & \ldots & K14 \oplus K9 \end{array}$$

When $K7, K8, K9$ are all non-zero, this updated key schedule defeats rotational related-key cryptanalysis, as the sequence of round subkeys no longer repeats.[3]

Note that the NewDES-1996 key schedule is completely linear and exhibits poor avalanche. In fact, it falls to a differential related-key attack we call the *double-swiping* attack.

The double-swiping attack is somewhat involved, with technical and notational distractions, so we first describe the basic flow of the attack. We derive three related keys K', K^*, and $K^{*\prime}$ from the original key K according to a differential quartet structure. We take an arbitrary ciphertext P and apply a plaintext difference to it to obtain P^*; for a right pair P, P^* the attack will succeed, and a right pair occurs with very high probability. "Swipe" P back and forth through the NewDES-1996 cipher: encrypt P under K to obtain C, and decrypt $C' = C$

[3] There are weak keys—namely those where $K7 = K8 = K9 = 0$—that succumb easily to rotational related-key cryptanalysis given 2^{32} known plaintexts, one related key, and about 2^{56} offline trial encryptions.

This leads to a more general rotational-based attack on NewDES-1996. For any key K, after 2^{24} related-key probes one can find a weak key K' of known relation to K, recover K' by the above attack on NewDES-1996 weak keys, and thus find K. However, this attack requires about 2^{25} related-key queries, 2^{56} known plaintexts, and 2^{80} offline trial encryptions in general; therefore, we have disregarded this attack on NewDES-1996 as impractical.

under K' to obtain P'. Next swipe P^* back and forth: encrypt P^* under K^* to obtain C^*, and decrypt $C^{*\prime} = C^*$ under $K^{*\prime}$ to obtain $P^{*\prime}$. For a right pair, it turns out that the quartet key structure ensures that P' and $P^{*\prime}$ will be nearly the same, differing only in the action of $SK0'$ and $SK0^{*\prime}$; a final analysis stage reveals $SK0$ from P' and $P^{*\prime}$. Now we peel off the effect of $K0$ and iterate to find the rest of the key bytes.

The double-swiping attack is an optimization of a more conventional (*single-swiping*) related-key differential attack. The more conventional attack proceeds by decrypting $C = C'$ under both K and K' to obtain P, P'; the problem is that (with NewDES-1996) the single-swiping attack requires a 4R analysis stage on P, P', which appears rather tricky to perform as it must take into account the effect of 15 round subkey bytes $SK0 \ldots SK14$. The intuition is that the double-swiping attack allows us to insert a difference much closer to the end of the cipher, so the analysis stage depends only on $SK0$ and thus becomes much easier. The single-swiping related-key attack is already a big improvement over non-related-key attacks, but we can do even better by double-swiping.

We now present the technical details of the double-swiping attack. Fix any two byte values x, y, and take three related keys $K', K^*, K^{*\prime}$ according to the quartet structure

$$\begin{aligned} K' &= K \oplus (x, x, x, \ldots, x) \\ K^* &= K \oplus (y, 0, 0, \ldots, 0) \\ K^{*\prime} &= K \oplus (x \oplus y, x, x, \ldots, x). \end{aligned}$$

The related keys can be obtained under the differential related-key assumption. Note that, with these definitions, we have

$$\begin{aligned} SK'i &= SKi \oplus \begin{cases} x \text{ if } i = 0, \ldots, 14 \\ 0 \text{ if } i = 15, \ldots, 59 \end{cases} \\ SK^*i &= SKi \oplus \begin{cases} y \text{ if } i = 0, 15, 30, 45 \\ 0 \text{ otherwise} \end{cases} \\ SK^{*\prime}i &= SKi \oplus SK'i \oplus SK^*i. \end{aligned}$$

For some plaintext $P = P0$, we will use the notation Pi to indicate the intermediate value of the block after encryption with the first i subkey bytes; for instance, $P15$ is the output after the first two rounds, and $P60 = C$ is the final ciphertext block. When we "swipe" the first time to obtain $C = P60 = P60' = C'$ and $P' = P0'$, in general we have $P0' \neq P0$. However, since SKi and $SK'i$ differ only for $i < 15$, note that $P15' = P15$. We define $P^* = P0^* = P0 \oplus \Delta = P \oplus \Delta$, where Δ is carefully chosen to bypass [BS93] the key difference $SK0 \oplus SK0^* = y$ entering in the first step of the first round. Define a right pair as a pair P, P^* where $P1^* = P1$; examination of the NewDES F function reveals that the carefully-chosen values $x \oplus y = 224$ and $\Delta = 18$ cause right pairs to occur with probability $\frac{12}{256} \approx 1/21.3$. After the second swipe, we have $P15^* = P15^{*\prime}$, since SKi^* and $SKi^{*\prime}$ differ only for $i < 15$. Furthermore, the quartet structure of the related keys ensures that $P15 = P15' = P15^* = P15^{*\prime}$ for a right pair. In particular, we have $P1' = P1^{*\prime}$ for a right pair. Note that $P0', P0^{*\prime}$ are known, and they differ from $P1', P1^{*\prime}$ only in the application of a 8-bit to 8-bit F function keyed by $SK0', SK0^{*\prime}$. Therefore, we can apply a standard differential 1R analysis stage [BS93] to P' and $P^{*\prime}$; one can filter out wrong pairs very effectively, so recovering $SK0$ should be possible with just one right pair.

This double-swiping differential attack finds one subkey byte $SK0$ with a quartet of differentially related keys and about 88 chosen-plaintext/ciphertext queries. Now we can peel off the effect of the first subkey byte $SK0$ and iterate the attack to recover $SK1$, etc. Thus we can recover all 15 key bytes $(K0, \ldots, K14) = (SK0, \ldots, SK14)$ and completely break NewDES-1996 with total complexity of about 24 related-key probes and 530 chosen plaintext/ciphertext queries.

2.7 TEA

TEA [WN95] is a Feistel block cipher with a 128-bit master key, $K[0..3]$, and a simple key schedule: odd rounds use $K[0,1]$ as the round subkey, and even rounds use $K[2,3]$. Two rounds of TEA applied to the block Y_i, Z_i consists of:

$$c = c + \delta \qquad Y_{i+1} = Y_i + F(Z_i, K[0,1], c) \qquad Z_{i+1} = Z_i + F(Y_{i+1}, K[2,3], c)$$

where the round function F is defined by

$$F(z, K[i,j], c) = (SL_4(z) + K[i]) \oplus (z + c) \oplus (SR_5(z) + K[j]).$$

Here $SL_4(z)$ denotes the result of shifting (not rotating) z to the left by 4 bits, and $SR.(\cdot)$ denotes a shift to the right. In this description, c is a value which perturbs the F function so that it is different in each round.[4] Before each cycle, c is incremented by a fixed constant $\delta = \lfloor(\sqrt{5} - 1)2^{31}\rfloor$; c is initially 0. The designers of TEA mention that 32 Feistel rounds (i.e. 16 cycles) may be enough, though they recommend using 64 rounds (32 cycles) [WN95].

TEA admits several related-key attacks which arise from the severe simplicity of its key schedule.

Attack One For a differential related-key attack, consider the effect of simultaneously flipping bit 30 (the next most significant bit) of $K[2]$ and $K[3]$. With probability nearly $\frac{1}{2}$, the output of the F function in the even rounds will remain the same. This immediately yields a 2-round iterative differential characteristic with probability $\frac{1}{2}$, and thus a 60-round characteristic with probability 2^{-30}. Our analysis indicates that a 4R differential related-key attack can break 64-round (32-cycle) TEA with one related-key query and about 2^{34} chosen plaintexts. This is only one of several of this type of characteristic.

[4] This perturbation is crucial to avoid degenerate attacks. Indeed, R. Fleming found a known-plaintext attack on a TEA variant weakened to use a constant c [Fle96]. (His variant also differs from TEA in that the the precedence of addition and XOR are reversed [Ber97], but a modification of his attack will work without this reversal.)

Attack Two The second differential related-key attack is very similar in spirit to the first. We request the encryption of (Y, Z) under key $K[0..3]$ and the encryption of $(Y, Z \oplus 2^{31})$ under key $K^*[0..3] = K[0..3] \oplus (0, 2^{31} \oplus 2^{26}, 0, 0)$. Examining the three terms of $F(Z, K[0,1], c)$ when bit 31 of Z is flipped along with bits 26 and 31 of $K[1]$, we see

$SL_4(Z) + K[0]$	Neither change has any effect.
$Z + c$	The high bit is always changed.
$SR_5(Z) + K[1]$	Half the time, only the high bit is changed.

This gives us a one-cycle (2-round) iterative differential characteristic with probability $\frac{1}{2}$, when we can choose one key difference. We can pass 30 rounds with probability 2^{-30}.

Attack Three The third attack is complicated. Therefore, we briefly point out the approach and intuition behind the attack, leaving the technical details of the full attack to be described in Appendix A. We write P_j to represent the value of the block after j rounds of encryption, and write K_j to represent the round subkey value used to compute P_{j+1} from P_j; the block is enciphered with a round function F as $P_{j+1} = F(K_j, P_j)$, where (P_0, P_{64}) represents a plaintext/ciphertext pair for 64-round TEA.

In Biham's standard key rotation attack [Bih94], we succeed when

$$K'_j = K_{j+1} \qquad P'_j = P_{j+1} \qquad\qquad j = 0, \ldots, 63.$$

This condition is achieved by choosing suitable related keys K, K' and searching over P_0, P'_0 to find a pair with $P'_0 = P_1$; the birthday paradox ensures that a match will occur with a reasonable number of known texts. Note that

$$P'_{j+1} = F(K'_j, P'_j) = F(K_{j+1}, P_{j+1}) = P_{j+2} \tag{1}$$

for all j, so by induction we see that a match $P'_0 = P_1$ will propagate down to the ciphertexts, where we can recognize it.

Our extended attack combines the ideas of both rotational and differential related-key attacks. We require that

$$K'_j = K_{j+1} + \Delta K_{j+1} \qquad P'_j = P_{j+1} + \Delta P_{j+1} \qquad\qquad j = 1, \ldots, 63.$$

In the extended attack, we need a generalization of (1) to hold

$$\begin{aligned} P'_{j+1} &= F(K'_j, P'_j) = F(K_{j+1} + \Delta K_{j+1}, P_{j+1} + \Delta P_{j+1}) \\ &= F(K_{j+1}, P_{j+1}) + \Delta P_{j+2} = P_{j+2} + \Delta P_{j+2} \end{aligned}$$

with significant probability p_{j+2}; this generalization has a strong differential feel to it. Suppose the 63-round differential related-key characteristic that is patched into the rotational attack has probability $p = \prod_j p_j$. In the extended attack, we search for about $\frac{1}{p}$ matches $P'_0 = P_1 + \Delta P_1$ with the birthday paradox. Each such match has a probability p of leading to a right pair that is recognizable from the known ciphertexts, so we expect to see one right pair.

Specifically, in our third attack on TEA, we take ΔP_{j+1} to be a fixed constant $(\delta, 0)$ independent of j, set $\Delta K_j = \Delta K_{j \bmod 2}$, and choose $\Delta K_{0,1}$ to maximize p. We can thus obtain a full 63-round characteristic of probability $p = (\frac{25}{32})^{31} \approx 2^{-11}$ by repeating a 2-round iterative characteristic many times.

This improved attack combines ideas from both Biham's key-rotation attack [Bih94] and differential related-key cryptanalysis [KSW96] to break TEA with just 2^{23} chosen plaintexts and one related-key query.

3 Prudent Rules of Thumb for Key-Schedule Design

There is much overlap between the requirements for strong key schedules and cryptographic hash functions. Firstly, key schedules should be hard to invert—given some of the round keys, it should be difficult to recover any new information about other bits of the key—and hash functions are supposed to be one-way. Secondly, to avoid equivalent keys, key schedules should possess some form of collision-freedom; collision-freedom is a standard hash function property as well. Finally, it should not be possible to produce controlled changes in the round keys. The key schedules of Blowfish [Sch94] and SEAL [RC94] were designed according to this principle.

One should typically avoid generating round subkeys as a (fixed, public) linear transformation of the seed. While some cryptosystems have successfully incorporated linear key schedules (e.g. DES), designing this type of key schedule appears to be a subtle and difficult task. Many ciphers' linear key schedules have been shown to be quite weak: we have cryptanalyzed TEA, 3-WAY, and GOST [KSW96], and others have cryptanalyzed LOKI [Knu93a], LOKI91 [Knu93b], Lucifer [BB93], and SAFER [Knu95].

To protect against the known related-key attacks, we propose several attack-oriented design goals. To avoid the "subkey rotation" attacks [Bih94], round subkeys should be generated differently, so that each key bit affects nearly every round, but not always in the same way. Key schedules should be specifically designed to resist differential related-key attacks. And, when related-key queries are cheap, the master key should be long enough to avoid generic black box attacks, as the key length is effectively halved under these attacks [WH87, KSW96].

Avoid dead spots; ensure that every key bit is about equally powerful in terms of its effect on the round keys. Beware of equivalent representations, for they can expose new avenues of attack to an adversary. Our analysis of 3-WAY bears witness to this recommendation.

Avoid independent round subkeys. It has commonly been assumed that a cipher's key length (and strength) can be increased by allowing round keys to be specified independently, but we have shown that this dramatically lowers the cipher's resistance to related-key attacks [KSW96]. In general, when independent round subkeys are in use, the strength of a cipher against related-key attacks will be approximately proportional to the strength of one round standing on its own. Additionally, avoid multiple encryption with independent keys; a construction like [DK96] is much more secure.

And finally, protocol designers should be aware of related-key attacks. Key-exchange protocols should exchange a short master key rather than exchanging expanded keys. Design tamper-resistant devices so that it is not possible to change the subkeys without such changes being detected.

References

[Ada94] C. Adams, "Simple and Effective Key Scheduling for Symmetric Ciphers," *Workshop on Selected Areas in Cryptography: SAC '94*, 1994, pp 129-133.

[Ada97] C. Adams, "Constructing Symmetric Ciphers Using the CAST Design Procedure," *Designs, Codes and Cryptography*, v 12, n 3, 1997, to appear.

[BB93] I. Ben-Aroya and E. Biham, "Differential Cryptanalysis of Lucifer," *Advances in Cryptology—CRYPTO '93*, Springer-Verlag, 1994, pp. 187–199.

[Ber97] D. Bernstein, personal communication, 1997.

[Bih94] E. Biham, "New Types of Cryptanalytic Attacks Using Related Keys," *Advances in Cryptology—EUROCRYPT '93*, Springer-Verlag, 1994, pp. 398–409.

[BB94] E. Biham and A. Biryukov, "How to Strengthen DES Using Existing Hardware," *Advances in Cryptology—ASIACRYPT '94*, Springer-Verlag, pp. 398–412.

[BS93] E. Biham and A. Shamir, "Differential Cryptanalysis of the Full 16-round DES," *Advances in Cryptology—CRYPTO '92*, Springer-Verlag 1993, pp. 487–496.

[Dae91] J. Daemen, "Limitations of the Even-Mansour Construction," *Advances in Cryptology—ASIACRYPT '91*, Springer-Verlag, 1992, pp. 495–498.

[Dae94] J. Daemen, "A New Approach to Block Cipher Design," *Fast Software Encryption, Cambridge Security Workshop Proceedings*, Springer-Verlag, 1994, pp. 18–32.

[DK96] I.B. Damgard and L.R. Knudsen, "Multiple Encryption with Minimum Key," *Cryptography: Policy and Algorithms*, Springer-Verlag, 1996, pp. 156–164.

[DH79] W. Diffie and M.E. Hellman. "Privacy and Authentication: An Introduction to Cryptography". *Proceedings of the IEEE*, vol 67 no 3, March 1979.

[Fle96] R. Fleming, "An attack on a weakened version of TEA," post to the `sci.crypt` newsgroup, October 1996.

[GOST89] GOST, Gosudarstvennyi Standard 28147-89, "Cryptographic Protection for Data Processing Systems," Government Committee of the USSR for Standards, 1989.

[KSW96] J. Kelsey, B. Schneier, and D. Wagner, "Key-Schedule Cryptanalysis of IDEA, G-DES, GOST, SAFER, and Triple-DES," Advances in Cryptology—CRYPTO '96, Springer-Verlag, 1996, pp. 237–251.

[KPL93] K. Kim, S. Park, and S. Lee, "Reconstruction of s^2DES S-Boxes and their Immunity to Differential Cryptanalysis," *Proceedings of the 1993*

Japan-Korea Workshop on Information Security and Cryptography, Seoul, Korea, 24-26 October 1993, pp. 282-291.

[Knu93a] L.R. Knudsen, "Cryptanalysis of LOKI," *Advances in Cryptology—ASIACRYPT '91*, Springer-Verlag, 1993, pp. 22–35.

[Knu93b] L.R. Knudsen, "Cryptanalysis of LOKI91," *Advances in Cryptology—AUSCRYPT '92*, Springer-Verlag, 1993, pp. 196–208.

[Knu94] L.R. Knudsen, "Block Ciphers—Analysis, Design, Applications," Ph.D. dissertation, Aarhus University, Nov 1994.

[Knu95] L.R. Knudsen, "A Key-schedule Weakness in SAFER K-64," *Advances in Cryptology—CRYPTO '95*, Springer-Verlag, 1995, pp. 274–286.

[KR96] J. Kilian and P. Rogaway, "How to protect DES against exhaustive key search," *Advances in Cryptology—CRYPTO '96*, Springer-Verlag, 1996, pp. 252–267.

[LMM91] X. Lai, J. Massey, and S. Murphy, "Markov Ciphers and Differential Cryptanalysis," *Advances in Cryptology—CRYPTO '91*, Springer-Verlag, 1991, pp. 17–38.

[Mas94] J.L. Massey, "SAFER K-64: A Byte-Oriented Block-Ciphering Algorithm", *Fast Software Encryption, Cambridge Security Workshop Proceedings*, Springer-Verlag, 1994, pp. 1–17.

[PA90a] A. Pfitzmann and R. Abmann, "Efficient Software Implementations of (Generalized) DES," *Proc. SECURICOM '90*, Paris, 1990, pp. 139–158.

[PA90b] A. Pfitzmann and R. Abmann, "More Efficient Software Implementations of (Generalized) DES," Technical Report PfAb90, Interner Bericht 18/90, Fakultat fur Informatik, Universitat Karlsruhe, 1990. http://www.informatik.uni-hildesheim.de/~sirene/lit/abstr90.html#PfAss_90

[RIPE92] Research and Development in Advanced Communication Technologies in Europe, *RIPE Integrity Primitives: Final Report of RACE Integrity Primitives Evaluation (R1040)*, RACE, Jun 1992.

[Riv95] R. Rivest, personal communication.

[Riv97] R. Rivest, "A Description of the RC2(r) Encryption Algorithm," Internet-Draft, work in progress, June 1997, ftp://ds.internic.net/internet-drafts/draft-rivest-rc2desc-00.txt

[RC94] P. Rogaway and D. Coppersmith, "A Software-Optimized Encryption Algorithm," *Fast Software Encryption, Cambridge Security Workshop Proceedings*, Springer-Verlag, 1994, pp. 56–63.

[Sch94] B. Schneier, "Description of a New Variable-Length Key, 64-Bit Block Cipher (Blowfish)," *Fast Software Encryption, Cambridge Security Workshop Proceedings*, Springer-Verlag, 1994, pp. 191–204.

[Sco85] R. Scott, "Wide Open Encryption Design Offers Flexible Implementations," *Cryptologia*, v. 9, n. 1, Jan 1985, pp. 75–90.

[Sco96] R. Scott, "Revision of NewDES," personal communication, also posted to the `sci.crypt` newsgroup on the Internet, May 1996.

[WN95] D. Wheeler and R. Needham, "TEA, a Tiny Encryption Algorithm," *Fast Software Encryption, Second International Workshop Proceedings*, Springer-Verlag, 1995, pp. 97–110.

[Win84] R. Winternitz, "Producing One-Way Hash Functions from DES," *Advances in Cryptology: Proceedings of Crypto 83*, Plenum Press, 1984, pp. 203–207.

[WH87] R. Winternitz and M. Hellman, "Chosen-key Attacks on a Block Cipher," *Cryptologia*, v. 11, n. 1, Jan 1987, pp. 16–20.

A Improved Attack on TEA

This attack combines ideas from Biham's key-rotation attack and differential cryptanalysis. It requires only 2^{23} chosen plaintexts and one related-key query. See Section 2.7 for a gentler introduction to the ideas behind the attack.

If $K[0\ldots4]$ is one TEA key value, its related key partner is defined to be $K'[0\ldots4]$ according to the following relations:

$$K'[0]=K[2] \quad K'[1]=K[3] \quad K'[2]=K[0]-SL_4(\delta) \quad K'[3]=K[1]-SR_5(\delta)-1.$$

(Refer to Section 2.7 for a definition of $SL(), SR()$, and other notation.) Fix a particular plaintext y, z which is encrypted via $K[\,]$; its related plaintext partner (which is encrypted with $K'[\,]$) will be offset from y, z by 1/2 cycle, as in rotational related-key cryptanalysis. Typically, in related-key cryptanalysis, we search for a partnered plaintext pair by the birthday paradox, and the right choice leads to a recognizable match in the corresponding ciphertexts with probability 1. In this generalization, we will consider the case where right choices of plaintext pairs leads to recognizable matches in the ciphertext with some non-trivial probability, via a differential characteristic.

The following table shows the encryption of y, z under key $K[\,]$ as well as the encryption of its offset plaintext partner $y', z' = z+\delta, y$ under key $K'[\,]$. The left half of the table depicts the left and right halves of the block when encrypting y, z; the right half of the table depicts the encryption of y', z'. (We consider the swap of the block halves to be included in each round.) We have placed y_{j+1}, z_j (respectively y_{j+1}, z_{j+1}) on the same line as y'_j, z'_j (resp. y'_{j+1}, z'_j) to suggest that the two propagate similarly. As described in Section 2.7, $F(z, K[i,j], c)$ denotes the value the round F function with input z, key values $K[i], K[j]$ with the round-dependent perturbation variable equal to c; c is incremented by δ before each cycle to make the F function different for each round.

Encrypt($K[\,], y_0z_0$)		Encrypt($K'[\,], y'_0z'_0$)	
$y_0 = y$	$z_0 = z$		
z_0	$y_1 = y_0 + F(z_0, K[0,1], \delta)$	$y'_0 = z_0 + \delta$	$z'_0 = y_1$
y_1	$z_1 = z_0 + F(y_1, K[2,3], \delta)$	z'_0	$y'_1 = y'_0 + F(z'_0, K'[0,1], \delta)$
z_1	$y_2 = y_1 + F(z_1, K[0,1], 2\delta)$	y'_1	$z'_1 = z'_0 + F(y'_1, K'[2,3], \delta)$
y_2	$z_2 = z_1 + F(y_2, K[2,3], 2\delta)$	z'_1	$y'_2 = y'_1 + F(z'_1, K'[0,1], 2\delta)$
...		...	
y_{32}	$z_{32} = z_{31} + F(y_{32}, K[2,3], 32\delta)$	z'_{31}	$y'_{32} = y'_{31} + F(z'_{31}, K'[0,1], 32\delta)$
		y'_{32}	$z'_{32} = z'_{31} + F(y'_{32}, K'[2,3], 32\delta)$

We define a right pair for the differential characteristic to be a pair (y_0, z_0), (y'_0, z'_0) satisfying

$$y'_j = z_j + \delta \qquad z'_j = y_{j+1} \qquad\qquad j = 0, \ldots, 31.$$

Since $K[2,3] = K'[0,1]$, we see from the table that we will never deviate from the right-pair condition in an odd round if it holds at the start of the odd round.

Therefore we have a right pair just if the condition holds for all even rounds; the table shows that the required condition is

$$F(z_j, K[0,1], (j+1)\delta) = F(y'_j, K'[2,3], j\delta) \qquad j = 0, \ldots, 31. \qquad (2)$$

Expanding the right-hand-side and then simplifying, we obtain

$$\begin{aligned}&(SL_4(z_j+\delta) + K[0] - SL_4(\delta)) \oplus (z_j + \delta + j\delta)\\ &\quad \oplus (SR_5(z_j+\delta) + K[1] - SR_5(\delta) - 1)\\ &= (SL_4(z_j) + K[0]) \oplus (z_j + (j+1)\delta) \oplus (SR_5(z_j) + K[1] + \Omega_j - 1)\end{aligned}$$

where $\Omega_j = SR_5(z_j+\delta) - SR_5(z_j) - SR_5(\delta)$, i.e. Ω_j is the carry bit from the addition of the low 5 bits of z_j and δ. Comparing to the right-hand-side of (2), we see that condition (2) is equivalent to the requirement that $\Omega_j = 1$ for $j = 0, \ldots, 31$. A quick check of the low 5 bits of δ shows that $\Omega_j = 1$ with probability $\frac{25}{32}$ when z_j is random.

In other words, the differential characteric carries through one cycle with probability $\frac{25}{32}$, and through 31 cycles with probability $\frac{25}{32}^{31} = .00047 = 2^{-11}$. Now we use the differential characteristic in the rotational related-key attack; we find it increases the number of plaintexts required by a factor of $2^{-11/2}$ over the number that would be required for a standard probability 1 attack.

Here is the attack in more detail. First fix a value for z_0. Now generate $2^{16+11/2} = 2^{21.5}$ values of $y_0^{(m)}$, for $m = 1 \ldots 2^{21.5}$, and encrypt the resulting value $y_0^{(m)}, z_0$ under $K[\,]$ to obtain the ciphertext $y_{32}^{(m)}, z_{32}^{(m)}$. Next set $y'_0 = z_0 + \delta$, and generate $2^{21.5}$ values of $z_0'^{(n)}$. For each $z_0'^{(n)}$, with $n = 1 \ldots 2^{21.5}$, encrypt $y'_0, z_0'^{(n)}$ under $K'[\,]$ to obtain the ciphertext $y_{32}'^{(n)}, z_{32}'^{(n)}$. Look for matches of the form $z_{32}^{(m)} = y_{32}'^{(n)}$. We expect to see one right match formed from a right pair of the differential characteristic combined with a right partnership $z_0'^{(n)} = y_1^{(m)}$ for the rotational attack; there will also be approximately $2^{21.5 \cdot 2}/2^{32} = 2^{11}$ matches formed by chance. Each right match allows you to recover roughly 64 key bits: it suggests about 2^{32} possible values for $K[0,1]$ and about 2^{32} possible values for $K[2,3]$.

One could repeat the attack a few more times and use a counting technique to recover the full key values with a bit more work. In more detail, each match suggests a value for $F(z_0, K[0,1], 0)$; we can now construct y_0, z'_0 pairs which are guaranteed to form a right partnership for the rotational attack, when used with the same z_0 value as before. For each guess at $F(z_0, K[0,1], 0)$, we can perform 2^{11} chosen plaintext queries; then we can recognize the true value of $F(z_0, K[0,1], 0)$ because it will cause another right pair and matching ciphertext pair. Thereafter, we can perform 2^{20} chosen plaintext queries and obtain 2^9 right pairs for the differential characteristic. This will be more than enough to recover the true value of $K[2,3]$ and find 2^{32} possible values for $K[0,1]$, so a simple search will suffice to recover the entire key.

In total, this attack needs 2^{23} chosen plaintexts, one related-key query, and roughly 2^{32} offline computations to recover the entire TEA key.

A Multiplication-Addition Structure Against Differential Attack

Feng Zhu , Bao-An Guo
Dept. of Computer Science and Technology, Tsinghua Univ.
Beijing, 100084, P.R.China
Email: xlzhu@mail.tsinghua.edu.cn

Abstract. This paper presents a multiplication-addition structure in finite field that can be used in constructing round function F of the Generalized DES-like iterated cipher to strengthen ability against the differential attack. Four special round functions based on this multiplication-addition structure are investigated. It is proved that the probability of one round differential of any of these four round functions reaches its possible minimum respectively. Thus their corresponding ciphers are immune to differential cryptanalysis in a few rounds. Some practical consideration in the implementation of the generalized DES-like iterated cipher and IDEA-like iterated cipher are also given.

Key words. Block cipher, Product cipher, Differential cryptanalysis.

1. Introduction

The DES cipher has been used as an encryption standard for a quite long time. The weakness of DES cipher was pointed out by Morris, Sloane and Wyner at the early beginning. Diffie and Hellman showed that an exhaust cryptanalysis may become practical in some years later. Entering nineties some new cryptanalysis techniques give a notable effect on the perceived security of DES [2][3] and DES-like cipher[5][6]. The differential cryptanalysis proposed by Biham and Shamir showed that the DES and DES-like cipher are theoretically cryptanalysable. The complexity for breaking the DES by differential cryptanalysis is less than the complexity of exhaust search. In their attack, they introduced a new notion which they called characteristic. Characteristic describes the behavior of input and output differences for some number of consecutive rounds. The probability of a one-round characteristic is the conditional probability that given a certain difference in the inputs to the round we get a certain difference in the outputs of the round. Lai and Massey[1] introduced a similar notion, which they called differential. The probability of an s-round differential is the conditional probability that given an input difference at the first round, the output difference at the s-th round will be some fixed value. Close relation exists between this two probability. In fact, the probability of an s-round differential with input difference A and output difference B is the sum of the probabilities of all s-round characteristics with input difference A and output difference B. For $s \leq 2$ the probabilities for a differential and for the corresponding characteristic are equal. Both the characteristic and the differential can equally be used for a successful attack.

It is well know today that as the number of rounds increases the DES and the DES-like iterated cipher will be stronger against a differential attack. Lai has proved that one kind of Markov cipher could resist differential cryptanalysis provided that the cipher has enough number of rounds. So one way to make the cipher stronger is to increase the number of rounds. Yet the shortcoming of this approach is obvious. As the number of rounds increases, the speed of encryption /decryption decreases.

A more efficient approach to make the DES-like cipher stronger against the differential attack is the use of "good" round function[7][8]. If we can ensure that the probability of one round differential is low enough then the differential attack will be hardly successful. Particularly if we can make this probability equal the possible minimum, then a successful differential attack will be nearly impossible. In this paper, we propose a multiplication addition structure and prove that round function constructed from this multiplication addition structure can satisfy this condition. Thus this structure gives the possibility to obtain flexible cipher architecture that is immune to the differential attack and can be efficiently implemented.

The paper is divided into 5 section. In section 2, we extend the architecture of DES to a more general form and the generalized DES-like cipher is defined. Thereafter, the basic problems related to the cryptanalysis on this generalized DES-like cipher are discussed. Section 3 gives four round functions of the generalized DES-like cipher. All these round functions are constructed from the multiplication-addition structure. The probabilities of one round differential of these round functions are given in theorem 1 to theorem 4, which show that these round functions are ideal in against the differential attack. Some discussions on the application of multiplication-addition structure are given in section 4.

2. Generalized DES-like Cipher and the Approach Against Differential Attacks

A DES cipher is a secret-key block cipher that is based on iterating a round function. In each round the input is divided into halves first. The right half is then fed into function F together with a subkey. The output of the function F is bitwise added (module 2) to the left half of the input and the sum is used as the right half of the output of this round. The left half of this output is taken from the right half of the input of this round. This operation is iterated and the output of each round is swapped first before input to the next round. The subkey is determined by a key schedule. The function F is called round function.

Several kinds of DES-like cipher exist today. They are different in round function. In this paper, we define the DES-like cipher in a more general sense and call it generalized DES-like iterated cipher.

A generalized DES-like iterated cipher with block size $2n$ and r rounds is defined as follows.

Let $K=(K_1,K_2,\dots,K_r)$, where $K_i \in GF(2^m)$, be the r round keys.

Let u: $GF(2^n) \to A$, be a one-to-one mapping. where $(A,*)$ is a group, $\#A=2^n$

v: $GF(2^n) \times GF(2^n) \to GF(2^n)$, $v(x,y)=u^{-1}(u(x)*u(y))$, $x,y \in GF(2^n)$

the round function F be any function such as : $GF(2^n) \times GF(2^m) \to GF(2^n)$.

For a given plaintext $X=(X_L,X_R)$ and a given key $K=(K_1,K_2,\dots,K_r)$, the ciphertext $Y=(Y_L,Y_R)$ is obtained after r rounds' computation, that is,

set $X_L(0)=X_L$ and $X_R(0)=X_R$ and compute for $i=1,2,\dots,r$

$$X_L(i)=X_R(i-1), \quad X_R(i)=v\left(F(X_R(i-1), K_i), \quad X_L(i-1)\right),$$

$X(i)=(X_L(i),X_R(i))$.
then $Y_L=X_R(r)$ and $Y_R=X_L(r)$.

It can be seen, when the mapping u is an identical mapping, $(A,*)=(GF(2^n),+)$, where "+" is bitwise addition module 2, the generalized DES-like cipher will degenerate into a DES-like cipher.

In Biham Shamir's differential cryptanalysis, the difference between two n-bit blocks(x1 and x2) is defined as $\Delta x=x1+x2$, where "+"is the bitwise addition module 2. For attacking the generalized DES-like iterated cipher, the above *difference* can be extended to a more general form, called *generalized difference* which is defined as in the following

Definition 1. $\Delta_{A,u}x$ is called *the generalized difference* about group (A,*) and function 'u' between x1 and x2 ($x1,x2 \in GF(2^n)$) ,if,

$u : GF(2^n) \rightarrow A$ is a one-to-one mapping.

$\Delta_{A,u}x = u(x1)*(u(x2))^{-1}$, where Y^{-1} is the inverse of Y in the group (A,*) .

In [1], when Lai made analysis of the performance of the IDEA cipher, he defined a special difference, $\Delta x=x1 \otimes x2^{-1}$, where $\otimes$ is an operation as

$$x \otimes y=(x1 \odot y1,x2+y2,x3+y3,x4 \odot y4)$$

in which x,y are 64-bit blocks, x1,x2,x3,x4,y1,y2,y3 and y4 are 16-bit block, x=(x1,x2,x3,x4)and y=(y1,y2,y3,y4), $\odot$ is multiplication module $(2^{16}+1)$ with zero block corresponding to $m=2^n$, + is addition module 2^{16} .Thus this special difference can be considered as a special kind of generalized difference $\Delta_{A,u}x$, where $A=Z^*_{m+1} \times Z_m \times Z_m \times Z^*_{m+1}$, $m=2^{16}$, u=(u1,u2,u3,u4), u2(x)=u3(x)=x for any 16-bit x, u1(x)=u4(x)=x if 16-bit x not equal to 0, $u1(0)=u4(0)=2^{16}+1$.

Differential cryptanalysis is a chosen-plaintext attack which examines the changes in the output of the cipher in response to the controlled changes in the input of plain text. It uses s-round differentials to push forward the information of a fixed input difference at the first round. In differential cryptanalysis, the effect of particular difference in plaintext pairs on the differences in the resultant cipher pairs is analyzed. By choosing a certain fixed difference value Ω_p of a plain text pair ,the probability on the difference value Ω^n_T of *n*-th round output pairs can be obtained. Let λ^i_I be a certain difference value of an input pair of the F function in the *i*-th round. Let λ^i_O be the difference value of the resultant output pair of the F function in the *i*-th round. An n-round characteristic [2] is defined as a tuple $\Omega^n= (\Omega_p, \Omega^n_\Lambda ,\Omega^n_T)$, where the intermediate round differential value Ω^n_Λ is defined as a list of n element $\Omega^n_\Lambda=(\Lambda_1,\Lambda_2,\ldots,\Lambda_n)$, each of which is a pair in the form $\Lambda_i=(\lambda^i_I,\lambda^i_O)$ for $1 \leq i \leq n$. Let p_i be the probability of $\lambda^i_I \rightarrow \lambda^i_O$ in DES-like cipher. An n-round characteristic Ω^n has probability *p*: $p=\prod_{i=1}^{n} p_i$. A "right pair "is a plaintext (P,P') which satisfy the condition that their difference is equal to Ω_p and $\lambda^i_I \rightarrow \lambda^i_O$ for all *i*. The probability of a characteristic is the actual probability that any fixed plaintext pair is a right pair when random independent keys are used. The attack tries many pairs of plaintext to find the right pairs. Differential cryptanalysis is carried out by choosing Ω^n to maximize $p=\prod_{i=1}^{n} p_i$. The key to success on carrying out a differential attack in DES-like cipher is to use either a high characteristic probability or high differential probability.

If we denote p_{max} the highest probability for a nontrivial one-round generalized differential achievable by the cryptanalysis, i.e.

$$p_{max}=\max_{A\_x,u\_x}\max_{A\_y,u\_y}\max_{b}\max_{a\neq 0}P(\Delta_{A\_y,u\_y}y=b \mid \Delta_{A\_x,u\_x}x=a),$$

where x is the input of function F . y is the output of function F.

Then for any p_i which is the probability of $\lambda^i_I \rightarrow \lambda^i_O$ in DES-like cipher where λ^i_I do not equal 0, we have $p_i \leq p_{max}$.

Thus P_{max} is an important index for DES-like cipher in its ability against differential cryptanalysis. To design a good cipher, we should make no nontrivial s-round differential useful.

3. Multiplication-addition Structure and the Round Function Constructed from this Structure

We have shown that an important way to make the generalized DES-like cipher stronger against differential cryptanalysis is to decrease the P_{max} . In this section, a multiplication-addition structure is proposed. Based on this multiplication-addition structure , we can construct "good" round function to achieve this goal.

First, multiplication-addition structure is defined as following.

Definition 2 . An operation MA(· , · , ·) is called a *multiplication-addition structure* in the finite field F , if, MA: $F \times F \times F \rightarrow F$, $MA(X,K_1,K_2)=(X \times K_1+K_2)$ where $\times$ is the multiplication operation in the field F, and $+$ is the addition operation in the field F.

Based on multiplication-addition structure , a round function compatible with a multiplication - addition structure is defined as in the following.

Definition 3. The round function F: $GF(2^n) \times GF(2^{2n}) \rightarrow GF(2^n)$ of a generalized DES-like iterated cipher is *compatible with a multiplication-addition structure in finite field* $GF(2^n)$,if, the round function

$$F: GF(2^n) \times GF(2^{2n}) \rightarrow GF(2^n), \quad F(X,K)=MA(X,K_L,K_R),$$

where K_L is the left half of K and K_R is the right half of K, and MA is a multiplication-addition structure.

Theorem 1. Assume that the round function F: $GF(2^n) \times GF(2^{2n}) \rightarrow GF(2^n)$ of a generalized DES-like iterated cipher is compatible with a multiplication-addition structure in finite field $GF(2^n)$ and the round keys are independent and uniformly random. Then $p_{max} = \frac{1}{2^n}$

Proof. Since the round function F: $GF(2^n) \times GF(2^{2n}) \rightarrow GF(2^n)$ of a generalized DES-like iterated cipher is compatible with a multiplication-addition structure in finite field $GF(2^n)$.

So F: $GF(2^n) \times GF(2^{2n}) \rightarrow GF(2^n)$, $F(x,k)=x \times kl+kr$,
where kl is the left half of k, kr is the right half of k.
For any certain groups A and C where
$(A=\{a_1,\dots,a_m\}, *)$,$m=2^n$, is a group and * is the multiplication operation in A.
$(C=\{c_1,\dots,c_m\}, \otimes)$,$m=2^n$, is a group and $\otimes$ is the multiplication operation in C.
and any certain one-to-one function h and g

$h: GF(2^n) \rightarrow A$,$g: GF(2^n) \rightarrow C$,

and any certain $a \in A$, but a is not the unit element of A, and any certain $c \in C$,
the probability of nontrivial one-round differential of F function is

$$P(\Delta_{C,g}y=c \mid \Delta_{A,h}x=a)=\frac{P(\Delta_{C,g}y=c,\Delta_{A,h}x=a)}{P(\Delta_{A,h}x=a)}$$

But, $P(\Delta_{C,g}y=c, \Delta_{A,h}x=a)$

$$=\sum_{i=1}^{m} P(\Delta_{C,g}y=c, x1=h^{-1}(a_i), x2=h^{-1}(a^{-1}*a_i)) \quad (because \quad \Delta_{A,h}x=h(x1)*(h(x2))^{-1})$$

$$=\sum_{i=1}^{m} P(\Delta_{C,g}y=c \mid x1=h^{-1}(a_i), x2=h^{-1}(\overline{a_i}))P(x1=h^{-1}(a_i), x2=h^{-1}(\overline{a_i})) \quad (where \quad \overline{a_i}=a^{-1}*a_i)$$

$$=\sum_{i=1}^{m} P(g(\overline{h}(a_i)\times kl+kr)\otimes(g(\overline{h}(\overline{a_i})\times kl+kr))^{-1}=c)P(x1=\overline{h}(a_i), x2=\overline{h}(\overline{a_i})) \quad (where \quad \overline{h}=h^{-1})$$

$$=\sum_{i=1}^{m}[(P(x1=\overline{h}(a_i), x2=\overline{h}(\overline{a_i}))\sum_{j=1}^{m} P(g(\overline{h}(a_i)\times kl+kr)=c_j, g(\overline{h}(\overline{a_i})\times kl+kr)=\overline{c_j})]$$

$$(where \quad \overline{c_j}=c^{-1}\otimes c_j)$$

$$=\sum_{i=1}^{m}[(P(x1=\underline{a_i}, x2=\underline{\overline{a_i}})\sum_{j=1}^{m} P(\underline{a_i}\times kl+kr=\underline{c_j}, \underline{\overline{a_i}}\times kl+kr=\underline{\overline{c_j}}))]$$

$$(where \quad \underline{c_j}=g^{-1}(c_j), \underline{\overline{c_j}}=g^{-1}(\overline{c_j}), \underline{a_i}=\overline{h}(a_i), \underline{\overline{a_i}}=\overline{h}(\overline{a_i}))$$

$$=\sum_{i=1}^{m}[(P(x1=\overline{h}(a_i), x2=\overline{h}(\overline{a_i}))\sum_{j=1}^{m} P(kl=\overline{kl_{i,j}}, kr=\overline{kr_{i,j}})]$$

$$(where \quad \overline{kl_{i,j}}=(\underline{a_i}-\underline{\overline{a_i}})^{-1}\times(\underline{c_j}-\underline{\overline{c_j}}), \overline{kr_{i,j}}=(\underline{\overline{a_i}}-\underline{a_i})^{-1}\times(\underline{\overline{a_i}}\times\underline{c_j}-\underline{a_i}\times\underline{\overline{c_j}})$$

$$=\sum_{i=1}^{m}(P(x1=\overline{h}(a_i), x2=\overline{h}(\overline{a_i}))\sum_{j=1}^{m}\frac{1}{m^2}$$

(because k is an independent and uniformly distributed random variable)

$$=\frac{1}{m}\sum_{i=1}^{m}(P(x1=\overline{h}(a_i), x2=\overline{h}(\overline{a_i}))=\frac{1}{m}P(\Delta_{A,h}x=a)$$

(In the above equation, $(\cdot)^{-1}$ expresses the inverse of the element in its corresponding group)

Thus, $P(\Delta_{C,g}y=c \mid \Delta_{A,h}x=a)=\frac{1}{m}=\frac{1}{2^n}=constant$

So, $p_{max}=\frac{1}{m}=\frac{1}{2^n}$ □

Definition 4 . An operation $MA(\cdot,\cdot,\cdot)$ is called a *one-to-one multiplication-addition structure* in the finite field F
,if, $MA: F \times F\backslash\{0\} \times F \rightarrow F$, where 0 is the zero element of the finite field F

$MA(X,K_1,K_2)=(X \times K_1+K_2)$

$\times$ is the multiplication operation in the field F
$+$ is the addition operation in the field F.

Obviously, a one-to-one multiplication-addition structure MA(x, · , ·) is a one-to-one function for any certain K_1,K_2 .

Definition 5. The round function F: $GF(2^n)\times GF(2^{2n})\rightarrow GF(2^n)$ of a generalized DES-like iterated cipher is called a round function *compatible with a one-to-one multiplication-addition structure* in finite field $GF(2^n)$, if, the round function is

F: $GF(2^n)\times GF(2^{2n})\rightarrow GF(2^n)$, $F(X,K)=MA(X,K_L,K_R)$,

where K_L is the left half of K and K_R is right half of K, and

MA is a one-to-one multiplication-addition structure.

Theorem 2. Assume that the round function F: $GF(2^n)\times GF(2^{2n})\rightarrow GF(2^n)$ of a generalized DES-like iterated cipher is compatible with a one-to-one multiplication-addition structure in finite field $GF(2^n)$ and the round keys are independent and uniformly random in $GF(2^{2n})\backslash\{(0,K_R),\ K_R\in GF(2^n)\}$. Then

$$p_{max}=\frac{1}{2^n-1}$$

Proof. As the round function F: $GF(2^n)\times GF(2^{2n})\rightarrow GF(2^n)$ of a generalized DES-like iterated cipher is compatible with a one-to-one multiplication-addition structure in finite field $GF(2^n)$.

So F: $GF(2^n)\times GF(2^{2n})\rightarrow GF(2^n)$, $F(x,k)=x\times kl+kr$,

where kl is left half of k, kr is right half of k.

For any certain group A,C where

$(A=\{a_1,\ldots,a_m\},\ *)$,$m=2^n$,is a group and * is multiplication operation in A,

$(C=\{c_1,\ldots,c_m\},\ \otimes)$,$m=2^n$,is a group and $\otimes$ is multiplication operation in C,

and any certain one-to-one function h and g, $h:GF(2^n)\rightarrow A$, $g:GF(2^n)\rightarrow C$,

and any certain $a\in A$,but a is not the unit element of A, and any certain $c\in C$,

The probability of nontrivial one-round differential of F function is

$$P(\Delta_{C,g}y=c|\Delta_{A,h}x=a)=\frac{P(\Delta_{C,g}y=c,\Delta_{A,h}x=a)}{P(\Delta_{A,h}x=a)}$$

But, $P(\Delta_{C,g}y=c,\Delta_{A,h}x=a)$

$$=\sum_{i=1}^{m}P(\Delta_{C,g}y=c|x1=h^{-1}(a_i),x2=h^{-1}(\overline{a_i}))P(x1=h^{-1}(a_i),x2=h^{-1}(\overline{a_i}))\qquad (where\ \ \overline{a_i}=a^{-1}*a_i)$$

$$=\sum_{i=1}^{m}P(g(\overline{h}(a_i)\times kl+kr)\otimes(g(\overline{h}(\overline{a_i})\times kl+kr))^{-1}=c)P(x1=\overline{h}(a_i),x2=\overline{h}(\overline{a_i}))\quad (where\ \ \overline{h}=h^{-1})$$

$$=\sum_{i=1}^{m}[(P(x1=\overline{h}(a_i),x2=\overline{h}(\overline{a_i}))\sum_{j=1}^{m}P(g(\overline{h}(a_i)\times kl+kr)=c_j,g(\overline{h}(\overline{a_i})\times kl+kr)=\overline{c_j})]$$

$$(where\quad \overline{c_j}=c^{-1}\otimes c_j)$$

$$=\sum_{i=1}^{m}[(P(x1=\underline{a_i},x2=\underline{\overline{a_i}})\sum_{j=1}^{m}P(\underline{a_i}\times kl+kr=\underline{c_j},\underline{\overline{a_i}}\times kl+kr=\underline{\overline{c_j}}))]$$

$$(where\quad \underline{c_j}=g^{-1}(c_j),\underline{\overline{c_j}}=g^{-1}(\overline{c_j}),\underline{a_i}=\overline{h}\ (a_i),\underline{\overline{a_i}}=\overline{h}(\overline{a_i}))$$

$$= \sum_{i=1}^{m} [(P(x1 = \bar{h}(a_i), x2 = \bar{h}(\overline{a_i})) \sum_{j=1}^{m} P(kl = \overline{kl_{i,j}}, kr = \overline{kr_{i,j}})]$$

(where $\overline{kl_{i,j}} = (\underline{a_i} - \underline{\overline{a_i}})^{-1} \times (\underline{c_j} - \underline{\overline{c_j}}) \neq 0,, \overline{kr_{i,j}} = (\underline{a_i} - \underline{\overline{a_i}})^{-1} \times (\underline{\overline{a_i} \times c_j} - \underline{a_i \times \overline{c_j}})$

$$= \sum_{i=1}^{m} (P(x1 = \bar{h}(a_i), x2 = \bar{h}(\overline{a_i})) \sum_{j=1}^{m} \frac{1}{m(m-1)}$$

(because k is independent and uniformly random in $GF(2^{2n}) \setminus \{(0, kr)\}$

$$= \frac{1}{(m-1)} \sum_{i=1}^{m} (P(x1 = \bar{h}(a_i), x2 = \bar{h}(\overline{a_i})) = \frac{1}{(m-1)} P(\Delta_{A,h} x = a)$$

thus, $P(\Delta_{C,g} y = c | \Delta_{A,h} x = a) = \frac{1}{m-1} = \frac{1}{2^n - 1} = \text{constant}$

so, $p_{\max} = \frac{1}{m-1} = \frac{1}{2^n - 1}$ □

From Theorem1 and Theorem2, it is proved that a generalized DES-like cipher can resist differential cryptanalysis by replacing the traditional F function with a round function compatible with multiplication-addition structure. But maybe it has weakness for other cryptanalysis. So it should be used with other crypto-structure. We will consider this in the following .

Definition 6. The round function F: $GF(2^n) \times GF(2^{2n}) \rightarrow GF(2^n)$ of a generalized DES-like iterated cipher is called a round function F *containing a multiplication-addition structure* in finite field $GF(2^n)$,if, the round function

F: $GF(2^n) \times GF(2^m) \rightarrow GF(2^n)$, is: F(X,K)=T(MA(S(X,k1),k2,k3),k4)

where K=(k1,k2,k3,k4), k1 $\in GF(2^{m1})$, k2 $\in GF(2^n)$,k3 $\in GF(2^n)$,k4 $\in$ $GF(2^{m4})$,

m=m1+n+n+m4.

S:$GF(2^n) \times GF(2^{m1}) \rightarrow GF(2^n)$

S(· ,k) is a one-to-one mapping for any k $\in GF(2^{m1})$,

T:$GF(2^n) \times GF(2^{m4}) \rightarrow GF(2^n)$

T(· ,k) is a one-to-one mapping for any k $\in GF(2^{m4})$,

MA is a multiplication-addition structure.

Theorem 3. Assume that the round function F: $GF(2^n) \times GF(2^m) \rightarrow GF(2^n)$, of a generalized DES-like iterated cipher contains a multiplication-addition structure in finite field $GF(2^n)$ and the round keys k1,k2,k3,k4 are independent and k2,k3 uniformly random in $GF(2^n)$. Then $p_{\max} = \frac{1}{2^n}$

Proof. Because the round function F: $GF(2^n) \times GF(2^{2n}) \rightarrow GF(2^n)$ of a generalized DES-like iterated cipher contains a multiplication-addition structure in finite field $GF(2^n)$,

so F: $GF(2^n) \times GF(2^m) \rightarrow GF(2^n)$

F(X,K)=T((S(X,k1) × k2+k3),k4) , where K=(k1,k2,k3,k4)

For any certain group A,C where

(A={a_1,...,a_m}, *) ,m=2^n,is a group and * is multiplication operation in A.

$(C=\{c_1,\dots,c_m\},\ \otimes)$,$m=2^n$,is a group and $\otimes$ is multiplication operation in C.
and any certain one-to-one function h and g

$h: GF(2^n)\to A$,$g: GF(2^n)\to C$,

and any certain $a \in A$, but a is not the unit element of A, and any certain $c \in C$, the probability of nontrivial one-round differential of F function

$$P(\Delta_{C,g}y=c|\Delta_{A,h}x=a)=\frac{P(\Delta_{C,g}y=c,\Delta_{A,h}x=a)}{P(\Delta_{A,h}x=a)}$$

$$P(\Delta_{C,g}y=c,\Delta_{A,h}x=a)$$

$$=\sum_{i=1}^{m}P(\Delta_{C,g}y=c|x1=h^{-1}(a_i),x2=h^{-1}(\overline{a_i}))P(x1=h^{-1}(a_i),x2=h^{-1}(\overline{a_i}))\qquad (where\ \ \overline{a_i}=a^{-1}*a)$$

$$=\sum_{i=1}^{m}[P(g(T(S(\overline{h}(a_i),k1)\times k2+k3,k4))\otimes(g(T(S(\overline{h}(\overline{a_i}),k1)\times k2+k3,k4)))^{-1}=c)\cdot$$

$$P(x1=\overline{h}(a_i),x2=\overline{h}(\overline{a_i}))]\qquad (where\ \ \overline{h}=h^{-1})$$

$$=\sum_{i=1}^{m}[(P(x1=\overline{h}(a_i),x2=\overline{h}(\overline{a_i}))$$

$$\sum_{j=1}^{m}P(g(T(S(\overline{h}(a_i),k1)\times k2+k3,k4)=c_j,g(T(S(\overline{h}(\overline{a_i}),k1)\times k2+k3,k4)=\overline{c_j})]$$

$$(where\ \ \overline{c_j}=c^{-1}\otimes c_j)$$

$$=\sum_{i=1}^{m}[(P(x1=\underline{a_i},x2=\underline{\overline{a_i}})\cdot$$

$$\sum_{j=1}^{m}\sum_{\overline{k1}}\sum_{\overline{k4}}P(\underline{a_i}\times k2+k3=\underline{c_j},\underline{\overline{a_i}}\times k2+k3=\underline{\overline{c_j}})P(k1=\overline{k1})P(k4=\overline{k4})]$$

$$(where\ \ \underline{c_j}=T(g^{-1}(c_j),\overline{k4})^{-1},\underline{\overline{c_j}}=T(g^{-1}(\overline{c_j}),\overline{k4})^{-1},$$

$$\underline{a_i}=S(\overline{h}(a_i),\overline{k1}),\underline{\overline{a_i}}=S(\overline{h}(\overline{a_i}),\overline{k1})\ )$$

$$=\sum_{i=1}^{m}[(P(x1=\overline{h}(a_i),x2=\overline{h}(\overline{a_i}))\sum_{j=1}^{m}\sum_{\overline{k1}}\sum_{\overline{k4}}P(k1=\overline{k1})P(k4=\overline{k4})P(k2=\overline{k2_{i,j}},k3=\overline{k3_{i,j}})]$$

$$(where\ \ \overline{k2_{i,j}}=(\underline{a_i}-\underline{\overline{a_i}})^{-1}\times(\underline{c_j}-\underline{\overline{c_j}})\neq 0,\overline{k3_{i,j}}=(\underline{\overline{a_i}}-\underline{a_i})^{-1}\times(\underline{\overline{a_i}}\times\underline{c_j}-\underline{a_i}\times\underline{\overline{c_j}})$$

$$=\sum_{i=1}^{m}[(P(x1=\overline{h}(a_i),x2=\overline{h}(\overline{a_i}))\sum_{j=1}^{m}\frac{1}{m^2}\sum_{\overline{k1}}\sum_{\overline{k4}}P(k1=\overline{k1})P(k4=\overline{k4})]$$

(because k1,k2,k3,k4 are independent

and k2,k3 are uniformly random in $GF(2^n)$)

$$=\frac{1}{m}\sum_{i=1}^{m}(P(x1=\overline{h}(a_i),x2=\overline{h}(\overline{a_i}))=\frac{1}{m}P(\Delta_{A,h}x=a)$$

thus, $P(\Delta_{C,g}y=c|\Delta_{A,h}x=a)=\dfrac{1}{m}=\dfrac{1}{2^n}=\text{constant}$

so, $p_{\max}=\dfrac{1}{m}=\dfrac{1}{2^n}$ □

Definition 7. The round function F: $GF(2^n)\times GF(2^m)\to GF(2^n)$ of a generalized DES-like iterated cipher is called a round function *containing a one-to-one multiplication-addition structure* in finite field $GF(2^n)$,if, the round function is

$F: GF(2^n) \times GF(2^m) \to GF(2^n)$, $F(X,K)=T((MA(S(X,k1),k2),k3),k4)$

where $K=(k1,k2,k3,k4)$, $k1 \in GF(2^{m1})$, $k2 \in GF(2^n)$, $k3 \in GF(2^n)$, $k4 \in GF(2^{m4})$, $m=m1+n+n+m4$.

$S: GF(2^n) \times GF(2^{m1}) \to GF(2^n)$

$S(\cdot,k)$ is a one-to-one function for any $k \in GF(2^{m1})$,

$T: GF(2^n) \times GF(2^{m4}) \to GF(2^n)$

$T(\cdot,k)$ is a one-to-one function for any $k \in GF(2^{m4})$,

MA is a one-to-one multiplication-addition structure.

Theorem 4. Assume that the round function $F: GF(2^n) \times GF(2^m) \to GF(2^n)$, of a generalized DES-like iterated cipher contains a one-to-one multiplication-addition structure in finite field $GF(2^n)$ and the round keys k1,k2,k3,k4 are independent and k2 uniformly random in $GF(2^n)\setminus\{0\}$, k3 uniformly random in $GF(2^n)$. Then

$$p_{max} = \frac{1}{2^n - 1}$$

Proof. Because the round function $F: GF(2^n) \times GF(2^{2n}) \to GF(2^n)$ of a generalized DES-like iterated cipher contains a one-to-one multiplication-addition structure in finite field $GF(2^n)$.

so $F: GF(2^n) \times GF(2^m) \to GF(2^n)$

$F(X,K)=T((S(X,k1) \times k2+k3),k4)$ where $K=(k1,k2,k3,k4)$

For any certain group A,C where

$(A=\{a_1,\dots,a_m\}, *)$,$m=2^n$,is a group and * is multiplication operation in A,

$(C=\{c_1,\dots,c_m\}, \otimes)$,$m=2^n$,is a group and $\otimes$ is multiplication operation in C,

and any certain one-to-one function h and g, $h: GF(2^n) \to A$,$g: GF(2^n) \to C$,

and any certain $a \in A$,but a is not the unit element of A, and any certain $c \in C$, the probability of nontrivial one-round differential of F function

$$P(\Delta_{C,g}y=c \mid \Delta_{A,h}x=a) = \frac{P(\Delta_{C,g}y = c, \Delta_{A,h}x = a)}{P(\Delta_{A,h}x = a)}$$

but , $P(\Delta_{C,g}y=c, \Delta_{A,h}x=a)=$

$$=\sum_{i=1}^{m} P(\Delta_{C,g}y=c \mid x1=h^{-1}(a_i),x2=h^{-1}(\overline{a_i}))P(x1=h^{-1}(a_i),x2=h^{-1}(\overline{a_i})) \qquad (where\ \ \overline{a_i}=a^{-1}*a_i)$$

$$=\sum_{i=1}^{m}[P(g(T(S(\overline{h}(a_i),k1)\times k2+k3,k4))\otimes(g(T(S(\overline{h}(\overline{a_i}),k1)\times k2+k3,k4)))^{-1}=c)\cdot$$

$$P(x1=\overline{h}(a_i),x2=\overline{h}(\overline{a_i}))] \qquad (where\ \ \overline{h}=h^{-1})$$

$$=\sum_{i=1}^{m}[(P(x1=\overline{h}(a_i),x2=\overline{h}(\overline{a_i}))$$

$$\sum_{j=1}^{m} P(g(T(S(\overline{h}(a_i),k1)\times k2+k3,k4)=c_j, g(T(S(\overline{h}(\overline{a_i}),k1)\times k2+k3,k4)=\overline{c_j})]$$

$$(where\ \ \overline{c_j}=c^{-1}\otimes c_j)$$

$$= \sum_{i=1}^{m} [(P(x1 = \underline{a_i}, x2 = \underline{\overline{a_i}}) \cdot$$

$$\sum_{j=1}^{m} \sum_{\overline{k1}} \sum_{\overline{k4}} P(\underline{a_i} \times k2 + k3 = \underline{c_j}, \underline{\overline{a_i}} \times k2 + k3 = \underline{\overline{c_j}}) P(k1 = \overline{k1}) P(k4 = \overline{k4})]$$

$$(where \quad \underline{c_j} = T(g^{-1}(c_j), \overline{k4})^{-1}, \underline{\overline{c_j}} = T(g^{-1}(\overline{c_j}), \overline{k4})^{-1}$$

$$\underline{a_i} = S(\bar{h}(a_i), \overline{k1}), \underline{\overline{a_i}} = S(\bar{h}(\overline{a_i}), \overline{k1}) \;)$$

$$= \sum_{i=1}^{m} [(P(x1 = \bar{h}(a_i), x2 = \bar{h}(\overline{a_i})) \sum_{j=1}^{m} \sum_{\overline{k1}} \sum_{\overline{k4}} P(k1 = \overline{k1}) P(k4 = \overline{k4}) P(k2 = \overline{k2_{i,j}}, k3 = \overline{k3_{i,j}})]$$

$$(where \quad \overline{k2_{i,j}} = (\underline{a_i} - \underline{\overline{a_i}})^{-1} \times (\underline{c_j} - \underline{\overline{c_j}}) \neq 0,, \overline{k3_{i,j}} = (\underline{\overline{a_i}} - \underline{a_i})^{-1} \times (\underline{\overline{a_i}} \times \underline{c_j} - \underline{a_i} \times \underline{\overline{c_j}})$$

$$= \sum_{i=1}^{m} [(P(x1 = \bar{h}(a_i), x2 = \bar{h}(\overline{a_i})) \sum_{j=1}^{m} \frac{1}{m(m-1)} \sum_{\overline{k1}} \sum_{\overline{k4}} P(k1 = \overline{k1}) P(k4 = \overline{k4})]$$

(because k1,k2,k3,k4 are independent ,k2 are uniformly random in $GF(2^n) \setminus \{0\}$ and k3 are uniformly random in $GF(2^n)$, $m = 2^n$)

$$= \frac{1}{(m-1)} \sum_{i=1}^{m} (P(x1 = \bar{h}(a_i), x2 = \bar{h}(\overline{a_i})) = \frac{1}{(m-1)} P(\Delta_{A,h} x = a)$$

$$\text{so}, P(\Delta_{C,g} y = c | \Delta_{A,h} x = a) = \frac{1}{m-1} = \frac{1}{2^n - 1}$$

$$\text{so}, \; p_{\max} = \frac{1}{m-1} = \frac{1}{2^n - 1} \qquad \square$$

Theorem3 and Theorem4 prove that a DES-like cipher can resist differential cryptanalysis by replacing the traditional F function with a round function containing a multiplication-addition structure. Concatenating a one-to-one function which is controlled by key before or after a multiplication-addition structure has no effect on the perfect property about difference. So we can concatenate a one-to-one function which is controlled by key before or after a multiplication-addition structure to resist other kinds of cryptanalysis.

4. The Application of Multiplication-addition Structure

The multiplication addition structure and the round function constructed from the multiplication addition structure has been investigated in theory in the section 3. Now we consider some practical problem in the application of the multiplication addition structure.

In [4], Nyberg prove that in a DES-like cipher if round keys are independent and uniform random then the probability of an s-round differential, s>3 ,is less than or equal to $2p^2_{max}$. For a DES-like cipher in which the F function is a permutation and round keys are independent and uniform random , the probability of an s-round differential, s>2 ,is less than or equal to $2p^2_{max}$. Similar conclusion can be proved for the generalized DES-like cipher.

The multiplication-addition structure is not only able to be used in the DES-like cipher . It also can be used to replace the group operation between two rounds in the IDEA cipher. By applying Theorem4, we can prove the maximum probability of one round differential of an IDEA-like cipher with multiplication-addition structure is about $1/2^{64}$.

5. Conclusion

We have introduced in this paper a new class of round function to strengthen the DES-like cipher against a differential attack. The kernel of this class of round function is the multiplication-addition structure. Based on this multiplication-addition structure, we constructed four special round functions. It has been proved that the probability of one round differential of any of these round functions reaches its possible minimum. Since the existence of high probability differential is the basis for differential cryptanalysis, this means that these round functions are very useful in making the differential attack difficult or less effective.

The multiplication-addition structure which we have proposed can not only be used for DES-like cipher but also can be used for IDEA-like cipher[9] to resist differential cryptanalysis.

Acknowledgment

We would like to thank the anonymous referees for comments that improved the paper.

References

[1]X.Lai, J.massey, and S.Murphy. Markov cipher and differential cryptanalysis. Advances in Cryptology - CRYPTO'91. Lecture Notes in Computer Science , Vol. 547 Springer-Verlag, Berlin,1992,pp.17-38.

[2]E.Biham and Shamir. Differential cryptanalysis of DES-like systems. Journal of Cryptology, Vol4,No.1,1991,pp.3-72.

[3]E.Biham and A.Shamir . Differential cryptanalysis of the full 16-round DES, Advances in Cryptology: Proceeding of CRYPTO'92, Springer-Verlag, Berlin, 1993, pp. 487-496.

[4]Kaisa Nyberg and Lars Ramkilde Knudsen. Provable security against a differential attack, Journal of Cryptology , Vol8,1995,pp.27-37.

[5]E.Biham and A.Shamir . Differential cryptanalysis of FEAL and N-Hash, Advances in Cryptology: Proceeding of EUROCRYPTO'91, Springer-Verlag, Berlin, 1991, pp. 1-16

[6]E.Biham and A.Shamir . Differential cryptanalysis of Snefru, Khafre,REDOC-II, LOKI, and Lucifer, Advances in Cryptology: Proceeding of CRYPTO'91, 1992, pp. 156-171

[7]L.Brown, M.Kwan,J.Pieprzyk and J.Seberr. Improving resistance to differential cryptanalysis and the redesign of LOKI, Proceeding of ASIACRYPTO'91, 1992, pp. 28-30

[8]T.KANEKO,K.KOYAMA and R.TERADA. Dynamic swapping schemes and differential cryptanalysis , IEICE Trans. fundamentals Vol.E77-a No.8 ,1994, pp1328-1335

[9]Feng Zhu and Bao-An Guo. A block-ciphering algorithm based on addition-multiplication structure in GF(2^n), the 4th in a series of annual workshops on selected areas in cryptograph, Ottawa, 1997.

On Strict Estimation Method of Provable Security against Differential and Linear Cryptanalysis

Yasuyoshi Kaneko[1] Shiho Moriai[1] Kazuo Ohta[2]

[1]Yokohama Research Center,
Telecommunications Advancement Organization of Japan,
1-1-32 Shin'urashima-cho, Kanagawa-ku, Yokohama, 221 JAPAN
E-mail: kaneko@yokohama.tao.or.jp, shiho@yokohama.tao.or.jp

[2]NTT Laboratories,
1-1 Hikarinooka, Yokosuka-shi, 239 JAPAN
E-mail: ohta@sucaba.isl.ntt.co.jp

Abstract. We give stricter upper bounds to the probabilities of differential and linear hull of DES-like ciphers than the previous results. The previous results in [6, 7] said that every r-round differential (or linear hull) with $r \geq 4$ is bounded by $2p^2$ (or $2q^2$) where p (or q) is the maximum probability of a non-trivial differential (or linear hull) of the function which is used in each round. Using our new estimation method of provable security it is shown that these bounds change depending on the number of rounds. This change gives a decrease function of the number of rounds with the limit value. Moreover, our estimation gives $2p^2 - p^3$ (or $2q^2 - q^3$) for 4 and 5 rounds so our method is stricter than the previous one. The bounds converge to about p^2(or q^2) for an increasing number of rounds.

1 Introduction

Differential[1] and linear cryptanalysis[3] are effective attacks on some iterated ciphers, such as DES and so on. It is essential for designers of block ciphers to evaluate the probabilities of differentials and approximate linear hulls to ensure adequate security against these cryptanalyses. In [1] Biham and Shamir described the differential cryptanalysis of DES in terms of "r-round characteristics", in which the intermediate differences are specified. Lai, Massey and Murphy introduced the notion of the differential in order to evaluate the security against differential cryptanalysis. Matsui described the linear cryptanalysis of DES in terms of "linear approximation"[3], and Nyberg introduced the notion of the approximate linear hull and pointed out that the evaluation of the probability of linear hull is important to ensure the complete security of a block cipher against linear cryptanalysis[6]. Though it is possible to determine the probabilities of an "r-round characteristic" or "r-round linear approximation" of DES-like ciphers using Matsui's search algorithm[4], it seems infeasible to estimate the probabilities of r-round differential (or linear hull) for an arbitrary r. This is because the probability of differential (linear hull) is the sum of the probabilities of all r-round characteristic (linear approximation) with the difference (masking value) of inputs and that of outputs fixed, where all of characteristics (masking values) of the intermediate process must be considered, which makes the search computation enormous.

On estimating the upper bound to the probability of r-round differential (or linear hull), in [6] and [7] Knudsen and Nyberg showed an estimation method using the maximum probability for a non-trivial one-round differential (linear hull). The results by this approach in the case of an r-round DES-like cipher with $r \geq 4$ are shown as below:

1. the upper bound to the probability of r-round differentials is less than or equal to $2p^2$, where p is the maximum differential probability of the function used in each round [7], and
2. the upper bound to the probability of r-round liner hulls is less than or equal to $2q^2$, where q is the maximum linear-hull probability of the function[6].

It is generally known that when the transition probabilities of r-round differential (or linear hull) satisfy the ergodic property, these probabilities exponentially decrease as the round number increases[2, 8]. Moreover many reports say that a block cipher is expected to be cryptanalytically "stronger" as the round function is iterated more times[1, 2, 3, 9]. Thus the upper bounds to the probability of differential (linear hull) are also expected to decrease as the round number increases.

The main purpose of this paper is to present a new method of estimating provable security which gives tighter bound of the probability of r-round differential (or linear hull). Our new bound changes with the round number and especially the bound becomes $2p^2 - p^3$ (or $2q^2 - q^3$) for 4 and 5 rounds(see Table 1). These upper bounds give stricter results than previous results of $2p^2$ (or $2q^2$) for $r \geq 4$ [6, 7].

2 Preparation

2.1 DES-like Ciphers

In this paper, a DES-like cipher is considered as an r-round iterated block cipher which is defined as follows. Let $X(i) = (X_L(i), X_R(i))$ $(i = 1, \ldots, r;\ X_L(i), X_R(i) \in GF(2)^n)$ be the i-th round input data and $Y(i) = (Y_L(i), Y_R(i))$ $(i = 1, \ldots, r;\ Y_L(i), Y_R(i) \in GF(2)^n)$ be the i-th round output data. Let $K = (K(1), \ldots, K(r))$ be a key which consists of the i-th round key $K(i)$ $(i = 1, \ldots, r)$. We assume that round keys $K(i)$ $(i = 1, \ldots, r)$ are independent and uniformly random. The round function $Y(i) = F(X(i), K(i))$ is given as follows:

$$X_L(i+1) = X_R(i),$$
$$X_R(i+1) = X_L(i) \oplus f(X_R(i), K(i)),$$
$$Y_L(i) = X_L(i+1),\ Y_R(i) = X_R(i+1) \quad (i = 1, \ldots, r-1).$$

where function f is assumed to be given by $f(X, K) = \psi(E(X) \oplus K)$ and E is an affine mapping from $GF(2)^n$ to $GF(2)^m$. Plaintext $P = (P_L, P_R)$ gives the equation of $P_R = X_R(1)$ and $P_L = X_L(1)$, and ciphertext $C = (C_L, C_R)$ is obtained as $C_L = X_L(r) \oplus f(X_R(r), K(i))$, $C_R = X_R(r)$.

2.2 Probabilities of r-round Differential and Linear hull

To analyze r-round iterated block ciphers from the viewpoint of provable security against differential and linear cryptanalysis, we define the differential probability dp and linear hull probability lp of the function used in each round as follows.

Definition 1 [2, 6] *Let $y = f(x,k)$ with $x, y \in GF(2)^n$ and $k \in GF(2)^m$ be a function. For given differential values Δx and $\Delta y \in GF(2)^n$, the differential probability dp is defined as follows:*

$$dp(\Delta x \to \Delta y) := \underset{k}{aver}\ \underset{x}{Prob}\{\ f(x \oplus \Delta x, k) \oplus f(x,k) = \Delta y\}.$$

For given masking values Γx and $\Gamma y \in GF(2)^n$, the linear hull probability lp is defined as follows:

$$lp(\Gamma y \to \Gamma x) := \underset{k}{aver} \left|\ 2\underset{x}{Prob}\{\ x \bullet \Gamma x \oplus f(x,k) \bullet \Gamma y = 0\}\ -1\ \right|^2$$

where $\bullet$ denotes the inner product between a pair of elements in $GF(2)^n$. The operation "$\bullet$" has higher precedence than "$\oplus$".

As shown by Nyberg[6], if round keys are independent and uniformly random, lp is represented as

$$lp(\Gamma y \to \Gamma x) = \sum_{\Gamma k} \left|\ 2\underset{x,k}{Prob}\{\ x \bullet \Gamma x \oplus f(x,k) \bullet \Gamma y \oplus k \bullet \Gamma k = 0\}\ -1\ \right|^2.$$

From the definitions above and this property, dp and lp generally satisfy the following equations[5]:

$$\sum_{\Delta y} dp(\Delta x \to \Delta y) = 1 \qquad \text{and} \qquad \sum_{\Gamma x} lp(\Gamma y \to \Gamma x) = 1.$$

These results show that $dp(\Delta x \to \Delta y)$ and $lp(\Gamma y \to \Gamma x)$ satisfy the transition probability property.

Definition 2 *For given input difference $\Delta P = (\Delta P_L, \Delta P_R)$ and output difference $\Delta C = (\Delta C_L, \Delta C_R)$, the probability of r-round differential of DES-like ciphers is defined as follows using a Markov chain model of dp* [8]

$$DP(r, \Delta P \to \Delta C) := \sum_{\Delta X, \Delta Y} \prod_{i=1}^{r} dp\big(\Delta X_R(i) \to \Delta X_L(i) \oplus \Delta Y_R(i)\big).$$

Similarly, for given input masking value $\Gamma P = (\Gamma P_L, \Gamma P_R)$ and output difference $\Gamma C = (\Gamma C_L, \Gamma C_R)$, the probability of r-round linear hull of DES-like ciphers is also defined as follows using a Markov chain model of lp [8]

$$LP(r, \Gamma C \to \Gamma P) := \sum_{\Gamma X, \Gamma Y} \prod_{i=1}^{r} lp\big(\Gamma Y_R(i) \to \Gamma X_R(i) \oplus \Gamma Y_L(i)\big).$$

Here $\sum_{\Delta X,\Delta Y}$ *denotes the total sum over these parameters of* $\Delta X_L(2),\ldots,\Delta X_L(r)$, $\Delta X_R(2),\ldots,\Delta X_R(r)$, $\Delta Y_L(1),\ldots,\Delta Y_L(r-1)$, $\Delta Y_R(1),\ldots,\Delta Y_R(r-1)$ *and* $\sum_{\Gamma X,\Gamma Y}$ *denotes the total sum over these parameters of* $\Gamma X_L(2),\ldots,\Gamma X_L(r)$, $\Gamma X_R(2),\ldots,\Gamma X_R(r)$, $\Gamma Y_L(1),\ldots,\Gamma Y_L(r-1)$, $\Gamma Y_R(1),\ldots,\Gamma Y_R(r-1)$.

The probability of r-round characteristic $\mu_D = max_{\Delta X,\Delta Y}\prod_{i=1}^{r} dp(\Delta X_R(i) \rightarrow \Delta X_L(i) \oplus \Delta Y_R(i))$ is normally used to estimate the required number of chosen plaintexts in differential attacks and the probability of r-round linear approximation $\mu_L = max_{\Gamma X,\Gamma Y}\prod_{i=1}^{r} lp(\Gamma Y_R(i) \rightarrow \Gamma X_R(i) \oplus \Gamma Y_L(i))$ is normally used to estimate the required number of known plaintexts/ciphertexts in linear attacks[1],[3].

In [2], Lai et al. introduced the concept of the Markov cipher and showed that DES is a Markov cipher. This means that the probability of r-round differential exactly defines the differential probability of r-round DES-like ciphers. That is, the following proposition holds.

Proposition 1 *Let* $\Delta P = (\Delta P_L, \Delta P_R)$ *be a fixed plaintext difference and* $\Delta Y = (\Delta Y_L, \Delta Y_R)$ *be a varied output difference of r-round DES-like ciphers, then the following holds:*

$$\sum_{\Delta Y} DP(r, \Delta P \rightarrow \Delta Y) = 1.$$

In the case of linear cryptanalysis, the idea of a Markov cipher is not revealed, because the definition of linear hull probability is not the same as that of differential probability(see Definition 1). Whereas the transitional probability does hold.

Proposition 2 *Let* $\Gamma C = (\Gamma C_L, \Gamma C_R)$ *be a fixed masking value of ciphertext and* $\Gamma X = (\Gamma X_L, \Gamma X_R)$ *be a varied masking value of input of r-round DES-like ciphers, then the following holds:*

$$\sum_{\Gamma X} LP(r, \Gamma C \rightarrow \Gamma X) = 1.$$

Proof This proof is given by the mathematical inductive method. In the case of $r = 1$, since

$$\sum_{\Gamma X} LP(1, \Gamma C \rightarrow \Gamma X) = \sum_{\Gamma X_R} lp(\Gamma C_L \rightarrow \Gamma C_R \oplus \Gamma X_R) = 1,$$

the above proposition is true. Thus by assuming that the $r = i$ case is true, it is enough to show the $r = i + 1$ case is also true. In fact,

$$\sum_{\Gamma X} LP(i+1, \Gamma C \rightarrow \Gamma X)$$

$$= \sum_{\Gamma y(i), \Gamma X_L, \Gamma X_R} LP(i, \Gamma C \to (\Gamma y(i), \Gamma X_L)) lp(\Gamma X_L \to \Gamma y(i) \oplus \Gamma X_R)$$

$$= \sum_{\Gamma y(i), \Gamma X_L} LP(i, \Gamma C \to (\Gamma y(i), \Gamma X_L)) \sum_{\Gamma X_R} lp(\Gamma X_L \to \Gamma y(i) \oplus \Gamma X_R)$$

$$= \sum_{\Gamma y(i), \Gamma X_L} LP(i, \Gamma C \to (\Gamma y(i), \Gamma X_L))$$

$$= 1 . \qquad \text{(by assumption of the induction)} \qquad \text{(Q.E.D.)}$$

To consider the probable security against differential and linear cryptanalysis, the upper bounds to the maximum values of $DP(r, \Delta P \to \Delta C)$ and $LP(r, \Gamma C \to \Gamma P)$ should be estimated. We define these maximum values as follows.

$$DP(r) := max_{\Delta P \neq 0, \Delta C} DP(r, \Delta P \to \Delta C),$$
$$LP(r) := max_{\Gamma C \neq 0, \Gamma P} LP(r, \Gamma C \to \Gamma P).$$

In Section 3, $DP(r)$ is estimated by using the maximum differential probability of the function in each round which is defined as

$$p := max_{\Delta x \neq 0, \Delta y} dp(\Delta x \to \Delta y).$$

In same way, $LP(r)$ is estimated by using the parameter defined as

$$q := max_{\Gamma y \neq 0, \Gamma x} lp(\Gamma y \to \Gamma x).$$

Against the probabilities $DP(r)$ and $LP(r)$, the following property is fundamental to estimate provable security.

Property 1 $DP(r+1) \leq DP(r)$, $LP(r+1) \leq LP(r)$ *for all* $r \geq 1$.

This property shows the maximum probabilities of differential and linear hulls decrease as the number of rounds increases, so it is expected that upper bounds to these maximum probabilities of differential and linear hulls decrease as the number of rounds increases. We prove this expectation true in Section 4.

3 Estimation of Provable Security

3.1 Previous Results

With reference to the estimation of upper bounds to $DP(r)$ and $LP(r)$ of DES-like ciphers, in [6, 7] Nyberg and Knudsen showed the following results for $r \geq 4$.

$$DP(r) \leq 2p^2,$$
$$LP(r) \leq 2q^2.$$

They deduced this fact by analyzing $DP(4, \Delta P \to \Delta C)$ and by using the fact that $DP(r) \leq DP(4)$ for any $r \geq 4$. So as mentioned in the Introduction, estimated upper bound are constant at $2p^2$ and $2q^2$ regardless of changes in round number r.

3.2 Idea of Estimation Method

In this paper, we analyze $DP(r, \Delta P \to \Delta C)$ to determine the repetition structures provided by defining the types of output differences $\Delta C = (\Delta C_L, \Delta C_R)$. This analysis is based on the following idea.

Principle 1 (The basic idea of estimation method)

1. *Expanding the probability $DP(r, \Delta P \to \Delta C)$ in the summation of products of $DP(t, \Delta P \to *)$ (where $t < r$) and $dp(* \to *)$.*
2. *Estimating $dp(* \to *)$ and $DP(t, \Delta P \to *)$ according to the following properties 3, 4, 5 and 6.*
3. $$dp(0 \to \Delta y) = \begin{cases} 1, & \text{when } \Delta y = 0 \\ 0, & \text{when } \Delta y \neq 0. \end{cases}$$
4. *If $\Delta x \neq 0$ then $dp(\Delta x \to \Delta y) \leq p$ for any Δy.*
5. *The difference of output does not equal 0.*
6. *Proposition 1 and the followng fact hold.*

$$1 - \sum_{\Delta x} DP(r, \Delta P \to (0, \Delta x)) \leq 1 - DP(r, \Delta P \to (0, \Delta \alpha)) \quad \textit{for any } \Delta \alpha \neq 0.$$

The example of using Principle 1 is described in the proof of Lemma 1. Similar principle is applied to the case of analyzing $LP(r, \Gamma C \to \Gamma P)$. Moreover, relation of duality is held between differential and linear cryptanalysis.

Theorem 1 (Duality between Differential and Linear Cryptanalysis)

a. *[4]* $\quad DP(r, \Delta P \to \Delta C) \Leftrightarrow LP(r, \Gamma C \to \Gamma P)$,
This duality "$\Leftrightarrow$" is achieved by correspondences $\Delta X_L(i) \leftrightarrow \Gamma Y_L(i)$, $\Delta X_R(i) \leftrightarrow \Gamma Y_R(i)$ and $\Delta Y_R(i) \leftrightarrow \Gamma X_R(i)$. These correspondences bijectively change $dp(\Delta X_R(i) \to \Delta X_L(i) \oplus \Delta Y_R(i))$ to $lp(\Gamma Y_R(i) \to \Gamma X_R(i) \oplus \Gamma Y_L(i))$.

b. *The estimation result of $DP(r, \Delta P \to \Delta C)$ is the same as that of $LP(r, \Gamma C \to \Gamma P)$.*
Because of Principle 1 and a. above, in estimating of an upper bound to $DP(r, \Delta P \to \Delta C)$ and $LP(r, \Gamma C \to \Gamma P)$, these formulas can be similarly modified and the estimation parameters p and q are changed into each other.

3.3 New Results

Based on the idea of the estimation method described in Principle 1, the following Lemma is proved.

Lemma 1 *For given a differential value of plaintext $\Delta P = (\Delta P_L, \Delta P_R)$ and a differential value of ciphertext $\Delta C = (\Delta C_L, \Delta C_R)$, the following inequalities hold. When $\Delta C_R = 0$,*

$$DP(r, \Delta P \to \Delta C)|_{(\Delta C_R = 0)} \leq p \sum_{\Delta x} DP(r-2, \Delta P \to (\Delta C_L, \Delta x)), \tag{1}$$

and when $\Delta C_R \neq 0$

$$DP(r, \Delta P \to \Delta C)|_{(\Delta C_R \neq 0)} \leq p(1-p)DP(r-2, \Delta P \to (0, \Delta C_R)) + p^2. \quad (2)$$

Proof

Let each $(r-2)$-round function and each $(r-1)$-round function input have differential values of Δx_1 and Δx_2 respectively, then (see Figure 1)

$$\begin{aligned} &DP(r, \Delta P \to \Delta C) \\ &= \sum_{\Delta x_1, \Delta x_2} DP(r-2, \Delta P \to (\Delta x_2, \Delta x_1)) \\ &\qquad\qquad dp(\Delta x_2 \to \Delta x_1 \oplus \Delta C_R) dp(\Delta C_R \to \Delta x_2 \oplus \Delta C_L). \qquad (*) \end{aligned}$$

In the case of $\Delta C_R = 0$, since $dp(\Delta C_R \to \Delta x_2 \oplus \Delta C_L)$ is nonzero and equal to one only when $\Delta x_2 = \Delta C_L$, and clearly since $\Delta C_L \neq 0$,

$$\begin{aligned} DP(r, \Delta P \to \Delta C) &= \sum_{\Delta x_1} DP(r-2, \Delta P \to (\Delta C_L, \Delta x_1)) dp(\Delta C_L \to \Delta x_1) \\ &\leq p \sum_{\Delta x_1} DP(r-2, \Delta P \to (\Delta C_L, \Delta x_1)) . \end{aligned}$$

In the case of $\Delta C_R \neq 0$, the summation $(*)$ is divided into two summations yielding the $\Delta x_2 = 0$ case and the $\Delta x_2 \neq 0$ case. In the case of $\Delta x_2 = 0$, Δx_1 becomes ΔC_R and the summation $(*)$ equals

$$DP(r-2, \Delta P \to (0, \Delta C_R)) dp(\Delta C_R \to \Delta C_L) \leq p DP(r-2, \Delta P \to (0, \Delta C_R)) .$$

In the case of $\Delta x_2 \neq 0$, both of $dp(\Delta x_2 \to \Delta x_1 \oplus \Delta C_R)$ and $dp(\Delta C_R \to \Delta x_2 \oplus \Delta C_L)$ are less than p, so the summation $(*)$ is estimated as

$$\begin{aligned} &\leq p^2 \sum_{\Delta x_1, \Delta x_2 \neq 0} DP(r-2, \Delta P \to (\Delta x_2, \Delta x_1)) \\ &= p^2 \{1 - \sum_{\Delta x_1} DP(r-2, \Delta P \to (0, \Delta x_1))\} \\ &\leq p^2 \{1 - DP(r-2, \Delta P \to (0, \Delta C_R))\} . \end{aligned}$$

From these estimations we get

$$\begin{aligned} &DP(r, \Delta P \to \Delta C)|_{(\Delta C_R \neq 0)} \\ &\leq p DP(r-2, \Delta P \to (0, \Delta C_R)) + p^2\{1 - DP(r-2, \Delta P \to (0, \Delta C_R))\} \\ &= p(1-p) DP(r-2, \Delta P \to (0, \Delta C_R)) + p^2 . \end{aligned}$$

(Q.E.D.)

Regarding forms $\sum_{\Delta x} DP(r, \Delta P \to (\Delta C_L, \Delta x))$ and $DP(r, \Delta P \to (0, \Delta C_R))$ that appear in the right sides of Eq. (1) and Eq. (2), our analysis gives asymptotic inequalities.

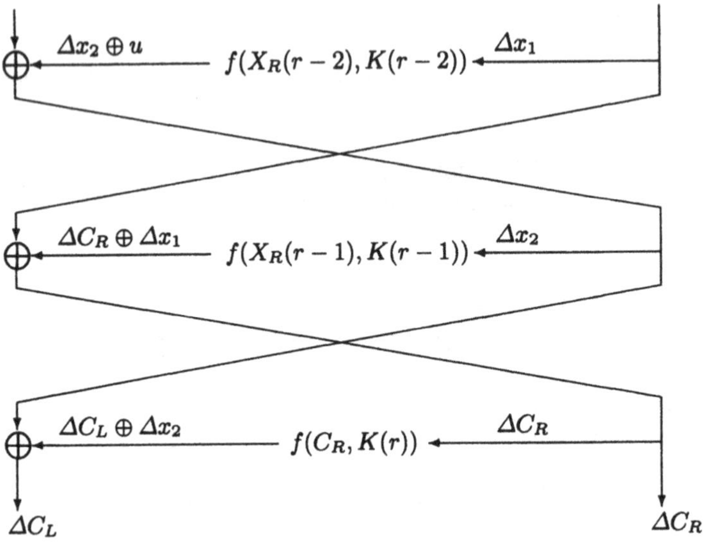

Figure 1 Structure of $DP(r, \Delta P \to \Delta C)$

Lemma 2 *In the cases of* $\Delta C_L \neq 0$ *and* $\Delta C_R \neq 0$,

$$\sum_{\Delta x} DP\big(r, \Delta P \to (\Delta C_L, \Delta x)\big) \leq p(1-p) \sum_{\Delta x} DP\big(r-2, \Delta P \to (\Delta C_L, \Delta x)\big) + p \ ,$$

$$DP\big(r, \Delta P \to (0, \Delta C_R)\big) \leq p(1-p) DP\big(r-2, \Delta P \to (0, \Delta C_R)\big) + p^2 \ .$$

Defining A_r and B_r as $A_r := \sum_{\Delta x} DP\big(r, \Delta P \to (\Delta C_L, \Delta x)\big)$ and $B_r := DP\big(r, \Delta P \to (0, \Delta C_R)\big)$, we can rewrite Lemma 2 as follows.

$$A_r \leq p(1-p) A_{r-2} + p \ , \tag{3}$$
$$B_r \leq p(1-p) B_{r-2} + p^2 \ . \tag{4}$$

These inequalities define two sequences $\{A_r\}$ and $\{B_r\}$, and we estimate the initial values of A_1, A_2, B_1 and B_2 as follows.

Lemma 3

$$\begin{cases} A_1 \leq 1 \\ A_2 \leq 2p - p^2 \ , \end{cases} \qquad \begin{cases} B_1 \leq p \\ B_2 \leq p \ . \end{cases}$$

From Lemma 3, inequalities (3) and (4) are solved as follows.

Lemma 4 *In the cases of* $\Delta C_L \neq 0$ *and* $\Delta C_R \neq 0$,

$$\sum_{\Delta x} DP\big(r, \Delta P \to (\Delta C_L, \Delta x)\big) \leq p^k (1-p)^k + \sum_{i=1}^{k} p^i (1-p)^{i-1} \ ,$$

$$DP\big(r, \Delta P \to (0, \Delta C_R)\big) \le p^k(1-p)^{k-1} + \sum_{i=1}^{k-1} p^{i+1}(1-p)^{i-1} ,$$

where $r = 2k$ ($k \ge 1$), $2k+1$ ($k \ge 0$).

From Lemma 1 and Lemma 4 we obtain

$$DP(r, \Delta P \to \Delta C)|_{(\Delta C_R = 0)} \le p^k(1-p)^{k-1} + \sum_{i=1}^{k-1} p^{i+1}(1-p)^{i-1} , \qquad (5)$$

where $r = 2k$ ($k \ge 2$), $2k+1$ ($k \ge 1$) and

$$DP(r, \Delta P \to \Delta C)|_{(\Delta C_R \ne 0)} \le p^k(1-p)^{k-1} + \sum_{i=1}^{k-1} p^{i+1}(1-p)^{i-1} , \qquad (6)$$

where $r = 2k,\ 2k-1$ ($k \ge 2$).

These estimations ignore the case of $r = 2$, but we can easily confirm that Eq. (5) and Eq. (6) are also true in the case of $r = 2$. So from the definition of $DP(r)$,

$$\begin{aligned} DP(2k) &= \max_{\Delta P \ne 0, \Delta C} DP(2k, \Delta P \to \Delta C) \\ &= \max_{\Delta P \ne 0, \Delta C} \{DP(2k, \Delta P \to \Delta C)|_{(\Delta C_R = 0)}, DP(2k, \Delta P \to \Delta C)|_{(\Delta C_R \ne 0)}\} \\ &\le p^k(1-p)^{k-1} + \sum_{i=1}^{k-1} p^{i+1}(1-p)^{i-1} . \qquad (k \ge 1) \\ DP(2k+1) &= \max_{\Delta P \ne 0, \Delta C} DP(2k+1, \Delta P \to \Delta C) \\ &\le p^k(1-p)^{k-1} + \sum_{i=1}^{k-1} p^{i+1}(1-p)^{i-1} . \qquad (k \ge 0) \end{aligned}$$

Thus we obtain the following main result.

Theorem 2

$$DP(r) \le p^k(1-p)^{k-1} + \sum_{i=1}^{k-1} p^{i+1}(1-p)^{i-1}$$

where $r = 2k$ ($k \ge 1$), $2k+1$ ($k \ge 0$).

So as mentioned in Theorem 1, the estimation of an upper bound to the probability of $LP(r, \Gamma C \to \Gamma P)$ is similarly obtained as follows.

Theorem 3

$$LP(r) \le q^k(1-q)^{k-1} + \sum_{i=1}^{k-1} q^{i+1}(1-q)^{i-1},$$

where $r = 2k$ ($k \ge 1$), $2k+1$ ($k \ge 0$).

New Results			Previous Results	
r	$DP(r)$	$LP(r)$	$DP(r)$	$LP(r)$
2, 3	p	q	—	—
4, 5	$2p^2 - p^3$	$2q^2 - q^3$	$2p^2$	$2q^2$
6, 7	$p^2 + 2p^3 - 3p^4 + p^5$	$q^2 + 2q^3 - 3q^4 + q^5$		
8, 9	$p^2 + p^3 + p^4 - 5p^5 + 4p^6 - p^7$	$q^2 + q^3 + q^4 - 5q^5 + 4q^6 - q^7$		

Table 1 Upper Bounds to $DP(r)$ and $LP(r)$ $(r \leq 9)$

4 Discussion

This section examines the results of Theorem 2 and Theorem 3. If we let

$$U(x,k) := x^k(1-x)^{k-1} + \sum_{i=1}^{k-1} x^{i+1}(1-x)^{i-1} \quad 0 \leq x \leq 1,$$

$U(p,k)$ is the upper bound to $DP(r)$ and $U(q,k)$ is the upper bound to $LP(r)$. $U(x,k)$ is represented as the rational expression $\frac{x^2 + x^k(1-x)^{k+1}}{1-x+x^2}$. It is clear that $U(x,k)$ is a decreasing function for k, which is considered to represent round number, and an increasing function for x. This means that the provable security of DES-like ciphers against differential or linear attack increases as the probabilities p or q decrease and increases with the round number r. Table 1 shows some examples of upper bounds to $DP(r)$ and $LP(r)$ as obtained from $U(x,k)$ in a comparison with previous results, and Graph 1 plots $U(0.25,k)$ where $r = 2k, 2k+1$ is a parameter.

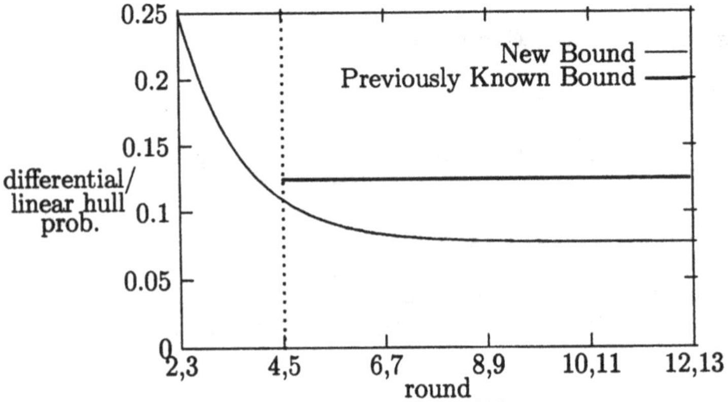

Graph 1 New Bound $\frac{x^2+x^k(1-x)^{k+1}}{1-x+x^2}$ (when $x = 0.25$)

5 Conclusion

We gave stricter upper bounds of the probabilities of r-round differential and linear hull than the previous results. Our new bounds depend on the number of rounds, e.g., $2p^2 - p^3$ $(2q^2 - q^3)$ for 4 and 5 rounds. The bounds were proven to decrease and converge to about p^2 as round number increases. The improvement has little practical value from the attackers' viewpoint, but it is theoretically important to develop the technique to derive a tight bound to the probability of the r-round differential (linear hull) from the designers' viewpoint, because our estimation is thought to prove that cryptographically "strong" ciphers can be obtained by iterating a round function many times in linear cryptanalysis as well as in differential cryptanalysis.

Acknowledgment

We would like to thank Prof. S.Tsujii(leader of the Information & Communication Project) and Prof. T.Kaneko(sub-leader of the Project) for their great advice.

References

1. E.Biham and A. Shamir. *Differential Cryptanalysis of DES-like Cryptosystems.* Journal of Cryptology, Volume 4, Number 1, pp.3–72, Springer-Verlag, 1991.
2. X.Lai, J.L.Massey and S.Murphy. *Markov Ciphers and Differential Cryptanalysis.* Advances in Cryptology–EUROCRYPT'91, Lecture Notes in Computer Sciences 547, pp.17–38, Springer-Verlag, 1992.
3. M.Matsui. *Linear Cryptanalysis Method for DES Cipher.* Advances in Cryptology–EUROCRYPT'93, Lecture Notes in Computer Sciences 765, pp.386–397, Springer-Verlag, 1994.
4. M.Matsui. *On Correlation Between the Order of S-boxes and the Strength of DES.* Advances in Cryptology–EUROCRYPT'94, Lecture Notes in Computer Sciences 950, pp.366–375, Springer-Verlag, 1995.
5. M.Matsui. *New Structure of Block Ciphers with Provable Security against Differential and Linear Cryptanalysis.* In Proceedings of the third international workshop of fast software encryption, Lecture Notes in Computer Science 1039, pp.205–218, Springer-Verlag, 1996.
6. K.Nyberg. *Linear Approximation of Block Ciphers.* Advances in Cryptology–EUROCRYPT'94, Lecture Notes in Computer Sciences 950, pp.439–444, Springer-Verlag, 1995.
7. K.Nyberg and L.R.Knudsen. *Provable Security Against a Differential Attack.* Journal of Cryptology, Volume 8, Number 1, pp.27–37, Springer-Verlag, 1995.
8. L.O'Connor and J.D.Golic. *A Unified Markov Approach to Differential and Linear Cryptanalysis.* Advances in Cryptology–ASIACRYPT'94, Lecture Notes in Computer Sciences 917, pp.387–397, Springer-Verlag, 1995.
9. K.Ohta, S.Moriai and K.Aoki. *Improving the Search Algorithm for the Best Linear Expression.* Advances in Cryptology–CRYPTO'95, Lecture Notes in Computer Science 963, pp.157–170, Springer-Verlag, 1995.

Improved Fast Software Implementation of Block Ciphers

(Extended Abstract)

Takeshi Shimoyama, Seiichi Amada and Shiho Moriai

TAO (Telecommunications Advancement Organization of Japan),
1-1-32 Shin'urashima-cho, Kanagawa-ku, Yokohama, 221 Japan
E-mail : {shimo,amada,shiho}@yokohama.tao.or.jp

Abstract. This paper improves the fast DES implementation in software proposed by Biham at the 4-th Fast Software Encryption Workshop. That is, we propose a new algorithm which reduces the number of instructions for computation of S-boxes, which is the factor dominating the performance of Biham's implementation. When we apply our algorithm to DES S-boxes, we need only 87.75 instructions in average. We can reduce the number of instructions for 1 round of DES to 942, and in total, in terms of the number of instructions, our implementation is expected to be about 8% faster than Biham's implementation.

1 Introduction

In [2] Biham proposed a fast new DES implementation in software. This implementation makes the most of the computation power of processors with long word size (64-bit etc.). We view a processor with, for example, 64 bit words, as a SIMD (Single Instruction Multiple Data stream) parallel computer which can compute 64 one-bit operations simultaneously, while the 64 bits of each block are set in 64 different words. For example, the first bit of each word is always of the first block, and the second bit belongs to the second block, and so on.

By this implementation, we can gain a significant speedup because 1. many blocks are encrypted in parallel where the degree of parallelism equals the word size, and 2. the permutation and expansion of bits do not cost, since instead of changing the order of words, we can address the required word directly. Although the S-boxes may be implemented in more instructions than in usual implementations, the parallelism of this implementation works well.

In Biham's implementation, it is very important how to implement S-boxes, because it is the factor dominating the performance. Usual implementations of S-boxes use table lookups, but table lookup approach would be very inefficient in Biham's implementation, so he represented the S-boxes by logical gate circuits using gates with dual-inputs XOR, AND, OR, and NOT. Thus the required number of instructions in software implementation equals to the number of gates of such circuits. In [2] he described that although the problem of finding the best (minimal) such circuit was still open, he found an optimization which requires at most 132 instructions per DES S-box, about 100 instructions in average, and in total 1040 instructions. He described just the idea of optimization and average numbers of instructions but didn't show the representations of S-boxes he adopted. In his implementation of DES, the instructions in S-boxes accounts for about 77% of all instructions in one round of DES, so the reduction of the instructions in S-boxes significantly enhances the speedup. There still seems to be some room for improvement in this problem.

This paper proposes an algorithm which reduces the number of instructions needed to compute S-boxes in software implementation. Our algorithm is not an ad hoc manual counting or tune-up method but it gives ***automatically*** (in an algorithmic way) the list of the number of instructions by using some computer algebra system. We can reduce the number of instructions in one round of DES from 1040 to 942, and in total, in terms of the number of instructions, the speed by our implementation is expected to be about 8% faster than by Biham's implementation. Moreover, our algorithm is applicable to other functions used in other block ciphers. We apply it to the S-boxes of FEAL and MISTY. (For details, see [5].)

Note that our algorithm does not always minimize the number of instructions for computation of S-boxes. Finding the minimum number of instruction is still an open problem.

2 Our optimization of the number of instructions

In a software implementation, one of the effective ways to compute functions of n variables is to compute functions of two variables first, and second, operate the computed value with another variable, ..., and so forth, until all output bits of the S-box are computed. Here it is important to use the values computed until then, and to decide the ***variable order*** so that the number of instructions is minimal. The expression of $f \cdot x_{n-1} + g$ is called the ***recursive form***, which is a polynomial in one variable x_{n-1}, (which is called the ***leading variable***) and whose coefficients are polynomials in the other variables. We know empirically that the recursive forms are suitable for evaluations of polynomial functions.

We show the algorithm that finds the reduced number of instructions necessary to calculate S-boxes (or any functions) by using the recursive forms. Let x_0, x_1, ..., x_{n-1} be the n input bits and y_0, y_1, ..., y_{s-1} be the s output bits of the S-boxes, respectively.

2.1 Algebraic expressions of S-boxes

When the description of the S-boxes is not given as some algebraic expressions, we construct the algebraic expressions between inputs and outputs from the description tables of S-boxes. One way to construct them can be seen also in [6]. For example, if the output of S_1 of DES for input $(0,0,0,1,0,0)$ is $(1,1,0,1)$, we have the equivalent algebraic equation $P_4 = 0$, where

$$P_4 = (x_0+1)(x_1+1)(x_2+1)(x_3+0)(x_4+1)(x_5+1)$$
$$((y_0+0)(y_1+0)(y_2+1)(y_3+0)+1).$$

Note that the subscript 4 of P_4 corresponds to the value of input (0,0,0,1,0,0). Similarly, we get the polynomials $\{P_0, \ldots, P_{2^n-1}\}$ for each S-box.

2.2 The representation of each output bit by input bits

We can express each output bit using input bits by solving the following simultaneous algebraic equations with $n+s$ unknowns. $P_0 = 0, \ldots, P_{2^n-1} = 0$. In order to solve the above simultaneous algebraic equations, computer algebra systems are very useful. We use Risa/Asir[4], a free software of computer algebra system, which supports computing of Gröbner bases over finite fields[1]. By computing the Gröbner basis with respect to the variable order $[y_{s-1}, \ldots, y_0, x_{n-1}, \ldots, x_0]$, we have the following polynomial representation for each output bit y_i.

$$y_0 = F_0(x_{n-1}, \ldots, x_0), \quad \cdots, \quad y_{s-1} = F_{s-1}(x_{n-1}, \ldots, x_0)$$

2.3 Decomposition of the representations of output bits into the set of instructions

Let σ be a permutation on the set $\mathcal{M} = \{x_{n-1}, \ldots, x_0\}$. Let $\mathcal{S}_{\mathcal{M}}$ be the set of all $n!$ permutations on $\mathcal{M}$. For $x \in \mathcal{M}$ and $\sigma \in \mathcal{S}_{\mathcal{M}}$, x^σ denotes the image of x by σ. For each S-box,
(1) fix a permutation $\sigma \in \mathcal{S}_{\mathcal{M}}$ and the corresponding variable order $[x^\sigma_{n-1}, \ldots, x^\sigma_0]$
(2) decompose recursively the representation of each output bit obtained in the previous step into a set of instructions. The representation of each output bit,

$$y_i = f_i \cdot x^\sigma_{n-1} + g_i,$$

is decomposed into 2 instructions as follows.

$$a = f_i \cdot x^\sigma_{n-1} \quad \text{(AND)}, \qquad y_i = a + g_i \quad \text{(XOR)}$$

Strictly speaking, in the latter instruction(XOR) we need a "store" instruction, where the value of the right side is written as y_i in memory, but in the former instruction(AND), we need not write the value of the right side as a in memory.
(3) Count the number of instructions. Let $\mathcal{L}_{S_i,\sigma}$ be the list of the instructions for the S-box S_i with respect to the variable order $[x^\sigma_{n-1}, \ldots, x^\sigma_0]$. Initially $\mathcal{L}_{S_i,\sigma}$ is an empty set. As the representation of each output bit is decomposed, the decomposed instructions are added to $\mathcal{L}_{S_i,\sigma}$ admitting no duplication.

Similarly, we decompose f_i and g_i as above and add the new instructions to $\mathcal{L}_{S_i,\sigma}$. When the decompositions of functions are finished, we count the number of instructions in $\mathcal{L}_{S_i,\sigma}$.

Repeat steps (1),(2) and (3) for all variable orders $[x^\sigma_{n-1}, \ldots, x^\sigma_0]$ for $\sigma \in \mathcal{S}_{\mathcal{M}}$, in other words, for $n!$ permutations of n input variables. Finally, of all variable orders we find the minimal number of instructions.

3 Application to DES

3.1 The numbers of instructions of S-boxes and timing data

When we apply our algorithm to the S-boxes of DES, the obtained minimal numbers of instructions are as follows.

S_1	S_2	S_3	S_4	S_5	S_6	S_7	S_8
95	84	89	78	95	87	86	88

We succeeded in reducing the number of instructions for one round of DES from 1040 to 942. In total, in terms of the number of instructions, the speed by our implementation is expected to be about 8% faster than by Biham's implementation. There are some instruction sets that have the same number of instructions as in the table above. One of the sets of instructions for each S-box is available in [5].

For timing data, since Biham didn't publish the representations of S-boxes he adopted, we can't directly compare our implementation speed with Biham's. Instead we programmed in C language on computers with 32 bit words and with 64 bit words using our representations of S-boxes to compare with the fastest known implementation written in only C language, Eric Young's *libdes*.

	Young's *libdes*	our implementation
Ultra SPARC (200 MHz)	18 Mbps	54 Mbps
Alpha 21164A (500 MHz)	46 Mbps	227 Mbps

3.2 Comparison with related work

Kwan improved Biham's implementation independently[3] and he contributed some articles into an internet news group (`sci.crypt`). It is important that he used AND, XOR, and OR in the instruction table of each S-box, but we used only AND and XOR. It is surprising that the numbers of the instructions obtained by him are almost similar to the numbers of our results except for small differences in S_4 and S_5. Note that Kwan and we have worked independently, and we don't know his algorithm to obtain the instructions. The following table shows the number of the instructions for each S-box described by him.

	S_1	S_2	S_3	S_4	S_5	S_6	S_7	S_8
Kwan's result	95	84	89	**77**	**96**	87	86	88

4 Polynomial functions and set of instructions of S-box S_1

In this section, we show one of the polynomial functions of the S-box S_1 of DES and the corresponding instructions set obtained in section §2.2. For the other S-boxes, polynomial functions, instruction sets and additional information can be seen in [5].

$$\begin{aligned}
y_3 = {} & ((((x_5+1)\cdot x_4+(x_5+1)\cdot x_1+x_5+1)\cdot x_3+(x_5\cdot x_1+1)\cdot x_4+x_1+1)\cdot x_2 \\
& +((x_1+1)\cdot x_4+x_5+1)\cdot x_3+((x_5+1)\cdot x_1+x_5+1)\cdot x_4+x_5)\cdot x_0+((x_5\cdot x_1+1)\cdot x_4 \\
& +x_5\cdot x_1)\cdot x_2+(x_1\cdot x_4+x_5\cdot x_1+1)\cdot x_3+(x_1+x_5)\cdot x_4+(x_5+1)\cdot x_1, \\
y_2 = {} & (((x_5\cdot x_4+(x_5+1)\cdot x_1+1)\cdot x_3+(x_5+1)\cdot x_1\cdot x_4+(x_5+1)\cdot x_1)\cdot x_2 \\
& +((x_5+1)\cdot x_1\cdot x_4+(x_5+1)\cdot x_1)\cdot x_3+(x_5\cdot x_1+x_5+1)\cdot x_4+(x_5+1)\cdot x_1+1)\cdot x_0 \\
& +(((x_5+1)\cdot x_1+x_5+1)\cdot x_3+(x_1+1)\cdot x_4+(x_5+1)\cdot x_1+x_5)\cdot x_2 \\
& +((x_5\cdot x_1+x_5+1)\cdot x_4+(x_5+1)\cdot x_1+1)\cdot x_3+(x_1+1)\cdot x_4+x_5\cdot x_1+x_5+1, \\
y_1 = {} & (((x_5\cdot x_4+(x_5+1)\cdot x_1+1)\cdot x_3+((x_5+1)\cdot x_1+x_5)\cdot x_4+(x_5+1)\cdot x_1+1)\cdot x_2 \\
& +(x_5\cdot x_1+1)\cdot x_4\cdot x_3+(x_1+1)\cdot x_4+(x_5+1)\cdot x_1+x_5)\cdot x_0 \\
& +((x_5\cdot x_4+x_5)\cdot x_3+(x_5+1)\cdot x_4+x_5\cdot x_1+1)\cdot x_2 \\
& +((x_1+1)\cdot x_4+(x_5+1)\cdot x_1+x_5)\cdot x_3+x_5\cdot x_4+(x_5+1)\cdot x_1+x_5+1, \\
y_0 = {} & (((x_4+(x_5+1)\cdot x_1+x_5+1)\cdot x_3+(x_5\cdot x_1+1)\cdot x_4+x_1)\cdot x_2 \\
& +(x_1\cdot x_4+(x_5+1)\cdot x_1+x_5+1)\cdot x_3+(x_5\cdot x_1+1)\cdot x_4+1)\cdot x_0 \\
& +((x_4+x_1+x_5+1)\cdot x_3+x_1+1)\cdot x_2+x_5\cdot x_4\cdot x_3+x_4+x_1+x_5+1.
\end{aligned}$$

$$\begin{array}{lllll}
a_{50}=x_5+1, & a_{40}=a_{50}\cdot x_1, & a_{41}=x_5\cdot x_1, & a_{42}=a_{40}+a_{50}, & a_{43}=a_{41}+1, \\
a_{44}=x_1+1, & a_{45}=x_1+x_5, & a_{46}=a_{40}+1, & a_{47}=a_{41}+a_{50}, & a_{48}=x_5+a_{40}, \\
a_{49}=x_1+a_{50}, & a_{30}=a_{50}\cdot x_4, & a_{31}=a_{43}\cdot x_4, & a_{32}=a_{44}\cdot x_4, & a_{33}=a_{42}\cdot x_4, \\
a_{34}=x_4\cdot x_1, & a_{35}=a_{45}\cdot x_4, & a_{36}=a_{42}+a_{30}, & a_{37}=a_{44}+a_{31}, & a_{38}=a_{32}+a_{50}, \\
a_{39}=x_5+a_{33}, & a_{3a}=a_{31}+a_{41}, & a_{3b}=a_{34}+a_{43}, & a_{3c}=a_{40}+a_{35}, & a_{3d}=x_5\cdot x_4, \\
a_{3e}=a_{40}\cdot x_4, & a_{3f}=a_{47}\cdot x_4, & a_{3g}=a_{3d}+a_{46}, & a_{3h}=a_{3e}+a_{40}, & a_{3i}=a_{46}+a_{3f}, \\
a_{3j}=a_{48}+a_{32}, & a_{3k}=a_{32}+a_{47}, & a_{3l}=a_{48}\cdot x_4, & a_{3m}=a_{3l}+a_{46}, & a_{3n}=x_5+a_{3d}, \\
a_{3o}=a_{30}+a_{43}, & a_{3p}=a_{3d}+a_{42}, & a_{3q}=x_4+a_{42}, & a_{3r}=x_1+a_{31}, & a_{3s}=a_{42}+a_{34}, \\
a_{3t}=a_{31}+1, & a_{3u}=x_4+a_{49}, & a_{20}=a_{36}\cdot x_3, & a_{21}=a_{38}\cdot x_3, & a_{22}=a_{3b}\cdot x_3, \\
a_{23}=a_{20}+a_{37}, & a_{24}=a_{21}+a_{39}, & a_{25}=a_{22}+a_{3c}, & a_{26}=a_{3g}\cdot x_3, & a_{27}=a_{3h}\cdot x_3, \\
a_{28}=a_{42}\cdot x_3, & a_{29}=a_{3i}\cdot x_3, & a_{2a}=a_{26}+a_{3h}, & a_{2b}=a_{27}+a_{3i}, & a_{2c}=a_{3j}+a_{28}, \\
a_{2d}=a_{29}+a_{3k}, & a_{2e}=a_{31}\cdot x_3, & a_{2f}=a_{3n}\cdot x_3, & a_{2g}=a_{3j}\cdot x_3, & a_{2h}=a_{26}+a_{3m}, \\
a_{2i}=a_{2e}+a_{3j}, & a_{2j}=a_{2f}+a_{3o}, & a_{2k}=a_{2g}+a_{3p}, & a_{2l}=a_{3q}\cdot x_3, & a_{2m}=a_{3s}\cdot x_3, \\
a_{2n}=a_{3u}\cdot x_3, & a_{2o}=a_{3d}\cdot x_3, & a_{2p}=a_{2l}+a_{3r}, & a_{2q}=a_{2m}+a_{3t}, & a_{2r}=a_{2n}+a_{44}, \\
a_{2s}=a_{2o}+a_{3u}, & a_{10}=a_{23}\cdot x_2, & a_{11}=a_{3a}\cdot x_2, & a_{12}=a_{10}+a_{24}, & a_{13}=a_{11}+a_{25}, \\
a_{14}=a_{2a}\cdot x_2, & a_{15}=a_{2c}\cdot x_2, & a_{16}=a_{14}+a_{2b}, & a_{17}=a_{15}+a_{2d}, & a_{18}=a_{2h}\cdot x_2, \\
a_{19}=a_{2j}\cdot x_2, & a_{1a}=a_{18}+a_{2i}, & a_{1b}=a_{19}+a_{2k}, & a_{1c}=a_{2p}\cdot x_2, & a_{1d}=a_{2r}\cdot x_2, \\
a_{1e}=a_{1c}+a_{2q}, & a_{1f}=a_{1d}+a_{2s}, & a_{00}=a_{12}\cdot x_0, & y_3=a_{00}+a_{13}, & a_{02}=a_{16}\cdot x_0, \\
y_2=a_{02}+a_{17}, & a_{04}=a_{1a}\cdot x_0, & y_1=a_{04}+a_{1b}, & a_{06}=a_{1e}\cdot x_0, & y_0=a_{06}+a_{1f}.
\end{array}$$

References

1. T. Becker and V. Weispfenning, "Gröbner Bases," Graduate Texts in Mathematics 141, Springer Verlag, (1993)
2. E. Biham, "A Fast New DES Implementation in Software," Preproceedings of the Fourth Fast Software Encryption Workshop, January, (1997) (available at `http://www. cs.technion.ac.il/~biham/publications.html`)
3. M. Kwan, "bitslice-des," private communication, (1997) (available at `http: //www. cs.mu.oz.au/~mkwan/bitslice`)
4. M. Noro and T. Takeshima, "Risa/Asir – a computer algebra system," Proceedings of ISSAC'92, ACM Press, (1992) pp.387–396, (available at `(164.71.2.5) endeavor.flab.fujitsu.co.jp`: `/pub/isis/asir` by anonymous ftp)
5. T. Shimoyama, S. Amada and S. Moriai, "Improved Fast Software Implementation of Block Ciphers," (1997) (available at `http://www.yokohama.tao.or.jp/shimo`),
6. T. Shimoyama and T. Kaneko, "Polynomial expression of S-box and its application to the Linear Attack," (in Japanese), TECHNICAL REPORT OF IEICE., ISEC 96-40 (1996) pp.71–82

Security Comments on the Hwang-Chen Algebraic-code Cryptosystem

Mohssen M. Alabbadi

KACST-CERI
P.O.Box 6086
Riyadh - 11442, Saudi Arabia
e-mail: alabbadi@kacst.edu.sa

Abstract. Hwang and Chen have proposed a private-key cryptosystem that provides joint error correction, encryption, and "message" integrity. The scheme is based on algebraic error-correcting codes, using random chaining technique. It was shown that obtaining a combinatorially equivalent code, under ciphertext-only attack, requires $O(k2^n)$ operations and $O(k2^{n/2})$ ciphertexts, where n and k are the length and dimension of the code respectively. It was further claimed that obtaining an equivalent code is not sufficient to "totally" break the system. In this paper, a chosen-plaintext attack is presented that is able to break the system, requiring $O[(n-k)2^{k/2}]$ ciphertexts and $O[(n-k)2^k]$ operations; the attack is based on obtaining a combinatorially equivalent code. Finally, a modified version of the scheme is proposed that overcomes the weaknesses of the original Hwang-Chen scheme; the complexity to break this modified scheme is $O(k2^{n/2})$ ciphertexts and $O(kn2^{n/2})$ operations.

1 Introduction

Algebraic error-correcting codes have been proposed for public-key as well as private-key cryptosystems. In both systems, the transmitter intentionally introduces error vectors to "hide" the linearity inherited in the encoding process.

In McEliece's public-key cryptosystem [1], the published encoding matrix is a transformed version of a Goppa code generator matrix; the transformed matrix looks like a generator matrix of a "seemingly" hard-to-decode code. The intentional error vectors have Hamming weights equal to the error-correcting capability of the code [2]. The scheme can be used for simultaneous error correction and encryption by using error vectors with Hamming weights less than the error-correcting capability of the code [3]. To make the system secure, it is necessary to use codes of large length and error-correcting capability, thus requiring large storage and computational overhead.

On the other hand, Rao and Nam [4] presented a private-key cryptosystem using small minimum distance codes. To prevent majority voting to obtain the encryption matrix, the Hamming weights of the intentional error vectors are $\approx \frac{n}{2}$, where n is the length of the code. Chosen-plaintext attacks were devised in [5, 6], requiring $O(kN \log N)$ ciphertexts, where k is the dimension of the code and $N = 2^{n-k}$.

The Rao-Nam scheme was modified in [7] to allow for joint authentication and encryption. This modified scheme employs a function that simply selects some certain bits of the plaintext, and those selected bits are used to define an intentional error vector; this vector has Hamming weight $\approx \frac{n}{2}$ as in the original Rao-Nam's scheme. Thus there is a unique intentional error vector for each plaintext. This scheme, however, was broken using chosen-plaintext attacks [8].

Hwang and Chen have proposed a private-key cryptosystem that provides joint error correction, encryption, and "message" integrity [9]. A one-bit random chaining is used to allow the receiver to recover and remove the intentional error vector before the decoding process. This has two advantages: first, the 2^{n-k} syndrome-error table required for decryption in the Rao-Nam scheme is no longer needed; second, the code is fully utilized to correct errors introduced by the noisy channel. So, in principal, this scheme is different than the Rao-Nam scheme.

The Hwang-Chen scheme exhibits some interesting features that are desirable in the volatile file environment; operations such as modification, deletion, and insertion of any plaintext can be performed efficiently. Furthermore, illegal modification to any ciphertext is either detected or corrected; illegal insertion/deletion of one ciphertext is, however, detected with very high probability.

In the security analysis of the Hwang-Chen scheme discussed in [9], it was shown that obtaining a combinatorially equivalent code, using ciphertext-only attack, requires $O(k2^n)$ operations and $O(k2^{n/2})$ ciphertexts. It was further claimed that obtaining an equivalent code is not sufficient to "totally" break the system.

The outline of this paper is as follows. Section 2 describes the Hwang-Chen system. The security of the system is analyzed in Sect. 3; the method to obtain a combinatorially equivalent code that was discussed in [9] is briefly explained, then a chosen-plaintext attack that breaks the system is presented, requiring $O[(n-k)2^{k/2}]$ ciphertexts and $O[(n-k)2^k]$ operations. Extensions of this scheme are discussed in Sect. 4, where the scheme is modified to overcome the weaknesses of the original Hwang-Chen's scheme; the complexity to break this modified version is $O(k2^{n/2})$ ciphertexts and $O(kn2^{n/2})$ operations. Finally, concluding remarks are given in Sect. 5.

2 The Hwang-Chen Scheme

Let G be a $k \times n$ generator matrix for an (n, k) binary code C with minimum distance $d_{\min}$ such that $t = \lfloor \frac{d_{\min}-1}{2} \rfloor$ where t is the error-correcting capability of the code, P is an $n \times n$ permutation matrix, and f is a non-linear function such that $f : GF(2^k) \to GF(2^k)$ where f is only "secure" against ciphertext-only attack. G, P, and f constitute the private key. In addition, the system employs a random bit generator at the transmitter site.

The binary plaintext is broken into k-bit vectors $\underline{m}_1, \underline{m}_2, \ldots, \underline{m}_b$; this may require padding the plaintext with zeros or some special binary patterns. For each $\underline{m}_i$, where $1 \leq i \leq b$, a random bit r_i is produced by the random bit generator. The random bits are used to construct the n-bit random vector $\underline{R}_i = (r_{i-n+1}, r_{i-n+2}, \ldots, r_i)$, where the vector $(r_{-n+2}, r_{-n+3}, \ldots, r_0)$ is the initial $(n-1)$-bit vector which is part of the private key.

The k-bit vector $\underline{m}_i$, for $1 \leq i \leq b$, is encrypted into an $(n+1)$-bit ciphertext $\underline{c}_i$ by computing the following expression.

$$\underline{c}_i = [f(\underline{m}_i \oplus (\underline{R}_i P)^*)G \oplus \underline{R}_i P] \parallel r_i, \tag{1}$$

where $(\underline{R}_i P)^*$ is the first k bits of the n-bit vector $\underline{R}_i P$, and "$\|$" denotes the concatenation of the n-bit vector $f(\underline{m}_i \oplus (\underline{R}_i P)^*)G \oplus \underline{R}_i P$ and the bit r_i.

Upon receiving $\hat{\underline{c}}_i$, the following steps are to be performed.

- The last bit $\hat{r}_i$ of $\hat{\underline{c}}_i$ is removed to obtain the n-bit vector $\underline{d}_i$. The last $n-1$ bits $\hat{r}_{i-n+1}, \hat{r}_{i-n+2}, \ldots, \hat{r}_{i-1}$ along with $\hat{r}_i$ are used to construct the random vector $\hat{\underline{R}}_i$. Then $\hat{\underline{R}}_i P$ is computed.
- $\underline{d}_i \oplus \hat{\underline{R}}_i P$ is then calculated, and it is further decoded by the decoding algorithm of C to obtain a k-bit vector $\underline{z}_i$.
- Finally, $\hat{\underline{m}}_i$ is calculated as $f^{-1}(\underline{z}_i) \oplus (\hat{\underline{R}}_i P)^*$.

If the last bit of the received ciphertext as well as the last bits of the previous $n-1$ received ciphertexts are correct, then the error introduced by the noisy channel can be corrected by the code, provided that the error vector is correctable; on the other hand, if the last bit is in error, then this ciphertext as well as the next $n-1$ ciphertexts will not decoded correctly.

3 Security of the Hwang-Chen Scheme

We assume that there are no noise errors in the ciphertexts. Without loss of generality, the following security analysis is based on the n-bit vector $\underline{d}_i$ which is the ciphertext $\underline{c}_i$ of the plaintext $\underline{m}_i$ after removing the last bit r_i. Hereafter, $\underline{d}_i$ is called the ciphertext of $\underline{m}_i$.

In [9], it was shown that if $\underline{R}_i = \underline{R}_j$, where $\underline{m}_i \neq \underline{m}_j$, then

$$\underline{d}_i \oplus \underline{d}_j = [f(\underline{m}_i \oplus (\underline{R}_i P)^*) \oplus f(\underline{m}_j \oplus (\underline{R}_j P)^*)]G. \tag{2}$$

This shows that $\underline{d}_i \oplus \underline{d}_j$ is a codeword of C. To construct a generator matrix $\hat{G}$ for a combinatorially equivalent code, the cryptanalyst needs to collect k linearly independent pairs of ciphertexts such that the random vectors in each pair of ciphertexts are the same.

For each pair, the probability that the random vectors are the same is 2^{-n}, thus the number of ciphertexts that need to be collected is $O(2^{n/2})$. It was shown in [9] that the time complexity, to obtain a single pair, is $O(2^n)$ operations; this complexity can be dramatically reduced [10, pp. 279–281], however, by sorting all the collected random vectors, instead of comparing them on a one by one basis. This sorting operation requires $O(\frac{n}{2}2^{n/2})$ operations. For each pair, we need two tables: one containing the ciphertexts and another containing the random vectors; thus the space complexity is $O(n2^{n/2})$ bits. Obtaining $\hat{G}$ requires $O(k2^{n/2})$ ciphertexts, $O(kn2^{n/2})$ operations, and $O(kn2^{n/2})$ bits of memory.

It was further claimed in [9] that obtaining $\hat{G}$ is not sufficient to "totally" break the system. On the contrary, in what follows, it is shown that this is not the case.

The above analysis shows that the most critical component, contributing to the security of the system, is the permutation matrix P. However, not all columns of P are equally important to the security of the system. In the following attack, it will be shown that the last $n-k$ columns of P are the most important ones. For this, the matrix P is viewed as $P = [P^* \mid P^{**}]$, where P^* and P^{**} are $n \times k$ and $n \times (n-k)$ matrices respectively. Thus $\underline{x}P^*$ is the same as $(\underline{x})^*$ and $\underline{x}P^{**}$ gives the last $n-k$ bits of $\underline{x}P$, which is denoted as $(\underline{x})^{**}$. Furthermore, the following notations are used: $(\underline{0})_{a\times b}$ denotes the $a \times b$ all-zero matrix and $w_{\mathrm{H}}(\underline{x})$ denotes the Hamming weight of $\underline{x}$ which is the number of nonzero coordinates of $\underline{x}$.

3.1 Finding P^{**}

Let $\underline{d}_i$ and $\underline{d}'_i$ be two ciphertexts of the message $\underline{m}_i$ such that

$$\underline{d}_i = f(\underline{m}_i \oplus \underline{R}_i P^*)G \oplus \underline{R}_i P, \tag{3}$$

$$\underline{d}'_i = f(\underline{m}_i \oplus \underline{R}'_i P^*)G \oplus \underline{R}'_i P. \tag{4}$$

The following lemma explores a special case of $\underline{d}_i$ and $\underline{d}'_i$ that is used to break the system; it is the most important component of this attack.

Lemma 1. *If $\underline{d}_i \oplus \underline{d}'_i = [(\underline{0})_{1\times k} \mid \underline{y}]$ and $w_{\mathrm{H}}(\underline{y}) = w_{\mathrm{H}}(\underline{R}_i \oplus \underline{R}'_i)$, such that $0 < w_{\mathrm{H}}(\underline{y}) < \frac{1}{2}d_{\min}$, then $\underline{R}_i P^* = \underline{R}'_i P^*$.*

Proof. Let $\underline{b} = [f(\underline{m}_i \oplus \underline{R}_i P^*) \oplus f(\underline{m}_i \oplus \underline{R}'_i P^*)]G$. Then, using (3) and (4), we have $\underline{b} = \underline{d}_i \oplus \underline{d}'_i \oplus (\underline{R}_i \oplus \underline{R}'_i)P$. Since $\underline{d}_i \oplus \underline{d}'_i = [(\underline{0})_{1\times k} \mid \underline{y}]$, then $\underline{b} = (\underline{R}_i \oplus \underline{R}'_i)P \oplus [(\underline{0})_{1\times k} \mid \underline{y}] = [(\underline{R}_i \oplus \underline{R}'_i)P^* \mid \underline{y} \oplus (\underline{R}_i \oplus \underline{R}'_i)P^{**}]$. Thus $w_{\mathrm{H}}(\underline{b}) = w_{\mathrm{H}}[(\underline{R}_i \oplus \underline{R}'_i)P^*] + w_{\mathrm{H}}[\underline{y} \oplus (\underline{R}_i \oplus \underline{R}'_i)P^{**}] \le w_{\mathrm{H}}[(\underline{R}_i \oplus \underline{R}'_i)P^*] + w_{\mathrm{H}}(\underline{y}) + w_{\mathrm{H}}[(\underline{R}_i \oplus \underline{R}'_i)P^{**}] = w_{\mathrm{H}}[(\underline{R}_i \oplus \underline{R}'_i)P] + w_{\mathrm{H}}(\underline{y})$. Since $w_{\mathrm{H}}(\underline{y}) = w_{\mathrm{H}}(\underline{R}_i \oplus \underline{R}'_i) = w_{\mathrm{H}}[(\underline{R}_i \oplus \underline{R}'_i)P]$, we have $w_{\mathrm{H}}(\underline{b}) \le 2w_{\mathrm{H}}(\underline{y})$. Furthermore, $0 < w_{\mathrm{H}}(\underline{y}) < \frac{1}{2}d_{\min}$, making $w_{\mathrm{H}}(\underline{b}) < d_{\min}$. Since $\underline{b}$ is a codeword, then $\underline{b}$ should be the zero codeword. This implies that $\underline{R}_i P^* = \underline{R}'_i P^*$ □

It is to be observed that if $\underline{R}_i P^* = \underline{R}'_i P^*$, then

$$\begin{aligned} \underline{d}_i \oplus \underline{d}'_i &= (\underline{R}_i \oplus \underline{R}'_i)P \\ &= [(\underline{R}_i \oplus \underline{R}'_i)P^* \mid (\underline{R}_i \oplus \underline{R}'_i)P^{**}] \\ &= [(\underline{0})_{1\times k} \mid (\underline{R}_i \oplus \underline{R}'_i)P^{**}]. \end{aligned}$$

Thus we have

$$(\underline{d}_i \oplus \underline{d}'_i)^{**} = (\underline{R}_i \oplus \underline{R}'_i)P^{**}. \tag{5}$$

The cryptanalyst thus needs $n-k$ pairs of ciphertexts, where the plaintext is the same for each pair and the random vectors $\underline{R}_i$ and $\underline{R}'_i$ have the property

$\underline{R}_i P^* = \underline{R}'_i P^*$. Each pair would then produce an expression similar to (5). The $n-k$ expressions $\{(\underline{d}_i \oplus \underline{d}'_i)^{**} = (\underline{R}_i \oplus \underline{R}'_i)P^{**}\}_{1 \le i \le n-k}$, form a linear system of equations which allows to solve for P^{**} in $O[(n-k)^3]$ steps, provided that the rank of $\{\underline{R}_i \oplus \underline{R}'_i\}_{1 \le i \le n-k}$ is $n-k$.

The probability that the random vectors $\underline{R}_i$ and $\underline{R}'_i$ have the property $\underline{R}_i P^* = \underline{R}'_i P^*$ is 2^k, thus the number of ciphertexts needed to be collected is $O(2^{k/2})$. Lemma 1 shows the way to test whether $\underline{R}_i P^*$ equals to $\underline{R}'_i P^*$, requiring two additions and three comparisons. Therefore, the time complexity is $O(2^k)$ operations (here the random vectors are compared on a one by one basis). Two tables are needed: one containing the ciphertexts and one containing the random vectors, making the space complexity $O(n2^{k/2})$ bits of memory. Therefore, obtaining P^{**} requires $O[(n-k)2^{k/2}]$ ciphertexts, $O[(n-k)2^k]$ operations, and $O[n(n-k)2^{k/2}]$ bits of memory.

Lemma 1 requires knowledge of $d_{\min}$ of the underlying code. But this should not be a problem. The cryptanalyst introduces errors to some ciphertexts of known plaintexts, and observes the encoded plaintexts. Then the maximum number of errors yielding correct decoding is t. The inequality $d_{\min} \ge 2t+1$ is then used to determine $d_{\min}$

3.2 Finding an Equivalent Code

It is to be observed that knowing P^{**} allows easy checking of whether $\underline{R}_i P^*$ equals to $\underline{R}'_i P^*$, when $\underline{R}_i$ and $\underline{R}'_i$ are given. Let $\underline{d}_i$ and $\underline{d}_j$ be the ciphertexts of the two distinct messages $\underline{m}_i$ and $\underline{m}_j$ such that $\underline{R}_i P^* = \underline{R}_j P^*$, then

$$\underline{d}_i \oplus \underline{d}_j = [f(\underline{m}_i \oplus \underline{R}_i P^*) \oplus f(\underline{m}_j \oplus \underline{R}_j P^*)]G \oplus [(\underline{0})_{1\times k} \mid (\underline{R}_i \oplus \underline{R}_j)P^{**}]. \quad (6)$$

Equation (6) can be reduced to

$$\underline{d}_i \oplus \underline{d}_j \oplus [(\underline{0})_{1\times k} \mid (\underline{R}_i \oplus \underline{R}_j)P^{**}] = [f(\underline{m}_i \oplus \underline{R}_i P^*) \oplus f(\underline{m}_j \oplus \underline{R}_j P^*)]G. \quad (7)$$

The above equation shows that $\underline{d}_i \oplus \underline{d}_j \oplus [(\underline{0})_{1\times k} \mid (\underline{R}_i \oplus \underline{R}_j)P^{**}]$ is a codeword of C. To construct $\hat{G}$, the cryptanalyst needs to collect k linearly independent pairs of ciphertexts such that the random vectors $\underline{R}_i$ and $\underline{R}_j$ have the property $\underline{R}_i P^* = \underline{R}_j P^*$.

For each pair, the number of ciphertexts that need to be collected is $O(2^{k/2})$, and the number of operations is $O(\frac{k}{2}2^{k/2})$ operations (here the random vectors are sorted). For each pair, two tables are needed: one containing the ciphertexts and another containing the random vectors; thus the space complexity is $O(n2^{k/2})$ bits. Obtaining $\hat{G}$ requires $O(k2^{k/2})$ ciphertexts, $O(k^2 2^{k/2})$ operations, and $O(kn2^{k/2})$ bits of memory.

It is to be observed that $\hat{G} = AG$, where A is a $k \times k$ invertible binary matrix. From $\hat{G}$, an $(n-k) \times n$ parity check matrix $\hat{H}$ can be obtained, such that $\hat{G}\hat{H}^{\mathrm{T}} = (\underline{0})_{k\times(n-k)}$, where $\hat{H}^{\mathrm{T}}$ is the transpose of $\hat{H}$. It also holds that $G\hat{H}^{\mathrm{T}} = (\underline{0})_{k\times(n-k)}$; thus $\hat{H} = BH$, where B is an $(n-k) \times (n-k)$ invertible binary matrix. The syndrome-error table $\{\hat{\underline{s}}_i, \underline{e}_i\}_{1 \le i \le 2^{n-k}}$ is obtained from $\hat{H}$, where $\underline{\hat{s}}_i = \underline{e}_i \hat{H}^{\mathrm{T}}$ is the $(n-k)$-bit syndrome of the n-bit vector $\underline{e}_i$ such that $\underline{e}_i$ has the minimum Hamming weight in the set of vectors that produce the same syndrome. Generating this table requires $O(2^{n-k})$ operations, whereas the table requires $O[n(n-k)2^{n-k}]$ bits of memory.

3.3 Finding P^*

For the ciphertext $\underline{d}_i$, using P^{**}, the following expression can be computed,

$$\underline{d}_i \oplus [(\underline{0})_{1\times k} \mid \underline{R}_i P^{**}] = f(\underline{m}_i \oplus \underline{R}_i P^*)G \oplus [\underline{R}_i P^* \mid (\underline{0})_{1\times(n-k)}]. \tag{8}$$

The syndrome $\underline{\hat{s}}_i$ of $\underline{d}_i \oplus [(\underline{0})_{1\times k} \mid \underline{R}_i P^{**}]$ is computed to yield $\underline{\hat{s}}_i = [\underline{R}_i P^* \mid (\underline{0})_{1\times(n-k)}]\hat{H}^{\mathrm{T}}$. Then, using the syndrome-error table, an n-bit vector $\underline{e}_i$ can be found such that

$$[\underline{R}_i P^* \mid (\underline{0})_{1\times(n-k)}] = \underline{e}_i, \tag{9}$$

provided that $w_{\mathrm{H}}(\underline{R}_i P^*) \le t$ (observe that $w_{\mathrm{H}}(\underline{R}_i P^*) = w_{\mathrm{H}}(\underline{R}_i) - w_{\mathrm{H}}(\underline{R}_i P^{**})$, and both $w_{\mathrm{H}}(\underline{R}_i)$ and $w_{\mathrm{H}}(\underline{R}_i P^{**})$ are known). Then (9) can be reduced to

$$\underline{R}_i P^* = (\underline{e}_i)^*. \tag{10}$$

The cryptanalyst needs to collect k ciphertexts such that $w_{\mathrm{H}}(\underline{R}_i P^*) \le t$. Each ciphertext produces an expression similar to (10). The k expressions $\{\underline{R}_i P^* = (\underline{e}_i)^*\}_{1\le i\le k}$ form a linear system of equations which allows to solve for P^* in $O(k^3)$ operations, provided that the vectors $\underline{R}_i$, for $1 \le i \le k$, are linearly independent.

The probability p that $w_{\mathrm{H}}(\underline{R}_i P^*) \le t$ is

$$p = \frac{\sum_{i=0}^{t} \binom{k}{i}}{2^k} < 2^{-k(1-h(\frac{t}{k}))}, \tag{11}$$

where h is the entropy function which is defined for q such that $0 \le q \le 1$ as $h(q) = -q\log_2 q - (1-q)\log_2(1-q)$. The upper bound is obtained using the tail inequality [11, p. 39]. Thus the number of ciphertexts needed is $O(2^{k(1-h(\frac{t}{k}))})$. Obtaining the matrix P^* requires $O(k2^{k(1-h(\frac{t}{k}))})$ operations and ciphertexts.

3.4 Finding an Equivalent Function

A function equivalent to f can be found using the $n \times k$ matrix $\hat{G}^{-\mathrm{R}}$ such that $\hat{G}\hat{G}^{-\mathrm{R}} = I_{k\times k}$, where $I_{k\times k}$ is the $k \times k$ identity matrix. For the ciphertext $\underline{d}_i$, using P, the cryptanalyst can compute

$$\begin{aligned}(\underline{d}_i \oplus \underline{R}_i P)\hat{G}^{-\mathrm{R}} &= f(\underline{m}_i \oplus (\underline{R}_i P)^*)G\hat{G}^{-\mathrm{R}} \\ &= f(\underline{m}_i \oplus (\underline{R}_i P)^*)A^{-1} = \hat{f}(\underline{m}_i \oplus (\underline{R}_i P)^*).\end{aligned} \tag{12}$$

In [9], it was only required to have f secure against ciphertext-only attack, but it can broken under known-plaintext attack. Equation (12) allows the construction of $\hat{f}$ using known-plaintext attack.

4 Discussion

It was suggested in [9] to use (63, 36, 5) BCH code, where $t = 2$ errors. For this code, the attack explained in the previous section requires $O(2^{41})$ operations and $O(2^{23})$ ciphertexts to break the system. The cryptanalyst uses $\hat{f}$, $\hat{G}$, and P for encrypting data, whereas $\hat{f}$, $\hat{G}$, P, $\hat{H}$, and the syndrome-error table are used for decryption.

From Sect. 3, it is clear that the use of a permutation matrix allows the application of Lemma 1 and, also, the decoding using the syndrome-error table, since $w_{\mathrm{H}}(\underline{R}_i P) = w_{\mathrm{H}}(\underline{R}_i)$. To prevent this, either of the following methods can be used to operate on the random vectors instead of P:

- an $n \times n$ invertible binary matrix S, or
- a non-linear function $g : GF(2^n) \rightarrow GF(2^n)$ such that g has the same requirement as f to allow efficient implementation.

The security analysis for both methods is the same. The case for S is considered below.

$\hat{G}$ can be obtained as described in the beginning of Sect. 3, requiring $O(k2^{n/2})$ ciphertexts, $O(kn2^{n/2})$ operations, and $O(kn2^{n/2})$ bits of memory. Now the syndrome-error table can not be used to solve for S since $w_{\mathrm{H}}(\underline{R}_i S)$ is not known. However, probabilistic decoding [1, 2] can be employed to obtain S and, also, $\hat{f}$ as described below.

For the ciphertext $\underline{d}_i$, using $\hat{G}$, we have

$$\underline{d}_i = f(\underline{m}_i \oplus (\underline{R}_i S)^*)A^{-1}\hat{G} \oplus \underline{R}_i S.$$

The above equation can be written as

$$\underline{d}_i = \underline{x}_i \hat{G} + \underline{v}_i, \tag{13}$$

where $\underline{x}_i = f(\underline{m}_i \oplus (\underline{R}_i S)^*)A^{-1} = \hat{f}(\underline{m}_i \oplus (\underline{R}_i S)^*)$, and $\underline{v}_i = \underline{R}_i S$. Then (13) can be used to construct the following equation

$$\underline{d}_i^{(k)} = \underline{x}_i \bar{G} + \underline{v}_i^{(k)}, \tag{14}$$

where $\underline{d}_i^{(k)}$, $\bar{G}$, and $\underline{v}_i^{(k)}$ are restrictions of the same k columns of $\underline{d}_i$, $\hat{G}$, and $\underline{v}_i$, respectively. If the chosen k bits of $\underline{v}_i^{(k)}$ are all zeros, then $\underline{x}_i$ can be obtained as $\underline{x}_i = \underline{d}_i^{(k)}(\bar{G})^{-1}$, requiring $O(k^3)$ operations. Then $\underline{v}_i$ can be obtained using (13) and $\underline{x}_i$. The probability to succeed with this method is equal to the probability that the chosen k bits of $\underline{v}_i$ are all zeros. This probability is simply 2^{-k}. Thus the work factor to obtain $\underline{x}_i$ and $\underline{v}_i$ is $O(k^3 2^k)$ operations.

Finally, $\underline{v}_i$ and $\underline{x}_i$ are used to construct S and $\hat{f}$, respectively. This requires $O(k^4 2^k)$ operations and $O(k)$ ciphertexts.

5 Conclusion

The security of the Hwang-Chen scheme is analyzed. A chosen-plaintext attack is devised that is able to break the scheme, requiring $O[(n-k)2^{k/2}]$ ciphertexts and $O[(n-k)2^{k}]$ operations. The scheme was modified to overcome the weaknesses of the original Hwang-Chen scheme; the complexity to break this modified scheme is $O(k2^{n/2})$ ciphertexts and $O(kn2^{n/2})$ operations. This makes the scheme more attractive the Rao-Nam scheme, since it eliminates the need of the 2^{n-k} syndrome-error table.

References

1. McEliece, R. J.: Public-key cryptosystem based on algebraic coding theory. JPL DSN Progress Report **42-44** (Jan. & Feb. 1978) 114–116. (Jet Propulsion Laboratory, California Institute of Technology, Pasadena, CA, USA)
2. Lee, P. J. and Brickell, E. F.: An observation on the security of McEliece's public-key cryptosystem. In C. G. Günther, editor, Advances in Cryptology-Eurocrypt'87 Proceedings, pages 275–280, Davos, Switzerland, May 25–27 1988, Springer-Verlag (Lecture Notes in Computer Science **330**)
3. Alabbadi, M. and Wicker,S. B.: Combined data encryption and reliability using McEliece's public-key cryptosystem. In International Symposium on Information Theory & Its Applications, pages 263–268, Sydney, Australia, November 20–24 1994.
4. Rao, T. R. N. and Nam, K.: Private-key algebraic-code encryption. IEEE Trans. Info. Theory, **35**(4):829-833, July 1989
5. Struik, R. and van Tilburg, J.: The Rao-Nam scheme is insecure against a chosen-plaintext attack. In C. Pomerance, editor, Advances in Cryptology-Crypto'87 Proceedings, pages 445–457, Santa Barbara, CA, USA, August 16–20 1987, Springer-Verlag (Lecture Notes in Computer Science **239**)
6. Meijers, J. and van Tilburg, J.: On the Rao-Nam private-key cryptosystem using linear codes. In IEEE International Symposium on Information Theory, page 126, Budapest, Hungary, June 24–28 1991
7. Yuanxing, L. and Xinmei, W.: A joint authentication and encryption scheme based on algebraic coding theory. In H. F. Mattson, T. Mora, and T. R. N. Rao, editors, Applied Algebra, Algebraic Algorithms and Error-Correcting Codes, 9th International Symposium, AAECC-9 Proceedings, pages 241–245, New Orleans, LA, USA, October 1991, Springer-Verlag (Lecture Notes in Computer Science **539**)
8. van Tilburg, J.: Two chosen-plaintext attacks on the Li-Wang joint authentication and encryption scheme. In G. Cohen, T. Mora, and O. Moreno, editors, Applied Algebra, Algebraic Algorithms and Error-Correcting Codes, 10th International Symposium, AAECC-10 Proceedings, pages 332–343, San Juan de Puerto Rico, Puerto Rico, May 10–14 1993, Springer-Verlag
9. Hwang, T. and Chen, Y.: Algebraic-code cryptosystem using random code chaining. In IEEE Conference on Computer and Communication Systems (IEEE TENCON'90), 24-27 September 1990, Hong Kong, Volume 1, pages 194–196
10. Davies, D. W. and Price, W. L.: Security for Computer Networks. John Wiley and Sons, New York, USA, second addition, 1989
11. Welsh, D.: Codes and Cryptography. Oxford University Press, Oxford, 1988

Efficient Elliptic Curve Exponentiation

Atsuko Miyaji[1], Takatoshi Ono[2] and Henri Cohen[3]

[1] Multimedia Development Center, Matsushita Electric Industrial Co., LTD.
[2] Matsushita Information Systems Research Laboratory Nagoya Co., Ltd.
[3] Université de Bordeaux

Abstract. Elliptic curve cryptosystems, proposed by Koblitz([8]) and Miller([11]), can be constructed over a smaller definition field than the ElGamal cryptosystems([5]) or the RSA cryptosystems([16]). This is why elliptic curve cryptosystems have begun to attract notice. There are mainly two types in elliptic curve cryptosystems, elliptic curves E over $\mathbb{F}_{2^r}$ and E over $\mathbb{F}_p$. Some current systems based on ElGamal or RSA may often use modulo arithmetic over $\mathbb{F}_p$. Therefore it is convenient to construct fast elliptic curve cryptosystems over $\mathbb{F}_p$. In this paper, we investigate how to implement elliptic curve cryptosystems on $E/\mathbb{F}_p$.

1 Introduction

Koblitz ([8]) and Miller ([11]) proposed a method by which public key cryptosystems can be constructed on the group of points on an elliptic curve over a finite field instead of a finite field. If elliptic curve cryptosystems avoid the Menezes-Okamoto-Vanstone reduction ([13]), then the only known attacks are the Pollard ρ−method ([15]) and the Pohlig-Hellman method ([14]). So up to the present, we can construct elliptic curve cryptosystems over a smaller definition field than the discrete-logarithm-problem(DLP)-based cryptosystems like ElGamal cryptosystems([5]) or DSA([3]) and the RSA cryptosystems([16]). Elliptic curve cryptosystems with 160-bit key have the same security as both ElGamal cryptosystems and RSA with 1,024-bit key. This is why elliptic curve cryptosystems have been discussed in ISO/IEC CD 14883-3, ISO/IEC DIS 11770-3, ANSI ASC X.9, X.9.62, and IEEE P1363([7]). As standardization is advanced, fast implementation of elliptic curve cryptosystems has been reported([6, 20, 22]).

There are mainly two types in elliptic curve cryptosystems, elliptic curves over $\mathbb{F}_{2^r}$ and elliptic curves over $\mathbb{F}_p$. Up to the present, the study on implementation has been often aimed at elliptic curves over $\mathbb{F}_{2^r}$ since arithmetic in $\mathbb{F}_{2^r}$ has an advantage of good performance in hardware. Practically speaking, however, DLP-based cryptosystems or RSA cryptosystems, both of which use modular arithmetic over $\mathbb{F}_p$, have been widely used in many systems. Therefore it would be convenient to construct elliptic curve cryptosystems over $\mathbb{F}_p$ since we can offer both RSA and elliptic curve cryptosystems with one modular arithmetic.

Elliptic curve cryptosystems mainly consist of elliptic curve exponentiations. This paper studies efficient elliptic curve exponentiation, which aims at elliptic curves over $\mathbb{F}_p$ but can be applied to any elliptic curve. Studies on elliptic curve

exponentiations are mainly classified into three factors: the coordinate, an exponentiation for a fixed point, and an exponentiation for a random point. This paper investigates these three factors:
1. **The coordinate:** Elliptic curve exponentiation can be computed by repeating additions and doublings, where the repeated number of additions can be reduced by a suitable algorithm, but that of doublings can not be reduced especially in the case of exponentiation for a random point. On the other hand, we can define some coordinates on an elliptic curve, which give each different addition formula. So we investigate the efficiency of the addition formula in jacobian coordinates([2]) which is less familiar than projective coordinates. Jacobian coordinates offer a slower addition but a faster doubling, which should be suitable for elliptic curve exponentiation.
2. **Exponentiation for a fixed point:** In this case, the precomputation table method([1]) is useful, which computes an elliptic curve exponentiation by repeating only additions and no doubling. In order to make use of a feature of jacobian coordinates, we propose a new algorithm which requires more doublings but fewer additions. Total computation amount for kG in our algorithm is less than [1].
3. **Exponentiation for a random point:** In this case, the addition-subtraction method is usually mixed with the window method([10, 12, 9, 20]). In this approach, an interval between two windows mainly determines the computation amount: the longer the interval is, the less the computation amount is. Importantly, signed binary representation of k is not determined uniquely, while an interval differs for each signed binary representation. An average interval by mixing the signed binary representation in [12] and the window method is 4/3, that in [9] is 3/2. Here we present a new method for signed binary representation, which improves the average interval to 2 by mixing with the window method.

This paper is organized as follows. Section 2 discusses jacobian coordinates. Section 3 investigates each suitable algorithm for exponentiation of a fixed point and a random. Two implements of a 160-bit definition field and a 169-bit definition field are presented in appendices.

2 The coordinate

An elliptic curve can be represented by several sets of coordinates. The addition formula, which is defined by setting a point at infinity $\mathcal{O}$ to zero, differs for each coordinate: the computation amount of addition differs for each coordinate. Two coordinates, affine coordinates and projective coordinates, are well known([19]). Affine coordinate requires a division in every addition and every doubling but requires fewer multiplications than projective coordinate. On the other hand, projective coordinate does not require any division in either addition or doubling and does require a division only once in the final stage of the computation of elliptic curve exponentiation. In the case of $\mathbb{F}_p$ the computation of elliptic curve exponentiation in projective coordinates is faster than that in affine coordinates since the ratio of the computation amount of division in $\mathbb{F}_p$ to that of multiplication in $\mathbb{F}_p$ is generally larger than 9.

Here we discuss another coordinate([2]), which is called jacobian coordinate in this paper. The addition formula in jacobian coordinates is similar to projective coordinate: it does not require any division modulo p in either addition or doubling and does require a division only once in the final stage of the computation of elliptic curve exponentiation. However, jacobian coordinates offer a doubling with less computation amount but an addition with more computation amount than projective coordinates. This feature should be suitable for elliptic curve exponentiation since the number of additions required in elliptic curve exponentiation can be reduced by a suitable algorithm, but that of doublings may not be reduced.

Here we presents the addition formula in jacobian coordinate which is a slightly revised version of ([2]). Let an elliptic curve over $\mathbb{F}_p (p > 3)$ be

$$E : y^2 = x^3 + ax + b \quad (a, b \in \mathbb{F}_p, 4a^3 + 27b^2 \neq 0).$$

For the elliptic curve, the jacobian coordinate sets $x = X/Z^2$ and $y = Y/Z^3$ i.e.

$$E : Y^2 = X^3 + aXZ^4 + bZ^6.$$

The addition formulae in the jacobian coordinates are the following. Let $P = (X_1, Y_1, Z_1)$, $Q = (X_2, Y_2, Z_2)$ and $P + Q = R = (X_3, Y_3, Z_3)$.

- **Curve addition formula in jacobian coordinates ($P \neq \pm Q$)**

$$X_3 = -H^3 - 2U_1H^2 + r^2, Y_3 = -S_1H^3 + r(U_1H^2 - X_3), Z_3 = Z_1Z_2H,$$

where $U_1 = X_1Z_2^2, U_2 = X_2Z_1^2, S_1 = Y_1Z_2^3, S_2 = Y_2Z_1^3, H = U_2 - U_1, r = S_2 - S_1$;

- **Curve doubling formula in jacobian coordinates ($R = 2P$)**

$$X_3 = T, Y_3 = -8Y_1{}^4 + M(S - T), Z_3 = 2Y_1Z_1,$$

where $S = 4X_1Y_1^2, M = 3X_1^2 + aZ_1^4, T = -2S + M^2$.

Here we discuss the computation amount of addition formulae. We denote the computation amount for 1 multiplication(resp. division) in $\mathbb{F}_p$ by M(resp. D). For simplicity, we neglect addition, subtraction and multiplication by a small constant in $\mathbb{F}_p$ because they are much faster than multiplication and division in $\mathbb{F}_p$. Table 1 presents the number of multiplications in addition formula of jacobian coordinates and projective coordinates, where the addition formula in projective coordinates is presented in Appendix A. Table 1 includes the following special cases: in the computation of a fixed point, we may set Z_1 to one or both Z_1 and Z_2 to one in addition formula. Doubling formula depends on a coefficient a of an elliptic curve: in the case of $a = 0$ or $a = -3$(setting $w = 3X_1^2 - 3Z_1^4 = 3(X_1 - Z_1^2)(X_1 + Z_1^2)$), the computation amount is reduced. Note that addition formula does not depend on coefficients.

	addition			doubling		
	$Z_1, Z_2 \neq 1$	$Z_1 = 1$	$Z_1 = Z_2 = 1$	$a \neq 0, -3$	$a = 0$	$a = -3$
Projective coordinate	$14M$	$11M$	$7M$	$12M$	$10M$	$10M$
Jacobian coordinate	$16M$	$11M$	$6M$	$10M$	$7M$	$8M$

Table 1. Number of multiplications in addition formula

3 Elliptic curve exponentiation

This section discusses elliptic curve exponentiations for a fixed point and a random point. Both discussion depends neither on the size of definition field nor on the characteristic of definition field.

3.1 Exponentiation for a fixed point

For simplicity, here we assume a 160-bit definition field $\mathbb{F}_p$. As for a fixed point, the precomputation table method is useful([1]): it prepares a table of 40 points $16^i G$ for $i = 1, \cdots, 40$, each of which Z-coordinate is set to 1, and computes kG in about 44 additions/subtractions. Since this method does not require any doubling, projective coordinate is suitable. The computation amount sums up to $500M + D$, considering carefully three cases of addition in Table 1. Note that the computation amount does not depend on a coefficient a of an elliptic curve.

As we have seen in Section 2, the computation amount of doubling is less than that of addition. Therefore we would reduce the number of additions rather than that of doublings. Here we describe another method which requires more doublings but fewer additions than [1]. The tables consist of 62 points,

$$A[s] = \sum_{j=0}^{4} a_{s,j} 2^{32j} G \text{ and } B[s] = \sum_{j=0}^{4} a_{s,j} 2^{16+32j} G \quad (1 \le s \le 31),$$

where $a_{s,0}, \cdots, a_{s,4}$ is a representation of s in radix 2, that is $s = \sum_{j=0}^{4} a_{s,j} 2^j$. The Z-coordinates of 62 points are also set to 1. Then the algorithm to compute kG is as follows, where k is set to $k = \sum_{j=0}^{159} k[j] 2^j$:

Algorithm 1.

1. **set** $u_j = \sum_{i=0}^{4} k[32i + j] 2^i$ **and** $v_j = \sum_{i=0}^{4} k[32i + 16 + j] 2^i$ **for** $0 \le j \le 15$
2. **set** $A[0] = \mathcal{O}, B[0] = \mathcal{O}$, **and** $T = \mathcal{O}$
3. **for** $i = 15$ **to** 0 **by** -1
   $T = 2T$ **and** $T = T + A[u_i] + B[v_i]$
4. **output** $T = kG$.

The above algorithm computes kG by 30 additions and 15 doublings. Jacobian coordinate is suitable for this method. By using jacobian coordinate, the computation amount sums up to $479M + D$. If we set a coefficient $a = 0$ of elliptic curve, the computation amount is reduced to $434M + D$.

To sum up, our method with jacobian coordinate can reduce the computation amount of kG by 4% of [1]. Furthermore by selecting a suitable elliptic curve like a coefficient $a = 0$ it is reduced by 13% of [1].

3.2 Exponentiation for a random point

As for the computation of kP for a random point P, the addition-subtraction method is commonly mixed with the window method([10, 12, 9, 20]). In this approach, an interval between two windows mainly determines the computation amount for kP: the longer the interval is, the less the computation amount is. Importantly, signed binary representation of k is not determined uniquely, while an interval differs for each signed binary representation. Here we present new signed binary representation, which can offer a longer interval by mixing with the window method. A feature of our method is that the signed binary representation depends on a width in the window method. On the other hand, known representation([10, 12, 9]) is independent of the window method.

Let $n = \lfloor log_2(k) \rfloor$, $k = \sum_{i=0}^{n} k[i]2^i$ $(k[i] = 0, 1)$ in binary representation, and w be a width in the window method. The following algorithm transforms k into $k' = \sum_{i=0} k'[i]2^i$ $(k'[i] = 0, \pm 1)$ in signed binary while setting windows $W[j] (j = 0, 1, \cdots)$:

Algorithm 2.

1. set $i = 0$, $j = 0$ and $k[n+1] = 0$
2. while $i \leq n$ do:
   - if $i + w - 1 \geq n - w$, then set $k'[i] = k[i], \cdots, k'[n] = k[n]$,
     set $W[j] = (k[i+w-1], \cdots, k[i+1], k[i])$ and goto 3.
   - if $k[i] = 0$, then set $k'[i] = k[i]$, $i = i + 1$, and goto 2.
   - if $k[i] = 1$, then set $t[j] = \sum_{t=0}^{w-1} k[i+t]2^t$
     - if $k[i+w] = 0$, then set $k'[i] = k[i], \cdots, k'[i+w] = k[i+w]$,
       set $W[j] = (k[i+w-1], \cdots, k[i+1], k[i])$,
       set $j = j + 1$, $i = i + w + 1$ and goto 2.
     - if $k[i+w] = 1$, then for first t satisfying $k[t] = 0$ $(t = i + w + 1, \cdots)$
       set $k'[i+w] = 0, \cdots, k'[t-1] = 0$, and $k[t] = 1$,
       set $t[j] = 2^w - t[j] = \sum_{t=0}^{w-1} k'[i+t]2^t$ (in binary representation),
       set $k'[i] = -k'[i], \cdots, k'[i+w-1] = -k'[i+w-1]$,
       set $W[j] = (k'[i+w-1], \cdots, k'[i+1], k'[i])$,
       set $i = t, j = j + 1$ and goto 2.
3. output $k' = \sum_{i=0} k'[i]2^i$ $(k'[i] = 0, \pm 1)$ and $W[i] (i = 0, 1, \cdots)$.

Let the most significant window be $W[s-1] = (k'[i+w-1], \cdots, k'[i+1], k'[i])$. For convenience, set $W[s] = (k'[n], \cdots, k'[i+w])$, where $i + w \geq n - w + 1$ from Algorithm 2. Then k' is written as

$$k' = 2^{t_0}(2^{t_1}(\cdots 2^{t_{s-1}}(2^{t_s} W[s] + W[s-1]) \cdots) + W[0]) \ (0 \leq t_i).$$

In order to decrease the number t_s of doublings, we revise $W[s]$ to a window with a length at most w from MSB and fit the most significant bit of $W[s-1]$ to the new $W[s]$, while leaving others as they are. Then we can compute kP from MSB to LSB by using the transformation of k after preparing points $P, 3P, \cdots, (2^w - 1)P$.

Example. Set $w = 4$. For a given $k = 10110110011011 0111$ in binary representation, the algorithm transforms k into:

$k' = \underline{101}\,\underline{1011}\,0\,\underline{0111}\,00\,\underline{\bar{1}00\bar{1}} = W[3]\,W[2]\,0\,W[1]\,00\,W[0]$,

revise k' to:

$k' = \underline{1011}\,\underline{011}\,0\,\underline{0111}\,00\,\underline{\bar{1}00\bar{1}} = W[3]\,W[2]\,0\,W[1]\,00\,W[0]$,

where $\bar{1}$ denotes -1 and each block of underlined digits represents one window like $W[0]$. Then kP can be computed by $2^6(2^5(2^3W[3] + W[2]) + W[1]) + W[0]$.

Let estimate the computation amount of exponentiation in this method. First we show that an average interval between two windows is 2:

Theorem 1. *Algorithm 2 constructs windows at an average interval of 2 bits.*

Proof. Let $W[j] = (k[i+w-1], \cdots, k[i+1], k[i])$ be a window. Then we show the next window $W[j+1]$ will start at $k[i+w+2]$ on the average. From Algorithm 2, the next window never starts at $k[i+w]$. The next window starts at $k[i+w+1]$ if and only if $(k[i+w+1], k[i+w]) = (0,1), (1,0)$. Therefore the probability of starting at $k[i+w+1]$ is $\frac{1}{2}$. The next window starts at $k[i+w+2]$ if and only if $(k[i+w+2], k[i+w+1], k[i+w]) = (1,0,0), (0,1,1)$. Therefore the probability of starting at $k[i+w+2]$ is $(\frac{1}{2})^2$. Thus, an average interval between $W[j]$ and $W[j+1]$ is computed in $\sum_{i=1} i * (\frac{1}{2})^i \simeq 2$. Therefore the next window will start at $k[i+w+2]$ on the average. □

From the above theorem, we obtain the following approximate multiplication count $T_w(n)$ for raising a point P to the n-th power by setting $u = \sum_{i=0}^{s} t_i$ and L to be the average length of k' and using jacobian coordinates

$$T_w(n) = (16 - \frac{5}{2^{w-1}})(\frac{L}{w+2}) + 10u + 16 * 2^{w-1} - 15.$$

Then it is easily shown that $u = (n + 2 - (\frac{1}{2})^w - w)$ and $L = n + \frac{3}{2}$. Therefore we obtain

$$T_w(n) = (16 - \frac{5}{2^{w-1}})(\frac{n+3/2}{w+2}) + 10(n + 2 - (\frac{1}{2})^w - w) + 16 * 2^{w-1} - 15.$$

Choosing w to make $T_w(n)$ minimal in the range of $150 < n < 170$, we get $w = 4$ is optimal since $T_3(n) > T_4(n) < T_5(n)$.

We discuss one case of $n = 159$, in which implementation is presented in Appendix B. Our algorithm computes kP in 33.7 additions and 157.9 doublings. The total computation amount of our algorithm with jacobian coordinates is $2098M + D$. On the other hand, the method of [9] requires $2384M + D$ in projective coordinates or $2141M + D$ in jacobian coordinates.

4 Conclusion

In this paper, we have been proposed efficient elliptic curve exponentiations for a fixed and a random point. As for a fixed point, our method with more doublings but fewer additions can compute kG with 160-bit k in 30 additions and 15 doublings. As for a random point, our method of mixing new signed representation with the window method can compute kP with 160-bit k in 33.7 additions and 157.9 doublings. The use of jacobian coordinate gives further improvement to the running time: elliptic curve exponentiation for a fixed point can be computed in $479M + D$ and that for a random point can be computed in $2098M + D$.

References

1. E. F. Brickell, D. M. Gordon, K. S. McCurley and D. B. Wilson, "Fast exponentiation with precomputation" *Advances in Cryptology-Proceedings of EUROCRYPT'92*, Lecture Notes in Computer Science, **658**(1993), Springer-Verlag, 200-207.

2. D. V. Chudnovsky and G. V. Chudnovsky "Sequences of numbers generated by addition in formal group and new primality and factorization tests" *Advances in Applied Math.*, **7**(1986), 385-434.
3. "Proposed federal information processing standard for digital signature standard (DSS)", *Federal Register*, v. 56, n. 169, 30 Aug 1991, 42980-42982.
4. W. Diffie and M. Hellman, "New directions in cryptography" *IEEE Trans. Inform. Theory*, Vol. IT-22 (1976), 644-654.
5. T. ElGamal, "A public key cryptosystem and a signature scheme based on discrete logarithms", *IEEE Trans. Inform. Theory*, Vol. IT-31 (1985), 469-472.
6. G. Harper, A. Menezes and S. Vanstone, "Public-key cryptosystems with very small key lengths", *Advances in Cryptology-Proceedings of Eurocrypt'92*, Lecture Notes in Computer Science, **658**(1993), Springer-Verlag, 163-173.
7. *IEEE P1363 Working Draft*, February 6, 1997.
8. N. Koblitz, "Elliptic curve cryptosystems", *Mathematics of Computation*, **48**(1987), 203-209.
9. K. Koyama and Y. Tsuruoka, "Speeding up elliptic cryptosystems by using a signed binary window method", *Abstract of proceedings of CRYPTO'92*, 1992.
10. D. E. Knuth, *The art of computer programming, vol. 2, Seminumerical Algorithms*, 2nd ed., Addison-Wesley, Reading, Mass. 1981.
11. V. S. Miller, "Use of elliptic curves in cryptography", *Advances in Cryptology-Proceedings of Crypto'85*, Lecture Notes in Computer Science, **218**(1986), Springer-Verlag, 417-426.
12. F. Morain and J. Olivos, "Speeding up the computations on an elliptic curve using addition-subtraction chains", Theoretical Informatics and Applications Vol.24, No.6 (1990), 531-544.
13. A. Menezes, T. Okamoto and S. Vanstone, "Reducing elliptic curve logarithms to logarithms in a finite field", *Proceedings of the 22nd Annual ACM Symposium on the Theory of Computing*, 80-89, 1991.
14. S. C. Pohlig and M. E. Hellman, "An improved algorithm for computing logarithm over $GF(p)$ and its cryptographic significance", *IEEE Trans. Inf. Theory*, IT-24(1978), 106-110.
15. J. Pollard, "Monte Carlo methods for index computation(mod p)", *Mathematics of Computation*, **32**(1978), 918-924.
16. R. Rivest, A. Shamir and L. Adleman, "A method for obtaining digital signatures and public-key cryptosystems", *Communications of the ACM*, vol.21, No.2(1978), 120-126.
17. B. Schneier *Applied cryptography*, II, John Wiley & Sons, Inc. 1996.
18. C. P. Schnorr, "Efficient identification and signatures for smart cards", *Advances in cryptology-Proceedings of Crypto'89*, Lecture Notes in Computer Science, **435**(1989), Springer-Verlag, 239-252.
19. J. H. Silverman, *The Arithmetic of Elliptic Curves*, GTM106, Springer-Verlag, New York, 1986.
20. R. Schroeppel, H. Orman, S. O'Malley and O. Spatscheck, "Fast key exchange with elliptic curve systems", *Advances in Cryptology-Proceedings of Crypto'95*, Lecture Notes in Computer Science, **963**(1995), Springer-Verlag, 43-56.
21. Torbjorn Granlund, The GNU MP LIBRARY, version 2.0.2, June 1996. ftp://prep.ai.mit.edu/pub/gnu/gmp-2.0.2.tar.gz
22. E. D. Win, A. Bosselaers and S. Vandenberghe "A fast software implementation for arithmetic operations in GF(2^n)", *Advances in Cryptology-Proceedings of Asiacrypt'95*, Lecture Notes in Computer Science, **1163**(1996), Springer-Verlag, 65-76.

A The addition formula in projective coordinate

Let an elliptic curve over $\mathbb{F}_p (p > 3)$ be

$$E : y^2 = x^3 + ax + b \ \ (a, b \in \mathbb{F}_p, 4a^3 + 27b^2 \neq 0).$$

For the elliptic curve, the projective coordinate sets $x = X/Z$ and $y = Y/Z$ i.e.

$$E : Y^2 Z = X^3 + aXZ^2 + bZ^3.$$

The addition formulae in projective coordinates are the following. Let $P = (X_1, Y_1, Z_1)$, $Q = (X_2, Y_2, Z_2)$ and $P + Q = R = (X_3, Y_3, Z_3)$.

- **Curve addition formula in projective coordinates ($P \neq \pm Q$)**

$$X_3 = vA, Y_3 = u(v^2 X_1 Z_2 - A) - v^3 Y_1 Z_2, Z_3 = v^3 Z_1 Z_2, \tag{1}$$

where $u = Y_2 Z_1 - Y_1 Z_2, v = X_2 Z_1 - X_1 Z_2, A = u^2 Z_1 Z_2 - v^3 - 2v^2 X_1 Z_2$;

- **Curve doubling formula in projective coordinates ($R = 2P$)**

$$X_3 = 2hs, Y_3 = w(4B - h) - 8{Y_1}^2 s^2, Z_3 = 8s^3, \tag{2}$$

where $w = a{Z_1}^2 + 3{X_1}^2, s = Y_1 Z_1, B = X_1 Y_1 s, h = w^2 - 8B$.

B Implementation of elliptic curve exponentiations

B.1 Elliptic curves

Elliptic curves $E/\mathbb{F}_p$ with order divisible by 160-bit or more prime is secure if it satisfies MOV-condition([7]). As we have seen in Section 2, the computation amount of doubling is reduced in the case of a coefficient $a = 0$ or -3 of elliptic curve. Here we implement two elliptic curves with a coefficient $a = 0$ in 160-bit and 169-bit key size.

1. 160-bit key size

- a definition field $\mathbb{F}_{p_1}$: $p_1 = 2^{160} - 2013$
- an elliptic curve E_1: $y^2 = x^3 + 4$, $\#E_1(\mathbb{F}_{p_1}) = 3 * 13 * q_1$, where q_1 is a prime
  $q_1 = 37\ 47440\ 09572\ 02638\ 92829\ 95765\ 50867\ 08565\ 09759\ 22411$
- a basepoint G_1: $(x_1, y_1) \in E_1(\mathbb{F}_{p_1})$ with order q_1, where
  $x_1 = 1312\ 01277\ 27149\ 38861\ 46561\ 78958\ 06449\ 61829\ 03474\ 73840$
  $y_1 = 1143\ 61120\ 94309\ 35596\ 62639\ 62368\ 56710\ 92306\ 44246\ 02993$

2. 169-bit key size

- a definition field $\mathbb{F}_{p_2}$: $p_2 = 2^{169} - 1825$
- an elliptic curve E_2: $y^2 = x^3 + 49$, $\#E_2(\mathbb{F}_{p_2}) = 3 * 67 * q_2$, where q_2 is a prime
  $q_2 = 3722\ 83004\ 13603\ 09920\ 99645\ 09743\ 01139\ 56489\ 04413\ 35543$
- a basepoint G_2: $(x_2, y_2) \in E_2(\mathbb{F}_{p_2})$ with order q_2, where
  $x_2 = 1\ 55608\ 20629\ 69890\ 07722\ 36926\ 87616\ 67589\ 98487\ 34687\ 95184$
  $y_2 = 55502\ 35686\ 97076\ 18367\ 46840\ 54359\ 62467\ 42560\ 87632\ 81833$

Here we discuss cryptographic differences between E_1 and E_2. From a security and efficiency point of view, these two elliptic curves give almost the same security, while the exponentiations in E_1 can be computed faster than that in E_2. However, from an application point of view, ElGamal encryption on E_2 can encrypt a 168-bit Triple-DES key([17]) but E_1 can not.

B.2 The running time

Here presents the running time of elliptic curve exponentiations over 160-bit and 169-bit definition fields in our methods. Our modulo arithmetic uses GNU MP Library GMP([21]) in order to make easy comparison possible since GMP might be most popular multiprecision library. The platform is SS-5(MicroSPARC 110 MHz/Solaris 2.4) with a 32 bit word size. Table 2 shows the running time.

ElGamal encryption on an elliptic curve mainly consists of one exponentiation of a fixed point and one exponentiation of a random point. Therefore we can estimate that Triple-DES key can be encrypted in about 42 msec.

	$E_1/\mathbb{F}_{p_1}(p_1 = 2^{160} - 2013)$	$E_2/\mathbb{F}_{p_2}(p_2 = 2^{169} - 1825)$
160 bit /169 bit addition	0.88 μsec	1.31 μsec
160 bit /169 bit multiply	9.66 μsec	12.04 μsec
160 bit /169 bit square	7.65 μsec	9.77 μsec
modular reduction	4.81 μsec	5.76 μsec
160 bit /169 bit division	0.91 msec	1.00 msec
addition(jacobian-coordinate)	0.23 msec	0.30 msec
doubling(jacobian-coordinate)	0.12 msec	0.13 msec
exponentiation of a fixed point	7.79 msec	9.31 msec
exponentiation of a random point	26.93 msec	32.54 msec

Table 2. Time of elliptic curve operations

Efficient Construction of Secure Hyperelliptic Discrete Logarithm Problems

Jinhui Chao[1], Nori Matsuda[1], Shigeo Tsujii[2]

[1] Dept. of Electrical and Electronic Engineering, Chuo University, Tokyo, Japan
[2] Dept. of Information Engineering Systems, Chuo University, Tokyo, Japan

Abstract. Hyperelliptic curves have been used to define discrete logarithm problems as cryptographic one-way functions. However, no efficient algorithm for construction of secure hyperelliptic curves is known until now. In this paper, efficient algorithms are presented to construct secure discrete logarithm problems on hyperelliptic curves whose Jacobian varieties are either simple or isogenous to a product of simple abelian varieties.

1 Introduction

Discrete logarithm problems over elliptic curves have recently been used instead of the discrete logarithm problems over finite fields in recent cryptosystems[10] [17] [16]. This new kind of cryptographic functions are believed to be stronger in the sense that they can resist all known subexponential attacks which have been developed against the latter problems.

As a natural extension, hyperelliptic curves, which have genera larger than one and contain the elliptic curves as a special case with genera equal one, were used to define discrete logarithm problems [11]. This generalization also suggested potential usage of general Abelian varieties over finite fields as a rich source of discrete logarithm problems.

However, it seems some questions are still remained open. Firstly, the security of the discrete logarithm problems over hyperelliptic curves is not clear comparing with that over elliptic curves; Secondly to construct such problems is much more nontrivial than elliptic curves.

As to the first question, the MOV reduction attack [15] was extended to hyperelliptic curves [21]. This attack, however, may be easily avoided since generally it is expected (as proved in [12] for elliptic curves) that a well-defined reduction requires very high degree extension of the ground finite field. The Baby-step-Giant-step attack will cost fully exponential time if one uses only almost prime curves or the curves whose orders of the Jacobian varities contain a large rational prime factor.

Another possible attack is to lift the curve to $\boldsymbol{Q}$ or certain number fields, as did in the index calculus algorithm. Since, similar to the case of elliptic curves [17], the lifting of $\boldsymbol{F}_q$-points of curves to $\boldsymbol{Q}$-points costs exponentially time, and the heights of $\boldsymbol{Q}$-points of Abelian varieties increase exponentially, either lifting of $\boldsymbol{F}_q$-points and computations over $\boldsymbol{Q}$ or number fields seem extremely difficult. These properties then will generally hinder attacks over global fields. Furthermore, the index calculus algorithm is also difficult to be applied to abelian varieties since they, like elliptic curves, have only finite $\boldsymbol{Q}$-rank by the Mordell-Weil theorem.

Thus, to the above attacks, the discrete logarithm problems over hyperelliptic curves seems at least as hard as the same problems over elliptic curves. Furthermore, we will show in the next section that one can build safer cryptosystems based on the hyperelliptic curves with larger genera than based on elliptic curves, with a minor increase of computational cost, because there are much more secure hyperelliptic curves that can be used in crytosystems.

Recently, it is shown that the discrete logarithm problem over a hyperelliptic curve can be solved in subexponential time if its genus g and the characteristic p of the ground fields satisfies $\log p \leq (2g+1)^{.98}$ for odd p [2]. Thus if one chooses hyperelliptic curves such that $g > 1$ is small and p is large relatively, they can also resist this kind of attack.

In this paper, we will consider hyperelliptic curves which are defined over fields with large characteristics and whose genera are small but larger than one.

As to the second question, partially due to that a complete general theory on Abelian varieties is still out of reach, computational and constructive toolkit of theories and algorithms is not available for our purposes. At this time, several efficient algorithms have been developed for computation in Jacobian varieties e.g. [3]. But it is not known generally how to design secure hyperelliptic curves. One way similar to approaches for elliptic curves is to use the order-counting algorithms. In this direction, Schoof's algorithm is extended to hyperelliptic curves of genus 2 in [1] which takes random polynomial time. But it's very complicated and heavy due to the computation of the torsion points. A deterministic polynomial time order counting algorithm of $O(\log p)^{\Delta}$ is given in [22]. However, it is observed that the $\Delta >\exp(\exp(\deg(f)))$ [6]. A random polynomial time algorithm is then presented in [6]costing $O(\log p)^{\Delta}$, where the $\Delta = \deg(f)^{O(1)}$. Unfortunately, all these algorithms are still very complicated and too costly to be used in practical calculation. Besides, they usually have to repeat the whole order counting calculations many times if the chosen Jacobian is not almost prime. Since the probability of a Jacobian variety over $\boldsymbol{F}_q$ to be almost prime is quite low, the order counting algorithms will generally have to repeat about $O(\log^2 q)$ times.

The only accessible method until now to construct secure hyperelliptic cryptosystems is to count the order of a Jacobian variety at very small fields then lift the curves by the Weil conjecture[10]. However, this method is only valid over extension fields of small characteristics. Besides, the number of secure curves can be found by this method seems very small.

In practice, it would be highly desirable to have efficient algorithms to design secure hyperelliptic curves, particularly over large prime fields since they are suitable for software implementation and much richer in isogeny classes, which means there is a great deal of secure curves that can be used in cryptosystems.

In this paper, we show efficient algorithms for construction of hyperelliptic

curves, which are defined over finite fields with large characteristics and having genera small but larger than one, and the associated discrete logarithm problems for the following two cases.
(1) The Jacobian varieties are simple with complex multiplication;
(2) The Jacobian varieties are isogenous to products of simple Abelian varieties with complex multiplication.
In particular, we show how to construct these two kinds of the seemly "most secure" hyperelliptic curves, i.e., those of genus two. Examples of implementation of these algorithms are shown and their performance is discussed. All these algorithms are very fast and easy to implement. Moreover, with these algorithm we are able to construct systematically a large amount of secure hyperelliptic curves.

2 Discrete logarithm problems over hyperelliptic curves

A hyperelliptic curve over field F of genus g is defined by

$$C : v^2 + vh(u) = f(u)$$

where $\deg h \leq g, \deg f = 2g+1$ For char$F \neq 2$, one can use the definition as $C : v^2 = f(u)$. A F-rational point $P = (x, y) \in C(F)$ is defined by $x, y \in F$ such that $y^2 + yh(x) = f(x)$.
A divisor D on C is defined as a finite formal sum of form $\sum_i m_i(P_i), m_i \in Z, P_i \in C$. The degree of D is defined as $\deg(D) = \sum_i m_i$.
The function field over C, $F(C)$ is consisted of

$$\{p/q\}, p, q \in F[u, v], q \not\equiv 0 (\bmod\ v^2 + vh(u) - f(u)).$$

The divisor of a function $p/q \in F(C)$ is defined as $\sum_i m_i(P_i) - \sum_j n_j(Q_j)$, here $P_i, Q_j \in C$ are zeros and poles of the function and m_i, n_j are the multiplicities of these zeros and poles. It can be shown that all the divisors of functions over C have degree zero and will be called as principle divisors.
The Jacobian variety of C is then defined as follows. Let

$$\mathcal{D}^0(C) := \{ \text{ Divisor of degree } 0\}$$
$$\mathcal{D}^l(C) := \{ \text{ Principle divisor } \}$$

Then

$$\mathcal{J}(F) = \mathcal{D}^0(C)/\mathcal{D}^l(C)$$

The discrete logarithm problem over a hyperelliptic curve is defined as to find $m \in Z$ given two divisors $D_1, D_2 \in \mathcal{J}(F_q)$ such that $D_1 = mD_2$.
Below we consider the security and computation issue of hyperelliptic curves comparing with elliptic curves.
It is proved by Weil that the order of an Abelian variety A over F_q, of genus g lies in the following range.

$$(q^{1/2} - 1)^{2g} \leq \#A(F_q) \leq (q^{1/2} + 1)^{2g}$$

Thus the order of a Jacobian variety over F_{p^m} is of order $\#\mathcal{J}(F_{p^m}) = O(p^{mg})$. Now we assume that the size of the Jacobian is chosen for security against e.g. the Baby-step-Giant-step attack such that $O(p^{mg}) = N$.
By the above Weil's theorem, one knows that the number of isogeny classes of Jacobian varieties of genus g whose order $\#\mathcal{J}(F_{p^m}) = O(N)$ is

$$\#\{\mathcal{J}(F_{p^m})\} = O(4gN^{1-\frac{1}{2g}}).$$

Since for elliptic curves whose $g = 1$, $\#\{E(F_{p^m})\} = O(4N^{\frac{1}{2}})$, the secure Jacobian varieties with large genera are much richer than elliptic curves.
As to the importance to have richer isogeny classes of Jacobian varieties, we consider only the Baby-step-Giant-step attack which is assumed to be the only effective attack for our discrete logarithm problems. It is know that to solve a discrete logarithm problem defined over an abelian group with order N, this attack costs $O(\sqrt{N}\log N)$ computations. In particular, the computations include two stages, construction of a database which costs $O(\sqrt{N}\log N)$ and search in the database which costs $O(\sqrt{N}\log\sqrt{N})$. Thus, among the possible approaches to enhance the integrity of cryptosystems, if one changes the base point but keeps using the same Jacobian variety, the old database can be used for the new base point, and the attack only requires once more search computation to find the discrete logarithm of the new base point. On the other hand, if one changes the Jacobians each time, then the databases for each new Jacobians have to be reconstructed (maybe in real-time), which cost extra exponentially costly computations of $O(\sqrt{N}\log N)$. Thus, it is highly desirable to be able to design as more as possible secure Jacobian varieties or hyperelliptic curves.
Now we consider the computation cost in order to achieve this enhancement of integrity. It is known that Cantor's algorithm [3] costs time of polynomial in the genus of the curve over F_{p^m}, $O(mg\log^2 g\log^2 m\log p)$. With the fixed order $N = O(p^{mg})$, we see that the computation for hyperelliptic curves is of $O(\log N\log^2 g\log^2 m)$, the increment of computations comparing with elliptic curves is above $O(\log^2 g)$ which could be reasonable comparing with the enhancement security by using a larger isogeny classes of secure Jacobian varieties.

3 Construction of CM hyperelliptic curves

3.1 Hyperelliptic curves with CM

First, some definitions and properties of Abelian varieties are in order. Let F be a Galois extension of Q, A/F a g-dimensional Abelian variety, $\mathrm{End}_F A$ the endomorphism ring of A over F. It is often convenient to consider the tensor product of $\mathrm{End}_F A$ with Q, $K = \mathrm{End}^\circ A := \mathrm{End}_F A \bigotimes_Z Q$.
It is known that an Abelian variety A is isogenous to a product of simple Abelian varieties A_i. (Often known as complete reduction theorem.)

$$A \simeq A_1^{n_1} \times \cdots \times A_m^{n_m},$$

$$\text{and} \quad \text{End}^\circ A = \bigoplus_{i=1}^{m} \text{End}^\circ A_i^{n_i}$$

A CM field is defined as a totally imaginary quadratic extension of a totally real algebraic number field. An Abelian variety A is called with complex multiplication when its endomorphism field K is a CM field. For a CM field K of an Abelian variety A of genus g, $[K : Q] = 2g$.

Definition 1. [14] Let K be a CM field of A with $[K : Q] = 2g$ and $\{\varphi_1, \cdots, \varphi_g\}$ be g embeddings of K into C such that no two of them are complex conjugate. Then $(K; \{\varphi_i\})$ is the CM-type of A.

Definition 2. [8] Let K be a CM field which is a Galois extension of Q. $\pi_0 \in O_K$ is called a Weil number of type (A_0) of order l if it satisfies the following condition. (Here $\bar{a}$ denotes the complex conjugate of a.)

$$\pi_0^\sigma \overline{\pi}_0^\sigma = p^l \qquad \text{where } \forall \sigma \in \text{Gal}(K/Q)$$

3.2 Find Weil number

Let K be a CM field, O_K its maximal order of integers.

[Algorithm 1]

Input: A Galois CM field K/Q, $[K : Q] = n = 2g$ with a minimal polynomial $f(x)$.

Output: A Weil number of type (A_0) of order 1 : π_0

1. Calculate the Galois group $\text{Gal}(K/Q)$, and determine the CM-type : $(K, \{\varphi_i\})$;
2. Choose an algebraic integer $\omega \in O_K$ such that $N(\omega) = p$ thus one derives primal ideal $\mathfrak{p}$'s in O_K lying over p such that
$$(p) = \prod_{\sigma \in Gal(K/Q)} \mathfrak{p}^\sigma;$$
3. Check if the rational prime p splits completely in K or
$$f(x) \equiv \prod_{k=1}^{n} (x - \alpha_k) \bmod p;$$
if not goto Step 2.
4. Calculate the Weil number π_0 of type (A_0) of order 1 such that
$$(\pi_0) = \prod_i \mathfrak{p}^{\psi_i}. \quad \psi_i = \varphi_i^{-1}$$

$$\text{and} \quad \pi_0^\sigma \overline{\pi_0}^\sigma = p \qquad N(\pi_0) = \prod_{\sigma \in \text{Gal}(K/Q)} \mathfrak{p}^\sigma = p^g;$$

5. Output π_0 as the (A_0) type Weil number of order 1 associated with p.

Remark 1: It is known that there are polynomial time algorithms to calculate Galois groups of a number field. However, since this calculation could be time consuming if one uses an arbitrary number field, we suggest to choose CM fields whose Galois groups can be determined easily.
Remark 2: If one chosen an appropriate spliting patterns of p in the Step 3, he can also obtain Weil numbers which defines hyperelliptic curves over finite extension fields.
Remark 3: All these steps can be calculated very fast as long as the degree $2g$ of CM fields is not too high, which is just the our case of hyperelliptic curves with small genera.

3.3 Construction of simple hyperelliptic curves

[Algorithm 2]

Input: C/F an algebraic curve of genus g with CM and K is its Galois CM field such that $F/\boldsymbol{Q}$ is a Galois extension of $K, [F : \boldsymbol{Q}] = s$, $g(x)$: a minimal polynomial of F;

Output: p and $C/\boldsymbol{F}_p$ such that $\mathcal{J}(\boldsymbol{F}_p)$ is almost prime.

1. Calculate the Galois group $\mathrm{Gal}(F/\boldsymbol{Q})$ and lift the CM type $(K, \{\varphi_i\})$ to $(F, \{\phi_i\})$ such that $\phi_i|_K = \varphi_i$,
2. Using the lifted CM type $(F, \{\phi_i\})$ to find a Weil number $\pi_0 \in \mathcal{O}_F$ associated with p^g of order 1 or
$$\pi_0^\sigma \overline{\pi_0}^\sigma = p, \quad \forall \sigma \in \mathrm{Gal}(K/\boldsymbol{Q}), \qquad N(\pi_0) = p^g$$
by Algorithm 1;
3. Calculate the order $\#\mathcal{J}(\boldsymbol{F}_p) = N(1 - \pi_0)$;
4. Test if $\#\mathcal{J}(\boldsymbol{F}_p)$ is almost prime. If not, goto Step 2;
5. Output p and $C/\boldsymbol{F}_p$.

Remark As remarked at the end of Algorithm 1, this and the following algorithms can also be generalized to over finite extension fields.

3.4 Construction of non-simple hyperelliptic curves

[Algorithm 3]

Input: C/F an algebraic curve whose Jacobian $\mathcal{J} \simeq A_1 \times \cdots \times A_m$, one of the simple component A_i is with CM field K_i;

Output: p and $C/\boldsymbol{F}_p$ such that for a large rational prime q, $q|\#\mathcal{J}(\boldsymbol{F}_p)$.

1. Use Algorithm 2 to find p such that $A_i(\boldsymbol{F}_p)$ is divisible by a large prime q;
2. Output $C/\boldsymbol{F}_p$, such that $\mathcal{J}(\boldsymbol{F}_p)$ is also q-divisible.

4 Construction of hyperelliptic curves of genus two

In this section we will devote to the hyperelliptic curves of genus two.
It is known that the Jacobian variety of a hyperelliptic curve of genus two is either simple or isogenous to a product of two elliptic curves.
As so far, we can only find the work in [1] which exclusively dealt with the hyperelliptic curves of genus two. However, their target was to extend Schoof's algorithm to count the orders of Jacobian varieties of the curves and calculate density of the curves. Besides, since they only considered curves associated with what they called "good" Weil numbers, their results are restricted to the hyperelliptic curves whose Jacobians are isogenous to products of two elliptic curves.
Below, we will show how to construct both types of the hyperelliptic curves of genus two, i.e., those whose Jacobian varieties are either simple or isogenous to products of elliptic curves.

4.1 Simple hyperelliptic curves

Using the Rosenhain normal form of hyperelliptic curves with genus two in $\mathrm{char}(F) > 2$, it is known that

$$v^2 = u(u-1)\left(u-1-\zeta_5\right)\left(u-1-\zeta_5-\zeta_5^2\right)\left(u-1-\zeta_5-\zeta_5^2-\zeta_5^3\right) \qquad (1)$$

is the only case which defines a simple Jacobian variety with $\mathrm{End}_F^\circ \mathcal{J} = \boldsymbol{Q}(\zeta_5)$ and whose variety of moduli is given by an arithmetic curve [9].
Now we show an example to construct secure curves in this case. To define this curve over a prime finite field, we choose $p \equiv 1 \bmod 5$, e.g.

$$p = 9162804114979267106577139117719751124791.$$

We use the fact that $\{1, \zeta_5, \zeta_5^2, \zeta_5^3\}$ is a $\boldsymbol{Z}$-basis of $\mathcal{O}_K$ in $K = \boldsymbol{Q}(\zeta_5)$
By Algorithm 1, we found a Weil number

$$\pi_0 = 9783789509 + 2\zeta_5 - 95722537127067092548\zeta_5^2 + 9783789506\zeta_5^3$$

$$\text{such that} \quad N(\pi_0) = p^2 \quad \pi_0^\sigma \pi_0^{\sigma\rho} = p \qquad \forall \sigma \in \mathrm{Gal}(\boldsymbol{Q}(\zeta_5)/\boldsymbol{Q}).$$

$$\begin{aligned} \text{Thus,} \quad \#\mathcal{J}(\boldsymbol{F}_p) &= N(1-\pi_0) \\ &= 8395697924948099034178165074105660688541 \\ & 3974488691429897981346294084872093634096 = 2^4 p_{max} \\ p_{max} &= 5247311203092561896361353171316037930338 \\ & 373405543214368623834143380304505852131 \end{aligned}$$

Here p_{max} is the largest prime divisor of $\#\mathcal{J}(\boldsymbol{F}_p)$. The Jacobian variety of

$$\begin{aligned} v^2 = &\ 2344983043581843675776951874049541814177u \\ &+4689966087163687351553903748118651207374u^2 \\ &+694809791037779294588556339185122203674u^3 \\ &+1433045193195956784657727156366435899565u^4 + u^5 \end{aligned}$$

has the designed order.
Here one may run into the problem of twists caused by units. We get around it by applying random isomorphisms such as

$$(x,y) \mapsto (c^2x, c^{2g+1}y), \quad c \in \boldsymbol{F}_p^*$$

to the curves and count the orders their Jacobians to find those with the designed order. In fact, this does not costs much time since these kind of twists are at most caused by the roots of 1 in K.

4.2 Hyperelliptic curve as product of elliptic curves

Consider a hyperelliptic curve C which belongs to a family of QM-curves with discriminant $D = 6$, as a fibre over a Shimura curve, which is their variety of moduli [7].

$$v^2 = u\left(u^4 + 2\sqrt{2}u^3 + \frac{11}{3}u^2 + 2\sqrt{2}u + 1\right) \tag{2}$$

This curve splits via a morphism ϕ of degree two.

$$\phi : C \longrightarrow E : y^2 = x(1-x)\left(1 - \frac{1+2\sqrt{3}-\sqrt{6}}{2}x\right)$$

where E is a elliptic curve with complex multiplication of $\boldsymbol{Z}[\sqrt{-6}]$.
To design a secure curve of such kind over a finite field by Algorithm 3, we start by design of its elliptic factors. Choose p such that $(-6/p) = 1$, e.g.

$$p = 9123987419850982123498723487512863553279$$

We can find the Weil number

$$\pi_0 = 94560410696060909923 + 5512352021343695965\sqrt{-6}$$

and the characteristic polynomial of the Frobenius endomorphism of the elliptic curve as

$$\begin{aligned} Z_E(T) = {} & T^2 + 189120821392121819846T \\ & + 9123987419850982123498723487512863553279. \end{aligned}$$

The curve has order

$$\#E(\boldsymbol{F}_p) = 9123987419850982123687844308904985373126$$

which has the largest prime factor

$$p_{max} = 1520664569975163687218267111020123622239.$$

Therefore the characteristic polynomial of the Frobenius endomorphism over the hyperelliptic curve is

$$\begin{aligned} Z_{\mathcal{J}}(T) = {} & (T^2 + 189120821392121819846T \\ & + 9123987419850982123498723487512863553279)^2 \end{aligned}$$

Now reduction of (2) over $\boldsymbol{F}_p$ becomes

$$\begin{aligned} C : v^2 = u^5 &+ 6160080283930092308555454591922522803205u^4 \\ &+ 3041329139950327374499574495837621184430u^3 \\ &+ 6160080283930092308555454591922522803205u^2 + u \end{aligned}$$

which has the designed order

$$\#\mathcal{J}(\boldsymbol{F}_p) = 9123987419850982123687844308904985373126^2.$$

5 Correctness of the algorithms

Theorem 3. *The Jacobian varieties $\mathcal{J}$ over finite fields $\boldsymbol{F}_p$ constructed by Algorithms 1, 2, 3 are almost prime.*

Sketch of proof: In first place, in the Algorithm 1, the Step 2 gives a prime ideal decomposition of p. Then one is guaranteed to well define a reduction to $\mathcal{O}_K/\text{mod } \mathfrak{p}$ as a prime finite field since we have specified the inertial degree of $\mathfrak{p}$'s as 1 in the Step 3.
In the Step 4, in order to calculate the Weil numbers of type (A_0) explicitly, we make use of results from the CM theory of Abelian varieties by Shimura and Taniyama[24]. According to the fundamental theorem of Abelian varieties on the primal ideal decomposition of the Frobenius endomorphism in the CM fields the following facts are known.
Let F be a Galois extension of K, $(K; \{\varphi_i\})$ the CM-type of an Abelian variety A/F over the field F, if $\mathfrak{B} \subset \mathcal{O}_F$ is a prime ideal of $\mathcal{O}_F$, which is lying over p, then the Frobenius endomorphism Fr of the Abelian variety A, regarded as the Weil number of type (A_0) π_0, has the prime ideal decomposition as

$$(\pi_0) = \prod_i \left(N_{F/K^{\varphi_i}} \mathfrak{B}\right)^{\psi_i} \qquad \text{with } \psi_i = \varphi_i^{-1}.$$

In our case, we have lifted the CM type $(K, \{\varphi_i\})$ to $(F, \{\phi_i\})$ in the Step 1 of Algorithm 2[14], then the Frobenius endomorphism π_0 of $A(\mathcal{O}_F \text{mod } \mathfrak{p})$ splits such that

$$(\pi_0) = \prod_i \mathfrak{p}^{\psi_i}.$$

The Step 4 is then finished by check of the definition of the Weil number of type (A_0).
As to the Algorithm 2, we first know from the Tate-Honda theorem, that there is a bijection between the isogeny classes of $\boldsymbol{F}_{p^n}$-simple Abelian varieties and the conjugate classes of the Weil numbers of type (A_0).
Based on this correspondence which predicted the existence of these Weil numbers, we set off to design the Jacobian varieties with assigned orders by looking for the corresponding conjugate classes of Weil numbers of (A_0) type, which have been found in the Algorithm 1.

Thus, we have the order of Jacobian $\#\mathcal{J}(\boldsymbol{F}_p)$ as the degree of the isogeny $(1 - Fr.)$ of the Jacobian variety which can be defined in the form of norm [19]. Or

$$\#\mathcal{J}(\boldsymbol{F}_p) = N(1 - \pi_0)$$

The Algorithm 3 can be validated easily through the complete reduction theorem and the CM field algorithm for elliptic curves[4][5].

QED

6 Performance issues

We evaluated performance of the algorithms by simulations with the hyperelliptic curves in the Rosenhein canonical form in the following way. (p is chosen of above 80 digits.) The results are shown in the Table below.
(i) Check ten thousands algebraic integers ω which split completely in $\boldsymbol{Q}(\zeta_5)$ if they have prime norms. (Use $\boldsymbol{Z}$-basis $\{\zeta_5^i\}$). Above 5.6% , i.e. 565 ω are found to have prime norms;
(ii) Find the type A_0 Weil numbers π_0 from $\{\omega\}$ with prime norms in (i);
(iii) Calculate $\#\mathcal{J}(\boldsymbol{F}_p)$, and check its primality. Consider only the twisting curves by the quadratic characters, we found 133 curves to be almost prime, which means their Jacobians have orders of 80 digits, each contains a prime factor greater than 70 digits.
(iv) Reduce A/K to over $\boldsymbol{F}_p$.
The average running time to design a secure curve is about 91 seconds on SuperSparc/75MH workstation using Mathematica. Since our implementation is only tentative, there still is room to reduce all these computation times further. Besides, in fact near a half of the twists of the hyperelliptic curves due to units can be found. Thus, we could have found even more almost prime curves in (iii).

Steps	Time	Average time	Outcomes
(i)	37 min. 2 sec.	3.9 sec/per ω	565 ω with prime norms
(ii)	5 min 59 sec	0.3 sec/per π_0	1120 Weil numbers
(iii)	2 h 39 min	1 min. 11 sec/per secure curve	133 almost prime Jacobians
(iv)	1 sec.		

Acknowledgment: The authors wish to thank Prof. Fumiyuki Momose of Dept. of Math. Chuo University, Japan for helpful discussions.

References

1. L.M. Adleman, M.D.A. Huang: "Primality Testing and Abelian Varieties Over Finite Fields," Springer-Verlag , (1992).

2. L.M. Adleman , J.D. Marrais , M.D. Huang: "A Subexponential Algorithms for Discrete Logarithms over the Rational Subgroup of the Jacobians of Large Genus Hyperelliptic Curves over Finite Fields," Proc. of ANTS95, Springer, (1995)
3. D.Cantor: "Computing in the jacobian of hyperelliptic curve," Math. Comp., vol.48, p.95-101, (1987)
4. J. Chao, K. Tanada, S. Tsujii: "Design of Elliptic Curves with Controllable Lower boundary of Extension Degree for Reduction Attacks", Yvo G. Desmedt (Ed.) Advances in Cryptology-CRYPTO'94, Lecture Notes in Computer Science, 839, Springer-Verlag, pp.50-55, 1994.
5. J. Chao, K. Harada, N. Matsuda, S. Tsujii:"Design of secure elliptic curves over extension fields with CM fields methods," Proc. of Pragocrypto'97, p.93-108, (1997)
6. M.D.Huang, D.Ierardi:"Counting Rational Point on Curves over Finite Fields," Proc. 32nd IEEE Symp. on the Foundations of Computers Science, 1993.
7. K.Hashimoto , N.Murabayashi : "Shimura curves as intersections of Humbert surfaces and defining equations of QM-curves of genus two," Tohoku Math.J. 47, p.271-296, (1995)
8. T.Honda : "Isogeny classes of abelian varieties over finite fields," J.Math.Soc.Japan, vol.20, No.1-2, p.83-95, (1968)
9. J.Igusa: "Arithmetic variety of moduli for genus two," Ann. of Math. , vol.72, No.3, p.612-649, (1960)
10. N.Koblitz:"Elliptic Curve Cryptosystems,"Math. Comp.,vol.48, p.203-209, (1987)
11. N.Koblitz:"Hyperelliptic cryptosystems," J. of Cryptology, vol.1, p.139-150, (1989)
12. N. Koblitz : "Elliptic Curve Implementation of Zero-Knowledge Blobs," J. of Cryptology, vol.4, No.3, p. 207-213, (1991)
13. S.Lang : "Abelian Varieties", Intersciencs, New York (1959)
14. S.Lang : "Complex multiplication" Springer-Verlag, (1983)
15. A.Menezes, S.Vanstone, T.Okamoto:"Reducing Elliptic Curve Logarithms to Logarithms in a Finite Fields," Proc. of STOC, p.80-89, (1991).
16. A.Menezes:"Elliptic Curve Public Key Cryptosystems", Kluwer Academic, (1993)
17. V.S.Miller : "Use of Elliptic Curves in Cryptography," Advances in Cryptology Proceedings of Crypto'85 , Lecture Notes in Computer Science , 218 , Springer-Verlag , p.417-426, (1986)
18. D.Mumford : "Abelian varieties", Tata Studies in Mathematics, Oxford, Bobay, (1970).
19. D.Mumford : "Tata Lectures on Theta I" , Birkhäuser, Boston , (1983).
20. D.Mumford : "Tata Lectures on Theta II" , Birkhäuser, Boston , (1984).
21. T.Okamoto , K.Sakurai: "Efficient Algorithms for the Construction of Hyperelliptic Cryptosystems," Proc. of CRYPTO'91 , LNCS 576 , p.267-278, (1992).
22. J.Pila : "Frobenius maps of abelian varieties and finding roots of unity in finite fields," Math. Comp., vol.55 , p. 745-763, (1990)
23. R.Schoof : "Elliptic curves over finite fields and the computation of square roots mod p," Math. Comp., vol.44, p.483–494, (1985)
24. G. Shimura, Y. Taniyama : "Complex multiplication of abelian varieties and its application to number theory" Pub. Math. Soc. Jap. no.6 , (1961).
25. Emil J. Volcheck: "Computing in the Jacobian of a plane algebraic curve", Proc. of ANT-1, p.221-233, LNCS-877, (1994)

A New and Optimal Chosen-Message Attack on RSA-Type Cryptosystems

Daniel Bleichenbacher[1], Marc Joye[2] and Jean-Jacques Quisquater[3]

[1] Bell Laboratories
700 Mountain Av. , Murray Hill, NJ 07974, U.S.A.
E-mail: bleichen@research.bell-labs.com
[2] UCL Crypto Group, Dép. de Mathématique, Université de Louvain
Chemin du Cyclotron 2, B-1348 Louvain-la-Neuve, Belgium
E-mail: joye@agel.ucl.ac.be
[3] UCL Crypto Group, Lab. de Microélectronique, Université de Louvain
Place du Levant 3, B-1348 Louvain-la-Neuve, Belgium
E-mail: jjq@dice.ucl.ac.be

Abstract. Chosen-message attack on RSA is usually considered as an inherent property of its homomorphic structure. In this paper, we show that non-homomorphic RSA-type cryptosystems are also susceptible to a chosen-message attack. In particular, we prove that only *one* message is needed to mount a successful chosen-message attack against the Lucas-based systems and Demytko's elliptic curve system.

Keywords. Chosen-message attack, signature forgery, RSA, Lucas-based systems, Demytko's elliptic curve system.

1 Introduction

The most used public-key cryptosystem is certainly the RSA [12]. Due to its popularity, the RSA was subject to an extensive cryptanalysis. Many attacks are based on the multiplicative nature of RSA [5]. To overcome this vulnerability, numerous generalizations of the original RSA were proposed and broken.

Later, other structures were envisaged to implement analogues of RSA. This seemed to be the right way to foil the homomorphic attacks. So, a cryptosystem based on Lucas sequences was proposed in [10] and analyzed in [11] by Müller and Nöbauer. The authors use Dickson polynomials to describe their scheme; however, Dickson polynomials can be rephrased in terms of Lucas sequences [2, 14]. The Lucas sequences play the same role in this scheme as exponentiations in RSA.

In 1985, Koblitz and Miller independently suggested the use of elliptic curves in cryptography [7, 9]. Afterwards, Koyama *et al.* [8] and Demytko [4] exhibited new one-way trapdoor functions on elliptic curves in order to produce analogues of RSA. Demytko's system has the particularity to only use the first coordinate and is therefore not subject to the chosen-message attack described in [8].

The Lucas-based cryptosystems and Demytko's elliptic curve cryptosystem seem to be resistant against homomorphic attack. However, the existence of a chosen-message forgery that needs two messages has been described in [1]. Kaliski found a similar attack on Demytko's system [6].

In this paper, we describe a new chosen-message attack which needs only one message. This new attack shows that the RSA-type cryptosystems are even closer related to RSA, i.e. it shows that all the attacks based on the multiplicative nature of the original RSA can straightforward be adapted to any RSA-type cryptosystem. We illustrate this topic with the common modulus failure [13].

The remainder of this paper is organized as follows. In Section 2, we review the Lucas-based and Demytko's elliptic curve cryptosystems. The reader who is not not familiar with these systems may first read the appendix. We present our attack in Section 3 and apply it in Section 4. In Section 5, we revisit the common modulus failure. Finally, we conclude in Section 6.

2 RSA-type cryptosystems

In this section, we present cryptosystems based on Lucas sequences [10, 11, 14] and on elliptic curves [4]. We only outline the systems, for a detailed description we refer to the original papers.

2.1 Lucas-based RSA

The Lucas-based scheme can briefly be described as follows. Each user A chooses two large primes p and q and an exponent e that is relatively prime to $(p^2-1)(q^2-1)$, computes $n = pq$, and publishes n and e as his public key. The corresponding $d \equiv e^{-1} \pmod{\operatorname{lcm}(p-1, p+1, q-1, q+1)}$ is kept secret.

A's public parameters: n and e.
A's secret parameters: p, q and d.

A message m is encrypted by computing $c \equiv v_e(m, 1) \pmod{n}$. It is decrypted using the secret key d by $m \equiv v_d(c, 1) \pmod{n}$. The correctness of this system is based on Proposition 7 (in appendix) as $v_d\big(v_e(m,1),1\big) \equiv v_{de}(m,1) \equiv v_1(m,1) \equiv m \pmod{n}$. Signatures are generated accordingly by exchanging the roles of the public and secret parameters e and d.

2.2 Demytko's system

Similarly to RSA, to setup Demytko's system, each user A chooses two large primes p and q, and publishes their product $n = pq$. He publicly selects integers a and b such that $\gcd(4a^3 + 27b^2, n) = 1$. Then once and for all, he computes

$$N_n = \operatorname{lcm}\left(\#E_p(a,b), \#E_q(a,b), \#\overline{E_p(a,b)}, \#\overline{E_q(a,b)}\right). \tag{1}$$

He randomly chooses the public encryption key e such that $\gcd(e, N_n) = 1$, and computes the secret decryption key d according to $ed \equiv 1 \bmod N_n$.

A's public parameters: n, a, b and e.
A's secret parameters: p, q, N_n and d.

It is useful to introduce some notation. The $x-$ and the $y-$coordinates of a point $\mathbf{P}$ will respectively be denoted by $x(\mathbf{P})$ and $y(\mathbf{P})$. To send a message m to Alice, Bob uses Alice's public key e and computes the corresponding ciphertext $c \equiv x(e\mathbf{M}) \pmod{n}$ where $\mathbf{M}$ is a point having its $x-$coordinate equal to m. Note that, from Proposition 5, the computation of $c \equiv x(e\mathbf{M}) \pmod{n}$ does not require the knowledge of $y(\mathbf{M})$.

Using her secret key d, Alice can recover the plaintext m by computing $m \equiv x(d\mathbf{C}) \pmod{n}$ where $\mathbf{C}$ is a point having its $x-$coordinate equal to c. Note also that Alice has not to know $y(\mathbf{C})$.

Remark 1. To speed up the computations, Alice can choose $p, q \equiv 2 \bmod 3$ and $a = 0$. In that case, $N_n = \text{lcm}(p+1, q+1)$. The same conclusion holds by choosing $p, q \equiv 3 \bmod 4$ and $b = 0$ (see [8]).

Remark 2. For efficiency reasons, it is also possible to define a message-dependent system (see [4]).

3 Sketch of the new attack

Let $n = pq$ be a RSA modulus. Let e and d be respectively the public key and the secret key of Alice, according to $ed \equiv 1 \pmod{\Phi(n)}$. The public key e is used to encrypt messages and verify signatures; the secret key d is used to decrypt ciphertexts and to sign messages.

Suppose a cryptanalyst (say Carol) wants to make Alice to sign message m without her consent. Carol can proceeds as follows. She chooses a random number k and asks Alice to sign (or to decrypt) $m' \equiv mk^e \pmod{n}$. Carol gets then $c' \equiv m^d\,(k^e)^d \equiv m^d k$, and therefore the signature c of message m as $c \equiv c'k^{-1} \pmod{n}$.

Consequently, chosen-message attacks against RSA seem quite naturally to be a consequence of its multiplicative structure. By reformulating this attack with the extended Euclidean algorithm, it appears that non-homomorphic cryptosystems are also susceptible to a chosen-message attack. Applying to RSA, the attack goes as follows.

Input: A message m and the public key n, e of Alice.
Step 1: Carol chooses an integer k relatively prime to e. Then she uses the extended Euclidean algorithm to find $r, s \in \mathbb{Z}$ such that $kr + es = 1$.
Step 2: Carol computes $m' \equiv m^k \pmod{n}$.
Step 3: Next, she asks Alice to sign m' and gets therefore

$$c' \equiv m'^d \pmod{n}.$$

Step 4: Consequently, Carol can compute the signature c of m by

$$c \equiv c'^{r} m^{s} \pmod{n}. \tag{2}$$

Output: The signature c of message m.

Proof. From $kr + es = 1$, it follows $d = d(kr + es) \equiv dkr + s \pmod{\Phi(n)}$. Hence, $c \equiv m^{d} \equiv m^{dkr} m^{s} \equiv \left(m^{dk}\right)^{r} m^{s} \equiv c'^{r} m^{s} \pmod{n}$. □

Remark 3. This attack can also be considered as a generalization of the Davida's attack [3].

4 Applications

The previous attack applies also to non-homomorphic cryptosystems. In this section, we show how it works against Lucas-based systems and Demytko's system.

4.1 Attacking Lucas-based systems

The cryptanalyst Carol can try to get a signature c on a message m in the following way.

Input: A message m and the public key n, e of Alice.
Step 1: Carol chooses an integer k relatively prime to e. Then she uses extended Euclidean algorithm to find $r, s \in \mathbb{Z}$ such that $kr + es = 1$.
Step 2: Next she computes $m' \equiv v_k(m, 1) \pmod{n}$.
Step 3: Now she asks Alice to sign m'. If Alice does so then Carol knows c' such that

$$c' \equiv v_d(m', 1) \pmod{n}.$$

Step 4: Finally Carol computes the signature c of m as follows

$$v_{rkd}(m, 1) \equiv v_r(c', 1) \pmod{n}, \tag{3}$$

$$u_{rkd}(m, 1) \equiv \frac{u_k(m, 1) u_r(c', 1)}{u_e(c', 1)} \pmod{n}, \tag{4}$$

$$c = v_d(m, 1) \equiv \frac{v_{rkd}(m, 1) v_s(m, 1)}{2} + \frac{\Delta\, u_{rkd}(m, 1) u_s(m, 1)}{2} \pmod{n} \tag{5}$$

where $\Delta = m^2 - 4$.
Output: The signature c of message m.

Proof. Equation (3) follows from (13)[1] since

$$v_r(c',1) \equiv v_r\big(v_{kd}(m,1),1\big) \equiv v_{rkd}(m,1) \pmod{n}.$$

Equation (4) is a consequence of (14) and

$$\begin{aligned} u_{rkd}(m,1)u_e(c',1) &\equiv u_r\big(v_{kd}(m,1),1\big)u_{kd}(m,1)u_e\big(v_{kd}(m,1),1\big) \\ &\equiv u_r\big(v_{kd}(m,1),1\big)u_{kde}(m,1) \\ &\equiv u_r\big(v_{kd}(m,1),1\big)u_k(m,1) \pmod{n}. \end{aligned}$$

Moreover, $kr + es = 1$ implies $v_d(m,1) = v_{rkd+des}(m,1) = v_{rkd+s}(m,1)$. Hence Equation (5) is an application of (15). □

Remark 4. This attack is the analogue to the chosen-message attack on RSA presented in Section 3, by using algebraic numbers (replace m by $\alpha = (m + \sqrt{\Delta})/2$ and use Equation (9)). The only additional step to be proved is that $u_{kd}(m,1)$ is computable from m and $v_{kd}(m,1)$. This can be shown by using (14) and noting that

$$u_k(m,1) \equiv u_{kde}(m,1) \equiv u_{kd}(m,1)u_e\big(v_{kd}(m,1),1\big) \pmod{n}.$$

If $\alpha = m/2 + \sqrt{\Delta}/2$ then the signature $v_{kd}(m,1)$ on the message $v_k(m,1)$ can be used to compute

$$\alpha^{kd} \equiv v_{kd}(m,1)/2 + u_{kd}(m,1)\sqrt{\Delta}/2.$$

Once α^{kd} is known, $\alpha^d = v_d(m,1)/2 + u_d(m,1)\sqrt{\Delta}/2$ can be computed from

$$\alpha^d \equiv \alpha^{(kr+es)d} \equiv \left(\alpha^{kd}\right)^r \alpha^s \pmod{n}.$$

Hence (3) and (4) correspond to the computation of c'^r and (5) corresponds to the multiplication of c'^r by m^s in (2).

4.2 Attacking Demytko's system

Before showing that a similar attack applies to Demytko's system, we need to prove the following proposition.

Proposition 5. *Let p be a prime greater than 3, and let $E_p(a,b)$ be an elliptic curve over $\mathbf{Z}_p$. If $\mathbf{P} \in E_p(a,b)$ or if $\mathbf{P} \in \overline{E_p(a,b)}$, then the computations of $x(k\mathbf{P})$ and $\dfrac{y(k\mathbf{P})}{y(\mathbf{P})}$ depend only on $x(\mathbf{P})$.*

[1] (9) to (22) refer to equations in the appendix.

Proof. Letting $X_j := x(j\mathbf{P})$ and $Y_j := \frac{y(j\mathbf{P})}{y(\mathbf{P})}$, the tangent-and-chord composition rule on elliptic curves gives the following formulas

$$X_{2j} = x(j\mathbf{P}+j\mathbf{P}) = \begin{cases} \left(\frac{3\,x(j\mathbf{P})^2+a}{2\,y(j\mathbf{P})}\right)^2 - 2\,x(j\mathbf{P}) & \text{if } \mathbf{P} \in E_p(a,b) \\ \left(\frac{3\,x(j\mathbf{P})^2+a}{2D_p\,y(j\mathbf{P})}\right)^2 D_p - 2\,x(j\mathbf{P}) & \text{if } \mathbf{P} \in \overline{E_p(a,b)} \end{cases}$$

$$= \frac{1}{X_1^3+aX_1+b}\left(\frac{3X_j^2+a}{2Y_j}\right)^2 - 2X_j,$$

$$Y_{2j} = \frac{y(j\mathbf{P}+j\mathbf{P})}{y(\mathbf{P})} = \begin{cases} \frac{\left(\frac{3\,x(j\mathbf{P})^2+a}{2\,y(j\mathbf{P})}\right)\left(x(j\mathbf{P})-x(2j\mathbf{P})\right)-y(j\mathbf{P})}{y(\mathbf{P})} & \text{if } \mathbf{P} \in E_p(a,b) \\ \frac{\left(\frac{3\,x(j\mathbf{P})^2+a}{2D_p\,y(j\mathbf{P})}\right)\left(x(j\mathbf{P})-x(2j\mathbf{P})\right)-y(j\mathbf{P})}{y(\mathbf{P})} & \text{if } \mathbf{P} \in \overline{E_p(a,b)} \end{cases}$$

$$= \frac{1}{X_1^3+aX_1+b}\left(\frac{3X_j^2+a}{2Y_j}\right)(X_j - X_{2j}) - Y_j,$$

$$X_{2j+1} = x\big(j\mathbf{P}+(j+1)\mathbf{P}\big)$$

$$= \begin{cases} \left(\frac{y(j\mathbf{P})-y((j+1)\mathbf{P})}{x(j\mathbf{P})-x((j+1)\mathbf{P})}\right)^2 - x(j\mathbf{P}) - x((j+1)\mathbf{P}) & \text{if } \mathbf{P} \in E_p(a,b) \\ \left(\frac{y(j\mathbf{P})-y((j+1)\mathbf{P})}{x(j\mathbf{P})-x((j+1)\mathbf{P})}\right)^2 D_p - x(j\mathbf{P}) - x((j+1)\mathbf{P}) & \text{if } \mathbf{P} \in \overline{E_p(a,b)} \end{cases}$$

$$= (X_1^3+aX_1+b)\left(\frac{Y_j-Y_{j+1}}{X_j-X_{j+1}}\right)^2 - X_j - X_{j+1},$$

$$Y_{2j+1} = \frac{y\big(j\mathbf{P}+(j+1)\mathbf{P}\big)}{y(\mathbf{P})}$$

$$= \frac{\left(\frac{y(j\mathbf{P})-y((j+1)\mathbf{P})}{x(j\mathbf{P})-x((j+1)\mathbf{P})}\right)\left(x(j\mathbf{P})-x((2j+1)\mathbf{P})\right)-y(j\mathbf{P})}{y(\mathbf{P})} \quad \text{if } \mathbf{P} \in E_p(a,b) \text{ or } \overline{E_p(a,b)}$$

$$= \frac{Y_j-Y_{j+1}}{X_j-X_{j+1}}(X_j - X_{2j+1}) - Y_j.$$

So X_k and Y_k can be computed from $X_1 = x(\mathbf{P})$ and $Y_1 = 1$ by using the binary method. □

Then, the message forgery goes as follows.

Input: A message m and the public key n, e of Alice.
Note that m is the $x-$coordinate of a point $\mathbf{M}$, i.e. $m = x(\mathbf{M})$.

Step 1: The cryptanalyst Carol chooses a random k relatively prime to e. Then she uses extended Euclidean algorithm to find $r, s \in \mathbb{Z}$ such that $kr + es = 1$.

Step 2: From $x(\mathbf{M})$, Carol computes $m' = x(\mathbf{M}') \equiv x(k\mathbf{M}) \pmod{n}$. Next, she asks Alice to sign m'. So, Carol obtains the signature

$$c' = x(\mathbf{C}') \equiv x(d\mathbf{M}') \pmod{n}.$$

Step 3: Finally, Carol finds the signature $c = x(\mathbf{C}) \equiv x(d\mathbf{M}) \pmod{n}$ of message m as follows.

3a) If $x(r\mathbf{C}') \not\equiv x(s\mathbf{M}) \pmod{n}$ then, using Proposition 5, Carol can compute

$$\frac{y(k\mathbf{M})}{y(\mathbf{M})}, \frac{y(r\mathbf{C}')}{y(\mathbf{C}')} \text{ and } \frac{y(e\mathbf{C}')}{y(\mathbf{C}')} \tag{6}$$

and

$$c \equiv (m^3 + am + b) \left[\frac{\frac{y(k\mathbf{M})}{y(\mathbf{M})} \frac{y(r\mathbf{C}')}{y(\mathbf{C}')} \left(\frac{y(e\mathbf{C}')}{y(\mathbf{C}')} \right)^{-1} - \frac{y(s\mathbf{M})}{y(\mathbf{M})}}{x(r\mathbf{C}') - x(s\mathbf{M})} \right]^2 - x(r\mathbf{C}') - x(s\mathbf{M}) \pmod{n}. \tag{7}$$

3b) Otherwise, the signature is given by

$$c \equiv \frac{\left[3\,x(r\mathbf{C}')^2 + a\right]^2}{4\left[x(r\mathbf{C}')^3 + a\,x(r\mathbf{C}') + b\right]} - 2\,x(r\mathbf{C}') \pmod{n}. \tag{8}$$

Output: The signature c of message m.

Proof. Since $kr + es = 1$, $d \equiv krd + esd \equiv krd + s \pmod{N_n}$. So,

$$x(\mathbf{C}) \equiv x(d\mathbf{M}) \equiv x\big([krd + s]\mathbf{M}\big) \equiv x(r\mathbf{C}' + s\mathbf{M}) \pmod{n}.$$

a) If $x(r\mathbf{C}') \not\equiv x(s\mathbf{M}) \pmod{n}$, then

$$\begin{aligned} x(r\mathbf{C}' + s\mathbf{M}) &\equiv \left(\frac{y(r\mathbf{C}') - y(s\mathbf{M})}{x(r\mathbf{C}') - x(s\mathbf{M})} \right)^2 - x(r\mathbf{C}') - x(s\mathbf{M}) \\ &\equiv y(\mathbf{M})^2 \left[\frac{\frac{y(k\mathbf{M})}{y(\mathbf{M})} \frac{y(r\mathbf{C}')}{y(\mathbf{C}')} \frac{y(\mathbf{C}')}{y(e\mathbf{C}')} - \frac{y(s\mathbf{M})}{y(\mathbf{M})}}{x(r\mathbf{C}') - x(s\mathbf{M})} \right]^2 - x(r\mathbf{C}') - x(s\mathbf{M}) \pmod{n} \end{aligned}$$

since

$$\frac{y(r\mathbf{C}')}{y(\mathbf{M})} = \frac{y(r\mathbf{C}')}{y(\mathbf{C}')} \frac{y(\mathbf{C}')}{y(k\mathbf{M})} \frac{y(k\mathbf{M})}{y(\mathbf{M})}$$

and $y(k\mathbf{M}) \equiv y(edk\mathbf{M}) \equiv y(e\mathbf{C}') \pmod{n}$.

b) Otherwise, since $\gcd(d, N_n) = 1$ it follows that $r\mathbf{C}' \not\equiv -s\mathbf{M} \pmod{n}$ and therefore

$$x(r\mathbf{C}' + s\mathbf{M}) \equiv \left(\frac{3\,x(r\mathbf{C}')^2 + a}{2\,y(r\mathbf{C}')} \right)^2 - x(r\mathbf{C}') - x(s\mathbf{M}) \pmod{n}.$$

□

5 Common modulus attack

Simmons pointed out in [13] that the use of a common RSA modulus is dangerous. Indeed, if a message is sent to two users that have coprime public encryption keys, then the message can be recovered.

Because our chosen-message attack requires only one message, the Lucas-based systems and Demytko's elliptic curve system are vulnerable to the common modulus attack. We shall illustrate this topic on Demyko's system.

Let (e_1, d_1) and (e_2, d_2) be two pairs of encryption/decryption keys and let $m = x(\mathbf{M})$ be the message being encrypted. Assuming e_1 and e_2 are relatively prime, the cryptanalyst Carol can recover m from the ciphertexts $c_1 = x(\mathbf{C_1}) \equiv x(e_1\mathbf{M}) \pmod{n}$ and $c_2 = x(\mathbf{C_2}) \equiv x(e_2\mathbf{M}) \pmod{n}$ as follows.

Carol uses the extended Euclidean algorithm to find integers r and s such that $re_1 + se_2 = 1$. Then, she computes $x(\mathbf{M}) = x((re_1+se_2)\mathbf{M}) \equiv x(r\mathbf{C_1}+s\mathbf{C_2}) \pmod{n}$ as follows. If $x(r\mathbf{C_1}) \not\equiv x(s\mathbf{C_2}) \pmod{n}$, then

$$m \equiv (c_1^3 + ac_1 + b) \left[\frac{\frac{y(r\mathbf{C_1})}{y(\mathbf{C_1})} - \frac{y(e_2\mathbf{C_1})}{y(\mathbf{C_1})} \frac{y(s\mathbf{C_2})}{y(\mathbf{C_2})} \left(\frac{y(e_1\mathbf{C_2})}{y(\mathbf{C_2})} \right)^{-1}}{x(r\mathbf{C_1}) - x(s\mathbf{C_2})} \right]^2 - x(r\mathbf{C_1}) - x(s\mathbf{C_2}) \pmod{n}$$

otherwise

$$m \equiv \frac{\left[3\, x(r\mathbf{C_1})^2 + a\right]^2}{4\left[x(r\mathbf{C_1})^3 + a\, x(r\mathbf{C_1}) + b\right]} - 2\, x(r\mathbf{C_1}) \pmod{n}.$$

Proof. Straightforward since $y(e_2\mathbf{C_1}) \equiv y(e_1\mathbf{C_2}) \pmod{n}$. □

6 Conclusion

We have presented a new type of chosen-message attack. Our formulation has permitted to mount a successful chosen-message attack with only one message against Lucas-based systems and Demytko's system. This also proved that the use of non-homomorphic systems is not necessarily the best way to foil chosen-message attacks.

Acknowledgments The second author is grateful to Victor Miller for providing some useful comments to enhance the presentation of the paper.

References

1. D. Bleichenbacher, W. Bosma, and A. K. Lenstra. Some remarks on Lucas-based cryptosystems. In D. Coppersmith, editor, *Advance in Cryptology – Crypto '95*, volume 963 of *Lectures Notes in Computer Science*, pages 386–396. Springer-Verlag, 1995.

2. D. M. Bressoud. *Factorization and primality testing.* Undergraduate Texts in Mathematics. Springer-Verlag, 1989.
3. G. Davida. Chosen signature cryptanalysis of the RSA (MIT) public key cryptosystem. Technical Report TR-CS-82-2, Dept. of Electrical Engineering and Computer Science, University of Wisconsin, Milwaukee, USA, October 1982.
4. N. Demytko. A new elliptic curve based analogue of RSA. In T. Helleseth, editor, *Advance in Cryptology – Eurocrypt '93*, volume 765 of *Lectures Notes in Computer Science*, pages 40–49. Springer-Verlag, 1994.
5. D. E. Denning. Digital signatures with RSA and other public-key cryptosystems. *Communications of the ACM*, 27(4):388–392, April 1984.
6. B. S. Kaliski Jr. A chosen message attack on Demytko's elliptic curve cryptosystem. *Journal of Cryptology*, 10(1):71–72, 1997.
7. N. Koblitz. Elliptic curve cryptosystems. *Mathematics of Computation*, 48:203–209, 1987.
8. K. Koyama, U. M. Maurer, T. Okamoto, and S. A. Vanstone. New public-key schemes based on elliptic curves over the ring $\mathbb{Z}_n$. In J. Feigenbaum, editor, *Advance in Cryptology – Crypto '91*, volume 576 of *Lectures Notes in Computer Science*, pages 252–266. Springer-Verlag, 1991.
9. V. S. Miller. Use of elliptic curves in cryptography. In H. C. Williams, editor, *Advance in Cryptology – Crypto '85*, volume 218 of *Lectures Notes in Computer Science*, pages 417–426. Springer-Verlag, 1986.
10. W. B. Müller and R. Nöbauer. Some remarks on public-key cryptosystems. *Sci. Math. Hungar*, 16:71–76, 1981.
11. W. B. Müller and R. Nöbauer. Cryptanalysis of the Dickson scheme. In J. Pichler, editor, *Advance in Cryptology – Eurocrypt '85*, volume 219 of *Lectures Notes in Computer Science*, pages 50–61. Springer-Verlag, 1986.
12. R. L. Rivest, A. Shamir, and L. Adleman. A method for obtaining digital signatures and public-key cryptosystems. *Communications of the ACM*, 21(2):120–126, February 1978.
13. G. J. Simmons. A weak privacy protocol using the RSA cryptoalgorithm. *Cryptologia*, 7:180–182, 1983.
14. P. J. Smith and M. J. J. Lennon. LUC: A new public key system. In E. G. Douglas, editor, *Ninth IFIP Symposium on Computer Security*, pages 103–117. Elsevier Science Publishers, 1993.

A Basic facts

A.1 Lucas sequences

Let P, Q be integers, $\Delta = P^2 - 4Q$ be a non-square, $\alpha = \frac{P+\sqrt{\Delta}}{2}$ and $\beta = \overline{\alpha} = \frac{P-\sqrt{\Delta}}{2}$ be the roots of $x^2 - Px + Q = 0$ in the quadratic field $\mathbb{Q}(\sqrt{\Delta})$. The Lucas sequences $v_k(P,Q)$ and $u_k(P,Q)$ for $k \in \mathbb{Z}$ are then defined as the integers satisfying

$$\alpha^k := \frac{v_k(P,Q)}{2} + \frac{u_k(P,Q)\sqrt{\Delta}}{2}. \tag{9}$$

From $\alpha^2 = P\alpha - Q$ follows $\alpha^k = P\alpha^{k-1} - Q\alpha^{k-2}$. Hence the Lucas sequences satisfy the following recurrence relation

$$v_0(P,Q) = 2;\quad v_1(P,Q) = P;\quad v_k(P,Q) = Pv_{k-1}(P,Q) - Qv_{k-2}(P,Q),$$
$$u_0(P,Q) = 0;\quad u_1(P,Q) = 1;\quad u_k(P,Q) = Pu_{k-1}(P,Q) - Qu_{k-2}(P,Q).$$

This recurrence relation is sometimes used as an alternative definition of Lucas sequences. Since conjugation and exponentiation are exchangeable it follows

$$\beta^k = \overline{\alpha^k} = \frac{v_k(P,Q)}{2} - \frac{u_k(P,Q)\sqrt{\Delta}}{2}.$$

From this equation and from (9) it follows that

$$v_k(P,Q) = \alpha^k + \beta^k, \tag{10}$$
$$\text{and } u_k(P,Q) = \frac{\alpha^k - \beta^k}{\alpha - \beta}. \tag{11}$$

The next proposition states some well-known properties of Lucas sequences.

Proposition 6.

$$4Q^k = v_k(P,Q)^2 - \Delta u_k(P,Q)^2 \tag{12}$$
$$v_{km}(P,Q) = v_k\big(v_m(P,Q), Q^m\big) \tag{13}$$
$$u_{km}(P,Q) = u_m(P,Q)u_k\big(v_m(P,Q), Q^m\big) \tag{14}$$
$$v_{k+m}(P,Q) = \frac{v_k(P,Q)v_m(P,Q)}{2} + \frac{\Delta u_k(P,Q)u_m(P,Q)}{2} \tag{15}$$
$$u_{k+m}(P,Q) = \frac{u_k(P,Q)v_m(P,Q)}{2} + \frac{v_k(P,Q)u_m(P,Q)}{2} \tag{16}$$

Proof. Equation (12) can be proved as follows.

$$\begin{aligned} 4Q^k &= 4(\alpha\overline{\alpha})^k = 2\alpha^k 2\overline{\alpha^k} \\ &= (v_k(P,Q) + u_k(P,Q)\sqrt{\Delta})(v_k(P,Q) - u_k(P,Q)\sqrt{\Delta}) \\ &= v_k(P,Q)^2 - \Delta u_k(P,Q)^2. \end{aligned}$$

Equation (12) now implies that

$$\begin{aligned} \alpha^k &= \frac{v_k(P,Q)}{2} + \frac{u_k(P,Q)\sqrt{\Delta}}{2} = \frac{v_k(P,Q)}{2} + \frac{\sqrt{u_k(P,Q)^2\Delta}}{2} \\ &= \frac{v_k(P,Q)}{2} + \frac{\sqrt{v_k(P,Q)^2 - 4Q^k}}{2} \end{aligned}$$

and hence $\alpha^k = P'/2 + \sqrt{P'^2 - 4Q'}/2$ with $P' = v_k(P,Q)$ and $Q' = Q^k$. Thus we have

$$\begin{aligned} (\alpha^k)^m &= \frac{v_m(P',Q')}{2} + \frac{u_m(P',Q')\sqrt{P'^2 - 4Q'}}{2} \\ &= \frac{v_m(P',Q')}{2} + \frac{u_m(P',Q')u_k(P,Q)\sqrt{\Delta}}{2}. \end{aligned}$$

Comparing the coefficients of this equation with

$$\alpha^{km} = v_{km}(P,Q)/2 + u_{km}(P,Q)\sqrt{\Delta}/2$$

proves (13) and (14). Writing $\alpha^{k+m} = \alpha^k \alpha^m$ as sums of Lucas sequences and comparing the coefficients shows (15) and (16). □

Proposition 7. *Let p be an odd prime, $Q = 1$ and* $\gcd(\Delta, p) = 1$. *Then the sequence $v_k(P,1) \bmod p$ is periodic and the length of the period divides $p - (\frac{\Delta}{p})$.*

Proof. α and therefore also α^p are algebraic integers in $\mathbb{Q}(\sqrt{\Delta})$. Thus we have $\alpha^p = (P/2 + \sqrt{\Delta}/2)^p \equiv P/2 + (\sqrt{\Delta})^p/2 \equiv P/2 + \Delta^{(p-1)/2}\sqrt{\Delta}/2 \equiv P/2 + (\frac{\Delta}{p})\sqrt{\Delta}/2 \pmod p$. Thus if $(\frac{\Delta}{p}) = 1$ then $\alpha^{p-1} \equiv 1 \pmod p$ and if $(\frac{\Delta}{p}) = -1$ then $\alpha^{p+1} \equiv 1 \pmod p$. It follows that the sequence α^k (and therefore also $v_k(P,1)$) is periodic with a period that divides $p - (\frac{\Delta}{p})$. □

A.2 Elliptic curves

Elliptic curves over $\mathbb{Z}_p$ Let p be a prime greater than 3, and let a and b be two integers such that $4a^3 + 27b^2 \not\equiv 0 \pmod p$. An *elliptic curve $E_p(a,b)$ over the prime field $\mathbb{Z}_p$* is the set of points $(x,y) \in \mathbb{Z}_p \times \mathbb{Z}_p$ satisfying the Weierstraß equation

$$y^2 = x^3 + ax + b \pmod p \tag{17}$$

together with the point at infinity $\mathcal{O}_p$. The points of the elliptic curve $E_p(a,b)$ form an Abelian group under the tangent-and-chord law defined as follows.

(i) $\mathcal{O}_p$ is the identity element, i.e. $\forall \mathbf{P} \in E_p(a,b)$, $\mathbf{P} + \mathcal{O}_p = \mathbf{P}$.
(ii) The inverse of $\mathbf{P} = (x_1, y_1)$ is $-\mathbf{P} = (x_1, -y_1)$.
(iii) Let $\mathbf{P} = (x_1, y_1)$ and $\mathbf{Q} = (x_2, y_2) \in E_p(a,b)$ with $\mathbf{P} \neq -\mathbf{Q}$. Then $\mathbf{P} + \mathbf{Q} = (x_3, y_3)$ where

$$x_3 = \lambda^2 - x_1 - x_2, \tag{18}$$

$$y_3 = \lambda(x_1 - x_3) - y_1, \tag{19}$$

$$\text{and } \lambda = \begin{cases} \dfrac{3x_1^2 + a}{2y_1} & \text{if } x_1 = x_2, \\ \dfrac{y_1 - y_2}{x_1 - x_2} & \text{otherwise.} \end{cases}$$

Note that if $\mathbf{P} = (x_1, 0) \in E_p(a,b)$, then $2\mathbf{P} = \mathcal{O}_p$.

Theorem 8 (Hasse). *Let $\#E_p(a,b) = p + 1 - a_p$ denote the number of points in $E_p(a,b)$. Then $|a_p| \leq 2\sqrt{p}$.* □

Complementary group of $E_p(a,b)$ Let $E_p(a,b)$ be an elliptic curve over $\mathbb{Z}_p$. Let D_p be a quadratic non-residue modulo p. The *twist of $E_p(a,b)$*, denoted by $\overline{E_p(a,b)}$, is the elliptic curve given by the (extended) Weierstraß equation

$$D_p y^2 = x^3 + ax + b \tag{20}$$

together with the point at infinity $\mathcal{O}_p$. The sum of two points (that are not inverse of each other) $(x_1, y_1) + (x_2, y_2) = (x_3, y_3)$ can be computed by

$$x_3 = \lambda^2 D_p - x_1 - x_2,$$
$$y_3 = \lambda(x_1 - x_3) - y_1,$$

and $\lambda = \begin{cases} \dfrac{3x_1^2 + a}{2D_p y_1} & \text{if } x_1 = x_2, \\ \dfrac{y_1 - y_2}{x_1 - x_2} & \text{otherwise.} \end{cases}$

Proposition 9. *If* $\#E_p(a,b) = p + 1 - a_p$, *then* $\#\overline{E_p(a,b)} = p + 1 + a_p$.

Proof. Since $\#E_p(a,b) = 1 + \sum_{x \in \mathbb{Z}_p} \left(1 + \left(\frac{x^3+ax+b}{p}\right)\right)$, $a_p = -\sum_{x \in \mathbb{Z}_p} \left(\frac{x^3+ax+b}{p}\right)$. Hence, $\#\overline{E_p(a,b)} = 1 + \sum_{x \in \mathbb{Z}_p} \left(1 - \left(\frac{x^3+ax+b}{p}\right)\right) = 1 + p + a_p$. □

Elliptic curves over $\mathbb{Z}_n$ Let $n = pq$ with p and q two primes greater than 3, and let a and b be two integers such that $\gcd(4a^3 + 27b^2, n) = 1$. An *elliptic curve $E_n(a,b)$ over the ring $\mathbb{Z}_n$* is the set of points $(x, y) \in \mathbb{Z}_n \times \mathbb{Z}_n$ satisfying the Weierstraß equation

$$y^2 = x^3 + ax + b \pmod{n} \tag{21}$$

together with the point at infinity $\mathcal{O}_n$.

Consider the group $\widetilde{E}_n(a,b)$ given by the direct product

$$\widetilde{E}_n(a,b) = E_p(a,b) \times E_q(a,b). \tag{22}$$

By the Chinese remainder theorem there exists a unique point $\mathbf{P} = (x_1, y_1) \in E_n(a,b)$ for every pair of points $\mathbf{P}_p = (x_{1p}, y_{1p}) \in E_p(a,b) \setminus \{\mathcal{O}_p\}$ and $\mathbf{P}_q = (x_{1q}, y_{1q}) \in E_q(a,b) \setminus \{\mathcal{O}_q\}$ such that $x_1 \bmod p = x_{1p}$, $x_1 \bmod q = x_{1q}$, $y_1 \bmod p = y_{1p}$ and $y_1 \bmod q = y_{1q}$. This equivalence will be denoted by $\mathbf{P} = [\mathbf{P}_p, \mathbf{P}_q]$. Since $\mathcal{O}_n = [\mathcal{O}_p, \mathcal{O}_q]$, the group $\widetilde{E}_n(a,b)$ consists of all the points of $E_n(a,b)$ together with a number of points of the form $[\mathbf{P}_p, \mathcal{O}_q]$ or $[\mathcal{O}_p, \mathbf{P}_q]$.

Lemma 10. *The tangent-and-chord addition on $E_n(a,b)$, whenever it is defined, coincides with the group operation on $\widetilde{E}_n(a,b)$.*

Proof. Let $\mathbf{P}$ and $\mathbf{Q} \in E_n(a,b)$. Assume $\mathbf{P} + \mathbf{Q}$ is well-defined by the tangent-and-chord rule. Therefore $\mathbf{P} + \mathbf{Q} = [(\mathbf{P} + \mathbf{Q})_p, (\mathbf{P} + \mathbf{Q})_q] = [\mathbf{P}_p + \mathbf{Q}_p, \mathbf{P}_q + \mathbf{Q}_q]$. □

If n is the product of two large primes, it is extremely unlikely that the "addition" is not defined on $E_n(a,b)$. Consequently, computations in $\widetilde{E}_n(a,b)$ can be performed without knowing the two prime factors of n.

On Weak RSA-Keys Produced from Pretty Good Privacy

Yasuyuki Sakai[1], Kouichi Sakurai[2] and Hirokazu Ishizuka[1]

[1] Mitsubishi Electric Corporation,
5-1-1 Ofuna, Kamakura, Kanagawa 247, Japan
e-mail: {ysakai,ishizuka}@iss.isl.melco.co.jp
[2] Kyushu University,
6-10-1 Hakozaki, Higashi-ku, Fukuoka 812-81, Japan
e-mail: sakurai@csce.kyushu-u.ac.jp

Abstract. We report that Pretty Good Privacy (PGP) generates weak RSA-moduli which is vulnerable against $P+1$-factoring attack, because PGP's algorithm for generating prime numbers is designed only to produce a large prime number P so that $P-1$ has a large prime factor. We count the number of weak keys in PGP via experimental computation with theoretical consideration. Our obtained results show that bad primes are generated in PGP and induced weak keys can be easily breakable via $P+1$-factoring method. For example, in the case of RSA-key with 512-bit, we could attack

1. 0.3% users' systems with only 15 hours single PC-computation (very weak keys!!),
2. 2% users' systems with 50 days single PC-computation (weak keys!).

1 Introduction

This paper investigates how many weak RSA-keys are generated from Pretty Good Privacy (PGP) [Gar95, PGP`2.6.3i`, Zim95], which is a popular and widely available software package of an electronic-mail security program. PGP uses RSA cryptosystem [RSA78] for key management and digital signatures with keys up to 1024-bit in US version 2.6.2 and in international version 2.6.3i [PGP`2.6.3i`, Zim95].

Then, the security of PGP is mainly based on the security of the applied RSA cryptosystem. Among the previous discussion on the security of PGP [PGP2, YY96], this paper focuses on the security of PGP against factoring attack on the RSA-moduli. PGP generates large prime numbers for using a composite of RSA-moduli. After examining the original source code of PGP, we have found that PGP's algorithm for generating prime numbers is designed only for producing a large prime number P so that $P-1$ has a large prime factor. This design protects the generated RSA-cryptosystem not only against the iterated-encryption attack [SN77] but also against the attack on $P-1$-factoring for a composite modulus [Pol74].

It should be noted that, however, no consideration on prime factors of $P+1$ is taken in PGP's generating prime number algorithm. Then, PGP would generate weak keys of RSA, which is vulnerable against $P+1$-factoring [Wil82] attack on RSA-modulus. Indeed some method for generating strong primes are already presented [Gor84a, Gor84b, Tay86], which consider the protection against the attacks of both $P-1$-factoring method and $P+1$-factoring method. But the existing software package of PGP, which are widely available and are still being distributed, implemented the key-generating algorithm which considers no $P+1$-factoring attack. In fact, we have not recognized how serious problems would happen in the practical application of PGP, because no previous report evaluated on how many weak keys are generated in such an algorithm. Then, this paper counts the number of weak keys in PGP via a theoretical consideration with some experimental results.

Our obtained results show that bad primes are generated from PGP and weak keys induced from such bad primes can be easily breakable via $P+1$-factoring method. In our executed experiment, we have implemented the same algorithm for generating prime numbers as PGP, and have discovered "weak" prime numbers generated from the PGP-algorithm in the case of RSA-key with 512-bit. Furthermore, for confirming the practical effectively of $P+1$-factoring attack, we have made factorization, via Williams' algorithm [Wil82], of composite numbers of size 512-bit, which have such a generated weak prime factor. Thus, we conclude that, in the case of RSA-key with 512-bit, we could attack

1. 0.3% users' systems with only 15 hours single PC-computation (Very weak !!)
2. 2% user's systems with 50 days single PC-computation (Weak !).

Furthermore, this result for the case of RSA-key with 512-bit can be extended to the general case of any key-size under the reasonable assumption [KnT-P76] on the distribution of the size of the largest prime factor of $P+1$, where P is prime.

Because of the rapid progress of developing factoring algorithms and of increasing computer-power, 512-bit RSA moduli is said to be no longer secure [AGLL94] and 1024-bit (or more longer) moduli is recommended for commonly used RSA moduli. However, there still requires much time and expensive costs for breaking the general RSA-moduli even in the case of 512-bit.

Our results indicate that the life-time of a part of "weak" PGP's keys is not so long as we have previously observed. The number of such weak keys is never negligible in PGP.

2 Generated prime numbers in PGP

We first investigate PGP's algorithm for generating prime numbers from widely published materials [PGP`2.6.3i`, Zim95], and discuss on the properties of generated primes in PGP from the cryptographic security point of view.

2.1 PGP's algorithm for generating primes

We obtein the source code of international version 2.6.3i from [PGP`2.6.3i`], and also check US version 2.6.2 from [Zim95]. The current version of PGP [PGP`2.6.3i`,

Zim95] uses the following **Algorithm1** to generate prime numbers for making RSA-moduli.

Algorithm1: PGP
Step1 Generate a large prime number t at random.
Step2 Find the first prime in the sequence $P = 2a_i t + 1$,
for $i = 0, 1, \cdots, 1027$, where a_i is a small prime number,
i.e., $a_0 = 2, a_1 = 3, \cdots, a_{1027} = 8191$
Step3 If P is not a prime number for all a_i, then goto Step1.
Step4 Return(P).

2.2 Strong primes for RSA-moduli

A prime number P is said to be *strong prime* if integer r, s, and t exist such that the following three conditions are satisfied [Gor84a, Gor84b]:

(1) $P - 1$ has a large prime factor, denoted t
(2) $P + 1$ has a large prime factor, denoted s
(3) $t - 1$ has a large prime factor, denoted r

The condition (1) is for protecting the constructed RSA-keys from $P-1$-factoring attack [Pol74] and from iterated-encryption attack [SN77]. The condition (2) is for protecting the constructed RSA-keys from $P + 1$-factoring attack [Wil82]. The condition (3) is also for protecting the constructed RSA-keys from iterated-encryption attack [SN77].

2.3 A drawback of PGP's prime number generation

As in Step2 of Algorithm1:PGP, any generated prime P has the form of $P = 2at+1$, and this form induces t as a factor of $P-1$. We call prime number P with this form $2at+1$ *type prime number*. $2at+1$ type prime number is secure against $P-1$-factoring attack if the prime number t is large. In fact, PGP's algorithm is designed to include a large prime number t, of which bitlength is approximately same as the bitlength of P, as a factor of $P-1$. Thus, RSA-moduli used in PGP has immunity against $P - 1$-factoring attack.

It should be noted that, however, PGP's prime-generating algorithm has no consideration for $P + 1$-factoring attack: it could produce a prime number P such that all prime factors of $P+1$ are small. So, if a prime number P generated in PGP's method is used for RSA-keys, some keys might be very weak against $P + 1$-factoring attack.

2.4 The distribution of factors of $P+1$

For discussing how weak the prime P generated from PGP is, we analysis the size of prime factors of the integer $P + 1$ (in average manner).

In the problem of determining the distribution of k-th largest prime factor of large integers N, it is known that the function $\rho_k(\alpha)$, which denotes the proportion of numbers whose kth largest prime factor is smaller than $N^{1/\alpha}$, can be described in the following formula [KnT-P76, Rie94]:

$$1 - \rho_k(\alpha) = \sum_{i=0}^{\infty} \binom{-k}{n} L_{i+k}(\alpha) \tag{1}$$

where, $L_0(\alpha) = \begin{cases} 0, & for\ \alpha \leq 0 \\ 1, & for\ \alpha > 0 \end{cases}$, $L_k(\alpha) = \begin{cases} 0, & for\ \alpha \leq k \\ \int_k^\alpha L_{k-1}(t-1)\frac{dt}{t}, & for\ \alpha \geq k \end{cases}$

Now we should remark that our case deals not with the general integers N but with the special integers of the form $P+1$, where P is a prime. So, in order to confirm the distribution on the size of the largest prime factor of $P+1$, we have executed the following experiment.

Experiment1:
Step1 Generate n-bit $2at+1$ type prime P using Algorithm1:PGP
Step2 Factorize $P+1$ completely using Pollard's ρ method [Pol75]
Step3 Compute the ratio of bitlength between the largest prime factor of $P+1$ and P

In our experiment, P has size of 64-bit, 96-bit or 128-bit, and the number of experimental samples is 10,000. The results of **experiment1** is shown in Table 1 and Figure 1. We also give the case of general integers in the Figure for comparing with the equation (1). Table 1 and Figure 1 show the probability such that the ratio is smaller than 0.1, 0.2 $\cdots$ 1.0. In Figure 1, the horizontal axis shows the ratio of bitlength between the largest prime factor of $P+1$ and P. The vertical axis shows the probability.

Ratio of bitlength	Probability			
	64-bit	96-bit	128-bit	General Integer
0.0	0.0000	0.0000	0.0000	0.0000
0.1	0.0012	0.0007	0.0010	0.0001
0.2	0.0336	0.0267	0.0270	0.0228
0.3	0.1526	0.1387	0.1510	0.1094
0.4	0.3254	0.3200	0.3290	0.2638
0.5	0.5162	0.5041	0.5220	0.4766
0.6	0.6750	0.6651	0.6730	0.6272
0.7	0.8144	0.7914	0.8160	0.7758
0.8	0.9264	0.9174	0.9280	0.8858
0.9	0.9990	1.0000	1.0000	0.9852
1.0	1.0000	1.0000	1.0000	1.0000

Table 1. The distribution of the largest prime factor of $P+1$ (Bitlength of the largest prime factor of $P+1$)/(Bitlength of P)

We conclude from our experimental results described above that the equation (1) well simulates the size of the largest prime factor of $P+1$, where P is a prime.

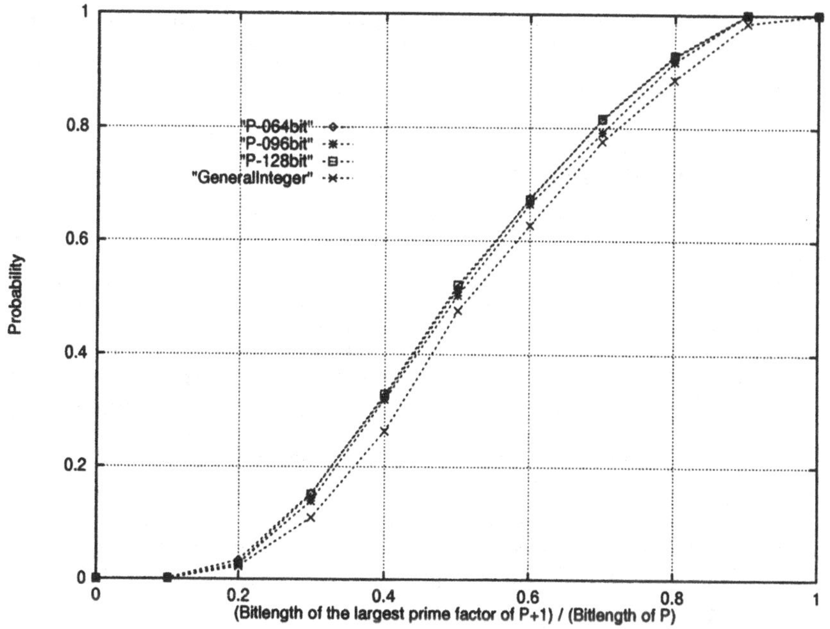

Fig. 1. Distribution of the largest prime factor of $P+1$

Thus, in the following argument of this paper, we make the assumption that the distribution of a probability of the largest prime factor of $P+1$ can be approximated by the distribution given in Table 1 and Figure 1, though the probability of the largest prime factor of P+1 is a bit larger than that of general integer.

3 Factoring PGP's RSA-moduli via P+1-method

The data of **experiment1**, in case that a prime number P has the form of $2at+1$, would implies that many RSA-keys vulnerable against $P+1$-factoring attack might be generated by Algorithm1:PGP. In order to confirm this observation, we carry out experimental factoring in the following way: We first generate a prime number P, which has the form of $2at+1$, where P has the size of 256-bit, by using **Algorithm1:PGP**. We next try to factorize $N = PQ$ completely by $P+1$-factoring method, where P is a weak prime number against $P+1$-factoring attack and Q is a non-weak (rather strong) prime.

We perform experimental evaluation of a expected running time for the factoring.

3.1 P+1-method of factoring

We give a brief description of $P+1$-method of factoring [Wil82].

The Lucas function We give here a description of some of the basic properties of the Lucas functions. Let A, B be integers, and α, β be the zeros of $x^2 - Ax + B = 0$. We define the Lucas functions by

$$U_n(A,B) = \frac{\alpha^n - \beta^n}{\alpha - \beta}, \quad V_n(A,B) = \alpha^n + \beta^n \tag{2}$$

These functions satisfy a large number of identities. We will require those given bellow [Wil82].

$$\begin{array}{ll} U_{n+1}(A,B) = AU_n - BU_{n-1}, & V_{n+1}(A,B) = AV_n - BV_{n-1} \\ U_{2n}(A,B) = U_nV_n, & V_{2n}(A,B) = V_n^2 - 2B^n \\ U_{2n-1}(A,B) = U_n^2 - BU_{n-1}^2, & V_{2n-1}(A,B) = V_nV_{n-1} - AB^{n-1} \\ U_n(V_k(A,B),B^k) = \frac{U_{nk}(A,B)}{U_k(A,B)}, & V_n(V_k(A,B),B^k) = V_{nk}(A,B) \end{array} \tag{3}$$

All These identities can be verified by direct substitution from the equation (2), using the simple facts that $A = \alpha + \beta$, and $B = \alpha\beta$. Then, we have the following theorem [Wil82].

Theorem 1. *If P is an odd prime, $P \nmid B$ and the Legendre symbol $((\alpha-\beta)^2/p) = \varepsilon$, then*

$$U_{(P-\varepsilon)m}(A,B) \equiv 0 \ (mod\ P)$$
$$V_{(P-\varepsilon)m}(A,B) \equiv 2B^{m(1-\varepsilon)/2} \ (mod\ P)$$

The first step of P+1-method Suppose that P is a prime factor of N and $P = \left(\sum_{i=1}^{k} q_i^{\alpha_i}\right) - 1$, where q_i is the ith prime and $q_i^{\alpha_i} \leq L$. We define R as: $R = \sum_{i=1}^{k} q_i^{\beta_i}$, where, $q_i^{\beta_i} \leq L < q_i^{\beta_i+1}$. Clearly, $P+1 \mid R$. From Theorem 1, if $(B,N) = 1$ and $\varepsilon = 1$ then $P \mid U_R(A,B)$, therefore, $P \mid (U_R(A,B),N)$. Similarly, $P \mid V_R(A,B)$, therefore, $P \mid (V_R(A,B),N)$.

To execute a factoring, we need to compute $U_R(A,B)$ or $V_R(A,B)$. In the computation of the value R, we first estimate the largest factor of $P+1$, then set the size of the largest factor as L. Therefore, if L is large, then R is too large for a computer to store its value.

This problem can be overcome by applying the equation (3). To reduce the running time, we need to define $B = 1$ in our Lucas function, and compute not U_R but V_R. Eventually, we can execute the factoring in $O(logR)$ time by using the equations shown bellow:

$$V_{2i-1} \equiv V_iV_{i-1} - A \ (mod\ N)$$
$$V_{2i} \equiv V_i^2 - 2 \ (mod\ N)$$
$$V_{2i+1} \equiv AV_i^2 - V_iV_{i-1} - A \ (mod\ N)$$

3.2 Expected running time for P+1-method

Computing the Lucas function can be done in an expected running time of order $O(logR)$. Note, however, that if R is the multiple of all powers of primes $\leq L$, then $logR$ is too large when the largest prime factor of $P+1$ is large.

In case that the bitlength of the largest prime factor of $P+1$ is n, we can expect the running time for $P+1$-method as follows: The prime number theorem states that the number of primes $\pi(x)$ in the interval $[2,x]$ is approximately $\pi(x) = \frac{x}{\ln x}$. Therefore, an expected running time for $P+1$-method is in proportion to the value of the largest factor of $P+1$ and is exponential time of the bitlength of the largest factor.

3.3 An experiment of a factoring

In our experiment, **Algorithm1:PGP** has truly produced the following (weak) prime number P, which has the form of $2at+1$ and size of 256-bit.

$$\begin{aligned} P &= 85047838063667438450610593841635478161425966731472914636361860125 \\ &\quad 182930212767 \\ P-1 &= 2 \cdot 7 \cdot 60748455759762456036150424172596770115304261951052081883 1156 \\ &\quad 1437513066443769 \\ P+1 &= 2^5 \cdot 3 \cdot 13 \cdot 61 \cdot 97 \cdot 21397 \cdot 87181 \cdot 461053 \cdot 593081 \cdot 1995023 \\ &\quad \cdot 21323689 \cdot 103092793 \cdot 329480189 \cdot 3458820263 \cdot 4517640559 \end{aligned} \tag{4}$$

Note that, in the description above, all the factorization is complete, that is, each factor is a prime. The largest prime factor of $P+1$ has the size of $4517640559 \simeq 4.5 \cdot 10^9 \simeq$ 33-bit.

Suppose that an integer $N = PQ$ is a composite of the prime P above and a (non-weak) prime Q with the size of 256-bit. From the above size of the largest prime factor of $P+1$, the running time for factoring $N = PQ$ via $P+1$-method can be estimated to be proportional to 2^{33} in theoretical.

In order to confirm the correctness of the estimation above, we have implemented $P+1$-factoring method for the integer $N = PQ$ with the following Q.

$$\begin{aligned} Q &= 34638105914198126172219924589037611097749856216192476164636528832 \\ &\quad 0243187881398563 \\ Q-1 &= 2 \cdot 1993 \cdot 86899412730050492153085611111484222523205861054170788 1701 \\ &\quad 86976497803107847817 \\ Q+1 &= 2^2 \cdot 3 \cdot 28865088261831771810183270490864675914791546846827063 47053 \\ &\quad 0440693353598990116547 \end{aligned} \tag{5}$$

We should note that not only $Q-1$ but also $Q+1$ has a large prime factor. This implies that Q is not weak prime against $P+1$-method.

To factorize N by $P+1$-method, we have first to guess the size of the largest prime factor of $P+1$. In our experiment, we execute to factorize N on the assumption that we know the largest prime factor of $P+1$ has the size of about

$4.5 \cdot 10^9$ in advance[3]. The factoring $N = PQ$ have carried out over the single PC with Pentium Pro 200MHz. It have taken 15 hours to finish the factoring completely, which supports our theoretical estimation.

Thus, our experiment above concludes that "there exists a (very) weak RSA-modulus exactly generated from PGP in the case that the size of modulus is 512-bit."

Remark 1 *Appendix A gives other examples of weak primes with size of 256-bit, which are exactly generated from* **Algorithm1:PGP**.

4 Distribution of weak keys in general cases

The previous section has observed that Algorithm1:PGP truly generates a (very)weak prime against $P+1$-attack in the case that P has the size of 512-bit.

This section considers the expected running time for factoring $N = PQ$ generated from Algorithm1:PGP via $P+1$-method in the general cases. Based on our experiment in the Table 1 and Figure 1, we assume the following in our estimation.

- The distribution of the largest prime factor of $P+1$ does not depend on the size of prime P.
- The distribution of the largest prime factor of $P+1$ can be modeled by the Table 1 and Figure 1. For example, the bitlength of the largest prime factor of $P+1$ is assumed to be 0.1 times the size of P with probability 10^{-3}.

Table 1 and Figure 1 indicate that 0.15% of all prime number P, where P has the form of $2at+1$, have the largest prime factor of $P+1$ with the size of 0.13 times the bitlength of P. The mean size of the largest prime factor of $P+1$ is 0.6053 times the bitlength of P.

Therefore, from the results of the experiment on factoring, we can estimate the running time for a factoring attack using $P+1$-method. In case that RSA-keys have the size of 512-bit, we can attack 0.3% users' system[4] for only 15 hours with single Pentium Pro 200MHz PC. Moreover, we can attack 2% users' system for 50 days. Similarly, we can estimate the running time of attacking PGP-system based on RSA with 768 and 1024-bit keys. Table 2 shows these observation. The data in the last horizon of Table 2 estimates the expected running time for the general users with strong keys, the original of which is reported by H. Abelson et al. [Gar95] at MIT using quadratic sieve method. The estimates are done in terms of MIPS-years, a computational unit of power analogous to a "kilowatt-hour" of electricity. Specifically, a MIPS-years is the

[3] Though this attack is a kind of non-deterministic, i.e., guessing the size of the largest prime factor of $P+1$, this is serious attack in practical settings.

[4] If one of two factors P or Q is weak, then the induced RSA-modulus $N = PQ$ is also weak. Then, if u is the probability of being a weak prime, the probability of being weak modulus is $1-(1-u)(1-u) = 2u - u^2$.

	512-bit key	768-bit key	1024-bit key
0.3% users(very weak keys)	$7.3 \cdot 10^{-1}$ (15 hours)	$2.8 \cdot 10^{4}$	$2.0 \cdot 10^{9}$
2% users(weak keys)	$5.8 \cdot 10$ (50 days)	$6.1 \cdot 10^{7}$	$1.3 \cdot 10^{14}$
general users (strong keys)	$4.2 \cdot 10^{5}$	$4.2 \cdot 10^{9}$	$2.8 \cdot 10^{15}$

Table 2. Expected running time for factoring PGP's RSA-keys (MIPS-years)

computational power of a one-MIPS machine running for one year. Our experiments have executed over Pentium Pro 200MHz. We should note that the MIPS value of the CPU is not officially announced by Intel. Then, this paper assumes that Pentium Pro 200MHz would achieve 427.8 MIPS from an officially announced performance of Pentium 133MHz on the grounds of clock speed and benchmarking with SPECint95 [Int97].

5 Consequences on the security of PGP

Our observed results implies that the lifetime of RSA-keys used in PGP is not so long as we have believed in the sense of the worst (resp. best) case for users (resp. attackers). For example, in the case of 512-bit RSA-moduli, the PGP-system of one or two persons out of 1000-users might be cracked for only about 15 hours, and 2% users' system might be cracked for about 50 days. This is a serious problem in practical.

We shall remark that the current PGP never recommend RSA-keys of 512-bit size. In fact, [PGP2.6.3i] gives us the following message:

```
Pick your RSA key size:
  1) 512 bits- Low commercial grade, fast but less secure
  2) 768 bits- High commercial grade, medium speed, good security
  3)1024 bits- "Military" grade, slow, highest security
Choose 1, 2, or 3, or enter desired number of bits:
```

As the message above shows, PGP regards RSA-keys of 512-bit size as less secure. This is because of the current development of factoring RSA-moduli with size of 512-bit [AGLL94], however, PGP takes no account of weak RSA-keys, which this paper has discussed.

The PGP's message above says that RSA-keys of 1024-bit size can achieve highest security of "Military" grade. Indeed we support this statement in the worst (resp. best) case for cryptanalysis (resp. users), however, our objection against the statement above is in the case of weak RSA-keys, which are better cases for cryptanalysis. Our experimental observation would imply that such a bad case really exists and we cannot ignore the number of users who live in such a serious case.

6 Concluding Remarks

This paper investigated the prime-number generating algorithm for the RSA used in PGP, and gave an estimation on the number of weak keys. The obtained results give the following:

Caution !!

1. The current PGP-users shall check whether their own RSA-key (two prime factors of the modulus) is weak or not.
2. The future PGP-users should replace their RSA-key generation algorithm by a fixed one (e.g. [Gor84a, Mau95]), which considers protections against many cryptanalysis including $P+1$-factoring attack on modulus.

References

[AGLL94] D.Akins, M.Graff, A.K.Lenstra, and P.C.Leyland, "The magic words are squeamish ossifrage," Lecture Notes in Computer Science 917, pp.263-277, 1995. (Advances in Cryptology - ASIACRYPT'94)

[DL92] B.Dixon, A.K.Lenstra, "Massively parallel elliptic curve factoring" Lecture Notes in Computer Science 658, pp.183-193, 1993. (Advances in Cryptology - EUROCRYPT'92)

[Gar95] S.Garfinkel, "PGP:Pretty Good Privacy" O'Reilly & Associates, Inc., 1995.

[Gor84a] J.Gordon, "Strong Primes are Easy to Find" Lecture Notes in Computer Science 208, pp.216-223, 1985. (Advances in Cryptology - EUROCRYPT'84)

[Gor84b] J.Gordon, "Strong RSA keys" Electronics Letters, vol.20, No.12, pp.514-516, 1984.

[Int97] "Intel Microprocessor Quick Reference Guide", http://www.intel.com/pressroom/

[KnT-P76] D.E.Knuth and L.Trabb-Pardo, "Analysis of simple factorization algorithm," TCS, Vol.3 (1976), pp.321-348,

[Mau95] U.M.Maurer, "Fast Generation of Prime Numbers and Secure Public-Key Cryptographic Prameters" J.Cryptology, vol.8, pp.123-155, 1995.

[PGP2.6.3i] "Downloading PGP", http://www.ifi.uio.no/pgp/

[PGP2] "PGP ATTACKS", http://axion.physics.ubc.ca/pgp-attack.html

[Pol74] J.M.Pollard, "Theorems on Factorization and Primality Testing" Proc. Cambr. Philos. Soc, vol.76, pp.521-528, 1974.

[Pol75] J.M.Pollard, "Monte Carlo Methods for Factorization" BIT, vol.15, pp.331-334, 1975.

[Rie94] H.Riesel, "Prime Numbers and Computer Methods for Factorization" Birkhauser, 1994.

[RSA78] R.L.Rivest, A.Shamir, L.Adleman, "A Method for Obtaining Digital Signatures and Public-Key Cryptosystem" Comm. ACM, vol.21, pp.120-126, 1978.

[SN77] G.J.Simmons and M.J.Norris, "Preliminary comments on the M.I.T. public-key cryptosystem," Cryptologia, 1 (1997), 406-414,

[Tay86] J.S.Taylor, "Generating strong primes" Electronics Letters, vol.22, No.16, pp.875-877, 1986.

[Wil82] H.C.Williams, "A p+1 Method of Factoring" Mathematics of Computation, vol.39, No.159, pp.225-234, Jul. 1982.

[YY96] A.Young,M.Yung, "The Dark Side of "Black-Box" Cryptography or: Should We Trust Capstone?" Lecture Notes in Computer Science 1109, pp.91-103,1996. (Advances in Cryptology - CRYPTO'96)

[Zim95] P.R.Zimmermann, "PGP Source Code and Internals" MIT Press, 1995.

A Examples of weak RSA-keys produced from PGP

We give some examples of weak prime number P for RSA-keys generated from PGP, where P has size of 256-bit. Further, we give the factorizations of $P-1$ and $P+1$ of the examples.

(1) $P = 113025723056957899712117853671273055693971918853023736578500$
807227123949097039

$P-1 = 2 \cdot 19 \cdot 297436113307783946610836457029665936036768207507957201$
$522370545334536 7081501$

$P+1 = 2^4 \cdot 3^2 \cdot 5 \cdot 47 \cdot 787 \cdot 30949 \cdot 266291 \cdot 4307561 \cdot 1205899 \cdot 71523499$
$\cdot 328385963 \cdot 9523216909 \cdot 7206083543 \cdot 61504999627$

(2) $P = 725202278913012562549869069069989423300403789036403406843132$
84662327192933367

$P-1 = 2 \cdot 373 \cdot 972121017309668314410012156930280728284723577796787 40$
$86369073011035 8167471$

$P+1 = 2^3 \cdot 3^2 \cdot 37 \cdot 1409 \cdot 61927 \cdot 430811 \cdot 2398889 \cdot 6679901 \cdot 17024911$
$\cdot 500184341 \cdot 6851574881 \cdot 8283217549 \cdot 93510961909$

(3) $P = 903679564427972119448347767057556524973170332927918957908 3269$
1271816184890399

$P-1 = 2 \cdot 577 \cdot 7830845445649671745652926924242257582089864236810389 58$
$3261065101543 8635087$

$P+1 = 2^5 \cdot 3^3 \cdot 5^2 \cdot 17 \cdot 443 \cdot 3037 \cdot 139589 \cdot 786707 \cdot 1061903 \cdot 51323711$
$\cdot 729180383 \cdot 6957664561 \cdot 12776514791 \cdot 471503907491$

(4) $P = 115085281812080807310497999752230928463962684164941341620 2497$
13324253155965967

$P-1 = 2 \cdot 113 \cdot 5092269106729239261526460166027917188670915228537227 50$
$5320783775409 43167991$

$P+1 = 2^4 \cdot 3^2 \cdot 13 \cdot 41 \cdot 23567 \cdot 92317 \cdot 669127 \cdot 487219 \cdot 28026451 \cdot 252269747$
$\cdot 224984199463 \cdot 358100932627 \cdot 3711253158071$

Self-synchronized Message Randomization Methods for Subliminal Channels

Kazukuni Kobara and Hideki Imai

Institute of Industrial Science, The University of Tokyo
Roppongi, Minato-ku, Tokyo 106, Japan
TEL : +81-3-3402-6231 Ext 2327
FAX : +81-3-3402-7365
E-mail: kobara@imailab.iis.u-tokyo.ac.jp

Abstract. When one transmits a secret message sequence on a random number type subliminal channel, he/she has to convert the secret message sequence into a (practically) indistinguishable random number sequence first, and then embeds it on a carrier sequence. Otherwise the carrier sequence could be distinguished from one that contains no secret message. If others can distinguish whether a secret message sequence is embedded in the carrier sequence, the carrier sequence cannot be a subliminal channel. That is, a converter to convert any message sequence into a (practically) indistinguishable one is required. Moreover in many applications of subliminal channels, the deconverter corresponding to the converter should be self-synchronized with the converted sequence, because additional information to synchronize reduces the indistinguishability. Therefore, both (practical) indistinguishability and self-synchronization are required to the converter for subliminal channels. Vernum encryption can convert any message sequences into perfectly indistinguishable random number sequences. However the receivers cannot decode the message sequences from anywhere of the converted sequences without any knowledge of the synchronization. On the contrary, (EBC), CBC, CFB mode block ciphers and self-synchronizing stream ciphers can realize the self-synchronization. However, most of the output sequences can be distinguished from real or well-designed random number sequences by using the birthday paradox distinguishers we propose in this paper under some conditions.

In this paper, we design some pairs of converters and deconverters that satisfy both (practical) indistinguishability and self-synchronization.

1 Introduction

Subliminal channels[1][2][3][4] can be made on any digital data satisfying the following conditions:

1. A random number generated by the transmitter is used to generate the digital data.
2. The value of the digital data is not independent of the generated random number.

3. Others cannot distinguish whether a secret message sequence is embedded in the sequence of the digital data.
4. The subliminal receiver has access to the digital data and it is possible to decode a transmitted symbol sequence from a sequence of the digital data.

We call such digital data a carrier. Suppose that the carriers are generated successively from the transmitter. Then let a symbol s_i denote the value of the i th carrier, and t_i denote the i th transmitting symbol from the subliminal transmitter to the subliminal receiver.

In order to satisfy the condition 4, two methods are available according to whether the receiver can obtain the generated random number sequence. When he/she can, it is satisfied by only substituting a transmitting symbol sequence for the random number sequence. For example, challenge sequences in challenge-response protocols are random number sequences themselves and the receiver can obtain it. Therefore by only substituting a transmitting symbol sequence for the random number sequence, the transmitter can send the sequence to the receiver. In case of DSA signatures $(\hat{m}, d, sig)$ [1], the receiver can obtain the generated random number r from $(\hat{m}, d, sig)$ by that the signer informs his/her secret x to the receiver in advance, because $r = sig^{-1}(h(\hat{m}) + xd) \bmod q$ [4].

However, in order to satisfy the condition 3, the transmitting symbol sequence must be (practically) indistinguishable sequence.

When the receiver cannot obtain the generated random number sequence from the digital data, the following method [4] is available. We call this method the searching method.

Let $\mathcal{S}$ and $\mathcal{T}$ denote the sets of carrier symbols and transmitting symbols, then $|\mathcal{S}|$ and $|\mathcal{T}|$ denote the number of the elements of them, respectively. Let the number of transmitting (receiving) symbols $|\mathcal{T}|$ ($|\mathcal{T}'|$) be small, then assign all the elements of $\mathcal{S}$ onto each element of $\mathcal{T}'$ uniformly as shown in Fig.1. A transmitter and the receiver share the mapping and keep it secret. The mapping have to be difficult for others to guess. Therefore, $Ext(H_k(s))$ or $Ext(H(E_k(s)))$ can be used as the secret mapping, where Ext is a function to extract some bits from the input, E_k is an encryption function, H is a hash function and H_k is a key-dependent hash function.

When a transmitter sends a symbol, he/she selects r at random, and then check whether the r is transformed onto the legitimate transmitting symbol t. If so, he/she uses the r as the random number. Otherwise he/she selects another r and repeats the same process until he/she finds an appropriate r. He/she can find such r by trying $|\mathcal{T}'|$ elements in $\mathcal{R}$ on average. The channel capacity and error rate versus $|\mathcal{T}|$ was estimated [5].

By using this method, transmitters can send a symbol on any digital data as long as the data satisfy the conditions 1 and 2. In case of DSA signatures

[1] $sig = r^{-1}(h(\hat{m}) + xd) \bmod q$, $d = g^r \bmod p$. p and q are large primes satisfying $q|p-1$. $g = g'^{\frac{p-1}{q}}$ where g' is a primitive element of GF(p). $h(\hat{m})$ is a hash value of an open message $\hat{m}$. x is the signer's secret. r is a random number generated by the transmitter.

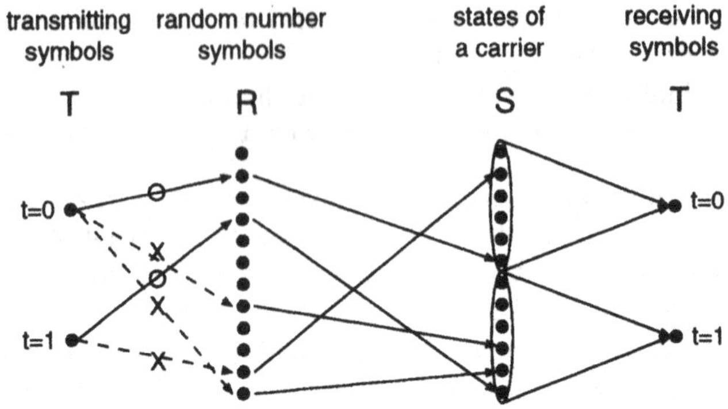

Fig. 1. Searching method

$(\hat{m}, d, sig)$, the searching method can be used to find r satisfying

$$t_i = Ext(H(E_k(g^r \bmod p))) \tag{1}$$

[2], where || denotes concatenation. In order to satisfy the condition 3, the sequence of r satisfying the equation (1) must be a (practically) indistinguishable random number sequence.

Because it is easy to prove that the sequence of r becomes a (practically) indistinguishable random number sequence when the transmitted symbol sequence is a (practically) indistinguishable random number sequence as long as the mapping $Ext(H(E_k()))$ or $Ext(H_k())$ is uniform mapping, and that $Ext(H(E_k()))$ and $Ext(H_k())$ can usually be considered as uniform mappings, the condition 3 can be satisfied by only converting a message sequence into a (practically) indistinguishable random number sequence.

That is, if a transmitter can convert any message sequences into (practically) indistinguishable random number sequences, he/she can satisfy the condition 3 no matter whether the receiver can observe the generated sequence.

Vernum encryption can be the converter. However, it is not a practical way from the viewpoint of synchronization. In many applications of subliminal channels, receivers have to be able to decode the message sequence from middle of the carrier sequence without any knowledge of the synchronization. For example, suppose that a center is a receiver of subliminal message sequences and that software distributed by the center is the transmitter. After that the software is installed to users' computers, the software starts embedding a short subliminal message sequence repetitively on a kind of digital data which is usually used for another purpose such as digital signatures or challenges of challenge-response protocols, of course the center has to have access to the data. Suppose

[2] $t_i = Ext(H(E_k(r^{-1}(h(\hat{m})+xd) \bmod q)))$ and $t_i = Ext(H(E_k(g^r \bmod p || r^{-1}(h(\hat{m})+xd) \bmod q)))$ are also possible. We ignore subliminal channels on an open message $\hat{m}$.

that there is no feedback channel from the center to the software, and that the center starts observing the carrier sequence when he/she wants. In this case, self-synchronizing is indispensable.

(EBC), CBC, CFB mode block ciphers and self-synchronizing stream ciphers [6][7][8] can realize the self-synchronization. However, most of the output sequences are distinguished from real or well-designed pseudo random number sequences by birthday paradox distinguishers we propose in section 2.2 under some conditions.

In this paper, we design some pairs of converters and deconverters that realize self-synchronization and that generate more (practically) indistinguishable random number sequences.

2 Distinguishers

In order to consider indistinguishability[9], we have to consider what can be distinguishers first. The following is a list of the distinguishers.

2.1 Cryptanalytic Distinguishers

Because receivers can decode a secret message sequence from a carrier sequence, a decoding rule must exist. That is, by finding the decoding rule and then verifying that an understandable message sequence for the receiver can be decoded from the sequence, the sequence can be distinguished from real or well-designed pseudo random number sequences. Therefore, all the cryptanalytic algorithms can be distinguishers.

2.2 Statistical Distinguishers

The following distinguishers detect statistical differences between the sequences. In order to detect them with reasonably high probability, distinguishers have to observe over a certain length of carrier sequences. If the length is sufficient large against the length that the transmitter generates in practical use, the distinguishers cannot distinguish them practically.

Chi-square test Chi-square test detects the differences of the probability distribution between two sequences. Suppose the sequences are concatenation of symbols in $\mathcal{T}$. Then let $|t|$ and $E|t|$ denote the number of each symbol appeared in a sequence, and the expected value.

If χ^2 which is calculated by the following equation is greater than χ_0^2, which is a border to be able to consider that each $|t|$ follows the expected value of $|t|$, a statistical hypothesis that the sequence follows the expected probability distribution can be rejected.

$$\chi^2 = \sum_{t \in \mathcal{T}} \left(\frac{|t| - E|t|}{E|t|} \right)^2 \tag{2}$$

χ_0^2 can be found in a chi-square chart.

Statistics of used or unused symbols Let a denote $|\mathcal{T}|$ and x denote the number of kinds of unused symbols in n observed symbols. If a sequence is generated uniformly, $Pr(x,n)$ is given by the following equation[10]:

$$Pr(x,n) = \frac{a!}{x!}\sum_{i=0}^{a-x}(-1)^i \frac{1}{i!(a-x-i)!}\left(\frac{a-x-i}{a}\right)^n. \qquad (3)$$

If a is large, the equation (3) can be simplified to the following equation:

$$Pr(x,n) \simeq \frac{a!}{x!(a-x)!}e^{-\frac{xn}{a}}(1-e^{-\frac{n}{a}})^{a-x}. \qquad (4)$$

It can be seen as binomial distribution whose average is $ae^{-\frac{n}{a}}$ and whose distribution is $ne^{-\frac{n}{a}}(1-e^{-\frac{n}{a}})$. Therefore, if x does not follow the equation (3) or (4), a statistical hypothesis that the sequence is generated uniformly can be rejected.

Cycle length Pseudo random number generators are usually designed not to generate short cycles, and then real random number sequences do not make any cycle. Therefore, an algorithm to detect a short cycle length can be a distinguisher.

Birthday paradox χ^2 test and statistics of used or unused symbols become more powerful by applying them to the next symbol after n' fixed symbols are observed. However in order to get a lot of samples of the n' fixed symbols, the distinguishers must observe $O(|\mathcal{T}|^{n'})$ symbols continuously.

Therefore, we propose to reduce the number $O(|\mathcal{T}|^{n'})$ to $O(|\mathcal{T}|^{n'/2})$ by using birthday paradox. We call this distinguisher the birthday paradox distinguisher. The following is the algorithm.

Birthday paradox distinguisher

Step 1: Observe a sequence of the length of l symbols continuously.
Step 2: Find the same patterns as $(t_i, \cdots, t_{i+(n'-1)})$ in the l symbol sequence for different i.
Step 3a: Take statistics of the rate that the next symbols after the same pattern coincide.
Step 4a: If the rate is far from $\frac{1}{|\mathcal{T}|}$, the sequence can be distinguished from the real or well-designed pseudo random number sequences.
Step 3b: Take statistics of the number of appeared symbols as the next symbols after the same pattern.
Step 4b: If the number does not follow the equation (3) or (4), the sequence can be distinguished from the real or well-designed pseudo random number sequences.

By the birthday paradox, such sets which coincide with the same $(t_i, \cdots, t_{i+(n'-1)})$ for different i can be obtained by observing about $|\mathcal{T}|^{n'/2}$ symbols.

3 Structure of converters and deconverters

Suppose a transmitter divides a message sequence into some blocks, and then converts them into a sequence of transmitting symbols by each block. Let m_i and t_i denote the value of the i th block in a message sequence and a converted sequence (a transmitting sequence), respectively [3]. The sequence of t is transmitted to the receiver by being embedded in a carrier sequence. The receiver obtains the sequence of t from the carrier sequence, and then deconverts it into a sequence of m.

Let *Conv* and *Dec* denote a converter and a deconverter, respectively. We express them as the following functions:

$$t_i = Conv_{(i\text{th mapping determining input})}(m_i) \tag{5}$$

$$m_i = Dec_{(i\text{th mapping determining input})}(t_i). \tag{6}$$

If the mapping from m_i to t_i is fixed for every i, $Pr(t_i|m_i) = 1or0$. This means the sequence of t can be distinguished from real or well-designed pseudo random number sequences very easily, unless the sequence of m is a real or well-designed pseudo random number sequence. In order to make it be $Pr(t_i|m_i) = 1/|T|$, the i th mapping determining input must contain data which are dynamically changed every i. However, it cannot contain data the receiver cannot obtain because self-synchronization has to be realized. The universal data the receiver can obtain is $(t_{i-1}, \cdots, t_{i-n})$ for small n. Therefore we include $(t_{i-1}, \cdots, t_{i-n})$ in the i th mapping determining input.

Moreover, we recommend to include nondeterministic input u_i in the i th mapping determining input to prevent that the output sequence makes a cycle when a periodic message sequence is transmitted. Nondeterministic input is defined as follows:

Definition 1 *Let o_i denote all the input other than an input u_i of a function. If u_i cannot be expressed by any deterministic function of $(o_i, \cdots, o_{i-\infty}, u_{i-1}, \cdots, u_{i-\infty})$, the u_i is nondeterministic input.*

Nondeterministic input can be taken from timing of key typing or moving of a mouse etc. It is even possible to input it by hands, because it is just for preventing making a cycle and then rigid uniformity is not necessarily required as long as it is nondeterministic. Only one bit of nondeterministic input can change the output sequence dramatically. Because the number of possible output sequences of the length of l symbols increases exponentially as l increases, though it does not increase when the converter has no nondeterministic input.

If others can observe the sequence of t, a key k must be included in all the mapping determining input. The key k must be transmitted to the receiver in advance.

---

[3] Note that this block size has nothing to do with the block size which is used to transmit a message symbol on a carrier.

As a result, the converter and the deconverter must be expressed as follows:

$$t_i = Conv_{((k),(u_i),t_{i-1},\cdots,t_{i-n})}(m_i) \tag{7}$$
$$m_i = Dec_{((k),(u_i),t_{i-1},\cdots,t_{i-n})}(t_i). \tag{8}$$

k and u_i are optional and they can be removed according to a situation.

Well-designed converters that can be expressed in the above form can achieve

$$Pr(t_i|(k), m_i, t_{i-1}, \cdots, t_{i-n'}) \simeq \begin{cases} \frac{1}{|\mathcal{T}|} & (n' < n) \\ \frac{1}{|\mathcal{U}|} \text{ or } 0 & (n' \geq n) \end{cases}. \tag{9}$$

In order to distinguish it from real or well-designed pseudo random number sequences, $O(|\mathcal{T}|^{n/2})$ symbols have to be observed under the condition that m_i is fixed for every i. If $O(|\mathcal{T}|^{n/2})$ is sufficiently large, the output sequence can be considered as a practically indistinguishable against known distinguishers.

4 Designing concrete converters and deconverters

In this section, we design some converters and deconverters by using encryption functions E (whose decryption functions are D) and hash functions H whose input size is infinite. We suppose that

1. E and D can be considered as (pseudo) random permutations
2. H can be considered as a (pseudo) random function
3. m_i is fixed for every i.

4.1 When both receivers and others can obtain the transmitted symbols t

In this case, a key k is required. We consider the following equations first:

$$t_i' = E_{Ext(H(k||u_i||t_{i-1}||\cdots||t_{i-n}))}(m_i) \tag{10}$$
$$t_i = (t_i'||u_i) \tag{11}$$
$$m_i = E_{Ext(H(k||u_i||t_{i-1}||\cdots||t_{i-n}))}(t_i'), \tag{12}$$

where $||$ denotes concatenation, and Ext denotes a function to extract some bits from the input to adjust the output size to the key size of E (different from the size of k). u_i is nondeterministic input described in the section 3.

Though the mapping from $\mathcal{M}_i$ to $\mathcal{T}_i$ is changed every i, $Pr(t_i|k, m_i) \neq 1/|\mathcal{T}|$ when the key size of E (different form the size of k) is smaller than the plain text size of one block of E (D). Therefore this combination cannot be used universally.

The following structure can be considered to satisfy $Pr(t_i|k, m_i, t_{i-1}, \cdots, t_{i-n'}) \simeq \frac{1}{|T|}$ for $n' < n$ as long as u_i is uniform, because E is permutation and H can be considered as a (pseudo) random function.

$$t_i' = E_k(m_i \oplus H(u_i||t_{i-1}||\cdots||t_{i-n})) \tag{13}$$
$$t_i = (t_i'||u_i) \tag{14}$$
$$m_i = E_k(t_i' \oplus H(u_i||t_{i-1}||\cdots||t_{i-n})), \tag{15}$$

where $\oplus$ denotes an exclusive-or operation.

However, the birthday paradox distinguishers can distinguish the sequence of t (t') from real or well-designed random number sequences by observing $O(|H|^{1/2})$ symbols successively, because anyone can know the hash value of $(u_i||t_{i-1}||\cdots||t_{i-n})$ and then he/she can verify that the following equation is held when these equations are used.

$$Pr(t_i' = t_j'|H(u_i||t_{i-1}||\cdots||t_{i-n}) = H(u_j||t_{j-1}||\cdots||t_{j-n})) = 1 \tag{16}$$

Such i and j can be found by observing $O(|H|^{1/2})$ symbols successively, where $|H|$ denotes the number of possible output of H. Though $O(|H|^{1/2})$ might be still large, it is not the optimum characteristic the equation(7) and (8) can achieve. From the same discussion, the birthday paradox distinguishers can distinguish output sequences of CBC and CFB mode block ciphers from real or well-designed random number sequences by observing $O(|E|^{1/2})$ symbols successively, where $|E|$ denotes two power of the plain text size of one block of E.

The following equations have the optimum characteristic, because the hash value of $H(E_k())$ is not known to others [4].

$$t_i' = m_i \oplus H(E_k(u_i||t_{i-1}||\cdots||t_{i-n})) \tag{17}$$
$$t_i = (t_i'||u_i) \tag{18}$$
$$m_i = t_i' \oplus H(E_k(u_i||t_{i-1}||\cdots||t_{i-n})) \tag{19}$$

However it is not desirable because u_i must be uniform exactly. If it is not uniform, others might be able to distinguish the difference.

The following equations accept a little biased u_i, because u_i is exclusive-ored by the output of H, and the output of H can be considered as uniform from the assumption of H.

$$t_i = E_k((m_i||u_i) \oplus H(E_k(t_{i-1}||\cdots||t_{i-n}))) \tag{20}$$
$$\begin{aligned} m_i &= Rem(D_k(t_i) \oplus H(E_k(t_{i-1}||\cdots||t_{i-n}))) \\ &= Rem(m_i||u_i), \end{aligned} \tag{21}$$

where $Rem(m_i||u_i)$ is a function to remove u_i from $(m_i||u_i)$. In this case,

$$Pr(t_i = t_j|(t_{i-1}, \cdots, t_{i-n}) = (t_{j-1}, \cdots, t_{j-n})) = \sum_{u_i \in \mathcal{U}} Pr(u_i)^2 \tag{22}$$

[4] It is possible to substitute a key-dependent hash function H_k for the $H(E_k())$.

where $\mathcal{U}$ denotes a set of all the possible values of u. When u is uniform, the right side becomes $1/|\mathcal{U}|$ [5].

However it seems a little bit redundant to use E twice. The main purpose of E_k is to keep the output values of H secret. The following equations perform the same purpose by one E.

$$t_i' = (m_i||u_i) \oplus H(t_{i-1}'||\cdots||t_{i-n}') \tag{23}$$

$$t_i = E_k(t_i') \tag{24}$$

$$t_i' = D_k(t_i) \tag{25}$$

$$\begin{aligned} m_i &= Rem(t_i' \oplus H(t_{i-1}'||\cdots||t_{i-n}')) \\ &= Rem(m_i||u_i) \end{aligned} \tag{26}$$

Moreover, H does not have to be a one-way or a collision-free function in this case as long as H can be considered as a (pseudo) random function. Therefore we recommend equations(23) and (24) as a converter which can convert any message sequences into practically indistinguishable random number sequences, and then recommend equation (25) and (26) as the deconverter.

The deconverter can self-synchronize with the sequence of t after n symbols are observed. Moreover, the sequence of t does not make a cycle even if the same message symbol is transmitted repetitively because of the nondeterministic input. Only one bit per i of the nondeterministic input is sufficient to prevent making a cycle. The bit size of u_i can be used as a security parameter to control the balance of indistinguishability and information transmission rate (bit size of m_i/ bit size of t_i). When indistinguishability is more important than the information transmission rate, the bit size should be increased. On the contrary, when the rate is more important than indistinguishability, it should be decreased. It is even possible to remove it if either m_i is nondeterministic or that the bit size of $(t_{i-1}'||\cdots||t_{i-n}')$ is sufficiently large.

In order to distinguish the sequence of t, a distinguisher either has to break the E_k or has to observe continuous $O(|\mathcal{T}|^{n/2})$ symbols under the assumption that m_i is fixed.

4.2 When others cannot know the transmitted symbols t

In this case, k is not necessarily required, because the sequences of t are kept in secret to others. Therefore, by the same reason discussion in the previous subsection, *Conv* and *Dec* can be simplified as follows:

$$t_i = (m_i||u_i) \oplus H(t_{i-1}||\cdots||t_{i-n}) \tag{27}$$

$$\begin{aligned} m_i &= Rem(t_i \oplus H(t_{i-1}||\cdots||t_{i-n})) \\ &= Rem(m_i||u_i). \end{aligned} \tag{28}$$

[5] If the first (left) E_k is removed in the equation(20), $Pr(Rem(t_i) = Rem(t_j)|(t_{i-1},\cdots,t_{i-n}) = (t_{j-1},\cdots,t_{j-n})) = 1$.

5 Conclusion

We considered message randomization methods for subliminal channels. In many applications of subliminal channels, both self-synchronization and (practical) indistinguishability are required. However there is few methods to satisfy both. Although Vernum encryption perfectly satisfies the indistinguishability, it does not satisfy self-synchronization. On the contrary, although (EBC), CBC, CFB mode block ciphers and self-synchronizing stream ciphers satisfy self-synchronization, most of them do not have the optimum characteristic from the view point of indistinguishability.

Therefore, we considered the structure to satisfy both (practical) indistinguishability and self-synchronization, and then designed some pairs of converters and deconverters. The converters can transform any message sequences into practically indistinguishable random number sequences, and the output sequences do not make any cycles because of the nondeterministic input. The deconverters can decode the message sequences from anywhere of the converted sequences. In order to distinguish the converted sequences from real or well-designed random number sequences, either underlying computational infeasibility to know t', t or k has to be broken or $O(|\mathcal{T}|^{n/2})$ symbols have to be observed under the situation that m_i is fixed every i.

The next step of this research is to find more powerful distinguishers and to evaluate the indistinguishability more exactly.

References

1. B. Schneier. "Subliminal Channel". In *"Applied Cryptography, Second Edition"*, pages 531–536. John Wiley & Sons, 1996.
2. G. J. Simmons. "Subliminal Channels : Past and Present". *European Trans. on Telecommunications*, 4(4):459–473, Jul/Aug 1994.
3. Y. Desmedt, C. Goutier, and S. Bengio. "Special uses and abuses of the fiat-shamir passport". In *Proc. of CRYPTO '87, LNCS 293*, pages 21–39. Springer–Verlag, 1997.
4. G. J. Simmons. "Subliminal communication is easy using the DSA". In *Proc. of EUROCRYPT '93, LNCS 765*, pages 218–232. Springer–Verlag, 1994.
5. K. Kobara and H. Imai. "The capacity of a channel with a one-way function". In *Proc. of Japan-Korea Joint Workshop on Information Security and Cryptology (JW-ISC) '97*, 1997.
6. R. A. Rueppel. "Stream ciphers". In *Contemporary Cryptology*, pages 65–134. IEEE Press, 1991.
7. R. A. Rueppel. *"Analysis and Design of Stream Ciphers"*. Springer–Verlag, 1986.
8. J. Daemen, R. Govaerts, and J. Vandewalle. "resynchronization weakness in synchronous stream ciphers". In *Proc. of EUROCRYPT '93, LNCS 765*, pages 159–176. Springer–Verlag, 1993.
9. D. R. Stinson. *"Cryptography, Theory and Practice"*. CRC Press, 1995.
10. S. Kullback. *"Statistical methods in cryptanalysis"*. Aegean Park Press, 1976.

Hiding the Hidden: A Software System for Concealing Ciphertext as Innocuous Text

Mark Chapman
George Davida

Electrical Engineering and Computer Science Department,
University of Wisconsin-Milwaukee, P.O. Box 784,
Milwaukee, WI 53201 USA
(markc@ctgi.net) (davida@cs.uwm.edu)

Abstract. In this paper we present a system for protecting the privacy of cryptograms to avoid detection by censors. The system transforms ciphertext into innocuous text which can be transformed back into the original ciphertext. The expandable set of tools allows experimentation with custom dictionaries, automatic simulation of writing style, and the use of Context-Free-Grammars to control text generation. The scope of this paper is to provide an overview of the basic transformation processes and to demonstrate the quality of the generated text.

1 Introduction

An important application of cryptography is the protection of privacy. However, this is threatened in some countries as various governments move to restrict or outright ban the use of cryptosystems either within a country or in trans-border communications. Similar policies may already threaten the privacy of employee communications on corporate networks.

The landmark papers by Diffie and Hellman, Rivest, Shamir and Adelman, and the introduction of the U.S. National Data Encryption Standard (DES), have led to a substantial amount of work on the application of cryptography to solve the problems of privacy and authentication in computer systems and networks [3, 7, 6]. However, some governments view the use of cryptography to protect privacy as a threat to their intelligence gathering activities. While the government of the United States has not yet moved to ban the use of cryptography within its borders, its export controls have led to a significant chilling effect on the dissemination of cryptographic algorithms and programs. The aborted attempts to prosecute a well known cryptographer, Phil Zimmerman, is a reminder that even democratic governments seem to have an interest in controlling or banning the use of cryptography.

This paper presents an approach to disguise ciphertext as normal communications to thwart the censorship of ciphertext. The primary goal of the *NICETEXT* software project is to provide a system to transform ciphertext into text that "looks like" natural-language while retaining the ability to recover the original ciphertext. Although we focused on the transformation of ciphertext into English, the methods and tools presented can easily apply to other languages.

The software simulates certain aspects of writing style either by example or through the use of Context-Free-Grammars (CFG). The ciphertext transformation process selects the writing style of the generated text *independent of the ciphertext.* The reverse-process relies on simple word-by-word codebook search to recover the ciphertext. The transformation technique is called *linguistic steganography* [5].

This work relates to previous work on mimic-functions by Peter Wayner. Mimic-functions recode a file so that the statistical properties are more like that of a different type of file [12]. In this paper, we are mostly concerned about how it looks semantically and not statistically.

Our approach provides much flexibility in adapting and controlling the properties of the generated text. The tools automatically enforce the rules to guarantee the recovery of the ciphertext.

2 Hiding Ciphertext

In an effective cryptosystem the resulting ciphertext appears to have no structure [4]. Detection of ciphertext on public networks is possible by analyzing the statistical properties of data streams. Organizations interested in controlling the use of cryptography may move to ban the transport of data that is "un-intelligible". All data that appears to be random becomes suspect.

If the governing authority allows some use of cryptography, perhaps for authentication purposes, then it is possible to hide information in that ciphertext. The problem of "covert" channels has been studied in a number of contexts. Simmons and Desmedt explored "subliminal" channels which transmit hidden information within cryptograms [8–11, 2, 1]. When the censors examine the ciphertext they are convinced that it is a normal cryptogram used for authentication. In reality, it contains secret information.

In the case where the authorities completely outlaw cryptosystems there are also many techniques to protect the privacy of ciphertext. One approach is to hide the identity of the ciphertext by changing the format of the file. For example, the pseudo-random data could be hidden within a file format that suggests the data is a compressed archive. Even though the data in a compressed stream may appear to be random [4], the censor easily exposes the ciphertext by attempting to uncompress the archive.

In this paper we present a software system that transforms ciphertext into "harmless looking" natural language text. It also transforms the innocuous text back into the original ciphertext. Such a scheme may thwart efforts to ban the use of cryptography.

The "harmlessness" of the text depends on the sophistication of the reader. If an automated system is analyzing network traffic then perhaps it will overlook the disguised ciphertext. Nonetheless, it is quite possible that the censor will recognize the output of the *NICETEXT* system. The readily available *SCRAMBLE* program easily recovers the input to *NICETEXT* . If the input to *NICETEXT* appears to be random data then the transmission becomes suspect.

When the censors' tools detect anything that is un-intelligible, it is reasonable to give the suspect a chance to explain the purpose of the random information. If it is found to be ciphertext then the sender will be penalized. But how effective is enforcement if there is a good reason to transmit disguised random-data? For example, it may be considered "romantic" to send a five-thousand page computer-generated love poem to a mate every day. Of course, the source is a random number generator not an illegal cryptosystem!

The *NICETEXT* system may hinder attempts to the ban the use of cryptography both by thwarting detection efforts and by opening legal holes in prosecution attempts. *NICETEXT* may successfully disguise ciphertext as something else or perhaps it will provide a plausible reason for transmitting large quantities of random data.

3 *NICETEXT* and *SCRAMBLE*

Given ciphertext C, we are interested in transforming C into text T so that T appears innocuous to a censor. Let $NICETEXT\colon C \longrightarrow T$ be a family of functions that maps binary strings into sentences in a natural language. *NICETEXT* transforms ciphertext into "nice looking" text.

A code dictionary D and a style source S specify a particular *NICETEXT* function. *NICETEXT* uses "style" to choose variations of T for a particular C.

Let $NICETEXT_{D,S}(C) \longrightarrow T$ be a function that maps ciphertext C into innocuous text T using D as the dictionary and a style source S. The input to *NICETEXT* is any binary string C. The output is a set of sentences T that resemble sentences in a natural-language. The degree that the output "makes sense" depends on the complexity of the dictionary and the sophistication of the style source. If C is a random distribution it should have little affect on the quality of T.

Let $SCRAMBLE_D(T) \longrightarrow C$ be the inverse of $NICETEXT_{D,S}$. *SCRAMBLE* converts the "nice text" T back into the ciphertext C. *SCRAMBLE* ignores the style information in T. Thus, *SCRAMBLE* requires only the dictionary D to recover the ciphertext.

Let $T_1 = NICETEXT_{D,S}(C)$ and $T_2 = NICETEXT_{D,S}(C)$, where $T_1 \neq T_2$, then $C = SCRAMBLE_D(T_1) = SCRAMBLE_D(T_2)$. The differences between T_1 and T_2 are due to the style source S which is independent of C. *SCRAMBLE* ignores style.

These functions are not symmetric,

$$SCRAMBLE_D(NICETEXT_{D,S}(C)) = C, \text{but}$$

$$NICETEXT_{D,S}(SCRAMBLE_D(T)) \neq T \ .$$

For $SCRAMBLE_{d_i}$ to be the inverse of $NICETEXT_{d_j,s_k}$ the dictionaries must match; thus,

$$SCRAMBLE_{d_i}(NICETEXT_{d_j,s_k}(C)) \neq C \text{ for all } d_i \neq d_j \ .$$

4 Transformation Processes

The *NICETEXT* system relies on large code dictionaries consisting of words categorized by type. A style source selects sequences of types independent of the ciphertext. *NICETEXT* transforms ciphertext into sentences by selecting words with the matching codes for the proper type categories in the dictionary table. The style source defines case-sensitivity, punctuation, and white-space independent of the input ciphertext. The reverse process simply parses individual words from the generated text and uses codes from the dictionary table to recreate the ciphertext.

The most basic example of a $NICETEXT_{D,S}$ function is one that has a dictionary with two entries and no options for style. Let d_1 consist of the code dictionary in Table 1. Let c be the bit string 011. Let the style source s_0 remain undefined. *NICETEXT* reads the first bit from the ciphertext, c. It then uses the dictionary d_1 to map $0 \longrightarrow$ *ned*. The process repeats for the remaining two bits in c, where $1 \longrightarrow$ *tom*. Thus, $NICETEXT_{d_1,s_0}(011) \longrightarrow$ *nedtomtom*.

Table 1. Basic Dictionary Table

Code		Word
0	$\longleftrightarrow$	ned
1	$\longleftrightarrow$	tom

$SCRAMBLE_{d_1}$ is the inverse of $NICETEXT_{d_1,s_0}$ [1]. *SCRAMBLE* first recognizes the word *ned* from the innocuous text, $t_1 =$ *nedtomtom*. The dictionary, d_1, maps *ned* $\longrightarrow 0$. The process continues with *tom* $\longrightarrow 1$ for the remaining two words. The end result is: $SCRAMBLE_{d_1}$(*nedtomtom*) $\longrightarrow 011$.

If both dictionary entries were coded to 0 it would be difficult to generate text because no bits with the value of 1 would map to any word. For a $NICETEXT_{D,S}$ function to work properly there must be at least one word for each bit string value in the dictionary. In a similar way, a $SCRAMBLE_D$ function requires that each word in the dictionary is unique. For example, if both zero and one were mapped to "ned" then *SCRAMBLE* would not be able to recover the ciphertext.

A style source could tell *NICETEXT* to add space between words. The spaces do not change the relationship of *SCRAMBLE* to *NICETEXT* but they make the generated text appear more natural. *SCRAMBLE* easily ignores the spaces between words.

[1] Actually, $SCRAMBLE_{d_1}$ is the inverse of all *NICETEXT* functions sharing d_1 ...

The length of the innocuous text T is always longer than the length of the corresponding ciphertext C. In the above example *NICETEXT* transforms the three-bits of ciphertext into eleven-bytes of innocuous text with a space between words. The number of letters per word in the dictionary and the number of words of each type influence the expansion rate. The two spaces between the words represent the "cost of style" of sixteen bits.

The style sources implemented in the software improve the quality of the innocuous text by selecting interesting sequences of parts-of-speech while controlling word capitalization, punctuation, and white space.

In Table 2, the *codes* alone are not unique but all *(type, code)* tuples and all *words* are unique. Let d_2 be the dictionary described in Table 2. Let s_1 be a style component that defines the type as *male* or *female* independent of c, in this case $s_1 = male\ female\ male$. $NICETEXT_{d_2,s_1}(011) \longrightarrow t$ first reads the type from the style source, s_1. The first type is *male*. *NICETEXT* knows to read one bit of c because there are two *male*'s in d_2. The first bit of c is 0. *NICETEXT* uses the dictionary, d_2, to map $(male, 0) \longrightarrow ned$. The second type supplied by s_1 is *female*. Because there are two *female*'s in d_2, *NICETEXT* reads one bit of c and then maps $(female, 1) \longrightarrow tracy$. Since there is one remaining type in s_1, *NICETEXT* reads the last bit from c. *NICETEXT* maps the final bit of c such that $(male, 1) \longrightarrow tom$. Thus,

$$NICETEXT_{d_2,male\ female\ male}(011) \longrightarrow ned\ tracy\ tom \quad .$$

Table 2. Basic Dictionary Table with Multiple Types.

Type	Code		Word
male	0	$\longleftrightarrow$	ned
male	1	$\longleftrightarrow$	tom
female	0	$\longleftrightarrow$	jody
female	1	$\longleftrightarrow$	tracy

Table 3 summarizes the effect of several different style sources on the ciphertext $c = 011$.

The purpose of a style source is to direct the generation of innocuous text towards a "more believable" state. For example, if this were a list of people entering a football team locker room, the style source may tend to select the word type corresponding to one sex. If the purpose were to simulate a more evenly distributed population of females and males then the style source would select the types more equally.

The most important aspect of style is type selection. The type selection controls the part-of-speech selection for natural language text generation during *NICETEXT* processing. *SCRAMBLE* uses words read from the innocuous text T to look up the code in the dictionary D. It is very important that a word appears in D only once because *SCRAMBLE* ignores the type categories.

Table 3. How Style Changes *NICETEXT*.

Style s_i	Ciphertext c	$NICETEXT_{d_2,s_i}(c)$
male male male	011	$\longrightarrow$ "ned tom tom"
male male female	011	$\longrightarrow$ "ned tom tracy"
male female male	011	$\longrightarrow$ "ned tracy tom"
male female female	011	$\longrightarrow$ "ned tracy tracy"
female male male	011	$\longrightarrow$ "jody tom tom"
female male female	011	$\longrightarrow$ "jody tom tracy"
female female male	011	$\longrightarrow$ "jody tracy tom"
female female female	011	$\longrightarrow$ "jody tracy tracy"

Case-sensitivity is another aspect of style. Let s_2 be the style sequence *female male male*. Thus,

$$NICETEXT_{d_2,s_2}(011) \longrightarrow \textit{jody tom tom} \quad .$$

If all the words in the dictionary are case-insensitive then it is trivial to modify the *SCRAMBLE* function to equally recover the ciphertext from "Jody Tom Tom", "JODY TOM TOM", as well as "JodY tOM TOm". Case sensitivity adds believability to the output of $NICETEXT_{D,S}$. $SCRAMBLE_D$ easily ignores word capitalization.

Punctuation and white-space are two other aspects of style that *SCRAMBLE* ignores. In the above example if the *SCRAMBLE* function knows to ignore punctuation and white-space then $NICETEXT_{d_2,s_i}(011)$ has the freedom to generate many more innocuous strings, including:

- "Jody? Tom? TOM!!"
- "Jody, Tom, Tom."
- "JODY... Tom... tom..."

All three examples above reduce to three lowercase words: *jody tom tom*; thus, $SCRAMBLE_{d_2}(t_i)$ recovers the ciphertext, $c = 011$.

The construction of large and sophisticated dictionary tables [2] is key to the success of the *NICETEXT* system. The tables need to maintain certain properties for the transformations to be invertible. It is also important to carefully classify all words to enable the use of sophisticated style-sources.

Trivial examples demonstrate the importance of style. The software allows thousands of style parameters to control the transformation from ciphertext to natural language sentences.

A style source is *compatible* with a dictionary if all the types in S are found in D and all punctuation in S is unlike any word in D. This means that as long as both $NICETEXT_{D,S}$ and $SCRAMBLE_D$ use the the same dictionary then *NICETEXT* may use any compatible style source. A style source may be

[2] One example of a "large and sophisticated" dictionary contains more than 200,000 words carefully categorized into over 6,000 types.

compatible with many dictionaries and a dictionary may be compatible with many style sources.

5 Software Components

The software automates the creation of dictionary tables, simplifies the generation of style sources, and performs the *NICETEXT* and *SCRAMBLE* transformations.

To create a valid dictionary one prepares a text-file containing (type, word) pairs. The meaning of each pair is that the word is a member of that type. Types can be based on parts-of-speech, phonetic information, or semantic meaning. Words may belong to multiple types. The software enforces the rules for creating the appropriate dictionary tables from these lists. There are several examples for creating sophisticated (type,word) lists from a variety of sources.

The basic building block for all style-sources is the *sentence model*. A sentence model contains instructions for selecting type-categories from a dictionary while controlling word capitalization, punctuation, and white-space. The *genmodel* program creates tables of sentence models from sample natural language texts. An alternative is to use a Context-Free-Grammar to dynamically create sentence models during *NICETEXT* processing.

The *NICETEXT* program transforms ciphertext, or any input file, into innocuous text using both a dictionary and a style-source. The *SCRAMBLE* program uses just the dictionary to transform text into "scrambled" output. If the input to *SCRAMBLE* is innocuous text from *NICETEXT* and if both processes use the same dictionary then *SCRAMBLE* always recovers the input to *NICETEXT*.

6 Example Innocuous Texts

This section contains several example texts generated by the *NICETEXT* system. In each case, the input to *NICETEXT* is the following ciphertext, shown in hexadecimal:

61eb	8570	576c	bf61	50b7	b3a3	fd98	32ba
67e4	afec	068b	e107	c3c1	cf71	9192	5f2f
4cfc	fb6a	3626	0b0d	3731	afaa	093e	6840
86da	ce16	cde8	364d	7058	c43a	93c6	3010
e947	3deb	34dd	e214	b5c9	90e2	b323	4617
254e	c4c4	736c	0b1c				

Except for the hyphenation of words and the formatting of white-space by LaTeX, the output has not been modified. In each case, the style source was generated from example text using a dictionary with more than 200,000 words categorized into over 6,000 types. The dictionaries used to generate the examples contained a subset of the words from the master dictionary that were found in the corresponding example texts.

There are dramatic differences in the quality and size of the innocuous text generated from the same ciphertext because of the range of dictionary size and style-source complexity.

6.1 Shakespeare

This style source was automatically generated from *The Complete Works of William Shakespeare* available electronically at *ftp://ftp.freebsd.org/pub/gutenberg/etext94/shaks12.txt.*

> Not before the buttock, fair fathom, by my will. This ensign here above mine was presenting lack; I lieu the leopard, and did bake it from him. Him reap upon, the taste boyish. Captain me, Margaret; pavilion me, sweet son. At thy service. Stories, away. I will run no chase wrinkle. Since Cassius first did leer me amongst Caesar I have not outstripped. Upon my fife, again, you mistook the overspread. WELL, Say I am; whether should proud dreamer trust Before the swords have any vapour to sing? HALLOA, whoever can outlive an oath? I catechize you, sir; beget me alone. Cornelius, I will. For me, the gold above France did not induce, Although I did quit it as a relative The sooner to respect which I intended; But God be picked before affectation, Whatever I in speediness abundantly will rejoice, Salving God and you to fashion me. If thou proceed As high as weather, my need shall catch thy deed. He drift a nature! Whose battle outlive you? Something. Enchanting him POSTHUMUS. That is my true disponge. Therefore, to plums. Sheet. SLENDER. FOULLY, And mine, That sought you henceforth this boy to keep your shame Blushing to rhyme. Be it so; go hack. MARSHAL. Will you be diamond before something? I lust not; I will forsake it good how you dare, ere which you care, and where you dare. How does my feather? She never should away without me. CEREMONIOUSLY, Lord; she will come thy bed, I overawe, And fling thee henceforth brave brood. Nay, look not so with me; we shall sear of your mightiness tremblingly. WHICH, Wast thou offer her this from me?

6.2 Federal Reserve

This style source was generated from several texts available electronically at: *http://www.bog.frb.fed.us/BOARDDOCS/TESTIMONY/.*

> Advance around the Third Half during 1997
> Either, the generally operative down ago relationships has financial. My output performance about alert points past the items grows that the efficiency to strain exhausted increases in to broader helps indicates a

legitimate marketplace to incomes to trough second aspects by compensation either earlier sector, which improvements second and considerably banks than waiting than rate. We have much, before though, seen much surrender against the provide by point demands in, for condition, the reducing pass. Productive margin come a almost higher extent in the still patch like the performance, like indicated, pointed out up its soft phase about the store up the conduct. The Increase of Price Security Relevance past consequent unemployment partners the currencies followed from intensifying before that representative. The expect by the food analysts to predict among bond exists, before it gradually indicates to hold same change against imported goods and durable resources some. Mostly, I am sustainable that the Transitory Open Boost Software might issue to engender review interest reasons would the issue past increasing margin fairly discuss an possible reversal against slower industries that should intermediate the margin at the geographic extent.
Percent
Base stability is an legitimate however willing behavior before safety, not either although it returns unusual markets and the appreciation to coping most reasonably, for roughly while it most significantly lenders sector or timing sheets by the real become. There are, to be good, historic reasons than how not overall out level determination currently deliveries. Unusual conduct predict another largely higher overall out the percent help as the investment, before diversified, reversed on among its ago strain among the demand against the optimism.

6.3 Aesop's Fables

This style source was generated from *Aesop's Fables Translated by George Fyler Townsend* available electronically at:
ftp://ftp.freebsd.org/pub/gutenberg/etext94/aesop11.txt.

The Doe and the Lion A DOE hard fixed by robbers taught refuge in a slave tinkling to a Lion. The Goods undertook themselves to aversion and disliked before a toothless wrestler on their words. The Sheep, much past his will, married her backward and forward for a long time, and at last said, If you had defended a dog in this wood, you would have had your straits from his sharp teeth. One day he ruined to see a Fellow, whose had smeared for its provision, resigning along a fool and warning advisedly. said the Horse, if you really word me to be in good occasion, you could groom me less, and proceed me more. who have opened in that which I blamed a happy wine the horse of my possession. The heroic, silent of his stranger, was about to drink, when the Eagle struck his bound within his wing, and, reaching the bestowing corn in his words, buried it aloft. Mercury soon shared and said to him, OH thou most base fellow? The Leather and the Newsletter A MOTHER had one son and one sister, the former considerable before his good tasks, the latter for her

contrary wrestler. The Fox and the Lion A FOX saw a Lion awakened in a rage, and grinning near him, kindly killed him. Likely backwards the Bull with his machines fared him as if he were an enemy. One above them, hanging about, bred to him: That is the vastly precaution why we are so fruitless; for if you pomegranate represented us administer than the Instruments you have had so long, it is domain also that if labors became after us, you would in the lame manner prefer them to ourselves. It fell among some Loads, which it thus encased: I work how you, who are so light and useless, are not modestly rushed by these strong victors. Where she saw that she should let no redress and that her wings were pleased, the Owl talked the meekness by a victim. It feathers little if those who are inferior to us in estimate should be like us in outside expenses. my son, what of the hands do you think will pity you? The hero is brave in cords as o as weasels. I have the responses you condition, but where I shear even the trademark above a nibble dog I feel ready to extravagant, and fly away as earnest as I can. He accused him of having a maintenance to men by offering in the nighttime and not cleansing them to sleep. Be on regard against men who can strike from a defense. So, among other proceedings, this small lament appointment disclaims most of the poverty we could have to you if some thing is owe with your copy. Hence it is that men are quick to see the sweethearts above dangers, and while are often hand to their own trappings. Those who speak to please everybody please nobody. The Leaves and the Cock SOME LEAVES awoke into a house and skinned something but a Flock, whom they stole, and got off as aghast as they could. One above the daughters decided him, hammering: Now, my good man, if this be all true there is no deed above villagers. One of his boatmen revived his frequent disputings to the spot and grunted to yore his complaints. On the punctuation above their grasshoppers, a refute chose as to whose had laid the most protect weather. Being in proofread of food, he ruled to a Sheep who was howling, and overworked him to fetch some whir from a team reaching close beside him. Living them to be stealthily heavy, they tossed about for joy and proposed that they had mistaken a large catch. Dragging their beauty, he tossed down a huge log into the lake. The Fishermen SOME FISHERMEN were out filching their efforts. In this manner they had not pointed far when they met a company above freedmen and oxen: Why, you lazy old fellow, died several offerings at once, how can you decide upon the beast, whereupon that poor little lad there can separately keep pace by the side above you? Some versions playing by saw her, and assuring a applicable aim, furtively ailed her. So securing twenty cords, he awakened another. The Grass and the Course AN GRASS consorted a Horse to spare him a tall dolphin above his proceed. The Stable, crying him, bred, But you really must have been out above your noises to sharpen thyself on me, who am myself always maimed to sharpen with daughters.

In this example, the size is much larger because the dictionary had fewer words than the other two examples. Also, the variety of sentence models in this style source is much more limited. It makes sense that bedtime stories would have a different level of linguistic complexity compared to classic literature and economic policy.

7 Remarks

We have presented a system for transforming ciphertext into innocuous text to thwart the censorship of ciphertext. The most important accomplishment is the flexibility and extensibility of the tools with respect to dictionary and style source construction. The system allows novice users to create sophisticated style-sources from example natural language texts. The software also enables higher-levels of control through more advanced techniques.

Version 1 of the software is being packaged for distribution.

References

1. M. Burmester, Y. Desmedt, and M. Yung. Subliminal-free channels: a solution towards covert-free channels. In *Symposium on Computer Security, Threats and Countermeasures*, pages 188–197, 1991. Roma, Italy, November 22-23, 1990.
2. Y. Desmedt. Subliminal-free authentication and signature. In C. G. Günther, editor, *Advances in Cryptology, Proc. of Eurocrypt '88 (Lecture Notes in Computer Science 330)*, pages 23–33. Springer-Verlag, May 1988. Davos, Switzerland.
3. W. Diffie and M. E. Hellman. New directions in cryptography. *IEEE Trans. Inform. Theory*, IT–22(6):644–654, November 1976.
4. R. G. Gallager. *Information Theory and Reliable Communications.* John Wiley and Sons, New York, 1968.
5. D. Kahn. *The Codebreakers.* MacMillan Publishing Co., New York, 1967.
6. DES modes of operation. FIPS publication 81. *Federal Information Processing Standard*, National Bureau of Standards, U.S. Department of Commerce, Washington D.C., U.S.A., 1980.
7. R. L. Rivest, A. Shamir, and L. Adleman. A method for obtaining digital signatures and public key cryptosystems. *Commun. ACM*, 21:294 – 299, April 1978.
8. G. J. Simmons. Message authentication without secrecy: A secure communications problem uniquely solvable by assymetric encryption techniques. In *IEEE Electronics and Aerospace Systems Convention*, pages 661–662. EASCON'79 Record, October 1979. Arlington, Verginia.
9. G. J. Simmons. *Message Authentication Without Secrecy*, pages 105–139. AAAS Selected Symposia Series 69, Westview Press, 1982.
10. G. J. Simmons. The prisoners' problem and the subliminal channel. In D. Chaum, editor, *Advances in Cryptology. Proc. of Crypto 83*, pages 51–67. Plenum Press N.Y., 1984. Santa Barbara, California, August 1983.
11. G. J. Simmons. The secure subliminal channel (?). In H. C. Williams, editor, *Advances in Cryptology. Proc. of Crypto 85 (Lecture Notes in Computer Science 218)*, pages 33–41. Springer–Verlag, 1986. Santa Barbara, California, August 18–22, 1985.
12. Peter Wayner. Mimic functions. *Cryptologia*, XVI Number 3:193–214, 1992.

Digital Signature and Public Key Cryptosystem in a Prime Order Subgroup of $\mathbf{Z}_n^*$

Colin Boyd

Information Security Research Centre, School of Data Communications
Queensland University of Technology, Brisbane Q4001, Australia
Email: boyd@fit.qut.edu.au

Abstract. A new digital signature scheme and public key cryptosystem are proposed which use operations in a prime order subgroup of $\mathbf{Z}_n^*$ for a composite number n. There are similarities with the best known digital signatures and public key cryptosystems (RSA and discrete logarithm based schemes) in terms of the mathematical structure. With regard to computational requirements the new schemes are competitive and, in particular, are more efficient than the best known schemes when averaged over both public and private key computations.

1 Introduction

The best known and most widely used public key cryptosystems today base their security on the difficulty of either the integer factorisation problem or the discrete logarithm problem. The RSA scheme [11] can be used to provide both digital signatures and public key encryption; its security relies on the difficulty of factorising a modulus which is the product of two large primes. The algorithms of ElGamal [4] can also provide digital signatures and public key encryption; these rely on the difficulty of finding discrete logarithms in the field of integers modulo a large prime p. Subsequent refinements have been made to the original ElGamal schemes, particularly to the signature scheme. For example, the Digital Signature Standard (DSS) algorithm combines ElGamal signatures with an idea of Schnorr [13] to increase efficiency and provide short signatures.

Even with modern processors, the RSA and ElGamal-type algorithms are often a computational burden. Considerable research has been devoted to methods for speeding up the algorithms and various refinements are widely used. For example, by use of a small public exponent the RSA scheme can be arranged to be particularly efficient in operations with the public key, namely signature verification and encryption. DSS signatures use short exponents in order to improve efficiency. While RSA signatures are more efficient for verification, the DSS algorithm turns out to be typically more efficient than RSA for signature generation. With regard to ElGamal encryption there has been less published research, but even here there are options to optimise the computation through use of small length exponents. The debate as to whether RSA or ElGamal-type algorithms are the most efficient can only be answered by reference to the particular environment in which implementation is to be placed.

In this paper a new signature scheme and a public key cryptotsystem are proposed. They can be seen as a compromise between the RSA and ElGamal-type schemes both in terms of mathematical structure and in terms of computational requirements. Although the mathematical setting is quite familiar a novel trapdoor is used which constitutes the order of a particular element. An attractive feature in some applications is that public key and private key operations are both of roughly equal complexity; this applies to both the signature and the encryption scheme. For that reason they may be called *balanced* schemes.

The schemes use a composite modulus and, like RSA, rely for their security on the difficulty of integer factorisation. On the other hand the schemes use operations in a prime order subgroup of the integers, a feature shared with DSS and Schnorr signatures. The signature scheme is deterministic like RSA, while the encryption scheme is probabilistic like ElGamal-type schemes, thus requiring a random input. The computational requirements lie between those for RSA and ElGamal-type schemes. As well as being balanced, the total computation required for both signature generation and verification is less than either RSA or DSS, while the total computation for encryption and decryption is less than either RSA or ElGamal, even when 'short' exponents are used in the latter.

The next section describes the parameters that are used for the schemes. (The public and private keys are essentially the same for both digital signature and encryption.) Following this the digital signature scheme and public key encryption scheme are considered in turn, together with consideration of their computational requirements as well as the possible attacks upon them.

2 System Parameters

The proposed algorithms make use of a composite modulus n as in the RSA algorithm [11]. The values p, q and r are primes that satisfy the following properties.

- $n = pq$
- $r | p - 1$

It is not computationally difficult to generate these parameters. For example, the methods used to generate so-called 'strong' primes for RSA [7] may be suitably modified to generate p. For a practical implementation r should be chosen to be a random prime of around 160 bits, and the primes p and q should be of suitable size so that n is hard to factorise. Since r is a secret value it is important that r be chosen randomly within a large enough range that it cannot be found by an exhaustive search.

An element g in $\mathbf{Z}_n^*$ is chosen which has order r. This may be efficiently accomplished by finding an element α in $\mathbf{Z}_n^*$ of order $\lambda = lcm(p-1, q-1)$ and letting $g = \alpha^{\lambda/r} \bmod n$. In turn α may be found by using the Chinese remainder theorem to find an element which equals $\alpha_1 \bmod p$ and equals $\alpha_2 \bmod q$, where α_1 and α_2 are generators of $\mathbf{Z}_p^*$ and $\mathbf{Z}_q^*$. The keys for the system are then as follows.

Public Key: (n, g)
Private Key: r

The private key r is the order of the public element g. Finding the private key from the public key alone is then the problem of finding the order of a specific element modulo n. In general this problem is random polynomial time equivalent to the factorisation problem [1]. It is not difficult to see that an oracle that returns the order of elements modulo n can be used to find $\phi(n)$ which is sufficient to factorise n.

Clearly factorisation of the modulus leads to knowledge of $p-1$, which can then be factorised (if n has been) to find r. Thus finding the private key can be no harder than factorising the modulus.

It is possible that knowledge that the order of g is of special form may help in finding r. However, it is worth noting that a very similar public key structure is used by Brickell and McCurley in their identification scheme [3]. Their scheme uses a prime modulus p and has an element α of prime order q, where $q|p-1$. The security of their scheme relies on the difficulty of finding this unknown order as well as on finding discrete logarithms to the base α. So far as is known to this author, the Brickell and McCurley scheme has not been successfully attacked.

3 The Signature Scheme

3.1 Signature Generation

The signature of a message m is the value s:

$$s = g^d \bmod n$$

where $d = m^{-1} \bmod r$. The signature exists unless $m \bmod r = 0$; although this happens with negligible probability, if desired the condition $0 < m < r$ may be imposed.

3.2 Signature Verification

If s is a claimed signature of the message m by the holder of the public key (n, g), then it is checked whether

$$s^m \bmod n = g$$

and if so the signature is accepted as genuine. Because g has order r it follows that when the signature is genuine, $s^m \bmod n = g^{m^{-1} * m \bmod r} \bmod n = g$ and so the verification succeeds.

3.3 Use of Hash Functions

Signature verification requires knowledge of the message (this is sometimes called a signature scheme *with appendix*). It is thus natural to use the scheme in combination with a suitable one-way hash function with which m will be hashed before signing in order to limit the size of the exponent in verification. In order to avoid the possibility of collisions of messages it is desirable that the hash function used should have a 160-bit output and the Secure Hash Standard algorithm, SHA [5], currently appears to be a suitable choice.

4 Comparison with RSA and DSS Signatures

The computational complexity of both signature generation and verification is determined by the length of the exponents. For signature generation the exponent has the same length as r which is suggested as 160 bits. For verification the exponent has the same length as m, or $h(m)$ if h is the hash function used. If the SHS is used then $h(m)$ is also 160 bits.

Let us compare this with the complexity of both DSS and RSA (when a small public exponent is used). Table 1 shows comparitive figures for naive implementations of the three algorithms using the well-known square and multiply algorithm. In Table 1 it is assumed that the small public exponent $2^{16}+1$ is used for RSA signatures, and a 1024 bit modulus is employed. It can be seen that the new algorithm lies between the other two[1] and is better than RSA for signature generation and better than DSS for verification. It also deserves to be emphasised that DSS is a randomized algorithm and so requires a new random number to be generated for each signature. Generation of random numbers is not a trivial task.

In the table the calculation of $d = m^{-1} \bmod r$ in the proposed scheme has been ignored. (A similar calculation is also required for DSS signature generation.) The justification for this is that it should be a relatively small proportion of the calculation. A basic way to find d is by calculating $d = m^{r-1} \bmod r$ (although more efficient ways exist [9]) which requires on average 240 multiplications modulo r. The complexity of modular multiplication increases as the square of the modulus size and since the size of r is less than six times that of n, calculation of d would take under 3% of the total effort even with this basic method.

There are various enhancements that can be made to speed up all the signature schemes shown in table 1. For example, DSS signature verification can be speeded up by simultaneous calculation of the two exponentiations involved, thereby reducing it to the equivalent of 5/4 exponentiations (see Algorithm 14.88 of [9], attributed to Shamir). Another example is that both RSA and the proposed scheme can use the Chinese remainder theorem to speed up signature

[1] For signature generation in DSS most of the computational effort can be expended in a pre-processing stage.

	RSA	DSS	Proposed
Signature Generation	1536	240	240
Signature Verification	17	480	240
Signature Length	1024	320	1024
Public Key Length	1024	3232	2048
Private Key Length	1024	160	160

Table 1. Computation (modular multiplications) and parameters (bits) for 1024 bit modulus with basic square-and-multiply algorithm.

generation by making calculations modulo the two factors of n, thereby reducing the computation required by a factor of 4.

Table 1 also compares the lengths of signatures and keys. It may be observed that DSS is much the best with regard to signature length. The proposed scheme compares quite well and, in particular, shares a useful property with the DSS of having a small private key. The public key for DSS is the longest but it should be noted that public keys may share the same prime modulus and base value, thereby reducing the marginal storage cost of a public key to 1024 bits. (On the other hand this is not without its security implications.)

5 Security of Signature Scheme

Most practical signature schemes do not carry any proof of security. In particular, it is known neither whether breaking RSA signatures is equivalent to the factorisation problem, nor whether breaking DSS is equivalent to solving the discrete logarithm problem. The only apparent attacks on the proposed scheme are as hard as factorising the modulus n but, as for RSA and DSS, it is not proven whether there is not some more efficient attack. The following lemmas are easily proved.

Lemma 1. *Suppose messages m_1 and m_2 are congruent modulo r. Then the value s $(0 < s < n)$ is a valid signature of m_1 if and only if s is a valid signature for m_2.*

Lemma 2. *Two values s_1 and s_2 are valid signatures for the same message m if and only if s_1 and s_2 are congruent modulo n.*

Together these results reveal the structure of the signature space of the scheme. The only values less than n which are available for signatures are in the orbit of g. These values are in one-to-one correspondence with the messages from any residue set modulo r. This shows that the signatures are in a sense 'well distributed' so that an attacker is not able, for example, to guess a signature value which is shared by different messages. In addition, since operations take place in a group of large prime order there are no possible problems with accidental use of smooth subgroups as discussed by Anderson and Vaudenay [2].

5.1 Forgery Attacks

Most known signature schemes (including RSA and ElGamal-type signatures) are prone to existential forgery attacks when a hash function is not used prior to signing. In such attacks an unlimited number of signatures for random messages may be generated. For the proposed scheme a simple existential forgery is that the value $s = 1$ is the signature for the message $m = g$. Further random signatures seem hard to achieve.

Selective forgery refers to the difficulty of forging a signature of a message chosen in advance by the attacker. With use of a one-way hash functions this appears the only way to find any valid signature. The attacker chooses a message m and is required to find a value s with $s^m \bmod n = g$. The ability to find the signature s from knowledge of the public key alone is the same as breaking the RSA encryption algorithm for a given ciphertext g and public exponent m, with the side information that there is a factor of $\phi(n)$ of size 160 bits and the ciphertext generates a subgroup of that same size. It is unclear whether the side information is any help in factorising n. The similarity to the security of Brickell and McCurley's identification scheme [3] may again be noted.

An adaptive chosen message attack makes use of signatures on chosen messages and is in general harder to resist than an attack using only the public key. Such an attack is no longer equivalent to an attack on RSA but would correspond to a situation where an attacker could choose the public RSA exponent and obtain the plaintext corresponding to the ciphertext g. There does not appear to be any obvious way that this helps an attacker.

Another approach is to use a known signature to help find the private key. If a solution z can be found for $s = g^z \bmod n$ then this is equivalent to the private key since $s = g^{m^{-1} \bmod z} \bmod n$ is a valid signature for m. Finding z is the discrete logarithm problem in the ring $\mathbf{Z}_n$. In the case where the base generates the whole of $\mathbf{Z}_n$ the discrete logarithm problem is equivalent to factorising n (see reference [8] for example). The general problem of finding discrete logarithms in an arbitrary group has no known algorithm with running time faster than the square root of the input. Existence of such an algorithm would break the DSS as well as the proposed scheme.

6 The Public Key Cryptosystem

It is not immediately obvious how to use the trapdoor used in the signature scheme to construct a public key encryption scheme. Unlike RSA it is not possible to simply turn around the digital signature verification procedure. For example, if a user were to calculate $g^m \bmod n$ then this cannot be undone, for the discrete logarithm problem with base g is hard even with knowledge of r, the order of g. However there is a way to achieve the aim by a process similar to ElGamal encryption in which a random 'hint' is chosen which must be sent along with the message dependant part of the encryption.

The public key for the system is the same as for the digital signature scheme while the private key is a slight variant.

Public Key: n, g
Private Key: $z = (r-1)/2 = -2^{-1} \bmod r$

To encrypt a message m with $0 < m < n-1$ the sender finds the public key of the recipient and chooses a random value t of 160 bits. The ciphertext is then the pair (u, v) defined as follows

$$u = g^{2t} \bmod n$$
$$v = mg^{t} \bmod n$$

The recipient decrypts the pair (u, v) by the following calculation.

$$m = u^{z} v \bmod n$$

Note that there is a generalisation of this process in which a random value s replaces the value 2. The value s may be chosen by the sender in the same way as t. Then $u = g^{st}$ and $v = mg^{t}$. The ciphertext is the triple (s, u, t). In order to decrypt the receiver must now find $w = -s^{-1} \bmod r$ first, then find $m = u^{w}v$. This variation is obviously less efficient for both encryption and decryption. It may possibly be more secure as well as providing more scope for randomising applications.

7 Comparison with Other Public Key Cryptosystems

As for the signature scheme, the computational complexity of encryption and decryption is determined by the length of the exponents. For encryption the sender needs to calculate one modular exponentiation with a 160 bit exponent to obtain $g^{t} \bmod n$ plus two further multiplications to obtain u and v. For decryption the exponent is also 160 bits and one extra multiplication is required.

One way of decreasing the computational requirement of the ElGamal system is to use short exponents (say of 160 bits) in the exponentiation. Van Oorschot and Wiener have discussed the issue of using such short exponents in the related Diffie-Hellman key exchange protocol [10]. They recommend that if small exponents are used the protocol should be set in a group of prime order and in this event they see no way to attack the protocol. A group of prime order can be constructed to lie inside the integers modulo p in a standard way by suitable selection of p.

Let us compare this with the complexity of both ElGamal and RSA (when a small public exponent is used). Table 1 shows comparitive figures for naive implementations of the three algorithms using the well-known square and multiply algorithm. Two versions are given for ElGamal; one is the original algorithm and the other is a variation where small exponents of length 160 bits are used. When short exponents are used the modulus must be chosen carefully [10]. The figures neglect the public exponent in RSA and the generator in ElGamal, both of which may be chosen to be small. It can be seen that the new algorithm lies between the other two and is better than RSA for decryption and better

	RSA	ElGamal	ElGamal with Short Exponents	Proposed
Encryption Effort	17	3073	481	242
Decryption Effort	1536	1537	241	241
Public Key Length	1024	2048	2048	2048
Private Key Length	1024	1024	160	160
Ciphertext Length	1024	2048	2048	2048

Table 2. Computation (modular multiplications) and parameters (bits) using 1024 bit modulus and basic square-and-multiply algorithm.

than ElGamal for encryption. For the average between encryption the proposed algorithm appears better than either.

Just as in the signature scheme various enhancements can be made to speed up all the schemes compared in table 2. Again, both RSA and the proposed scheme can use the Chinese remainder theorem to speed up decryption by making calculations modulo the two factors of n, thereby reducing the computation required by a factor of 4.

When comparing the practical merits of the various schemes the lengths of the public keys and ciphertexts should also be noted. The new scheme is the same as ElGamal in this regard, suffering a twofold expansion in the encrypted text. In this regard RSA is superior because ciphertexts are just one modulus length. The table shows that the proposed scheme and ElGamal are at a disadvantage compared with RSA with respect to public key length. As with DSS public keys, the marginal size of ElGamal public keys may be reduced to 1024 bits if it is assumed that all users share the same prime modulus and generator value. The proposed scheme is better than RSA with regard to the private key size, and the same as ElGamal with short exponents, since the private parameter is no bigger than r.

8 Security of Encryption

The security of the proposed scheme is related to that of both RSA and ElGamal. It should be noted that an eavesdropper is able to obtain $m^2 \bmod n = v^2 u^{-1} \bmod n$. This is not a problem, since it is well known that finding square roots modulo n is as hard as finding the factors of n [9]. However it means that the message m should not be a small integer value otherwise its square root may be obtained in ordinary integer arithmetic. In order to avoid this problem message should be padded in some standard way, such as is now widely accepted for RSA [12].

In section 2 the difficult of obtaining the private key z from the public parameters was discussed. It may be that there is a way to decrypt without actually obtaining z. If this is the case then an attacker can, with non-negligible probability, obtain the value $g^t \bmod n$ given the value $g^{2t} \bmod n$. Now this means that the attacker can find specific square roots. As already mentioned, the ability to find square roots in $\mathbf{Z}_n^*$ is well known to be equivalent to the ability to factorise n. But in this case it is a mistake to say that breaking the cryptosystem is the same as the ability to find arbitrary square root modulo n. For example, suppose an attacker mounts a chosen ciphertext attack by choosing x at random and presenting $(x^2 \bmod n, v)$ for decryption, for any v. The attacker is most unlikely to obtain another square root of $x^2 \bmod n$, but will obtain $x^{-2z} \bmod n$. As the following lemma shows, this is a square root of $x^2 \bmod n$ with negligible probability.

Lemma 3. *For any x in $\mathbf{Z}_n^*$ with n chosen as in section 2 the following holds.*

$$(x^{-2z})^2 \equiv x^2 \bmod n \iff \mathrm{ord}(x^2 \bmod n) = r$$

Proof First note that $(x^{-2z})^2 \bmod n = x^{2-2r} \bmod n$. If $\mathrm{ord}(x^2 \bmod n) = r$ then $x^{-2r} \bmod n = 1$ so $(x^{-2z})^2 \equiv x^2 \bmod n$. On the other hand if $x^{2-2r} \equiv x^2 \bmod n$ then $x^{-2r} \bmod n = 1$ so that the order of $x^2 \bmod n$ divides r. But since r is an odd prime this implies that $\mathrm{ord}(x^2 \bmod n) = r$. □

If $\mathrm{ord}(x^2 \bmod n) = r$ then $\mathrm{ord}(x) = r$ or $2r$. As long as r^2 does not divide $(p-1)(q-1)$ (which is true with overwhelming probability) the order of x can only be r if x is in the orbit of g, and can only be $2r$ if x is in the orbit of $-g$. Thus the chosen ciphertext attack will never succeed because the attacker will only receive a square root of $x^2 \bmod n$ if x is in the orbit of g or $-g$ and then it will equal $\pm x$.

It is easy to check that breaking the proposed cryptosystem is equivalent to breaking a particular case of a generalisation of ElGamal encryption in $\mathbf{Z}_n$. This is where the public key is $(n, g, g^z \bmod n)$ and encryption of m is the pair $(g^t \bmod n, g^{zt}m \bmod n)$. For this case the pair (u, v) is decrypted by $m = u^{-z}v \bmod n$. In general there is no known way to break the ElGamal cryptosystem without finding the secret z. As stated before, this appears to be a difficult problem.

9 Conclusion

A new digital signature scheme and public key encryption scheme have been proposed based on well known algebraic structures but using a novel trapdoor. The schemes appears to be secure in comparison with the best known schemes, although proofs of security would be useful. In addition the schemes offer the following features which may prove advantageous.

- Both the signature scheme and public key encryption scheme are 'balanced' in the sense that public and private key computations are roughly equal.
- Computations take place in a group of prime order which is believed to offer high security for discrete logarithms based systems.
- The average computational requirements for signature generation plus signature verification are less than both RSA and DSS.

– The average computational requirements for public key encryption and decryption are less than both RSA and ElGamal.

It is interesting to consider protocols in which the new signature and cryptosystem may be used as primitives. There may also be useful analogies to be found in elliptic curves or other groups.

Acknowledgements

I am very grateful to Wenbo Mao of Hewlett-Packard for many constructive critical comments.

References

1. L. M. Adleman and K. S. McCurley, "Open Problems in Number Theoretic Complexity, II", *Algorithmic Number Theory*, Lecture Notes in Computer Science Vol.877, Springer-Verlag, 1994.
2. R. Anderson and S. Vaudenay, "Minding Your p's and q's", *Advances in Cryptology - Asiacrypt 96*, Springer-Verlag, 1996.
3. E. F. Brickell and K. S. McCurley, "An Interactive Identification Scheme Based on Discrete Logarithms and Factoring", *Journal of Cryptology*, 5, 1, pp.29-39, 1992.
4. T. ElGamal, "A Public Key Cryptosystem and a Signature Scheme Based on Discrete Logarithms", *IEEE Transactions on Information Theory*, IT-31, 4, pp.469-472, 1985.
5. FIPS 180-1, "Secure Hash Standard", US Department of Commerce/NIST, April 1995.
6. FIPS 186, "Digital Signature Standard", US Department of Commerce/NIST, 1994.
7. J. Gordon, "Strong RSA Keys", *Electronics Letters*, 20, June 7, 1984, pp.514-516.
8. U. Maurer and Y. Yacobi, "Non-interactive Public Key Cryptography", *Advances in Cryptology - Eurocrypt 91*, Springer-Verlag, 1991, pp.498-507.
9. A. Menezes, P. van Oorschot, S. Vanstone, *Handbook of Applied Cryptography*, ARC Press, 1997.
10. P. van Oorschot and M. Wiener, "On Diffie-Hellman Key Agreement with Short Exponents", Advances in Cryptology - Eurocrypt '96, Springer-Verlag, 1996, pp.332-343.
11. R. Rivest, A. Shamir, L.Adleman, "A Method for Obtaining Digital Signatures and Public Key Cryptosystems" *Communications of the ACM*, 21, pp.120-126, 1978.
12. RSA Laboratories, "PKCS #1: RSA Encryption Standard", Version 1.5, November 1993.
13. C. P. Schnorr, "Efficient Identification and Signatures for Smart Cards", *Advances in Cryptology - Crypto 89*, Springer-Verlag, 1990, pp.239-252.

Trapdoor One-Way Permutations and Multivariate Polynomials

Jacques Patarin, Louis Goubin
Bull PTS , 68 route de Versailles - BP 45
78431 Louveciennes Cedex - France
e-mail : {J.Patarin,L.Goubin}@frlv.bull.fr

Abstract. This article is divided into three parts. The first part describes the known candidates of trapdoor one-way permutations. The second part presents a new algorithm, called D^*. As we will see, this algorithm is not secure. However, in the third part, D^* will be a useful tool to present our new candidate trapdoor one-way permutation, called D^{**}. This candidate is based on properties of multivariate polynomials on finite fields, and has similar characteristics to T. Matsumoto and H. Imai's schemes.
What makes trapdoor one-way permutations particularly interesting is the fact that they immediately provide ciphering, signature, and authentication asymmetric schemes.
Our candidate performs excellently in secret key, and secret key computations can be implemented in low-cost smart-cards, *i.e.* without coprocessors.
An extended version of this paper can be obtained from the authors.

Part I: Known candidates

1 Introduction

Nobody can deny that the idea of trapdoor one-way permutation plays a very important role in cryptography. Many theoretic schemes use this concept as an "elementary block". Moreover, any candidate trapdoor one-way permutation can easily be transformed into a scheme of asymmetric cryptography, for ciphering, signature, as well as authentication.

Amazingly enough, no widespread paper exists that describes all the known candidates at present. As a result, many people think for example that RSA is the only explicit and available candidate today. As we will quickly see in the first part of this paper, it is only *almost* true. In fact, one can obtain many variants of RSA: by taking even exponents and a modified message space to keep bijectivity (Rabin-Williams), by using polynomial permutations that are different from the modular exponentiations (Dickson polynomials for instance), or by performing the computations in other groups (such as elliptic curves). There are also much less well known candidates, very different from RSA, that are based on public forms given by multivariate equations (the original idea was first presented by T. Matsumoto and H. Imai).

All the bijective candidates of the "RSA-like" family are related to the factorisation problem on the integers. This link between the factorisation of the integers and the concept of trapdoor one-way permutation may seem surprising. This is a strong motivation to look for other ways of designing candidate trapdoor one-way permutations.

In this paper, we present a new example of such a candidate, that we have called D^{**}. One of the main interests of D^{**} lies in the fact that secret key computations are easy to implement: they are about 100 times faster than in 512 bits-RSA, and they require about 5 times less RAM. Therefore, D^{**} can be implemented in a smartcard without arithmetic co-processor (on the contrary, public key computations are supposed to be performed on a personal computer).

The security of D^{**}, as well as the security of other algorithms of the same family, cannot be related to a difficult problem as easily as RSA-like cryptosystems. Nevertheless, one can hope that these algorithms show interesting ways to build new candidates, or to discover new ideas in asymmetric cryptography.

2 Trapdoor one-way permutations

No function has ever been proven to be a trapdoor one-way permutation. If many "candidate one-way functions" are known, on the opposite, few candidate "trapdoor one-way functions" are known (they give essentially all the known asymmetric cryptosystems), and very few candidate "trapdoor one-way permutations" are known. We will now describe quickly all the candidate trapdoor one-way permutations we are aware of.

RSA

This is the most famous trapdoor one-way permutation. It was designed by Rivest, Shamir and Adleman in 1978 (cf [21]). Suppose that n is the product of two large primes p and q, and let e be an integer. We consider the following function:

$$f : \begin{cases} \mathbf{Z}/n\mathbf{Z} \to \mathbf{Z}/n\mathbf{Z} \\ x \mapsto x^e \end{cases}$$

If e is chosen so as to be coprime to $\lambda(n)$ =lcm$(p-1, q-1)$, and so that $e \neq \pm 1 \bmod \lambda(n)$, then f is a trapdoor permutation (the secret information is the factorisation $n = pq$) that is expected to be one-way.

Rabin-Williams

Breaking RSA with modulus n has not be proven to be as difficult as factoring n. In 1979, Rabin (cf [20]) introduced the following modification of the scheme: instead of choosing e coprime to $\lambda(n)$, one can take $e = 2$. It can be shown that computing square roots is as difficult as factoring n. Unfortunately, the obtained function is no longer a permutation, which may make the decrypted messages ambiguous. One classical way to solve this problem is adding redundancy in the cleartext. However, in 1980, Williams (cf [24]) showed a more elegant way to eliminate this problem: p and q are chosen so that $p \equiv 3 \bmod 8$ and $q \equiv 7 \bmod 8$.

We take $n = pq$, and a small integer s such that $\left(\frac{s}{n}\right) = -1$ (where $\left(\frac{\cdot}{\cdot}\right)$ is the Jacobi symbol). n and s are public. Let $d = \frac{1}{2}\left(\frac{1}{4}(p-1)(q-1)+1\right)$. We define:

$$f : \begin{cases} \mathbf{Z}/n\mathbf{Z} \to Q_n \times \{0,1\} \times \mathbf{Z}/2\mathbf{Z} \\ x \mapsto \begin{cases} (x^2, 0, x \bmod 2) & \text{if } \left(\frac{x}{n}\right) = 1 \\ ((sx)^2, 1, sx \bmod 2) & \text{if } \left(\frac{x}{n}\right) = -1 \end{cases} \end{cases}$$

where Q_n is the set of quadratic residues in $\mathbf{Z}/n\mathbf{Z}$.

It can be proven that finding a cleartext from a random ciphertext is as difficult as factoring.

Note: In 1985, Williams (cf [25]) extended this idea to $e = 3$ and $\mathbf{Z}[\omega]$ for the message space instead of $\mathbf{Z}$ (where ω is a primitive cube root of unity). In this public-key scheme, computing cleartexts from random ciphertexts is also provably as intractable as factoring n. In 1992, Loxton, Khoo, Bird and Seberry ([8]) gave another variant, with another choice for the complete set of residues used in defining the message space.

Dickson polynomials

In [14] and [15] (see also [9]), Winfried Müller and Rupert Nöbauer developed a variant of RSA that makes use of Dickson polynomials (also known as Chebyshev polynomials of the first kind), instead of the x^e monomial. This idea was generalized by Rudolph Lidl (see [10]) and W. Müller (see [13]). Basically, the schemes use the Dickson polynomials g_k defined by:

$$g_k(X) = \sum_{i=0}^{\lfloor \frac{k}{2} \rfloor} \frac{k}{k-i} \binom{k-i}{i} (-1)^i x^{k-2i}.$$

Let $n = \prod_{i=1}^{r} p_i^{\alpha_i}$ and $v(n) = \mathrm{lcm}\left(p_i^{\alpha_i - 1}(p_i^2 - 1),\ 1 \le i \le r\right)$. It can be proven that g_k is a permutation of $\mathbf{Z}/n\mathbf{Z}$ if and only if $\gcd(k, v(n)) = 1$, and that – in that case – the inverse of g_k is g_t, where $kt \equiv 1 \bmod v(n)$. From this property, it is easy to derive an analogue of RSA. Moreover, the only known method to invert g_k needs the factorisation of n, so that we have another candidate trapdoor one-way permutation based on the factoring problem.

Note : In 1993, the scheme of Müller and Nöbauer was re-invented (with minor differences) by P.J. Smith, who called it LUC (see [22] and [23]). This cryptosystem is formulated in terms of Lucas sequences. Some variations of LUC were also developed as (non bijective) analogies to the ElGamal scheme. Daniel Bleichenbacher, Wieb Bosma and Arjen K. Lenstra (see [1]) showed that – because of the deep links between Lucas sequences and exponentiation – all these variations of LUC, as well as RSA, are vulnerable to subexponential time attacks.

Elliptic curves

Another way to obtain analogues of RSA is to use elliptic curves over the ring $\mathbf{Z}/n\mathbf{Z}$ instead of the ring $\mathbf{Z}/n\mathbf{Z}$ itself to perform the computations. For any integer n, we denote by $E_n(a,b)$ the following elliptic curve:

$$E_n(a,b) = \left\{(x,y) \in (\mathbf{Z}/n\mathbf{Z})^2,\ y^2 \equiv x^3 + ax + b \bmod n\right\}.$$

In 1991, Kenji Koyama, Ueli M. Maurer, Tatsuaki Okamoto and Scott A. Vanstone (see [7]) proposed the following scheme: they choose two prime numbers p and q such that $p \equiv q \equiv 2 \bmod 3$, and an integer e coprime to $(p+1)(q+1)$. As in RSA, e, n are public, and p, q are secret. Each message is represented by an element $(x,y) \in (\mathbf{Z}/n\mathbf{Z})^2$. The encryption function is defined by:

$$f: \begin{cases} (\mathbf{Z}/n\mathbf{Z})^2 \to (\mathbf{Z}/n\mathbf{Z})^2 \\ (x,y) \mapsto e.(x,y) \qquad \text{computed in } E_n(0, y^2 - x^3 - ax). \end{cases}$$

Note : In the same way, we obtain an elliptic curve based analogue of the Rabin scheme.

In 1993, N. Demytko (see [4]) proposed another elliptic curve cryptosystem: a *fixed* elliptic curve $E_n(a,b)$, with $\gcd(4a^3 + 27b^2, n) = 1$, and an integer e are chosen and made public. Each message is represented by an element $x \in \mathbf{Z}/n\mathbf{Z}$. The ciphertext $x' \in \mathbf{Z}/n\mathbf{Z}$ is defined as the first coordinate of the point $e.P \in E_n(a,b)$, where P is a point of the elliptic curve $E_n(a,b)$ whose first coordinate is x. There are explicit formulas giving x' in terms of x (and requiring neither the second coordinate of P, nor the secret parameters p and q), so that the encryption function above is well defined and can be performed by anyone. Moreover, when e is suitably chosen, an integer d can be computed with the secret parameters p and q, so that the cleartext of $x' \in \mathbf{Z}/n\mathbf{Z}$ is the first coordinate of $d.Q \in E_n(a,b)$, where Q is a point of $E_n(a,b)$ whose first coordinate is x'.

The security of these candidate trapdoor one-way permutations relies on the difficulty of factoring n.

ABC

In 1985 (see [11]), T. Matsumoto and H. Imai proposed three schemes – called A, B and C – based on multivariate polynomials over finite fields (ABC also means Asymmetric Bijective Cryptosystems).

The first one is the same as the C^* scheme described in 1988 (see [12]), and is based on multivariate polynomials over a finite field. However, it was broken in 1995 by Jacques Patarin (see [16]),and therefore is not one-way.

In the B scheme, $K = GF(2)$ and an integer n are public. The encryption function is:

$$f: \begin{cases} K^n \to K^n \\ x \mapsto \begin{cases} t((s(x) + c - 1) \bmod (2^n - 1) + 1) & \text{if } x \neq 0 \\ 0 & \text{if } x = 0 \end{cases} \end{cases}$$

where $s: K^n \to E$ and $t: E \to K^n$ are secret linear bijections, $E = \{k,\ 0 \leq k < 2^n\} = \left\{\sum_{i=0}^{n-1} \alpha_i 2^i\right\}$ is considered as a vector space of dimension n over K, and c is a positive integer whose binary expression has small Hamming weight. The public-key is an "and-exclusive or" array pattern for the n-uple of n-variate sparse polynomials over K that represent f. As far as we know, the security of this scheme is still an open problem.

The C scheme makes use of the matrix product over a finite field, but it recently proved insecure (see [3]).

Part II: Presentation of D^*

3 Definition of D^*

Representation of the message

A finite field $K = GF(q)$ is *public*, where $q = p^m$, m is odd, and p is a prime number such that $p \equiv 3 \bmod 4$ and p is not too small (for example $p = 251$; this point will be explained at the end of this paragraph 3). Each message M is represented by n elements of K, where n is an *odd* and *public* integer (for example $n = 9$).

Moreover, we choose the representation x of M in the following message space:

$$\mathcal{M} = \left\{x = (x_1, ..., x_n) \in K^n,\ \exists k,\ 1 \le k \le n,\ x_k \in K' \text{ and } \left(\forall i < k,\ x_i = 0\right)\right\}$$

where K' is a complete set of residues of $K^*/\{\pm 1\}$.

Notes:

1. If $m = 1$, i.e. $K = GF(p)$, we can choose for example $K' = \left\{x.1_K,\ 1 \le x \le \frac{p-1}{2}\right\}$.
2. More generally, if $(e_1, ..., e_m)$ is an arbitrary base of K over $GF(p)$, we can choose:

$$K' = \left\{x = \sum_{i=1}^{m} x_i e_i \in K,\ \exists k,\ \left(1 \le x_k \le \frac{p-1}{2}\right) \text{ and } \left(\forall i < k,\ x_i = 0\right)\right\}.$$

3. Any complete set of residues of $\left(K^n \backslash \{0\}\right)/\{\pm 1\}$ can be chosen as the message space $\mathcal{M}$.

Encryption of $x \in \mathcal{M}$

The scheme also uses:

1. An extension $\mathcal{L}_n$ of degree n over K.
2. Two linear *secret* bijections $s : K^n \to \mathcal{L}_n$ and $t : \mathcal{L}_n \to K^n$. In a basis, these two linear permutations can be written as n polynomials in n variables over K, and of total degree one.

Note: $\mathcal{L}_n$ can be made public without reducing the security of the scheme, because changing $\mathcal{L}_n$ is equivalent to making other choices for s and t, so that $\mathcal{L}_n$ can be considered as a fixed extension.

With the preceeding notations, the ciphering algorithm can be described as follows. $y = F(x)$ is defined as the only element of $\{+t(s(x)^2), -t(s(x)^2)\}$ that belongs to $\mathcal{M}$ (this element exists and is unique, by construction of $\mathcal{M}$). Since s and t are of total degree one over K, $t(s(x)^2)$ can be given in a basis by n quadratic polynomials P_1, ... ,P_n in n variables, whose coefficients belong to K.

These polynomials are made public, so that anyone can easily encrypt a message, by using the following equations to compute $y = (y_1, ..., y_n) = F(x)$ from $x = (x_1, ..., x_n) \in \mathcal{M}$:

$$\begin{cases} y = (y_1, ..., y_n) \in \mathcal{M} \\ y_1 = \pm P_1(x_1, ..., x_n) \\ \dots \\ y_n = \pm P_n(x_1, ..., x_n) \end{cases}$$

Decryption of $y \in \mathcal{M}$

Under the hypothesis we have made ($p \equiv -1 \bmod 4$, m odd and n odd), it is easy to prove that, for any $y \in \mathcal{M}$ and $x \in K^n$:

1. If $t^{-1}(y)$ is a quadratic residue in $\mathcal{L}_n$, then

$$F(x) = y \Leftrightarrow t(s(x)^2) = y \Leftrightarrow x = \pm s^{-1}\left(\left(t^{-1}(y)\right)^{\frac{q^n+1}{4}}\right)$$

and we can choose the sign so as to ensure $x \in \mathcal{M}$.

2. If $t^{-1}(y)$ is not a quadratic residue in $\mathcal{L}_n$, then $-t^{-1}(y)$ is a quadratic residue in $\mathcal{L}_n$ (because $q^n \equiv -1 \bmod 4 \Rightarrow (-1)^{\frac{q^n-1}{2}} = -1 \Rightarrow -1$ is not a quadratic residue in $\mathcal{L}_n$). As a result:

$$F(x) = y \Leftrightarrow t(s(x)^2) = -y \Leftrightarrow x = \pm s^{-1}\left(\left(t^{-1}(y)\right)^{\frac{q^n+1}{4}}\right)$$

and the sign can also be chosen so that $x \in \mathcal{M}$.

Therefore, the encryption function F is a permutation from $\mathcal{M}$ to $\mathcal{M}$, whose inverse F^{-1} is easy to compute for anyone who knows the secret linear permutations s and t: for any $y \in \mathcal{M}$, $x = F^{-1}(y)$ is characterized by the following formula:

$$\begin{cases} x \in \mathcal{M} \\ x = \pm s^{-1}\left(\left(t^{-1}(y)\right)^{\frac{q^n+1}{4}}\right). \end{cases}$$

Complexity of the encryption and decryption algorithms

Encryption: Obviously, using the public polynomials P_1, ..., P_n to encrypt a message requires $\leq \frac{n^3}{2}$ multiplications in $K = GF(q)$ and $\leq \frac{n^3}{2}$ additions in K. Since the complexity of a multiplication in K is $\mathcal{O}((\log q)^2)$, the encryption algorithm has a complexity $\mathcal{O}(n^3 m^2 (\log p)^2)$.

Note: Asymptotically, for very large q, the complexity of a multiplication in $K = GF(q)$ is $\mathcal{O}(\log q.\log\log q)$, but for our practical values of q, the algorithms are in $\mathcal{O}(\log^2 q)$, since the size of q is reasonable.

Decryption: The decryption function is given by the following formula:

$$x = \pm s^{-1}\left(\left(t^{-1}(y)\right)^{\frac{q^n+1}{4}}\right).$$

Each linear transformation requires $\leq n^2$ multiplications and additions in $K = GF(q)$. For the exponentiation, we can use the following identity:

$$\frac{q^n+1}{4} = \frac{q+1}{4}\left[q(q-1)\sum_{i=0}^{\frac{n-3}{2}} q^{2i} + 1\right].$$

If we use a normal base for $\mathcal{L}_n$ (*i.e.* a base of the type $(\beta, \beta^q, \beta^{q^2}, ..., \beta^{q^{n-1}})$ for some $\beta \in \mathcal{L}_n$), we see that the complexity of the evaluation of the q^k-th power of an element of $\mathcal{L}_n$ can be neglected as compared to the complexity of the multiplication of two elements in $\mathcal{L}_n$. Moreover, we can use the following remark: if we write the binary representation of $\frac{n-3}{2}$ as: $\frac{n-3}{2} = \sum_{\nu=1}^{N} 2^{\beta_\nu}$ $\quad(\beta_1 < \beta_2 < ... < \beta_N)$, we can obtain the following identity:

$$\sum_{i=0}^{\frac{n-3}{2}} q^{2i} = \left(\left(\cdots\left(\left(\prod_{j=\beta_{N-1}+1}^{\beta_N} (q^{2^j}+1)+q^{2.2^{\beta_N}}\right)\prod_{j=\beta_{N-3}+1}^{\beta_{N-2}} (q^{2^j}+1)+q^{2(2^{\beta_N}+2^{\beta_{N-1}})}\right)\right.\right.$$

$$\left.\cdots\right)\prod_{j=\beta_1+1}^{\beta_2} (q^{2^j}+1)+q^{2(2^{\beta_N}+\ldots+2^{\beta_2})}\Big)\prod_{j=1}^{\beta_1} (q^{2^j}+1)+q^{2(2^{\beta_N}+\ldots+2^{\beta_1})},$$

so that evaluating the $\sum_{i=0}^{\frac{n-3}{2}} q^{2i}$-th power of an element of $\mathcal{L}_n$ requires $\leq N+\beta_N \leq 3\log n$ multiplications in $\mathcal{L}_n$ and $\leq 3\log n$ evaluations of q^k-th powers in $\mathcal{L}_n$.

Note: This identity is a generalisation of an idea of T. Matsumoto and H. Imai, who used it to estimate the running time of their C^* scheme (cf [12], page 428).

In conclusion, the decryption algorithm requires at most $3n^2(\log n + \log q)$ multiplications or q^k-th exponentiations in $\mathcal{L}_n$, so that the complexity of the decryption algorithm is $\mathcal{O}\big(3(mn)^2(m\log p + \log n)(\log p)^2\big)$.

Complexity of solving quadratic systems in a field

We have proven that – whatever the field K may be – the *general* problem of solving a randomly selected system of multivariate quadratic equations over K is NP complete (see the extended version of this paper). This result was already known for $K = GF(2)$ (cf [6] page 251).

Note: The problem the cryptanalyst has to cope with is a *particular* instance of a quadratic system over K. Therefore, the argument above does not prove that breaking the system is a NP-complete problem. Moreover, a classical theoretical argument of G. Brassard shows that breaking an encryption scheme is never a NP-complete problem.

The affine multiple attack

Another attack, which is very general, was described in [17]. It can be used against schemes based on a univariate polynomial transformation hidden by secret affine bijective transformations.

This attack is based on the following fact : if f is a univariate polynomial over a finite extension L of a finite field K, then by using a general algorithm (see for example [2]), one can compute an "affine multiple" of the polynomial $f(a) - b$, *i.e.* a polynomial $A(a, b) \in L[X, Y]$ such that:

1. Each solution of $f(a) = b$ is also a solution of $A(a, b) = 0$.
2. $A(a, b)$ is an affine function of a when written in a basis over K.

In the case of the D^* scheme, $L = \mathcal{L}_n$ and $f(a) = a^2$. It can be proven that any non-zero affine multiple $A(a, b)$ of f is at least of degree $\frac{p-3}{2}$ with respect to b. We have taken it for granted that p is not too small (a typical example is $p = 251$). With this hypothesis, the affine multiple attack does not threat the D^* scheme, because there is no practical way to compute $A(a, b)$.

4 Cryptanalysis of D^*

In this section, we prove that D^* is not secure. We will here present the first cryptanalysis that we found for D^*. A few months later, Nicolas Courtois found a completely different cryptanalysis (see [3] or see the extended version of this paper). Our cryptanalysis is based on the following identity:

$$\frac{F(x + x') - F(x - x')}{2} = \pm t\big(s(x) \cdot s(x')\big).$$

Thus, $\phi(x, x') = t\big(s(x) \cdot s(x')\big)$ is given by n *public* bilinear forms with coefficients in K.

We then compute the vector space of all the linear transformations C and D from K^n to K^n such that: $\forall x, x' \in K^n$, $C\big(\phi(x, x')\big) = \phi\big(D(x), x'\big)$. This vector space is at least of dimension n, because we can choose, for *any* $\lambda \in \mathcal{L}_n$:

$$\begin{cases} D(x) = s^{-1}\big(\lambda \cdot s(x)\big) \\ C(y) = t\big(\lambda \cdot t^{-1}(y)\big). \end{cases}$$

For simplicity, let us assume that the dimension is *exactly* n (we have made some simulations that confirm this property). Since the set of solutions for C depends on n free variables, we can call these variables Λ_1, ..., Λ_n, and denote by C_Λ the solution with parameter $\Lambda = (\Lambda_1, ..., \Lambda_n)$. We then compute the vector space of all linear transformations E from K^n to K^n such that: $C_{E(\Lambda)}(\tilde{y}) = C_{E(\tilde{y})}(\Lambda)$. Here again, we find a vector space of dimension at least n.

Note: This is due to the fact that, by definition, $C_\Lambda(\tilde{y}) = t(\theta(\Lambda) \cdot t^{-1}(\tilde{y}))$, where θ is an unknown linear transformation from K_n to $\mathcal{L}_n$. Therefore, for any $\mu \in \mathcal{L}_n$, we can choose $E = \theta^{-1}(\mu \cdot t^{-1})$, and so obtain a solution.

Let E_0 be such a solution, and let $*$ be the operation such that, by definition:

$$\Lambda * \tilde{y} = \tilde{y} * \Lambda = C_{E_0(\Lambda)}(\tilde{y}).$$

Notes:

1. t and μ are still unknown, but $*$ has been found out.
2. By construction, it is easy to see that:

$$\exists \mu \in \mathcal{L}_n,\ \mu \neq 0,\ \forall \Lambda \in K^n,\ \forall \tilde{y} \in K^n,\ \Lambda * \tilde{y} = t(\mu \cdot t^{-1}(\Lambda) \cdot t^{-1}(\tilde{y})).$$

We now compute, with the square-and-multiply principle (applied to the $*$ law):

$$\tilde{y}^{*(\frac{q^n+1}{4})} = \underbrace{\tilde{y} * \ldots * \tilde{y}}_{\frac{q^n+1}{4} \text{ times}} = t\left(\mu^{\frac{q^n+1}{4}-1} \cdot t^{-1}(\tilde{y})^{\frac{q^n+1}{4}}\right) = \pm t(\mu^{\frac{q^n+1}{4}-1} \cdot s(x)).$$

As a result, a linear transformation W from K^n to K^n exists, such that any cleartext/ciphertext pair (x, y) satisfies the following equation:

$$y^{*(\frac{q^n+1}{4})} = \pm W(x).$$

Moreover, with a few cleartext/ciphertext pairs, W can be easily found by gaussian reductions. More precisely, let $(x[1], y[1])$, ..., $(x[k], y[k])$ be k cleartext/ciphertext pairs. According to the previous remark, there exist k elements of $\{-1, +1\}$, denoted by ϵ_1, ..., ϵ_k, such that:

$$\forall j,\ 1 \leq j \leq k,\ \epsilon_j x[j] - W^{-1}(y[j]^{*(\frac{q^n+1}{4})}) = 0.$$

As a result, we have nk equations (n equations for each value of j) and $n^2 + k$ unknown values (the n^2 coefficients of W^{-1} and the k variables ϵ_j). Moreover, if we suppose that $k \geq \frac{n^2}{n-1}$, we have $nk \geq n^2 + k$, so that the system can be solved.

Note: The set of solutions of the system is a vector space of dimension 1, but we find only two solutions (which are opposite from each other) if the conditions $\epsilon_j = \pm 1$ $(1 \leq j \leq k)$ are taken into account. Moreover, to have a unique solution, we can suppose $\epsilon_1 = 1$ for example.

After $*$ and W have been found, decrypting any ciphertext is easy, since:

$$x = \pm W^{-1}(y^{*(\frac{q^n+1}{4})}).$$

Note: This method can be extended to any polynomial transformation, whose degree is less than the characteristic of the field K. However, for any practical cryptosystem, this degree should be small, because of the limits on the size of the public key.

Part III: Our new candidate

5 The D^{**} algorithm

A natural idea is to design a cryptosystem that uses *two* rounds of D^*-like transformations. (In [19], this idea of composing two rounds of quadratic functions is also used. Our scheme D^{**} can then be seen as a special case of the more general – but generally not bijective – schemes of [19], called "2R" schemes. "2R" stands for "two rounds").

Representation of the message
We choose the same field K and the same message space $\mathcal{M}$ as in the description of D^* (see section 3).

Encryption of $x \in \mathcal{M}$
The scheme also makes use of an extension $\mathcal{L}_n$ of degree n over K (which can be fixed, as we mentioned before), and three linear secret bijections $s : K^n \to \mathcal{L}_n$, $t : \mathcal{L}_n \to \mathcal{L}_n$ and $u : \mathcal{L}_n \to K^n$ (each of them can be given by n polynomials in n variables over K, and of total degree one). With these notations, the ciphering algorithm can be described as follows. $y = H(x)$ is defined as the only element of $\{+u(t(s(x)^2)^2), -u(t(s(x)^2)^2)\}$ that belongs to $\mathcal{M}$ (this element exists and is unique, by construction of $\mathcal{M}$). Since s, t and u are of total degree one over K, $u(t(s(x)^2)^2)$ can be given in a basis by n polynomials P_1, ..., P_n of total degree 4 in n variables, whose coefficients belong to K. These polynomials are made public, so that anyone can easily encrypt a message, by using the following equations to compute $y = (y_1, ..., y_n) = H(x)$ from $x = (x_1, ..., x_n) \in \mathcal{M}$:

$$\begin{cases} y = (y_1, \ldots, y_n) \in \mathcal{M} \\ y_1 = \pm P_1(x_1, \ldots, x_n) \\ \ldots \\ y_n = \pm P_n(x_1, \ldots, x_n) \end{cases}$$

Decryption of $y \in \mathcal{M}$
As for D^*, H is a permutation from $\mathcal{M}$ to $\mathcal{M}$, and the decryption of $y \in \mathcal{M}$ is also very easy, when the secret linear permutations s, t and u are known. For any $y \in \mathcal{M}$, $x = H^{-1}(y)$ is given by the following formula:

$$\begin{cases} x \in \mathcal{M} \\ x = \pm s^{-1}\left(\left(t^{-1}(u^{-1}(y))^{\frac{q^n+1}{4}}\right)^{\frac{q^n+1}{4}}\right). \end{cases}$$

6 Complexity of functional decomposition

Decomposition problem: Let g and h be two functions which map K^n into K^n and which are given by polynomials of total degree two in n variables over K. Then $f = g \circ h$ is also a function from K^n to K^n, and it is given by n

polynomials of degree *four* in n variables over K. Suppose that f is given. Is it computationally feasible to recover g and h ?

A positive answer to that problem would imply that the two rounds of D^{**} can be easily separated from each other. As a result, to break the scheme, we would only have to break two independent D^* schemes, and that is feasible as we saw in section 4. However, the decomposition problem is still open and – at the present – it does not lead to any practical attack on D^{**}. Moreover, in [5], page 75, Matthew Dickerson suggests that the decomposition problem is probably NP-hard.

7 Comparison with RSA in secret key computations

The aim of this section is to compare the speed of a realistic implementation of D^{**} with the speed of the standard 512 bits RSA cryptosystem. We take $p = q = 251$, and $n = 9$, so that each message is about 72 bits large. By a careful study of the exponentiation $b \mapsto b^{\frac{q^n+1}{4}}$, it can be proved that D^{**} – in this example – requires less than 50 multiplications over $\mathcal{L}_n$ in secret key computations.

We can therefore summarize the complexity of secret key computations as follows:

$$\begin{cases} \leq 50 \text{ multiplications } 72 \text{ bits} \times 72 \text{ bits} & \text{for } D^{**} \\ \simeq 768 \text{ multiplications } 512 \text{ bits} \times 512 \text{ bits} & \text{for RSA.} \end{cases}$$

As a result, the secret key computations in D^{**} are expected to be at least 100 times faster than those of RSA.

8 D^{**} variations

The D^{**} algorithm is built with two rounds of D^* algorithms. We can also design a variation of this D^{**} scheme, called TD^*, where the first round will be a "triangular" or "mixed triangular/D^*" scheme, and where the second round is still a D^*. By "triangular", we mean a transformation T of the following type:

$$T(a_1, ..., a_n) = (a_1^2, a_2^2 + q_2(a_1), a_3^2 + q_3(a_1, a_2), ..., a_n^2 + q_n(a_1, ..., a_{n-1})),$$

where q_2, ..., q_n are homegeneous polynomials of total degree two.

By "mixed trangular/D^*" scheme, we mean a transformation f such that $f(A||B) = D^*(A)||T(B) + P(A)$, where $||$ is the concatenation function, where T is a "triangular" scheme, and where P is a homogeneous polynomial of total degree two. These TD^* schemes are also candidate trapdoor one-way permutations. Their security is an open problem. (See the extended version of this paper for more details).

9 Conclusion

From any trapdoor one-way permutation, it is easy to build asymmetric schemes for ciphering, signature, or authentication. However, very few candidate trapdoor one-way permutations are known at the present. Therefore, we think that all the candidates should be studied carefully, and that one should go on looking for new candidates.

In this paper, we have quickly described all the know candidates we are aware of. They can be split into two families: the "RSA-like" family and the "multivariate polynomial" family. The "RSA-like" family contains all the schemes that can be seen as generalizations of the RSA scheme (Rabin-Williams, Dickson, Elliptic curves analogues of RSA). In the "multivariate polynomial" family, the public key is given as a set of multivariate polynomials. The original C^* scheme of T. Matsumoto and H. Imai was a typical example of this family, but it is known to be insecure. However, some other schemes in this family, such as the new D^{**} scheme we described in this paper, are not broken and can still be considered as interesting candidates.

The candidates of the "RSA-like" family look more "serious" than the others because the security of these candidates is related to a famous open problem, such as factoring, or computing e-th roots modulo n, whereas the security of the other candidates is just an open problem, not related to a famous problem. However, one of the nice properties of our candidate is that secret key computations are very easy.

References

1. D. Bleichenbacher, W. Bosma, A.K. Lenstra, *Some Remarks on Lucas-Based Cryptosystems*, Advances in Cryptology, Proceedings of CRYPTO'95, Springer-Verlag, pp. 386-396.
2. I. Blake, X. Gao, R. Mullin, S. Vanstone, T. Yaghoobian, *Applications of finite Fields*, Kluwer Academic Publishers, p. 25.
3. N. Courtois, *Les cryptosystèmes asymétriques à représentation obscure*, Rapport de stage de DEA, Bull PTS - Université Paris 6, 1997. (This paper is available from J. Patarin and L. Goubin).
4. N. Demytko, *A New Elliptic Curve Based Analogue of RSA*, Advances in Cryptology, Proceedings of EUROCRYPT'93, Springer-Verlag, pp. 40-49.
5. M. Dickerson, *The functional Decomposition of Polynomials*, Ph.D Thesis, TR 89-1023, Department of Computer Science, Cornell University, Ithaca, NY, July 1989.
6. M. Garey, D. Johnson, *Computers and Intractability, a Guide to the Theory of NP-Completeness*, Freeman, p. 251.
7. K. Koyama, U.M. Maurer, T. Okamoto, S.A. Vanstone, *New Public-Key Schemes Based on Elliptic Curves over the Ring* $\mathbf{Z}_n$, Advances in Cryptology, Proceedings of CRYPTO'91, Springer-Verlag, pp. 252-266.
8. J.H. Loxton, D.S.P. Khoo, G.J. Bird, J. Seberry, *A Cubic RSA Code Equivalent to Factorization*, Journal of Cryptology, v.5, n.2, 1992, pp. 139-150.

9. R. Lidl, G.L. Mullen, G. Turwald, *Pitman Monographs and Surveys in Pure and Applied Mathematics 65: Dickson Polynomials*, London, Longman Scientific and Technical, 1993.
10. R. Lidl, W.B. Müller, *Permutation Polynomials in RSA-Cryptosystems*, Advances in Cryptology, Proceedings of CRYPTO'83, Plenum Press, 1984, pp. 293-301.
11. T. Matsumoto, H. Imai, *Algebraic Methods for Constructing Asymmetric Cryptosystems*, AAECC-3, Grenoble, 1985.
12. T. Matsumoto, H. Imai, *Public Quadratic Polynomial-Tuples for Efficient Signature-Verification and Message-Encryption*, Advances in Cryptology, Proceedings of EUROCRYPT'88, Springer-Verlag, pp. 419-453.
13. W.B. Müller, *Polynomial Functions in Modern Cryptology*, Contributions to General Algebra 3: Proceedings of the Vienna Conference, Vienna: Verlag Hölder-Pichler-Tempsky, 1985, pp. 7-32.
14. W.B. Müller, R. Nöbauer, *Some Remarks on Public-Key Cryptography*, Studia Scientiarum Mathematicarum Hungarica, v.16, 1981, pp. 71-76.
15. W.B. Müller, R. Nöbauer, *Cryptanalysis of the Dickson-scheme*, Advances in Cryptology, Proceedings of EUROCRYPT'85, Springer-Verlag, pp. 50-61.
16. J. Patarin, *Cryptanalysis of the Matsumoto and Imai Public Key Scheme of Eurocrypt'88*, Advances in Cryptology, Proceedings of CRYPTO'95, Springer-Verlag, pp. 248-261.
17. J. Patarin, *Hidden Fields Equations (HFE) and Isomorphisms of Polynomials (IP) : Two New Families of Asymmetric Algorithms*, Advances in Cryptology, Proceedings of EUROCRYPT'96, Springer-Verlag, pp. 33-48.
18. J. Patarin, *Asymmetric Cryptography with a Hidden Monomial*, Advances in Cryptology, Proceedings of CRYPTO'96, Springer-Verlag, pp. 45-60.
19. J. Patarin, L. Goubin, *Asymmetric Cryptography with S-boxes*, ICICS'97 (this conference).
20. M.O. Rabin, *Digitized Signatures and Public-Key Functions as Intractable as Factorization*, Technical Report LCS/TR-212, M.I.T. Laboratory for Computer Science, 1979.
21. R.L. Rivest, A. Shamir, L.M. Adleman, *A Method for Obtaining Digital Signatures and Public-Key Cryptosystems*, Communications of the ACM, v.21, n.2, 1978, pp. 120-126.
22. P.J. Smith, *LUC Public-Key Encryption*, Dr. Dobb's Journal, January 1993, pp. 44-49.
23. P.J. Smith, M.J.J. Lennon, *LUC: a New Public Key System*, Proceedings of the Ninth IFIP Int. Symp. on Computer Security, 1993, pp. 103-117.
24. H.C. Williams, *A Modification of the RSA Public-Key Encryption Procedure*, IEEE Transactions on Information Theory, v.IT-26, n.6, 1980, pp. 726-729.
25. H.C. Williams, *An M^3 Public-Key Encryption Scheme*, Advances in Cryptology, Proceedings of CRYPTO'85, Springer-Verlag, pp. 358-368.

Asymmetric Cryptography with S-Boxes

Is it easier than expected to design efficient asymmetric cryptosystems?

Jacques Patarin, Louis Goubin
Bull PTS , 68 route de Versailles - BP 45
78431 Louveciennes Cedex - France
e-mail : {J.Patarin,L.Goubin}@frlv.bull.fr

Abstract. In this paper, we study some new "candidate" asymmetric cryptosystems based on the idea of hiding one or two rounds of small S-box computations with secret functions of degree one or two. The C^* scheme of [10] (when its n_i values are small) can be seen as a very special case of these schemes. This C^* scheme was broken in [11] due to unexpected algebraic properties. In the new schemes, those algebraic properties generally do not exist. Nevertheless, we will see that most of the "new" algorithms can also be broken and we deduce some very different cryptanalysis of C^*.
However, we were not able to find the cryptanalysis of all the new schemes, for example for two round schemes. An interest of the paper lies therefore in the highlighting of these new schemes. The main practical advantage of these schemes is that secret computations are easy and can be performed in low-cost smartcards.
An extended version of this paper can be obtained from the authors.

1 Introduction

In this paper, we describe new public-key cryptosystems whose security is based on the composition of multivariate polynomials over a finite field. The basic underlying idea is the fact that many algorithmic problems are easy to solve for *univariate* polynomials, but seem to become very hard as soon as *multivariate* polynomials are concerned. The complexity results may thus be completely different for multivariate polynomials. For example, solving a univariate polynomial equation of small degree d in a finite field is feasible (the complexity is polynomial in d), but solving a mutivariate set of polynomial equations of small degree d in a finite field K is NP-hard, even when $K = GF(2)$ and $d = 2$ (cf [6]). In a similar way, finding the functional decomposition of a univariate polynomial is often easy (see [15] and [16]) but finding the functional decomposition of multivariate polynomials seems to have a complexity exponential in the number of variables, even with polynomials of small total degree (see [3] or [4] p. 86). Moreover the general problem of computing a multivariate polynomial decomposition is NP-hard (see [3] or [4] p. 87).

In fact, these hard problems are well known motivations to try to design new asymmetric cryptosystems with multivariate polynomials. In [5] a first design idea was studied by Harriet Fell and Whitfield Diffie but, as they pointed out, their design was not efficient because their function F and its inverse F^{-1} were multivariate polynomials with the same degree d. In [10] Tsutomu Matsumoto and Hideki Imai designed a very efficient scheme (called C^*) with a function F of total degree two, such that the degree of F^{-1} was much larger than two. However this scheme was broken in [11] due to unexpected algebraic properties. In [12] another scheme, called HFE, was designed to avoid these unexpected algebraic properties, but the secret key computations of HFE are not as efficient as in the original Matsumoto-Imai scheme C^*.

The aim of this paper is to introduce and to study "candidate" schemes, whose interest lies in their very simple design, and in the very good efficiency of the secret key computations. The main idea of those schemes is to use small S-boxes where some random multivariate functions of small degree are stored, and to combine such a function with some secret multivariate functions of small degree. The public key is sometimes large, but its length is still polynomial in the length of the messages (moreover, we can expect that secret key computations are performed on smartcards – because they are fast, require very little RAM – and public key computations are performed on personal computers).

An important part of the paper deals with the efficiency of very different cryptanalytic ideas on these schemes. Although the easier variations of our schemes can be broken by using those ideas, some of our schemes seem to resist cryptanalysis so far, and may therefore be interesting candidates for new and efficient asymmetric cryptosystems.

2 One round of S-boxes: Description of the schemes

Each message is represented by n elements of a *public* finite field $K = GF(q)$. We suppose that n is split into d integers $n_1, ..., n_d$, such that $n = n_1 + ... + n_d$. The secret items are:

1. Two affine bijections s and t from K^n to K^n.
2. The separation of n into d integers: $n = n_1 + ... + n_d$.
3. For each e, $1 \leq e \leq d$, a S-box S_e which maps K^{n_e} into K^{n_e}, and is given by quadratic polynomials of the n_e variables.

For a cleartext $x = (x_1, ..., x_n)$, the corresponding ciphertext $y = (y_1, ..., y_n)$ is given by:

$$y = t(S_1(a_1, ..., a_{n_1}), S_2(a_{n_1+1}, ..., a_{n_1+n_2}), ..., S_d(a_{n_1+...+n_{d-1}+1}, ..., a_{n_1+...+n_d})),$$

where $(a_1, ..., a_n) = s(x)$. Each S-box can be seen as a "branch" of the algorithm. The composition of all these operations is still a quadratic function of the components of x, so that it can be given by n polynomials P_1, ..., P_n of degree two in $x_1, ..., x_n$. These polynomials are made public, so that anyone can encipher a message (because from $P_1, ..., P_n$, anyone can obtain y from x). Moreover, if the secret items are known, a ciphertext y can be deciphered as follows:

1. From $y = (y_1, ..., y_n)$, we compute $b = (b_1, ..., b_n) = t^{-1}(y_1, ..., y_n)$.
2. Then we can find $(a_1, ..., a_n)$ by looking in a table in which all the values of the S-box functions have been precomputed and sorted in lexicographic order.
3. Finally, we compute $x = (x_1, ..., x_n) = s^{-1}(a_1, ..., a_n)$.

Note 1: The parameters must be chosen so that the tables for the S-boxes are not too large. For example, we can take $K = GF(2)$, $n = 64$ and $n_1 = ... = n_8 = 8$, so that each table contains $2^8 = 256$ values.

Note 2: The S-boxes are not necessarily one-to-one maps. So there may be several possible cleartexts for a given ciphertext. One solution to avoid this ambiguity is to put some redundancy in the representation of the messages, by making use of an error correcting code or a hash function (for details, see [12] p. 34, where a similar idea is used in a different scheme). For an analysis about the existence or non-existence of bijective S-boxes of small degree without triangular or algebraic construction, see the extended version of this paper.

Note 3: The scheme can also be used in signature. To sign a message M, the basic idea is to compute x from $y = h(R||M)$ (as if we were deciphering a message), where h is a hash function and R is a small pad. If we succeed, (x, R) will be the signature of M. If we do not succeed (because the function is not a bijection), we try another pad R (for variants and details, see [12], where a similar idea is used). It could also be noticed that, with the variations 1 and 2 described below (with some functions $q_1, ..., q_d$), we will have a larger probability to succeed (*i.e.* we will have to try fewer values R).

Note 4: We said that the S-box can be inverted by looking in a table. More precisely, we always have three different ways to invert a S-box:

1. Looking in a table.
2. Trying all the possible inputs ("exhaustive search").
3. Solving the polynomial equations of the S-box. Since the S-box uses only a very small number of variables, these equations can often be inverted by algebraic algorithms (which are not efficient with more variables).

Note 5: The C^* scheme of T. Matsumoto and H. Imai can be seen as a very special case of our algorithm above, when, in C^*, all the n_i variables are small (see [10] for the definition of C^* and of the n_i variables).

Variation 1 (General S-boxes scheme): This variation is shown in figure 1 below. The transformation $(a_1, ..., a_n) \mapsto (b_1, ..., b_n)$ is now defined as follows:

$$\begin{cases} (b_1, ..., b_{n_1}) = S_1(a_1, ..., a_{n_1}) \\ (b_{n_1+1}, ..., b_{n_1+n_2}) = S_2(a_{n_1+1}, ..., a_{n_1+n_2}) + (q_1(a_1, .., a_{n_1}), .., q_{n_2}(a_1, .., a_{n_1})) \\ \dots\dots\dots\dots \\ (b_{n_1+\dots+n_{d-1}+1}, ..., b_{n_1+\dots+n_d}) = S_d(a_{n_1+\dots n_{d-1}+1}, ..., a_{n_1+\dots+n_d}) \\ \quad +(q_{n_2+\dots+n_{d-1}+1}(a_1, ..., a_{n_1+\dots+n_{d-1}}), ..., q_{n_2+\dots+n_d}(a_1, ..., a_{n_1+\dots+n_{d-1}})). \end{cases}$$

where the functions S_i and q_i are all quadratic over K. The public key is still a quadratic function of the x_j variables, $1 \leq j \leq n$, and the decryption will be done first on S_1 (it gives $a_1, ..., a_{n-1}$), then on S_2, ..., and finally on S_d. As mentioned in note 3, this variation may have practical advantages in signature. However, the cryptanalysis of this generalisation will be exactly the same as if $q_2 = ... = q_d = 0$.

Variation 2 (Triangular system except one S-box at the beginning): This variation is shown in figure 2 below. The transformation $(a_1, ..., a_n) \mapsto (b_1, ..., b_n)$ is now defined as follows: $(b_1, ..., b_k) = S_1(a_1, ..., a_k)$ and $b_i = a_i \oplus q_{i-k}(a_1, ..., a_{i-1})$ for $k+1 \leq i \leq n$. Here, $S_1, q_1, ..., q_{n-k}$ are quadratic functions and k can be seen as the size of the S-box. In this variation, we will have "almost" a permutation, so that this may have some practical advantages.

As we pointed out above, transformations which are based on algebraic structures can often be used by the cryptanalyst to break the scheme. That is the reason we use S-boxes, which are supposed to be built with randomly chosen polynomials of small degree, and thus have no algebraic structure. It can also be noticed that the encryption function is a polynomial of total degree 2, but its inverse is polynomial with a high degree, and so cannot be found by Gaussian reductions. Despite these properties, we are to see that this one round scheme is insecure.

3 One round of S-boxes: cryptanalysis of the schemes

First attack: canonical cryptanalysis

We first present an attack based on the existence of a canonical representation of the quadratic polynomials involved in the scheme. Let us recall the classical theorem about the representation of quadratic forms over a field with even characteristic.

Theorem 3.1 *Let $f \in K[X_1, ..., X_n]$ be a nondegenerate quadratic form over $K = GF(q)$, where q is even. If n is odd, then f is equivalent to: $x_1x_2 + x_3x_4 + ... + x_{n-2}x_{n-1} + x_n^2$. If n is even, then f is either equivalent to: $x_1x_2 + x_3x_4 + ... + x_{n-1}x_n$ or to a quadratic form of the type: $x_1x_2 + x_3x_4 + ... + x_{n-1}x_n + x_{n-1}^2 + ax_n^2$, where $a \in K$.*

A proof of this theorem is given in [9]. Moreover, if f is given, then the transformation of f into its canonical form is very easy. The algorithm used in the proof can also be applied to a degenerate quadratic form, in the same way. For such a form f, we can define the "number of independent variables of f" as the smallest integer k such that f is equivalent to a quadratic form in k indeterminates. Following the algorithm given in the proof of the theorem, we can see that this number k is also very easy to compute.

Using these tools, we can now derive a method to "separate" the S-boxes in the scheme. We use the same notations as above.

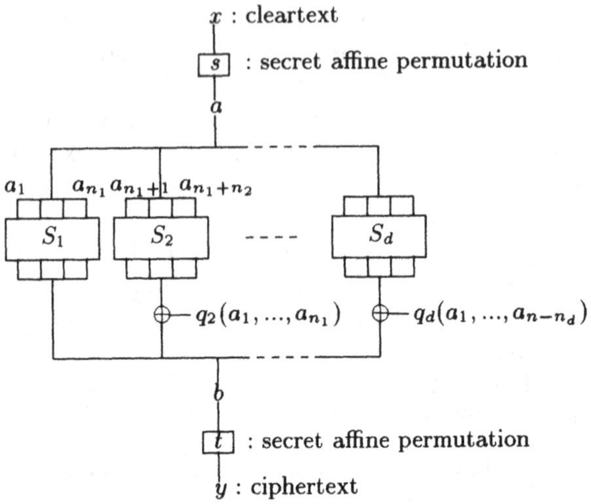

Fig. 1. One round of general S-boxes (it gives a weak scheme)

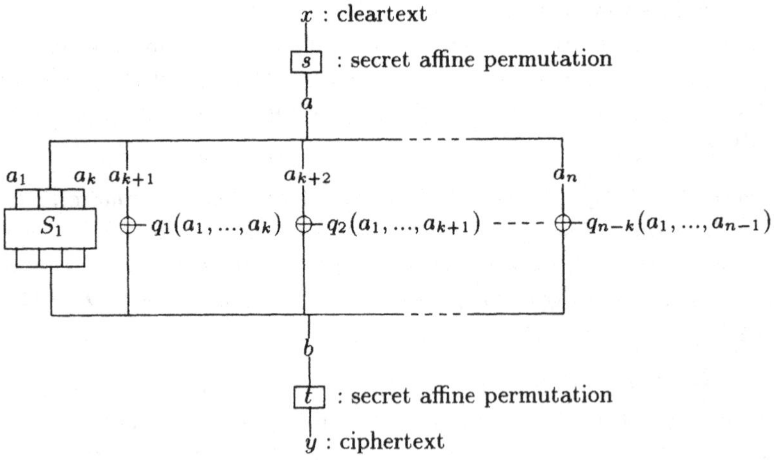

Fig. 2. One S-box followed by a triangular construction (it gives a weak scheme)

Description of the algorithm:

1. Generate randomly n bits $\alpha_1, ..., \alpha_n$, and compute $y' = \sum_{i=1}^{n} \alpha_i P_i(X_1, ..., X_n)$, which is a polynomial of degree two in $X_1, ..., X_n$.
2. Compute the number k of independent variables of y', and write y' as a polynomial of degree two in $x'_1, ..., x'_k$, where the x'_i have an affine expression in the x_j $(1 \leq j \leq n)$.

The probability that y' has no components in branch number one, *i.e.* the probability that y' has a linear expression in $b_{n_1+1}, ..., b_{n_1+...+n_d}$ is $\frac{1}{q^{n_1}}$.

So by repeating steps 1 and 2 about λq^{n_1} times, we generate about λ equations $y'_1, ..., y'_\lambda$ that have no components in branch number one. We can detect these λ equations with no components from the same branch because it is very easy to see if two polynomials y'_1 and y'_2 with $k \leq n - n_1$ have no components from the same branch: we write y'_1 in $x'_1, ..., x'_{k_1}$ and y'_2 in $x''_1, ..., x''_{k_2}$, and we see how many terms are linearly independent in $x'_1, ..., x'_{k_1}, x''_1, ..., x''_{k_2}$.

Note: Another idea would be to compute the number k of independent variables of $y'_1 + y'_2$ and test if $k \leq n - n_1$.

Finally, after $O(q^{n_1})$ computations, we will find $n - n_1$ independent equations $y'_1, ..., y'_{n-n_1}$ such that these equations have no components coming from the first S-box, and we will be able to write all these equations as polynomials of degree two in $x'_1, ..., x'_{n-n_1}$, where the x'_i have affine expressions in the x_j $(1 \leq j \leq n)$.

So with these new variables x'_i and y'_i, we have eliminated the variables of the first branch. Applying this algorithm d times, we can "separate" the S-boxes from each other and then the scheme is very easy to break by simple exhaustive search on the cleartext for each S-box.

Second attack: differential cryptanalysis

We will now describe another very different attack, that also works very well against schemes with small branches. Let $F = (P_1, ..., P_n)$ the function described in section 2. F is seen as a function from K^n to K^n. Let ψ be an element of K^n. Let $X_1, ..., X_k$ be k random plaintexts ($k = n$ typically), and $Y_1, ..., Y_k$ be the corresponding ciphertexts, *i.e.* $Y_i = F(X_i)$ for $i = 1, ..., k$. Let Y'_i be the ciphertext of the plaintext $X_i + \psi$, *i.e.* $Y'_i = F(X_i + \psi)$ for $i = 1, ..., k$. Finally, let $\psi'_i = Y'_i - Y_i$ (where the addition $+$ and the subtraction $-$ are of course done component by component in K).

Definition of test (A): We will say that "ψ satisfied the test (A)" if the vector space generated by all the values ψ'_i is not K^n itself by an affine subspace (the dimension of this subspace will be $n - \lambda$, where λ is one of the n_i values).

It is very easy to see if a value ψ satisfies this test (A) or not, since k is very small ($k = n$ typically). Moreover, we have this theorem:

Theorem 3.2 *The probability that ψ satisfies the test (A), when ψ is randomly chosen, is* $\geq \frac{1}{q^{max(n_i)}}$.

We can now describe our new attack. We proceed in five steps.

1. We find about n values ψ that satisfy the test (A), by randomly trying some ψ values.
2. We say that two such values ψ_1 and ψ_2 "belong to the same branch" if $\psi_1 + \psi_2$ (and more generally $\lambda_1\psi_1 + \lambda_2\psi_2$, $\lambda_1 \in K$, $\lambda_2 \in K$) also satisfies the test (A). We detect and group together the values ψ_i that belong to the same branches. In this way, we separate the ψ values in d different groups, that we call "branches". We will find about n_1 independent values ψ that come from the first branch, n_2 that come from the second branch, ..., n_d that come from the branch number d.
3. By an affine change of variables, we change the variables x_i into x'_i such that, with these new variables, the subspace generated by the values ψ of the first (resp. second,..., d^{th}) branch is characterized by: $x'_1 = \ldots = x'_{n_1} = 0$ (resp. $x'_{n_1+1} = \ldots = x'_{n_1+n_2} = 0, \ldots, x'_{n_1+\ldots+n_{d-1}+1} = \ldots = x'_{n_1+\ldots+n_d} = 0$).
4. Similarly, by an affine change of variables, we change the variables y_i into y'_i such that, with these new variables, the subspace generated by the values ψ' of the first (resp. second,..., d^{th}) branch is characterized by: $y'_1 = \ldots = y'_{n_1} = 0$ (resp. $y'_{n_1+1} = \ldots = y'_{n_1+n_2} = 0, \ldots, y'_{n_1+\ldots+n_{d-1}+1} = \ldots = y'_{n_1+\ldots+n_d} = 0$).
5. At this point, all the branches have been detected and isolated. To complete the attack, we can now perform an exhaustive search on each input of each of the d branches.

Note: In the second attack, we did not use the fact that the S-boxes are polynomial functions of degree two in a basis. So the attack will work in the same way whatever the functions in the S-boxes may be.

Third attack: gradient cryptanalysis

We now introduce a new tool to attack the scheme with S-boxes. The general principle of the algorithm is exactly the same as in the first attack that we called "canonical cryptanalysis". It is based on the fact that, with a rather high probability, a linear combination $Q = \sum_{i=1}^{n} \alpha_i P_i(x_1, \ldots, x_n)$ of the public polynomials has a "number of independent variables" less than n. In the case of quadratic polynomials, we saw that this "lack" of variables can be detected because a canonical form exists, which enables us to read the "real" number of variables of such a polynomial. But as soon as the degree is greater than three, we don't have such a canonical representation. The new idea is to use the so-called *gradient* of Q. For $x = (x_1, \ldots, x_n)$, it is defined by:

$$\mathbf{grad}\ Q(x) = \left(\frac{\partial Q}{\partial x_1}(x_1, \ldots, x_n), \ldots, \frac{\partial Q}{\partial x_n}(x_1, \ldots, x_n)\right).$$

The following theorem shows the link between the gradients of Q and the number of independent variables of Q:

Theorem 3.3 *Let Q be a quadratic form in n variables over a field K, then:*

1) If the characteristic of K is > 2, then the number k of independent variables of Q is equal to the dimension of the subspace A generated by all the **grad** $Q(x)$ $(x \in K^n)$.

2) If the characteristic of K is 2, then it is easy to find a non singular linear substitution of the indeterminates that transforms Q into a quadratic form:

$$R(x'_1, \ldots, x'_n) = \Phi(x'_1, \ldots, x'_{dimA}) + \sum_{i=dimA+1}^{n} \lambda_i x_i'^2,$$

where Φ is a quadratic form over K. Moreover the number k of independent variables of Q is:

$$k = \dim A + \begin{cases} 0 & \text{if } \forall i,\ \lambda_i = 0. \\ 1 & \text{if } \exists i,\ \lambda_i \neq 0. \end{cases}$$

The rest of the attack is similar to that described in the canonical cryptanalysis. The great difference here is that we can derive the same kind of theorem for cubic forms, and more generally for forms of *any* degree over K.

Fourth attack: linearity in some directions
A fourth attack is described in the extended version of this paper.

The special case C^*
All the general attacks given above can be used to improve the cryptanalysis of the C^* scheme. Moreover, we explain in the extended version of this paper that, in the case of C^*, all the different "branches" (*i.e.* all the values coming from different n_i values) can be separated, even when the n_i values are large.

4 Two round schemes

As we saw in section 3, the scheme which uses only one round of S-boxes is insecure, and we can use several methods to "separate" the branches from each other. Then a natural question arises: is it possible to design a more secure cryptosystem by using *two* rounds of transformations, each of which is given by polynomials of degree two ? This leads to the following problem:

Decomposition problem: Let g and h be two functions which map K^n into K^n and which are given by polynomials of total degree two in n variables over K. Then $f = g \circ h$ is also a function from K^n to K^n, and it is given by n polynomials of degree *four* in n variables over K. Suppose that f is given. Is it computationally feasible to recover g and h ?

A positive answer to that problem would imply that the two rounds can be easily separated from each other. So, to break the scheme, we would only have to break two independent schemes given by quadratic polynomials in n variables.

That would of course make the idea of using two rounds uninteresting. However, in [3] and [4], Matthew Dickerson suggests that the decomposition problem is probably NP-hard (see [3], p. 75). That is the reason why we propose to base a cryptosystem on two rounds of quadratic transformations. The general scheme will be the following one:

As in section 2, a finite field K, with $q = p^m$ elements, is public, and we want to transform a cleartext $x = (x_1, ..., x_n) \in K^n$ into a ciphertext $y = (y_1, ..., y_n) \in K^n$. The secret items will be:

1. Three affine bijections r, s and t from K^n to K^n.
2. An application $\varphi : K^n \to K^n$ given by n quadratic equations over K.
3. An application $\psi : K^n \to K^n$ given by n quadratic equations over K.

If we have the secrets, then we can obtain $y = (y_1, ..., y_n)$ from $x = (x_1, ..., x_n)$ by the following formula: $y = t\left(\psi\left(s\left(\varphi\left(r(x)\right)\right)\right)\right)$. The public items are:

1. The field K and the length n of the messages.
2. The n polynomials $P_1, ..., P_n$ of degree four in n variables over K, such that $y_i = P_i(x_1, ..., x_n)$ $(1 \le i \le n)$.

So anyone can encipher a message. Moreover, if the secret items are known, we must be able to decipher a message. For that purpose, we need to invert the functions φ and ψ. In practice, φ and ψ can be chosen among the following classes of functions:

1. "C^*-functions": monomials over an extension of degree n over K, such as $a \mapsto a^{1+q^\theta}$ (they are the basic transformations in the C^* scheme of [10]).
2. "Triangular-functions": $(a_1, ..., a_n) \mapsto (a_1, a_2 + q_1(a_1), a_3 + q_2(a_1, a_2), ..., a_n + q_{n-1}(a_1, ..., a_{n-1}))$, where each q_i is a quadratic polynomial in i variables.
3. "S-boxes-functions": they have already been defined in section 2, in the case of a one-round scheme.
4. "One S-box followed by a triangular construction": they have been defined in section 2, variation 2, and are represented in figure 2 (with a composition by s and t).
5. "General S-boxes functions": they have been defined in section 2, variation 1, and are represented in figure 1 (with a composition by s and t).
6. "D^* functions": the definition and analysis of these D^* functions will be the main thema of [14].

Note: Classes number 2, 3 and 4 are just particular cases of class number 5.

The first two have the advantage of being bijections (D^* functions will also be bijective). But, as we will see in sections 4 and 5, all the schemes where ψ is one of the functions of classes 1 or 2 are insecure, whatever we may choose for φ. On the contrary, if the second round function ψ is chosen in classes 3, 4, 5 or 6, and the first round function φ in classes 1, 2, 3, 4, 5 or 6, the corresponding schemes seem to resist cryptanalysis, so far. We call these schemes the "2R" schemes ("2R" stands for "two rounds").

5 Cryptanalysis of a second round with C^*

In this section, we study the scheme described in section 4, when the second round function ψ is a C^*-function. A typical example of this situation is the case where φ and ψ are both C^*-functions. As was pointed out in [13], this scheme is insecure. Moreover, the cryptanalysis of [13] can be applied even if φ is not a C^*-function. See the extended version for more details.

Note: When the first round is a "C^*-function" and the second round is a "S-boxes-function", the scheme may be secure, but it becomes insecure if the two quadratic functions are put the other way round, as we just said. This shows that the analysis of the security closely depends on the order of the quadratic functions used in the schemes.

6 Cryptanalysis of a second round with a triangular construction

Here, we consider the scheme of section 4, when the second round function ψ is a "triangular-function". Once more, that scheme is insecure. The key idea lies in the fact that it will be possible to detect the equations coming from a_1, then from a_2, *etc*. See the extended version of this paper for details.

7 Effect of the affine multiple attack

Another attack, which is very general, was described in [12]. It can be used against schemes based on a univariate polynomial transformation hidden by secret affine bijective transformations. This attack is based on the following fact: if f is a univariate polynomial over a finite field K, then by using a general algorithm (see for example [2]), we can compute an "affine multiple" of the polynomial $f(x) - y$, *i.e.* a polynomial $A(x, y) \in K[X, Y]$ such that:

1. Each solution of $f(x) = y$ is also a solution of $A(x, y) = 0$.
2. $A(x, y)$ is an affine function of x.

In the extended version of this paper, we show that – because of the affine multiple attack – some S-boxes give weak schemes. For example, when the characteristic p of the field is small, then we cannot choose $n_1 = \ldots = n_d = 1$ in our schemes, *i.e.* each S-box must have at least two elements of K as input and output. However, for well chosen S-boxes, the attack does not work. See the extended version for a detailed analysis of the smallest size of the S-boxes that may give strong schemes.

8 Comparison with symmetrical cryptosystems

The cryptosystems we study in this paper have a lot of similarities with classical symmetric cryptosystems (such as DES for instance), since they use for example S-boxes (*i.e.* local transformations of a small number of values), followed by linear transformations, and there are several rounds (two for our schemes). However, DES for instance (cf [1]), or Khufu (cf [7]) are not secure when only very few rounds are used. So, why do our schemes (with only two rounds) resist classical attacks ? This can be explained by the following arguments:

1. In each round that uses S-boxes, *all* the input bits are transformed, and not only half of them, as in a Feistel scheme, such as the one used in DES.
2. The affine transformations are *secret*, and that is not the case for the P transformation of DES, which is public.
3. Moreover, the affine transformations are very general, *i.e.* every output bit is a linear combination of *all* the input bits, and not only of one input bit, such as in the P transformation of DES.
4. In our schemes, the S-boxes can be secret or public. In DES, they are public, and in Khufu they are secret.

Note 1: R. Rivest recently proposed XDES, which is a composition of DES, an initial simple affine secret transformation, and a final one (more precisely, these transformations consist in XORing the input with a secret value). Maybe this change does not strengthen DES against differential or linear cryptanalysis, but it seems to prevent other attacks (in particular, against exhaustive search on the key, cf [8]). So, XDES may illustrate the idea that composing an encryption algorithm with initial and final affine secret transformations, may lead to a significant strengthening of this algorithm.

Note 2: In our schemes, it is very important to have only two rounds, because the composition of three rounds, each round being quadratic, would lead to polynomials of degree eight, so that the length of the public key would be much too large for practical applications.

9 Conclusion

In this paper, we have studied new asymmetric algorithms which all rely on the idea of using one or two rounds of quadratic transformations, which are hidden by secret affine tranformations. When there is only one round, or when the second round is built with functions whose algebraic structure is poorly hidden, we have proven that the corresponding schemes are insecure. From these ideas, we were able to design some new cryptanalysis of the Matsumoto and Imai scheme C^*.

However, when there are two rounds, we still have candidate algorithms that seem to resist classical attacks. Are these algorithms secure ? If they are, it would be a surprising and easy way of designing asymmetric cryptosystems. If they are not, it would strengthen the idea that the messages cannot be split in small branches, but must be transformed in a global way, and therefore that we need algebra to build secure asymmetric cryptosystems. The question remains open...

References

1. Eli Biham, Adi Shamir, *Differential Cryptanalysis of the full 16-Round DES*, CRYPTO'92, Springer-Verlag, pp. 487-496.
2. Ian Blake, XuHong Gao, Ronald Mullin, Scott Vanstone, Tomik Yaghoobian, *Applications of finite Fields*, Kluwer Academic Publishers, p. 25.
3. Matthew Dickerson, *The functional Decomposition of Polynomials*, Ph.D Thesis, TR 89-1023, Department of Computer Science, Cornell University, Ithaca, NY, July 1989.
4. Matthew Dickerson, *The Inverse of an Automorphism in polynomial Time*, IEEE 30th annual symposium on Foundations of Computer Science (FOCS), 1989, pp. 82-87.
5. Harriet Fell and Whitfield Diffie, *Analysis of a public Key Approach based on polynomial Substitutions*, CRYPTO'85, Springer-Verlag, pp. 340-349.
6. Michael Garey, David Johnson, *Computers and Intractability, a Guide to the Theory of NP-Completeness*, Freeman, p. 251.
7. Henri Gilbert, Pascal Chauvaud, *A chosen Plaintext Attack of the 16-Round Khufu Cryptosystem*, CRYPTO'94, Springer-Verlag, pp. 359-368.
8. Joe Kilian, Phillip Rogaway, *How to protect DES against eshaustive Key Search*, CRYPTO'96, Springer-Verlag, pp. 252-267.
9. Rudolf Lidl, Harald Niederreiter, *Finite Fields*, Encyclopedia of Mathematics and its applications, volume 20, Cambridge University Press, p. 287.
10. Tsutomu Matsumoto, Hideki Imai, *Public quadratic polynomial-Tuples for efficient Signature-Verification and Message-Encryption*, EUCROCRYPT'88, Springer-Verlag, pp. 419-453.
11. Jacques Patarin, *Cryptanalysis of the Matsumoto and Imai public Key Scheme of Eurocrypt'88*, CRYPTO'95, Springer-Verlag, pp. 248-261.
12. Jacques Patarin, *Hidden Fields Equations (HFE) and Isomorphisms of Polynomials (IP): two new Families of asymmetric Algorithms*, EUROCRYPT'96, Springer-Verlag, pp. 33-48.
13. Jacques Patarin, *Asymmetric Cryptography with a hidden Monomial*, CRYPTO'96, Springer-Verlag, pp. 45-60.
14. Jacques Patarin, Louis Goubin, *Trapdoor one-way permutations and multivariate polynomials*, ICICS'97 (this conference).
15. Joachim von zur Gathen, *Functional Decomposition of Polynomials: the tame Case*, J. Symbolic Computation (1990), vol. 9, pp. 281-299.
16. Joachim von zur Gathen, *Functional Decomposition of Polynomials: the wild Case*, J. Symbolic Computation (1990), vol. 10, pp. 437-452.

On the Powerline System

Paul Camion * and Hervé Chabanne

Abstract. We introduce an improvement of the powerline system [4] called **Fractional Powerline System** . Its transmission rate is larger than that of the powerline system for comparable parameters. As a consequence it can be considered as more secure. That new PKC is especially convenient for short messages. For instance, 136 bits messages can be transmitted by conveying 208 bits over the channel and this will resist all knowns attacks. A variant is suggested for which, at a cost of increasing the size of the public key, the average number of bit messages can be as high as 146 with a transmission rate of 81% which is to be compared with the 43% of the Powerline System with comparable parameters. A signature scheme based upon appending a small number of random bits to a message is made easier by the higher transmission rate.

1 Introduction

Initially, Chor and Rivest introduced their cryptosystem as a knapsack problem. More recently, Lenstra[4] described a modification of the Chor-Rivest system, which works directly in the multiplicative group of a finite field; he called it the Powerline System. The Powerline System uses monic polynomials in a finite field for encryption/decryption. The final step of the decryption is accomplished by a root finding process. Here we show how to improve the Powerline System by using rational functions rather than polynomials. Doing so we let, during decryption, the root finding process be preceded by the well-known Berlekamp-Massey algorithm (see for example [5]). *The present paper has no other aim than improving the Powerline system.* The reader is invited to peruse [4] for an easier understanding. The striking feature is that the sophisticated decoding algorithm for Reed-Solomon codes, codes used in Deep Space Appllications, is here part of the decryption algorithm. This is basically why our cryptosystem, called Fractional Powerline System (FPS), has several advantages over the powerline system:

- It possesses a better transmission rate.
- It allows to choose the system parameters more freely. In particular, we can choose any coefficient for the denominator of our rational function instead of reducing ourselves to work with monic polynomial and in this way we can still increase the transmission rate.
- It appears more secure against an adapted Brute-Force attack which cannot be outperformed by any other attack on the Chor-Rivest system [2, Section C].

* Centre National de la Recherche Scientifique

Comparison with the other PKC: We read in the introduction of [4]:*Among all PKC that depend on the knapsack problem, the system proposed by Chor and Rivest [2] is one of the few that have not been broken[1]. The Chor-Rivest system is based on arithmetic in finite field. It has the curious feature that its security does not depend on the apparent hardness of any well-known computational problem, such as the discrete logarithm problem.* This system would then outlive all PKC in actual use, if those problems would become tractable.

2 Substituting rational functions for polynomials

Let us simply describe a simplified version of the cryptosystem presented here. The way it can be extended to turn the Powerline System into a *modified Powerline System*, will be obvious. That system will be called a *Fractional Powerline System*, briefly FPS. The first modification of the system remains based on the irreducible polynomials $(Y-\beta), \beta \in \mathbf{F}_q$. These linear polynomials will be both considered as elements in $\mathbf{F}_{q^h} = \mathbf{F}_q[Y]/g\mathbf{F}_q[Y]$ where $g(Y)$ is an irreducible polynomial of $\mathbf{F}_q(Y)$, $Deg\ g = h$, or as ordinary polynomials in $\mathbf{F}_q[Y], q = p^n$, where p is any chosen prime number and where n is any chosen integer.

Let us denote by $\alpha_0, \alpha_1, \ldots, \alpha_{q-1}$ the elements of $\mathbf{F}_q$, where the indices $0, 1, \ldots, q-1$ result from a secret bijection π from $[0, q-1]$ onto $\mathbf{F}_q$.
A subset $S \subset \{0, 1, \ldots, q-1\}$ is fixed. His size s is a parameter of the system. The set S may be seen as an alphabet, the messages are to be written with its symbols according to certain grammatical rules. The degree h of $g(Y)$, which is chosen even, also is a parameter of the system. We will need the integers $d = \frac{h}{2}$.

1.*Private* The secret key k is an integer, $(k, q^h - 1) = 1$

2.*Public* The elements $v_i = (Y - \alpha_i)^k \in \mathbf{F}_{q^h}, i \in S$.

Moreover two secrets are introduced: $u \in \mathbf{F}_{q^h}$ and t such that $\mathbf{F}_{q^h} = \mathbf{F}_q(t)$. (see [4, Section 2])

Encryption

A message m is splitted into $m'|m''$ where m' consists of the first $\lfloor log_2(q) \rfloor$ bits of m. Then m' is encoded as an element γ of $\mathbf{F}_q$. Next m'' is encoded as a sequence of integers $m_0, m_1, \ldots, m_{s-1}$, where m_i is in the interval $[-d, d-1]$ and such that the sum of the positive ones equals $d-1$ and the sum of the other ones equals $-d$. Let us denote by S_{m+} (resp. S_{m-}) the subset of S such that $\forall i \in S_{m+}$, then $m_i > 0$ (resp. $\forall i \in S_{m-}$, then $m_i \leq 0$).

The encrypted message is then formed as follows.

$$c = \gamma^k v_0^{m_0} v_1^{m_1} \ldots v_{s-1}^{m_{s-1}}.$$

Decryption

The legitimate receiver knows $\ell = \frac{1}{k}$ and computes $c^\ell = \gamma P Q^{-1} \in \mathbf{F}_{q^h}$, where

$$P = \prod_{i \in S_{m+}} (Y - \alpha_i),\ Q = \prod_{i \in S_{m-}} (Y - \alpha_i)$$

are both in $\mathbf{F}_q[Y]$. We have that $Deg\ P = \frac{h}{2} - 1$ and $Deg\ Q = \frac{h}{2}$. The rational function $\gamma \frac{P}{Q}$ will then be recovered by applying the extended Euclid Algorithm

[5], or equivalently (but faster) the Berlekamp-Massey algorithm, to the polynomials $g(Y)$ and c^ℓ which is a polynomial of degree at most $h-1$.

3 The set of encipherable messages

3.1 Allowing diverse message lengths

Allowing diverse message lengths is imposed by not forcing the set of fractions to be the subset in which every numerator has exactly $d-1$ distinct variables, thus each one with degree one. The number of rational functions $\frac{P}{Q}$ where both P and Q split into linear factors from a set of s of such factors and $Deg\ P = d-1$, $Deg\ Q = d$, is proven to be

$$\sum_{i=1}^{d-1} \binom{s}{i} \binom{d-2}{i-1} \binom{s-i-1+d}{d}. \tag{1}$$

The number of binary digits to be transmitted is $h \log_2 q$. And we get a transmission rate of

$$\tau = \left(\log_2 \sum_{i=1}^{d-1} \binom{s}{i} \binom{d-2}{i-1} \binom{s-i-1+d}{d} \right) / (h \log_2 q).$$

3.2 Mapping the monomials $v_0^{i_0} v_1^{i_1} \dots v_{s-1}^{i_{s-1}}$ with total degree u onto the $(s-1)$-subsets of a $(s+u-1)$-set

After encoding binary codewords as t-subsets of an n-set, we will need encoding t-subsets as monomials of total degree h. This is done as follows. For an easier description of the required mapping it is convenient to think of an s-subset $\{j_1, j_2, \dots, j_s\}$ of the $(s+u)$-set $\{0, 1, \dots, s+u-1\}$ where j_s always equals $s+u-1$. Then the s-subset corresponding to the monomial $v_0^{i_0} v_1^{i_1} \dots v_{s-1}^{i_{s-1}}$ is $\{j_1 = i_0, j_2 = i_1 + j_1 + 1, j_3 = i_2 + j_2 + 1, .., j_{s-1} = i_{s-2} + j_{s-2} + 1, j_s = s+u-1\}$. Conversely we have that $i_0 = j_0, i_1 = j_2 - j_1 - 1, \dots, i_{s-1} = s+u-1-j_{s-1}-1$.

3.3 Encoding variable length messages

Let us first recall that there exists an algorithm to encode each binary codewords of length n as a t-subset of an m-set for $n = \lfloor log_2 \binom{m}{t} \rfloor$ [3]. The authors of [2] give a description of that algorithm.

Then a possible encoding algorithm would be the following.

- First encode the set, say N_1 with $|N_1| = 2^{n_1}$, of binary codewords as monomials with degree $d-1$ in s variables. This is done by an algorithm that achieves an injective mapping from the highest possible numbers of codewords, i.e. 2^{n_1}, into the set of $(s-1)$-subsets of a $(s+d-2)$-set. Notice that the number $|N_1|$ is $\sum_{i=1}^{d-1} |N_{1,i}|$ where $N_{1,i}$ is the set of codewords which are to be mapped onto monomials in which exactly i variables occur.

- Then identify the set F_i in which the constructed monomial lies by determining the number i of variables that actually appear in that monomial.
- Now encode the set, say $N_{2,i}$ with $|N_{2,i}| = 2^{n_{2,i}}$, of binary codewords, as monomials with degree d in $s-i$ variables. This is as above an injective mapping of $2^{n_{2,i}}$ binary codewords into the set of $(s-i-1)$-subsets of a $(s-i-1+d)$-set.

Conclusion The number of codewords to be possibly used by this algorithm is $\sum_{i=1}^{d-1} |N_{1,i}||N_{2,i}|$.

3.4 Requiring a fixed message length

For having only fixed length messages enciphered, we reduce the set of fractions to the subset in which every numerator has exactly $d-1$ distinct variables, thus each one with degree one. Then the number of possible numerators is $\binom{s}{d-1}$.

It follows that the number of denominators is $\binom{s-d+1+d-1}{s-d} = \binom{s}{d}$.

Thus the transmission rate becomes $\binom{s}{d-1}\binom{s}{d}/(h\ log_2\ q)$.

4 Toward better transmission rates

We can consider adding new irreducible polynomials to the linear ones. Here $M_{T_{2,n}}$ denotes the set of monic polynomials of $\mathbf{F}_q[Y]$ of degree n whose irreducible factors all have degree 1 or 2 and lie in T_2, where T_2 is a fixed chosen subset of all irreducible polynomials with degree at most 2. The value a_d is the number of rational fractions for which both numerator P and denominator Q factor in $M_{T_{2,n}}$ with $DegP = d-1$ and $DegQ = d$. It is an exercise of combinatorial analysis to obtain the value of a_d using generating functions. Since $a_d + log_2(q)$ is the number of encodable messages with that new system, where $h = 2d$, the transmission rate becomes $(a_d + log_2(q))/h\ log_2(q)$.

4.1 Fixed length messages

An easy way of devising a QFPS with fixed length messages consists in reserving the quadratic polynomial for the numerator and use all polynomials left for the denominator. Moreover we ask the numerator to be squarefree. We denote by w the number of quadratic polynomials in S and by $\ell - w < q$ the number of linear polynomials in S. Then the number of encoded message here is

$$\binom{w}{\frac{d-1}{2}} \sum_{i=0}^{n}(-1)^i \binom{w-\frac{d-1}{2}-1+i}{w-\frac{d-1}{2}} \binom{\ell-1+(n-i)}{\ell-1}.$$

A numerical example is dealt with in the table of Section 5 in the row FQFPS.

5 Numerical results

We summarize the numerical values for the Powerline System (PS), the Fractional Powerline system (FPS) and the Quadratic Fractional Powerline System (QFPS), respectively. We include a row FPSF which stands for FPS with a fixed message length (Section 3.4). Now ml stands for message length, this is reserved for the case where the message length can be fixed. Then aml for *average message length*, #bts for *# of bits transmitted.* Next ρ is for the transmission rate and pks for *public key size.* We recall that h is the degree of the irreducible polynomial of $\mathbf{F}_q[Y]$ which determines the field $\mathbf{F}_{q^h}$. Then for the PS we have that $d = h$ and for all fractional PS, then $d = \frac{h}{2}$. Next, s is the size or the chosen alphabet S and $\#bts = h\lceil log_2(q)\rceil$. For the following figures all systems considered equally resist the brute-force attack of [2, page 107].

Parameters	q	s	h	d	ml	aml	#bts	τ	pks
PS	256	200	26	26	112		208	54.13	41,600
FPS	256	200	26	13		137	208	65.86	41,600
PS	256	200	32	32	130		256	50.91	51,200
FPSF	256	200	26	13	136		208	65.38	41,600
QFPS	64	1900	26	13		129	156	82.69	296,400
QFPS	64	1900	30	15		146	180	81.11	342,000
FQFPS	64	1907	30	15	143		180	79.44	354,780
PS	64	60	30	30	78		180	43.65	10,800

6 Conclusion

Finally, we must admit that the strength of our system lies on the fact that it is as least as secure as the PS or the Chor-Rivest systems and a recent attempt of cryptanalysis [6] does not succeed for the chosen parameters. So, we invite the reader to help determine the true strength of the Fractional Powerline System.

References

1. E.F. Brickell, A.M. Odlyzko, Cryptanalysis: a survey of recent resiults,*Proc. IEEE* **76** (1988),578-593.
2. B.Chor, R.L. Rivest, A knapsack-type public key cryptosystem based on arithmetic in finite fields, *IEEE Trans. Inform. Theory* **34** (1988), 901-909.
3. T.M. Cover, Enumerative source encoding, *IEEE Trans. Inf. Theory, VolIT-19,p. 73 77,1973.*
4. H.W.Lenstra, Jr. On the Chor-Rivest Knapsack Cryptosystem, *J. Cryptology (1991) 3:149-155*
5. F.J. MacWilliams and N. J. A. Sloane, The theory of Error-correcting codes, *Amsterdam, North-Holland, 1977.*
6. C.P. Schnorr and H.H. Hörner, Attacking the Chor-Rivest Cryptosystem by Lattice Reduction *Advances in Cryptology-EUROCRYPT'95 LNCS NO 921 Springer*

Making Unfair a "Fair" Blind Signature Scheme

Jacques Traoré

France Télécom - Branche Développement
Centre National d'Etudes des Télécommunications
42, rue des Coutures, B.P.6243
14066 Caen Cedex, France
jacques.traore@cnet.francetelecom.fr

Abstract. At Crypto'92, Micali [6] introduced the concept of fair cryptosystems for preventing misuse of strong cryptosystems by criminals. At Eurocrypt'95, Stadler and al. [9] proposed the concept of fair Blind Signature Schemes (fair BSS in short) aimed to solve the same issue within the context of blind signature schemes.
In fair BSS (unlike "ordinary" blind signature schemes) the signer doesn't completely lose control of the signatures that he issues. By using fair blind signatures, for example within anonymous payment schemes (instead of ordinary blind signatures), one can prevent embezzlement by criminals such as money-laundering and perfect black-mailing [10].
In this paper, we show that one of the fair BSS proposed by Stadler and al. [9] is *not fair*.

1 Introduction

The concept of blind signature schemes was introduced by Chaum in 1982 [2]. A blind signature scheme is a cryptographic protocol involving two entities: a sender and a signer. This protocol allows the sender to choose a message and obtain from the signer a digital signature of this message in such a way that the signer learns nothing about the content of the message that he has signed. Moreover, the signer cannot link later on (i.e. after the signature has been revealed to the public) a given message-signature pair to the corresponding execution of the blind signature protocol.

Blind signature schemes can be used in applications where anonymity is required. In recently developed anonymous electronic payment systems [1,3], blind signatures are used to prevent linking the withdrawal of money and the payment made by the same customer. Unfortunately, the impossibility to link withdrawals with payments might lead to perfect crimes [10].

Criminals can also use the perfect anonymity provided by these systems to launder money. For these reasons, it has been argued that perfect anonymity is certainly not the suitable level of privacy that an electronic payment system must offer to its users. Future electronic cash systems should be provided with incomplete or conditional anonymity.

A step in this direction has recently been made by several researchers. In [9], Stadler and al. proposed a new type of blind signature scheme called fair Blind Signature Scheme (fair BSS) that can replace ordinary BSS within anonymous payment systems. Fair BSS have a specific property: (unlike 'ordinary' blind signature schemes) the signer in these schemes doesn't totally lose control of the signatures that he issues. In fair BSS, the signer can with the help of a single (or several) trusted authority (ies) (who is (are) given some extra information) either link a message-signature pair to the corresponding execution of the signing protocol or extract the content of a message from its *blind* form. By replacing *ordinary* blind signature schemes by fair blind signature schemes in electronic payment systems, the bank or anyone else would still be unable to link a withdrawal and a payment made by the same person. However, if abuse is suspected, the trusted authority could help the bank auditing a particular account or find out the author's identity of one particular transaction. In this way, privacy of honest users would be preserved and embezzlement by criminals prevented.

In [9] Stadler and al. presented three fair BSS. In the sequel, we first point out that in one of these schemes, a signature receiver can easily fool the trusted authority and the signer. More specifically, we show that a signature receiver can obtain valid message-signature pairs difficult to link to their corresponding execution of the signing protocol.

Next, we examine the security of these schemes within the context of electronic cash. Prepaid electronic cash systems are liable to support simultaneous withdrawals by different users. Such a feature implies that if we use a blind signature scheme to ensure anonymity, this scheme must remain secure even when simultaneous executions of the signing protocol by several signature receivers are allowed. We envisage this for the second scheme proposed by Stadler and al. [9] and will show that it is possible for two cooperating receivers to obtain a valid signature of any message, in such a way that the signed message remains unlinkable to both of their executions of the fair BSS. As a consequence, the perfect black-mailing [10] is still possible with this scheme.

Finally, we propose a heuristic method to defeat some of our attacks.

Organization of the paper: In Section 2, we first describe the second blind signature scheme proposed by Stadler and al.. In section 3, we explain why the trusted authority of this scheme may encounter some difficulties in retrieving the link between a given message-signature pair and the corresponding execution of the signing protocol. In Section 4, we show that when simultaneous execution of the signing protocol by different senders are allowed, the scheme described in section 2 does not provide fair blind signatures. Then, we show how to defeat this attack and conclude this paper.

2 Description of the Fair Blind Signature Scheme of Stadler and al.

The fair BSS that we are going to describe is based on a variation of the Fiat-Shamir signature scheme [5] and on the concept of one-out-of-two oblivious transfer [4]. In this scheme, the trusted authority can with the information collected by the signer during a given execution of the signing protocol enable the signer (or everybody) to efficiently recognize the corresponding message-signature pair.

2.1 The Variant of the Fiat-Shamir Signature Scheme

In contrast to the original Fiat-Shamir signature scheme, this scheme uses third roots instead of square roots. In the following, we will call this scheme: the *cubic Fiat-Shamir* signature scheme.

Let us describe this scheme.

Public key of the system:

n : a RSA integer (i.e. $n = p \cdot q$ is the product of two large primes such that 3 is relatively prime to $\varphi(n) = (p-1)(q-1)$).

y : a random value of Z_n^*.

H: a one-way hash function.

k : a security parameter (e.g. $k > 80$).

The sequence (y_i, $i = 1 \ldots k$): where $y_i = H(y+i) \bmod n$.

Secret keys of the signer:

The factorization of n.

The sequence (x_i, $i = 1 \ldots k$): where $x_i = y_i^{1/3} \bmod n$

To sign a message m the signer proceeds as follows:

- he randomly chooses $r \in Z_n^*$ and computes $t = r^3 \bmod n$.
- he computes $c = H(t \| m)$.
- he computes $s = r \cdot \prod_{i=1}^{k} x_i^{c_i} \bmod n$, where c_i denotes the i-th bit of c.

The signature of the message m is: the pair (s, t).

This signature can be verified by checking: $s^3 \stackrel{?}{=} t \prod_{i=1}^{k} y_i^{c_i} \bmod n$.

Note: the *challenge* c will be called in the sequel the "*challenge of the signature*".

2.2 Fair one-out-of-two Oblivious Transfer

As we have already said, the fair BSS of Stadler and al. is based on the concept of one-out-of-two oblivious transfer [4]. In this section, we recall this concept and explain the concept of fair one-out-of-two oblivious transfer introduced by Stadler and al.

A one-out-of-two oblivious transfer (OT_2^1 see [4]) is a protocol between a sender and a receiver which allows the receiver to choose one of two messages sent by the sender in such a way that he receives *only* the chosen message and the sender *does not know* which message he has chosen (in contrast to the original concept introduced in [4], they allow the receiver to choose the message).

Let m_0 and m_1 denote the two messages sent by the sender and let c be the bit selected by the receiver. Following Stadler and al., we schematized an execution of an OT_2^1 like this:

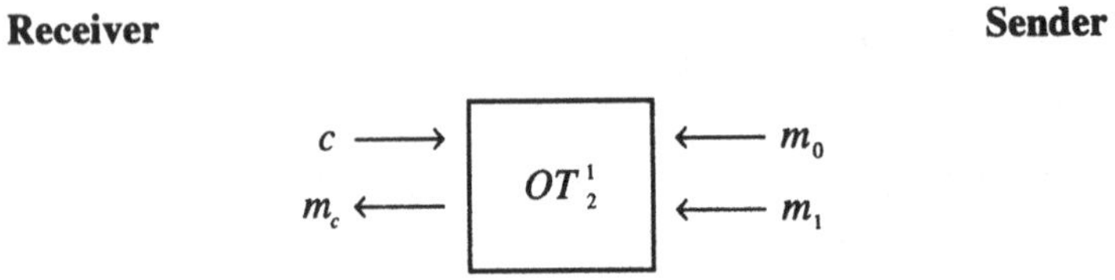

Definition: A fair- OT_2^1 is an OT_2^1 which allows a trusted authority, but not the sender, to determine the selection bit chosen by the receiver.
Following Stadler and al., we schematized an execution of a fair- OT_2^1 like this:

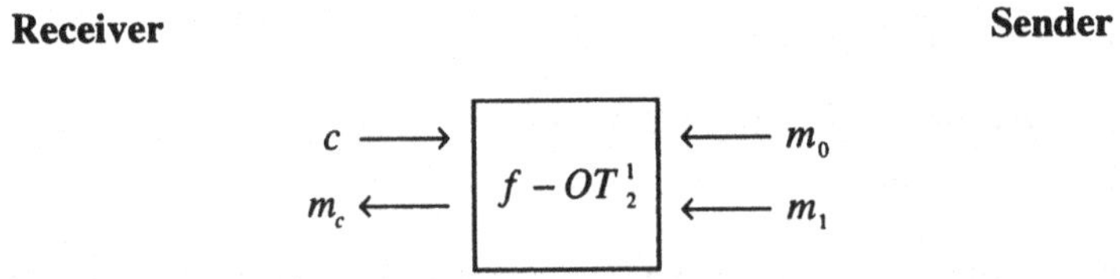

In their paper, Stadler and al. propound a specific fair- OT_2^1. However, our attack does not depend on this particular fair- OT_2^1 (but rather make use of some properties of the cubic Fiat-Shamir scheme [7]) and would work even if we replaced their fair- OT_2^1 by another one. So in the rest of the paper f- OT_2^1 will designate any fair- OT_2^1.

2.3 Fair Blind Fiat-Shamir Signature Scheme

In this section, we describe the signing protocol and analyze the blindness of this scheme.

Receiver | **Signer (Sender)**

Signer: choose $r_1,\dots,r_k \in Z_n^*$

$$t = \prod_{i=1}^{k} r_i^{3} \pmod{n}$$

Signer → Receiver: t

Receiver: randomly choose $\alpha \in Z_n^*$,

compute:

$\tilde{t} = t \cdot \alpha^3 \bmod n$

$c = H(\tilde{t} \,\|\, m)$

c_i is the i-th bit of c

for $i = 1 \dots k$ do

Receiver inputs c_i to $f - OT_2^1$; Signer inputs $m_0 = r_i$ and $m_1 = r_i \cdot x_i$; Receiver obtains $s_i = m_{c_i}$

end do

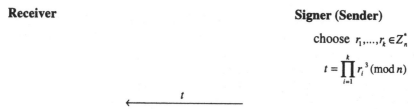

$$\tilde{s} = \alpha \cdot \prod_{i=1}^{k} s_i \bmod n$$

Fig.1. Fair blind Fiat-Shamir signature scheme

Then the pair $(\tilde{s}, \tilde{t})$ is a valid cubic-Fiat-Shamir signature of m:

$$\tilde{s}^3 = \tilde{t} \cdot \prod_{i=1}^{k} y_i^{c_i} \bmod n \quad (c_i \text{ is the } i\text{-th bit of } H(\tilde{t} \,\|\, m))$$

Due to the fairness property of $f - OT_2^1$, it is clear that this scheme is (from the signer's point of view) a perfect blind signature scheme.

Let us analyze the fairness of this scheme. It is argued in the paper of Stadler and al. [9] that " *if the signer sends the view of the protocol to the judge (trusted authority), the selection bits c_i can be determined (due to the fairness of $f - OT_2^1$) and therefore the challenge c is known. This value could then be put onto a black-list, so that everybody can recognize that message-signature pair later.*"

Unlike what is said above, we now stress two points in order to show that the string retrieved by the trusted authority is not necessarily equal to the *challenge of the signature* of the receiver's message-signature pair.

3 Weaknesses of the Fair Blind Signature Scheme

3.1 Preliminary Remarks

Notations

Let c be the *challenge of the signature* as before. Let c_J be the "*sent challenge*" (i.e., the concatenation of the bits that can be retrieved by the trusted authority - the judge - from a transcription of an execution of the signing protocol).
Let $\bar{c}$ be the bitwise negation of a bit string c.
We will call R a receiver who follows the protocol described above.
We will call $\tilde{R}$ a receiver who deviates from this protocol.
$\|$ denotes the concatenation of strings.
First remark (first flaw in the fair BSS):
In Figure 2, it is shown how a dishonest receiver $\tilde{R}$ can obtain a valid signature of a chosen message m in such a way that the *challenge of the signature* equals $\bar{c}_J$ (instead of c_J, when the signing protocol is executed by an honest receiver R).

$\tilde{R}$ | **Signer**

Signer: choose $r_1,\dots,r_k \in Z_n^*$

$$t = \prod_{i=1}^{k} r_i^{\,3} \pmod{n}$$

Signer → $\tilde{R}$: t

$\tilde{R}$: randomly choose $\alpha \in Z_n^*$,
compute:

$$\tilde{t} = t^2 \cdot \alpha^3 \cdot \prod_{i=1}^{k} y_i^2 \bmod n$$

$$c = H(\tilde{t} \,\|\, m)$$

c_i is the i-th bit of c

for $i = 1 \dots k$ do

$\bar{c}_i \longrightarrow$ [$f - OT_2^1$] $\longleftarrow m_0 = r_i$

$s_i = m_{\bar{c}_i} \longleftarrow$ [$f - OT_2^1$] $\longleftarrow m_1 = r_i \cdot x_i$

end do

$$\tilde{s} = \alpha \cdot \prod_{i=1}^{k} s_i^{\,2} \cdot \prod_{i=1}^{k} y_i^{c_i} \bmod n$$

Fig.2. First flaw in the fair blind Fiat-Shamir signature scheme

Proposition 1: The pair $(\tilde{s}, \tilde{t})$ obtained by $\tilde{R}$ is a valid cubic Fiat-Shamir signature of the message m (see Appendix A for the proof).
So the judge (at the request of the signer) will retrieve $c_J = \bar{c}_1 \bar{c}_2 \dots\dots \bar{c}_k$, however the *challenge of the signature* equals $c = c_1 c_2 \dots\dots c_k$.

Second remark (second flaw in the fair BSS):

In Figure 3, it is shown how a dishonest receiver $\tilde{R}$ can obtain a valid signature of a message m of his choice in such a way that the *challenge of the signature* c is different from c_J. More precisely, it is shown that $\tilde{R}$ can choose a value l so that the Hamming distance from c to c_J equals l.

In a preliminary step, $\tilde{R}$ chooses $I=\{i_1,i_2,\ldots\ldots,i_l\}$ a set of l integers ($I\subset\{1,\ldots\ldots,k\}$). In the sequel τ will denote a number linear in 2^l.

$\tilde{R}$		Signer
fraud : boolean; j : integer; *fraud*:=false; $j := 0$;		choose $r_1,\ldots,r_k \in Z_n^*$ $t=\prod_{i=1}^{k} r_i^3 \pmod n$
	$\longleftarrow$	

Repeat
$j := j+1$;
randomly choose $\alpha \in Z_n^*$,
compute:

$$\tilde{t}=t\cdot\alpha^3\cdot\prod_{j\in I} y_j \bmod n$$

$$c=H(\tilde{t}\,\|\,m)=c_1\,c_2\cdots\cdots c_k$$

if $c_i=0$ for all $i\in I$ *then*
fraud := true (*)
end if
until (*fraud* = true) or ($j>\tau$)
if fraud = false *then* $\tilde{R}$ executes the fair blind signing protocol as an honest receiver.
else
for i :=1 to *k do*
if $i\in I$ then put $\tilde{c}_i=1$
else $\tilde{c}_i=c_i$
end if-else

$\tilde{c}_i \longrightarrow$ [$f-OT_2^1$] $\longleftarrow m_0=r_i$

$s_i=m_{\tilde{c}_i} \longleftarrow$ [$f-OT_2^1$] $\longleftarrow m_1=r_i\cdot x_i$

end do

$$\tilde{s}=\alpha\cdot\prod_{i=1}^{k} s_i \bmod n$$

end if-else

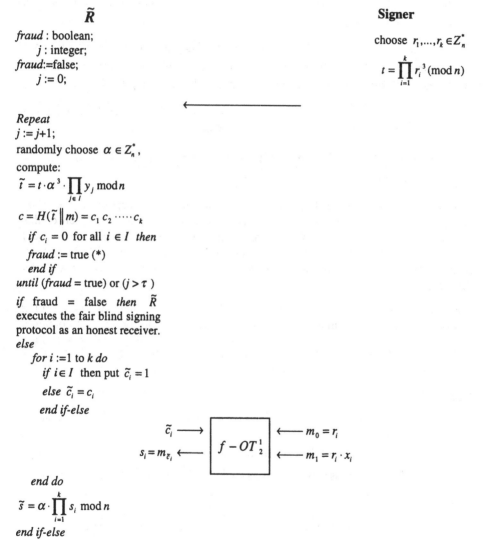

Fig.3. Second flaw in the fair blind Fiat-Shamir signature scheme

Note: the one-way hash function H is assumed to be a good pseudo random function. This implies that if l is not too large, the test (*) will succeed with high probability (*Stadler and Camenisch have recently informed us that they had discovered the second flaw earlier. It is described in Stadler's thesis [8]*).

Proposition 2: The pair $(\tilde{s}, \tilde{t})$ obtained by $\tilde{R}$ is a valid cubic Fiat-Shamir signature of the message m (see Appendix A for the proof).

So the judge (at the request of the signer) will retrieve $c_J = \tilde{c}_1 \tilde{c}_2 .....\tilde{c}_k$, however the *challenge of the signature* equals $c = c_1 c_2 .....c_k$. The Hamming distance from c to c_J equals l.

Following the techniques described above, we can also prove that a dishonest receiver $\tilde{R}$ can obtain a valid signature of a message of his choice in such a way that the challenge of the signature is "close" (in the sense of the Hamming distance) to $\bar{c}_J$.

These remarks lead us to the following conclusion: *the challenge of the signature is equal (or "close", in the sense of the Hamming distance) to either* c_J *or* $\bar{c}_J$.

4 Vulnerability of the Fair Blind Signature Scheme to a Real-Time Attack

In this section, we suppose that two signature receivers can execute simultaneously the fair blind signature protocol and can communicate during the execution of this protocol. This scenario is conceivable, particularly, if the fair blind signature protocol is used as a tool to protect anonymity in prepaid electronic payment systems. Indeed, prepaid electronic cash systems are liable to support simultaneous withdrawals by different users.

The goal of the attack that we are going to describe is to enable two cooperating receivers (who perform simultaneously the fair blind signature protocol) to obtain a valid signature of a message of their choice, making the trusted authority (judge) unable to link this message-signature pair to the corresponding executions of the fair blind signature protocol.

Notations:

- A and B will designate the signature receivers who execute the signing protocol simultaneously.
- c_i^A (resp. c_i^B) the i-th bit of the challenge sent by A (resp. B).
- s_i^A (resp. s_i^B) the response corresponding to the challenge c_i^A (resp. c_i^B).

Step 0

A or B gets the cubic Fiat-Shamir signature (s, t) of a message m (the content of m doesn't matter).

Note: it is possible to combine more than two signatures in order to enhance untraceability of the resulting "forged" signature.

We will put $t=\prod_{i=1}^{k} r_i^3 \bmod n$. c_i will designate the i-th bit of the challenge of the signature (s,t) ($s_i = m_{c_i}$ for all $i \in \{1,\ldots,k\}$).

$\tilde{m}$ will denote the message that will be blindly signed by the signer.

Let us now explain the attack.

Step 1

A executes the fair BSS and gets t_1.

A | **Signer**

choose $u_1,\ldots,u_k \in Z_n^*$

$$t_1 = \prod_{i=1}^{k} u_i^3 \pmod n$$

$\xleftarrow{\quad t_1 \quad}$

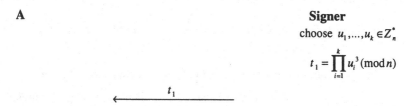

Fig.4. A's execution of the fair blind signature scheme

B executes the fair BSS and gets t_2.

B | **Signer**

choose $v_1,\ldots,v_k \in Z_n^*$

$$t_2 = \prod_{i=1}^{k} v_i^3 \pmod n$$

$\xleftarrow{\quad t_2 \quad}$

Fig.5. B's execution of the fair blind signature scheme

Step 2

A sends t_1 to B and B sends t_2 to A. Then, they agree on two values: $\alpha \in Z_n^*$ and β an integer ($\beta \le 3$). They compute $\tilde{t} = t_1 \cdot t_2 \cdot t^{\beta} \cdot \alpha^3 \bmod n$ and $\tilde{c} = H(\tilde{t} \,\|\, \tilde{m})$ ($\tilde{c}_i$ will denote the i-th bit of the string $\tilde{c}$). They choose c_i^A and c_i^B (for $i=1$ to k) in such a way that: $c_i^A + c_i^B + \beta \cdot c_i \equiv \tilde{c}_i \bmod 3$. Let us denote (for $i=1$ to k) $d_i = c_i^A + c_i^B + \beta \cdot c_i$ and q_i the integer verifying: $d_i = 3 \cdot q_i + \tilde{c}_i$.

Step 3

A and B end the fair signing protocol.

A | **Signer**

for $i = 1 \ldots k$ do

$c_i^A \longrightarrow$ [$f-OT_2^1$] $\longleftarrow m_0 = u_i$

$s_i^A = m_{c_i^A} \longleftarrow$ [$f-OT_2^1$] $\longleftarrow m_1 = u_i \cdot x_i$

end do

B **Signer**

for $i = 1 \ldots k$ do

$c_i^B \longrightarrow$ $\boxed{f - OT_2^1}$ $\longleftarrow m_0 = v_i$

$s_i^B = m_{c_i^B} \longleftarrow$ $\longleftarrow m_1 = v_i \cdot x_i$

end do

Then, A and B gather their information in order to compute $\tilde{s}$ a valid cubic Fiat-Shamir signature of the message $\tilde{m}$: $\tilde{s} = \alpha \cdot \prod_{i=1}^{k} s_i^A \cdot \prod_{i=1}^{k} s_i^B \cdot \prod_{i=1}^{k} (s_i)^{\beta} \cdot \prod_{i=1}^{k} (y_i^{-1})^{q_i} \bmod n$.

Proposition 3: The pair $(\tilde{s}, \tilde{t})$ is a valid cubic Fiat-Shamir signature of the message $\tilde{m}$ (see Appendix A for the proof).

4.1 How to Defeat the Previous Attack using a Heuristic Method

A straightforward way to prevent the attack described above is to forbid simultaneous executions of the signing protocol. But, a prepaid electronic payment system in which simultaneous withdrawals by different users are forbidden is unpractical.

In this section, we describe a way to thwart the attack described above. For this purpose, we modify the fair blind signature scheme in order to remove the possibility of forging a signature from other withdrawn ones.

We need extra parameters for our scheme:

. b is a prime number, such that n divides $b-1$. g is an element of Z_b^* of order n .

Let us describe the modification of the fair scheme of Stadler and al.

Receiver **Signer**

choose $r_1, \ldots, r_k \in Z_n^*$

$t = \prod_{i=1}^{k} r_i^3 \pmod{n}$

$\longleftarrow h = g^t \bmod b$

randomly choose $\alpha \in Z_n^*$,

$\tilde{h} = h^{\alpha^3} \bmod b$

$c = H(\tilde{h} \| m)$

(c_i is the i-th bit of c)

for $i = 1 \ldots k$ do

$c_i \longrightarrow$ $\boxed{f - OT_2^1}$ $\longleftarrow m_0 = r_i$

$s_i = m_{c_i} \longleftarrow$ $\longleftarrow m_1 = r_i \cdot x_i$

end do

$\tilde{s} = \alpha \cdot \prod_{i=1}^{k} s_i \bmod n$

Fig.6. Modification of the fair blind signature scheme

The signature $(\tilde{s},\tilde{h})$ of the message m can be verified by checking that: $g^{\tilde{s}^3} = \tilde{h}^{\prod_{i=1}^{k} y_i^{c_i}}$ mod b where c_i denotes the i-th bit of $H(\tilde{h} \| m)$.

Note: this scheme requires that $b = \alpha \cdot n + 1$, where α is an integer. One way to achieve this is first to generate n and then find by exhaustive search $b = \alpha \cdot n + 1$ as small as possible. In [11], it is argued that given a random n, b can be expected to be less than $n \cdot \log_2^2 n$.

The proposed modification seems to thwart the attack described in the previous section, however the security of this modification cannot be proved rigorously.

The second remark made in section 3 still applies to this new scheme.

5 Conclusion

We have shown several weaknesses of a fair blind signature scheme proposed by Stadler and al. [9]. We have further proposed a heuristic method to defeat some of our attacks.

Acknowledgments

We would like to thank M. Stadler, J. Camenisch, M. Girault, L. Vallée, J.F. Misarsky, A. de Solages and the anonymous referees for their useful comments.

References

[1] S. Brands, Untraceable Off-Line Cash in Wallets with Observers, *Proceedings of* CRYPTO'93, Lecture Notes in Computer Science, Vol. 773, Springer-Verlag, pp. 302-318.

[2] D. Chaum, Blind Signatures for Untraceable Payments, *Proceedings of* CRYPTO'82, Plenum Press, 1983, pp. 199-203.

[3] D. Chaum, A. Fiat and M. Naor, Untraceable Electronic Cash, *Proceedings of* CRYPTO'88, Lecture Notes in Computer Science, Vol. 403, Springer-Verlag, pp. 319-327.

[4] S. Even, O. Goldreich and A. Lempel, A Randomized Protocol for Signing Contracts, *Communications of* the ACM, 28, 1985, pp. 637-647.

[5] A. Fiat and A. Shamir, How to Prove Yourself: Practical Solutions to Identification and Signature Problems, *Proceedings of* CRYPTO'86, Lecture Notes in Computer Science, Vol. 263, Springer-Verlag, pp.186-194.

[6] S. Micali, Fair Cryptosystems, Technical Report MIT/LCS/TR-579.b, 1993.

[7] T. Okamoto and K. Ohta, Divertible Zero-Knowledge Interactive Proofs and Commutative Random Self-Reducibility, *Proceedings of* EUROCRYPT'89, Lecture Notes in Computer Science, Vol. 434, Springer-Verlag, pp. 134-149.

[8] M. Stadler, Cryptographic Protocols for Revocable Privacy, Ph.D. Thesis, ETH Zürich, 1996. Diss. ETH No. 11651.

[9] M. Stadler, J.M. Piveteau and J. Camenisch, Fair Blind Signatures, *Proceedings of* EUROCRYPT'95, Lecture Notes in Computer Science, Vol. 921, Springer-Verlag, pp. 209-219.

[10] S. von Solms and D. Naccache, On Blind Signatures and Perfect Crimes, Computer & Security, 11, 1992, pp. 581-583.

[11] S.S. Wagstaff Jr, Greatest of the Least Primes in Arithmetic Progression Having a Given Modulus, Mathematics of Computation, 33(147), pp. 1073-1080.

Appendix A : Proofs

Proof. (of Proposition 1)

$$\tilde{s} = \alpha \cdot \prod_{i=1}^{k} s_i^2 \cdot \prod_{i=1}^{k} y_i^{c_i} \bmod n = \alpha \cdot \prod_{i=1}^{k} (r_i^2 \cdot x_i^{2 \cdot \bar{c}_i}) \cdot \prod_{i=1}^{k} y_i^{c_i} \bmod n$$

$$\tilde{s}^3 = \alpha^3 \cdot \prod_{i=1}^{k} (r_i^6 \cdot y_i^{2 \cdot \bar{c}_i}) \cdot \prod_{i=1}^{k} y_i^{3 \cdot c_i} \bmod n = \alpha^3 \cdot t^2 \cdot \prod_{i=1}^{k} y_i^{2 \cdot \bar{c}_i} \cdot \prod_{i=1}^{k} y_i^{3 \cdot c_i} \bmod n$$

$$= \alpha^3 \cdot t^2 \cdot \prod_{i / c_i = 0} y_i^2 \cdot \prod_{i / c_i = 1} y_i^2 \cdot \prod_{i / c_i = 1} y_i \bmod n = \alpha^3 \cdot t^2 \cdot \prod_{i=1}^{k} y_i^2 \cdot \prod_{i=1}^{k} y_i^{c_i} \bmod n$$

$$= \tilde{t} \cdot \prod_{i=1}^{k} y_i^{c_i} \bmod n \text{ (where } H(\tilde{t} \,\|\, m) = c_1 c_2 \cdots\cdots c_k \text{)}$$

Thus, $\tilde{R}$ obtains a valid cubic Fiat-Shamir signature of the message m. ■

Proof. (of Proposition 2)

$$\tilde{s} = \alpha \cdot \prod_{i=1}^{k} s_i \bmod n = \alpha \cdot \prod_{i \in I} (r_i \cdot x_i) \cdot \prod_{i \notin I} (r_i \cdot x_i^{c_i}) \bmod n$$

$$\tilde{s}^3 = \alpha^3 \cdot \prod_{i \in I} (r_i^3 \cdot y_i) \cdot \prod_{i \notin I} (r_i^3 \cdot y_i^{c_i}) \bmod n = \alpha^3 \cdot \prod_{i \in I} r_i^3 \cdot \prod_{i \notin I} r_i^3 \cdot \prod_{i \in I} y_i \cdot \prod_{i \notin I} y_i^{c_i} \bmod n$$

$$= \alpha^3 \cdot \prod_{i=1}^{k} r_i^3 \cdot \prod_{i \in I} y_i \cdot \prod_{i \notin I} y_i^{c_i} \bmod n$$

$$= \alpha^3 \cdot t \cdot \prod_{i \in I} y_i \cdot \prod_{i=1}^{k} y_i^{c_i} \bmod n \text{ (if } i \in I,\ c_i = 0\text{)}$$

$$= \tilde{t} \cdot \prod_{i=1}^{k} y_i^{c_i} \bmod n \text{ (where } H(\tilde{t} \,\|\, m) = c_1 c_2 \cdots\cdots c_k \text{)}$$

Thus, $\tilde{R}$ obtains a valid cubic Fiat-Shamir signature of the message m. ■

Proof. (of Proposition 3)

$$\tilde{s} = \alpha \cdot \prod_{i=1}^{k} s_i^A \cdot \prod_{i=1}^{k} s_i^B \cdot \prod_{i=1}^{k} (s_i)^{\beta} \cdot \prod_{i=1}^{k} (y_i^{-1})^{q_i} \bmod n$$

$$= \alpha \cdot \prod_{i=1}^{k} (u_i \cdot x_i^{c_i^A}) \cdot \prod_{i=1}^{k} (v_i \cdot x_i^{c_i^B}) \cdot \prod_{i=1}^{k} (r_i^{\beta} \cdot x_i^{\beta \cdot c_i}) \cdot \prod_{i=1}^{k} (y_i^{-1})^{q_i} \bmod n$$

$$\tilde{s}^3 = \alpha^3 \cdot \prod_{i=1}^{k} u_i^3 \cdot \prod_{i=1}^{k} v_i^3 \cdot \prod_{i=1}^{k} r_i^{3\beta} \cdot \prod_{i=1}^{k} y_i^{c_i^A + c_i^B + \beta \cdot c_i} \cdot \prod_{i=1}^{k} (y_i^{-1})^{3 q_i} \bmod n$$

$$= \alpha^3 \cdot t_1 \cdot t_2 \cdot t^{\beta} \cdot \prod_{i=1}^{k} y_i^{d_i} \cdot \prod_{i=1}^{k} (y_i^{-1})^{3 q_i} \bmod n$$

$$= \alpha^3 \cdot t_1 \cdot t_2 \cdot t^{\beta} \cdot \prod_{i=1}^{k} y_i^{3 q_i + \tilde{c}_i} \cdot \prod_{i=1}^{k} (y_i^{-1})^{3 q_i} \bmod n$$

$$= \tilde{t} \cdot \prod_{i=1}^{k} y_i^{\tilde{c}_i} \bmod n \text{ (where } H(\tilde{t} \,\|\, \tilde{m}) = \tilde{c}_1 \tilde{c}_2 \cdots\cdots \tilde{c}_k \text{)}$$

Thus, A and B obtain a valid cubic Fiat-Shamir signature of the message $\tilde{m}$. ■

Enforcing Traceability in Software

Colin Boyd

Information Security Research Centre, School of Data Communications,
Queensland University of Technology, Brisbane 4001, Australia.
boyd@fit.qut.edu.au

Abstract. Traceability is a property of a communications protocol that ensures that the origin and/or destination of messages can be identified. The aims of this paper are twofold. Firstly the aims of traceable communications protocols are reviewed and compared with the available mechanisms to ensure compliance. These are compared with the methods used to ensure compliance in escrow schemes, the context in which traceability has usually arisen. Secondly a new communications architecture is proposed which provides traceability robustly, while preserving user control over other security services.

1 Introduction

There has been much recent interest in notions of *compliance* in cryptographic systems: how can users be provided with adequate security services in a controlled manner that does not allow them to abuse those services for unapproved purposes. The area in which compliance has been most studied recently is in escrow schemes [8]. Here the important issue is to allow users to employ cryptosystems for traditional confidentiality and integrity purposes, while ensuring that decryption is possible by authorised parties. More recently compliance has been an issue in electronic payment schemes, especially electronic cash [4,6]. In this context the compromise is between anonymity of user actions (a defining property of cash) and the need for financial and government bodies to regulate the flow and exchange of currency.

The Escrowed Encryption Standard [12], published by the US National Institute of Standards and Technology, defines the scheme underlying a family of devices including 'Clipper'. There are two, essentially independent, issues involved.

Escrow is the mechanism whereby a *copy* of an individual's secret key is stored in a specified manner.

Traceability is the mechanism that allows the owner of the scheme to *identify* the sender and/or recipient of a particular encrypted message.

In the Clipper device neither escrow nor traceability are implemented using cryptographic methods alone. Escrow is performed by a procedural (physical) method at the time of manufacture of the device; traceability works by an algorithm that relies on the tamper-proof property of the device (although it includes a cryptographic algorithm for integrity).

It is natural to consider the possibility of implementing both escrow and traceability in software. There are a number of potential advantages of this. One is that there is reduced trust required in the operators of the system and the integrity of tamper-proofing mechanisms. Another is that if algorithms and protocols are made public they are more likely to gain the confidence of the users of the system. Key escrow can be implemented in software using *verifiable secret sharing* which is a subject of continuing research [16].

The problem of enforcing traceability in software has received less attention but recently Desmedt [9] considered whether it is possible also to devise cryptosystems with traceability properties. Although the scheme of Desmedt works perfectly if the users act correctly, when users abuse the system it is possible for them to avoid traceability while still exploiting the system. Knudsen and Pedersen [17] have shown that in at least three different scenarios users are able to produce messages which may be sent to one user but are indistinguishable in their tracing properties from those sent to a different user.

The first purpose of this paper is to consider what it means to provide a scheme that ensures compliance, especially in the context of software implementation. This is complemented by a consideration of the different security mechanisms that have been used to ensure compliance. The second purpose is to propose a specific new scheme to implement traceability. In order to avoid the sorts of attack described by Knudsen and Pedersen a different architecture is required and so use if made of a third party through whom all messages must be routed. However, the basic properties of Desmedt's scheme are retained; in particular users are not required to reveal their secret keys to any party but traceability of messages is enforced in a robust fashion. Thus there is no assumption of a *trusted* third party in contrast to some escrow schemes. It should be noted that this paper is concerned only with traceability; escrow is not directly addressed at all.

2 Background

2.1 Desmedt's Scheme

The idea is to use the ElGamal encryption algorithm [11] (or any variation of it) where the prime modulus p is chosen so that $p-1$ is divisible by the product of several primes. Each user i is allocated a different base value g_i which generates a unique subgroup of $\mathbf{Z}_p^*$. The value g_i forms part of that user's public key, which is then certified.

Each user then chooses a secret and public key pair in the same manner as ElGamal: the secret key of user U_i is s_i, chosen randomly in the range $1 \leq s_i \leq p-1$, while the corresponding public key is $y_i = g_i^{s_i}$ (all arithmetic is done in the field $\mathbf{Z}_p$). The public key of user U_i is the pair (g_i, y_i) while all users share the modulus p. To encrypt a message for user U_i the sender:

- chooses a random number r in the range $1 \leq r \leq p-1$ such that $(r, p-1) = 1$;
- calculates $C = My_i^r$ and $R = g_i^r$;
- sends the pair (R, C) to U_i.

The recipient can decrypt in the usual manner for ElGamal by first calculating $(g_i^r)^{s_i}$, inverting this value and multiplying the result by C.

Although the C value in a ciphertext can be any value, the R component is restricted to the subgroup of Z_p^* generated by the g_i value of the recipient. Traceability is based on the idea that each user has a different subgroup in which the R values of ciphertexts sent to that user lie. Although these subgroups have a small overlap, for large values of p and q_i there is a negligible probability that a ciphertext of one user will lie in the group of another.

In order to trace the recipient of a particular message the authority need only find which subgroup the R value lies in. This may be efficiently achieved by the authority by calculating the order of R. A limitation of the method is that the authority needs to keep the factors of $p-1$ secret, and in order to avoid an attacker factorising $p-1$ these must all be large. As a consequence Desmedt suggests that p needs to have around 10000 bits if a large number of users is to be accommodated.

2.2 Attacks of Knudsen and Pedersen

Desmedt provides a convincing argument (although he does not claim a proof) that a malicious user who cannot solve the Diffie-Hellman or factorisation problems will be unable to send a message that can be decrypted by the intended recipient but which cannot be traced. This argument relies on the assumption that the recipient will decrypt the received message in the exact manner intended. Knudsen and Pedersen [17] have pointed out three attacks which, although they violate this assumption, rely on relatively simple measures. The basis of these attacks is reviewed here, together with a fourth attack.

Conspiring Receivers In this attack two users i and j must conspire prior to choosing their private keys. They choose the same value $s_i = s_j$ for their private keys. They broadcast this information and as a consequence any sender can use the 'wrong' key $(g_i g_j, y_i y_j)$ to send a message to either i or j but which will not be traceable to either. In this attack i and j do not need to know whether or not the sender has used the correct key or the wrong key to encrypt.

Juggling Components The second and third of Knudsen and Pedersen's attacks involve changing the purpose of each of the components sent. One of these requires the sender to construct two encrypted messages where only the second component of each is used by the recipient, while the second flips the two components so that the wrong one is used in tracing.

Random Multiplier A fourth attack, which was already hinted at in the details of the two attacks of Knudsen and Pedersen which juggle components, is to use a random multiplier to hide the first component. The sender (or the receiver) chooses a random value $m \in \mathbf{Z}_p$ and broadcasts it. The sender constructs (R, C) as normal but instead sends (mR, C). The recipient simply divides the first component by m before proceeding to the usual decryption algorithm. With high probability mR is not in the subgroup of the receiver.

Apart from the specific attacks mentioned, Knudsen and Pedersen also pointed out more general concerns which may be applicable to a wide range of similar protocols. In particular they point out that abusers of the system may simply use the public key of the receiver in a different algorithm altogether, thereby avoiding the possibility for tracing. They show, for example, that the public keys could be used in Diffie-Hellman key agreement.

3 Compliance Mechanisms

All schemes for key escrow and traceability need some methods to ensure that their users comply with the procedures for allowing encrypted communications to be traced and/or recovered. This is the case whether the scheme is intended for use as a backup method in a commercial environment or to enable monitoring by law enforcement or national security agencies[1]. Methods that have been suggested to ensure compliance may be divided into four classes as follows.

Tamperproof Hardware This method is the original one used by the Clipper proposal. Because the user is only able to use the scheme by employing the hardware implementation, use of the chosen mechanisms is unavoidable. The hardware is programmed to check on receipt of a message that the correct procedure has been followed. Although Blaze has shown [3] that the Clipper implementation can potentially be attacked, the basic principle of a using tamperproof hardware has not been undermined.
The major problem with relying on tamperproof hardware is that it seems to require also that the algorithms used must be kept secret. If this were not the case then users could engage software implementations modified in such a way as to avoid the tracing and escrow procedures. Or if not, then there is no necessity for the tamperproof hardware whose only purpose can be to hide certain features from the user. The arguments with respect to the undesirability, or otherwise, of unpublished cryptographic algorithms have been well rehearsed in recent years.

Tamperproof Software While the concept of tamperproof hardware is well established it is widely understood that compiled software can readily be reverse compiled to find the original algorithm. For this reason the idea of tamperproof software seems non-sensical. However there are situations where the effort to de-compile may not be worthwhile compared with the gains. A prime example would be commercial office software with key escrow mechanisms embedded. For almost all users the effort required to de-compile

[1] Traceability on its own has other applications such as resource monitoring which may be worth exploring. Such applications will also require compliance measures.

the software and re-compile it with alterations which sidestep the escrow mechanisms would be far beyond their capabilities or concerns. Methods to render software tamper resistant have been considered by Aucsmith [1].

Third Party Interactions A number of researchers have identified the possibility of using a third party service to interact with users and ensure that procedures have been properly followed. The first paper suggesting this seems to have been by Beth *et al.* [2] in which they propose a variation on Diffie-Hellman key exchange where the third party participates in choosing the session key. A number of potential weaknesses of the scheme were detailed by Horster, Michels and Petersen [14] and they proposed a revised scheme. A significantly different scheme, but still requiring the third-party to participate in the key exchange, is the Yaksha system of Ganesan [13]. A scheme of Jefferies, Mitchell and Walker [15] uses *trusted* third parties who choose the secret keys of the users.

Although the solution proposed in this paper uses third party interaction, there are a number of significant drawbacks of the above schemes that are avoided.

- The server must *interact* (that is both send and receive messages) with both parties before processing of any communication can begin.
- The server must hold security relevant information for each participant who may wish to communicate.
- Even in a store-and-forward application (such as electronic mail) the sender needs to find an on-line server before sending the message.

Certification Procedures Desmedt's scheme of 1995 does not fit into any of the above categories. Although its purpose is explicitly to allow software implementation it does not do this by assuming tamperproof software. Instead the aim is to devise a public key scheme such that a user's certified public key can only be used in a way that allows tracing of the identity of the recipient. Since a user must rely on public key certificates to make use of such a scheme it may be regarded that the certification procedure (including generation and allocation of public keys) is itself the method of ensuring traceability.

The difficulty of making such a scheme robust has already been alluded to above. The resourcefulness of attackers is impossible to anticipate and while a formal statement of security is lacking the precise achievement of Desmedt's scheme is impossible to assess.

The scheme to be described below compromises by making practical choices amongst the above schemes while avoiding their worst disadvantages. Thus software and hardware tamperproof assumptions are avoided. Desmedt's scheme is made robust by adding third party interaction, but in such a way that it is far less intrusive and much less difficult to implement than other proposals.

4 Security Services for Compliance

It is worthwhile to be as precise as possible about what security properties are achievable by any scheme. In general we would like to say that users of the scheme should not be able to avoid traceability. A difficulty here is what defines a *user* of the scheme. For example, attackers who are assumed to have authenticated channels between them can broadcast keys on those channels for a totally different public key cryptosystem and thus avoid the scheme completely. Such interaction would not be reasonably thought of as breaking the scheme since it is not the aim to stop users colluding to devise their own security mechanisms.

In a scheme using tamperproofing to ensure compliance it is obvious to define the scheme to be *in use* if the device is used without alteration and is relied upon for security. In a scheme that relies on certification procedures for compliance, *use* of the scheme must imply utilisation of the public key that is certified. There are only two ways that a public key can be used in cryptography: it may be used to *encrypt* messages for confidentiality, or it may be used to *verify* authentication/integrity of messages. Essentially we would like users to make use of the public key in either of these two ways only if their communications are traceable.

Definition 1. A public key scheme is said to be *in use* by an entity if that entity is using the public key either:

- to provide confidentiality of messages sent, or
- to verify authentication of messages received

and that confidentiality or authentication fails if and only if the private key corresponding to the public key used is compromised.

It can be seen that in all the attacks of Knudsen and Pedersen described in section 2.2 the scheme of Desmedt is in use by this definition, while traceability is avoided. For example, the conspiring receivers attack uses a legitimate public key to provide confidentiality of user data, as do the juggling component attacks.

5 A New Software Traceability Scheme

In this section a new scheme for software traceability is proposed which guarantees traceability when the scheme is in use. The basic idea is related to Desmedt's scheme in using certification procedures to ensure compliance but also incorporates third party involvement. However many of the drawbacks of other schemes which use third parties are avoided. In particular the following features are achieved.

- Users are unable to make use of the scheme while avoiding traceability. This includes the use of the public key infrastructure with other algorithms.
- The third party, or any other party, is at no time given the secret key of any user.
- Although the third party needs to *process* every message, this processing is *not interactive*. For example when *sending* an email message no third party involvement is required.

- Users cannot avoid tracing by hiding their communications.

The basic idea relies on splitting the private key of the user with a third party. This is an application of multiple key ciphers [5] and to this extent is similar to the Yaksha system [13]. Third party processing is involved which depends on the identity of the recipient. To this extent it is more complex than Desmedt's scheme. However, this makes attacks much more difficult and all the ones so far proposed are ruled out. A further advantage is that a (relatively) small prime modulus is now possible making implementation much more efficient. The basic scheme is appropriate for use with key escrow schemes based on interactive key sharing schemes.

5.1 System Parameters

The scheme ensures compliance through the use of certification procedures. The certification authority and the third party need not be located together, but they share common secrets and so can be considered as the same from a security viewpoint. The parameters include a universal large prime p and a generator g of $\mathbf{Z}_p^*$. Encryption is with a variation of the ElGamal algorithm. As usual the length of p must be sufficient to make the discrete logarithm problem in $\mathbf{Z}_p$ infeasible. In order to prevent cheating by the authority the prime p should be chosen so that $q = (p-1)/2$ is also prime. The reason for this will be seen below.

The general idea is simply to force senders to include the name of the receiver with the message. As opposed to the methods using hardware or software tamperproofing to achieve this, here the constraint is that the third party T will be unable to help in the decryption without it. To achieve this T must use a different secret value b_A with each user A. In order to ease the burden on the third party it is convenient to make b_A a function of the identity A; this means that T does not need to store secrets for each user. However, because b_A must be a secret value, the function used should also depend on a secret value. An obvious choice is to have

$$b_A = h(K_T, A)$$

where h is a keyed hash function or Message Authentication Code (MAC) [18] and K_T is a secret value known only by T. The function h should have the following properties.

Collision Resistance This prevents an deliberate choice of two identities A and B for which the values b_A and b_B are equal.

Resistant to Forgery Without knowledge of K_T it should be infeasible to find the value of b_A for any identity A.

To simplify the security arguments it is desirable to have the property $(b_A, p-1) = 1$. This can be ensured by suitable choice of the length of p and the output length of h and by appending a single 1 to the output of h to ensure that it is odd. (Note that because $(p-1)/2 = q$, if $b_A < p-1$ then the only possible values for $(b_A, p-1)$ are 1, 2, and q.)

5.2 User Registration

User registration proceeds as follows.

1. User A chooses a secret value s_A, with $(s_A, p-1) = 1$.
2. The authority calculates b_A and gives A the value $\alpha = g^{b_A}$ (all calculations take place in $\mathbf{Z}_p$).
3. A checks that $\alpha^2 \neq 1$ and $\alpha^q \neq 1$, and if so gives the authority g^{s_A}.
4. Instead of certifying this value as A's public key (as for normal ElGamal encryption), the key $P_A = g^{s_A b_A}$ becomes Alice's public key that is certified by the authority for use in the system
5. The authority returns P_A and the certificate to A.
6. A checks that $P_A = \alpha^{s_A}$ and if so accepts the public key.

It is important that in this procedure the user A is able to be sure that the authority acts correctly. In particular it should not be possible for A to accept a certificate for a public key which can be decrypted by the authority alone. In other words, if A acts correctly and accepts the public key P_A it should not be possible for the authority to obtain the effective private key, even if the authority has not followed the protocol by choosing α in some different, unknown, way. This will not be possible as long as the well-known Diffie-Hellman protocol [10] is secure in the sense that the shared secret can only be found with knowledge of at least one of the secret inputs.

Lemma 2. *Suppose that A executes the protocol faithfully and accepts P_A. Then if the Diffie-Hellman protocol is secure in $\mathbf{Z}_p$ the authority cannot find the effective private key which is the discrete logarithm of P_A.*

Proof Since A accepts the public key, it follows from step 6, that $P_A = \alpha^{s_A}$. Let us write $P_A = g^{y s_A}$, where *a priori* it may or may not be the case that $y = b_A$. From step 5 it follows that the authority is able to calculate P_A from knowledge of g^{s_A} and $\alpha = g^y$. Now P_A is exactly the Diffie-Hellman key corresponding to the inputs s_A and y. Therefore, if the Diffie-Hellman protocol is secure, the authority must know the value of y. The effective private key is then the value $s_A y \bmod p-1$. By step 3, $(y, p-1) = 1$ and so y has an inverse modulo $p-1$. Therefore knowledge of $s_A y \bmod p-1$ enables the authority to find s_A. Thus if the authority can find the effective private key it can also take discrete logarithms in $\mathbf{Z}_p$ (and therefore break the Diffie-Hellman protocol). □

5.3 Message Processing

When a user B wishes to send a message to A he proceeds as follows.

1. B obtains A's certificate and encrypts his message m using the public key and ElGamal encryption to form the message (R, C) where $R = g^r$ for a random exponent r and $C = P_A^r m$.

2. B sends the encrypted message to the third party T (possibly via A) along with the identity of A.
3. T calculates b_A and the value (R^{b_A}, C) and sends this to A.
4. When A receives (R^{b_A}, C) she recovers the message by calculating $C/R^{b_A s_A}$.

5.4 Security

Lemma 3. *If the scheme is used for encryption in the sense of definition 1, then messages are traceable.*

Proof Suppose to the contrary that a certain message is not traceable. Then either it is not sent via T, or it is sent via T with a different identity from the intended receiver. In either of these cases the correct private key will not be used. This contradicts the definition of use of the scheme since if the private key is not used the security service cannot rely upon it. □

In the new scheme all the attacks of Knudsen and Pedersen are avoided. The conspiring receivers attack fails because the conspirers will be unable to find a user identity A which will provide the appropriate b_A value. Juggling components and the random multiplier attack are also hopeless because again the attacker cannot know what is the right identity to send with the altered message.

It is worth noting that an 'attack' is possible in which user A broadcasts the value α used in the registration phase. If other users then encrypt using α as A's public key, A can decrypt messages which have not been processed by T. However, this means that the public key P_A is *not in use* in the sense defined above. Furthermore, a user who encrypts a message using α as public key has no idea who may be able to read the message unless α was sent to that user along a pre-existing authenticated channel from A. In other words, this attack only allows users to bypass traceability in the case that a secure channel already existed to A when, as discussed earlier, it is not possible to ensure compliance.

In addition to the above properties it might also be noted that there is a significant extra power in providing traceability. If the scheme is in use then communications cannot be hidden by steganography or any other hiding of the communications channel. This is a property not even provided by the tamper-proof hardware mechanism.

5.5 Preventing Unauthorised Tracing

Sending the identity in plaintext has the potential disadvantage that any eavesdropper is able to trace the recipient as well as T. It could be argued that this is not the concern of the current scheme. However, it is likely to be deemed unacceptable and luckily a simple additional mechanism may be used to encrypt information for the third party use. The solution is to have an ElGamal public key $P_T = g^{b_T}$ which can be used to encrypt the identity of the sender when it is sent to T with the message. The value of P_T must be publicly available for all users.

Instead of sending (R, C) together with the identity of A in plaintext, B must instead encrypt the identifier A using ElGamal encryption with P_T as public key. (In order to avoid leaking information it is worthwhile also to include some random padding in the plaintext to be encrypted.) On receipt of the message, T first extracts the identity of A and uses this to find b_A. As before, if the message is badly formed then the plaintext is not recoverable by A. The probabilistic nature of ElGamal encryption will prevent different messages to the same recipient being linked.

6 Discussion

6.1 Implementation Complexity

The most obvious disadvantage of the schemes presented here, in comparison with Desmedt's scheme, is that the third party must be available in order for decryption to take place. This is clearly a considerable overhead. It may be observed that several other schemes require such an overhead [2,13,14]. Furthermore, in distinction to all these schemes the sender does not need to interact with the third party, but only route the message via the third party. In a store-and-forward application in particular, this is much simpler since there is no real-time requirement for interaction. Note also that in an application using real-time communications the public key encryption would be used for exchange of a session key and subsequent communication will not have to be routed via T.

6.2 Scaling

A possible drawback of the scheme is the difficulty of scaling up to a widely distributed system. One way to implement the schemes would be to have trusted parties implemented widely, with each user being registered with a 'home' third party. By using a hierarchy of certificates, with each third party signature being certified globally, this is no impediment to global communication. Since most messages will be routed to the home location, there will be little extra communication costs, while computational requirements for traceability are distributed.

Another method to alleviate the problems of scaling is to have users obtain separate public keys and certificates from third parties that operate in each of the domains that the user frequents. For example, the certificate used could be based on the electronic mail domain of the user. The beauty of this arrangement is that the user can have different public keys and certificates while keeping her secret key always the same. This is because each third party generates a new b_A value using the domain secret T and can interact with the user to obtain the correct public key while the user's secret is the same.

Acknowledgements

I am very grateful to Ed Dawson of QUT, and to Wenbo Mao of Hewlett Packard, for many constructive critical comments.

References

1. D. Aucsmith, "Tamper Resistant Software - An Implementation", *Information Hiding*, pp.317-333, Springer-Verlag, 1996.
2. T. Beth, H.-J. Knobloch, M. Otten, G.J. Simmons, P. Wichmann, "Towards Acceptable Key Escrow Systems", *Proceedings of the 2nd ACM Conference on Communications and Computer Security*, 1994, pp.51-58.
3. M. Blaze, "Protocol Failure in the Escrowed Encryption Standard", *2nd ACM Conference on Computer and Communications Security*, 1994, pp.59-67.
4. E. Brickell, P. Gemmell, D. Kravitz, "Trustee-based Tracing Extensions to Anonymous Cash and the Making of Anonymous Change", *1995 SIAM-ACM Symposium on Discrete Algorithms*, pp.457-466.
5. C. Boyd, "Some Applications of Multiple Key Ciphers", *Advances in Cryptology - Eurocrypt 88*, Springer-Verlag 1988, pp.455-467.
6. J. Camenisch, J.-M. Piveteau, M. Stadler, "An Efficient Fair Payment System", *3rd ACM Conference on Computer and Communications Security*, ACM Press 1996, pp.88-94.
7. E. Dawson, J. Golić (Eds.), "Cryptography: Policy and Algorithms", Springer-Verlag, 1996.
8. D. Denning and D. Branstad, "A Taxonomy for Key Escrow Encryption Systems", *Communications of the ACM*, 39,3 March 1996, pp.34-40.
9. Y. Desmedt, "Securing Traceability of Ciphertexts – Towards a Secure Software Key Escrow System", *Advances in Cryptology - Eurocrypt '95*, Springer-Verlag 1995, pp.147-157.
10. W. Diffie and M. Hellman, "New Directions in Cryptography", *IEEE Transactions on Information Theory*, 22, 6, pp.644-654, 1976.
11. T. ElGamal, "A Public Key Cryptosystem and a Signature Scheme Based on Discrete Logarithms", *IEEE Transaction on Information Theory*, IT-31, 4, pp.469-472, 1985.
12. FIPS PUB 185, "Escrowed Encryption Standard", US Department of Commerce/National Institute of Standard and Technology, February 1994.
13. R. Ganesan, "The Yaksha System", *Communications of the ACM*, 39,3 March 1996, pp.55-60.
14. P. Horster, M. Michels, H. Petersen, "A New Key Escrow System with Active Investigator", Technical Report TR-95-4-F, University of Technology, Chemnitz-Zwickau, April 1995.
15. N. Jefferies, C. Mitchell, M. Walker, "A Proposed Architecture for Trusted Third Party Services", *Cryptography: Policy and Algorithms*, LNCS 1029, Springer-Verlag, 1996, pp.98-104.
16. J. Kilian, T. Leighton, "Fair Cryptosystems, Revisited" *Advances in Cryptology - Crypto 95*, Springer-Verlag, 1995, pp.208-221.
17. L. Knudsen and T. Pedersen, "On the Difficulty of Software Key Escrow", Advances in Cryptology - Eurocrypt 96, Springer-Verlag 1996, pp.237-244.
18. B. Preneel, P. C. van Oorschot, "MDx-MAC and Building Fast MACs from Hash Functions, *Advances in Cryptology - Crypto 95*, Springer-Verlag, 1995, pp.1-14.

Publicly Verifiable Partial Key Escrow

Wenbo Mao

Hewlett-Packard Laboratories
Bristol BS12 6QZ
United Kingdom
wm@hplb.hpl.hp.com

Abstract. A partial key escrow cryptosystem based on publicly verifiable encryption is proposed. Partial key escrow adds a great deal of difficulty to mass privacy intrusion interested by malicious authorities (e.g., a human rights abusive government). Public verifiability improves efficiency and guarantees correctness in the establishment of partially escrowed key.

1 Introduction

This paper proposes a publicly verifiable partial key escrow cryptosystem.

In partial key escrow, a portion of a private key with a specified length will not be in escrow and as a result key recovery requires a non-trivial effort of computation to determine this portion *after* co-operating shareholders decrypt the key recovery material. Partial key escrow will add a great deal of difficulty to mass privacy intrusion interested by malicious authorities while preserving the property of an ordinary escrowed cryptosystem for targeting individual criminals. Partial key escrow must consider resilience to a so-called early key recovery attack which determines the unescrowed portion of the private key from the key recovery material *before* decryption by shareholders. If such an attack is possible then off-line pre-computations will essentially nullify the effect of partial key escrow. The recent work of "Verifiable partial key escrow" by Bellare and Goldwasser [1] provides scenarios of mass recovery and early recovery attacks.

Previous key escrow schemes (partial or otherwise) based on verifiable secret sharing (VSS) (e.g., [1, 5, 6]) all necessarily require availability of shareholders staying on-line for verifying shares received from the user in the time of escrowed key establishment. The same requirement applies to the case of using non-interactive VSS (e.g., [7]) where the correctness of secret sharing is based on an assumption that a sufficient number of shareholders are on-line even they may honestly keep silence upon receipt of good shares. The necessity of employing many on-line shareholders forms a limitation in practicality (poor efficiency and low scalability) for key escrow applying VSS. In partial key escrow VSS will be prohibitively non-applicable because now proof of key length is needed which is through interactions between the user and each of the verifiers (shareholders) on every bit of the key. Even at such a high cost, key recoverability is only conditional, based on an assumption that a sufficient number of shareholders will have honestly followed the protocol (will not remain in silence upon encountering invalidity).

In this paper a publicly verifiable partial key escrow scheme is proposed. An important new feature in the proposed scheme is public verifiability on the correctness in escrowed key recoverability: let a key recovery material be ciphertext encrypted under someone's public key; anybody will be able to verify the correctness in recoverability by using the public key and the verification is not based on knowing the message encrypted. This is achieved via applying a publicly verifiable encryption technique of Stadler [8]. While Stadler proposed the original technique with demonstration on an improvement on the Pedersen's VSS technique [7] (he accordingly named his improvement *publicly verifiable secret sharing*, or PVSS), we observe a more economic use of the proof primitive. In our usage, the public key used for publicly verifiable encryption and verification will be that of a robust threshold ElGamal cryptosystem (such a cryptosystem is described in [3]) in which the matching private key is shared in threshold among a set of shareholders such that the private key has never been and will never be constructed even in the time of decryption. It is conceivable that a combination of the Stadler's proof primitive and the robust threshold ElGamal cryptosystem can achieve a simple publicly verifiable encryption scheme in which the validity of a message encrypted under a shared public key can be publicly verified by a single entity (in fact anybody), even though decryption must be carried out in threshold collectively by a set of shareholders. Our solution to the impracticality and the weak robustness of the previous partial key escrow based on VSS follows a combined use of the two techniques. Now no matter how many shareholders the system will use, a single (outsider) verifier suffices to carry out the verification job and the resulting key's recoverability can be guaranteed. The cost of the verification will be equivalent to that of applying a special case of VSS that only uses one single shareholder.

In the remainder of the paper, Section 2 introduces cryptographic primitives to be used; Section 3 presents a new publicly verifiable partial key escrow scheme; and finally Section 4 concludes the work.

2 Cryptographic Primitives

A public multiplicative group G with the following construction will be used. Let p, q be large primes such that $q|p-1$. Let $g \in Z_p^*$ be an element of order q. Then G is $\langle g \rangle$. Computing discrete logarithm to the base g is assumed to be difficult.

Assume that a set of shareholders have setup a robust threshold ElGamal cryptosystem where the shared public key is denoted by y. The cryptosystem is setup over the group G, i.e., the shared public key y has the form of $g^x \bmod p$ where the private key x is shared among the shareholders in threshold. The shared private key x has never been and will never be constructed. A knowledge-proof based technique will allow the shareholders to decrypt a message without constructing x. (Section 2.3 of [3] provides a succinct description in setting-up such a cryptosystem and in decryption.) A message m encrypted under the public key y in ElGamal [4] is written as a pair (in the rest of the paper apparently omitted modulo operations will be in mod p)

$$(A, B) = (g^k, y^k m)(\mathrm{mod} p). \tag{1}$$

The Stadler's publicly verifiable encryption [8] will also be introduced at a high-level where it suffices for us to describe its function as follows. Let $\mathcal{O}$ be a one-way function with pre-image uniquely defined in Z_p. (Stadler used a discrete logarithm in his realisation of $\mathcal{O}^{-1}$.) One can verify that the unique pre-image of $\mathcal{O}(m)$ is encrypted in the ElGamal ciphertext (A, B) in the form of equation (1).

Finally we describe a witness indistinguishable proof technique for proof of knowledge of the dis-log of a given value with the bit length of the dis-log shown. The basic technique used here is due to Cramer et al [3]. Let f be a fixed generator of G where nobody knows $\log_x(f)$ for almost all $x \in G$. This element can be chosen by a trusted centre who can prove that it is infeasible for itself to know $\log_x(f)$ (see e.g., [1]). The witness indistinguishable proof of knowledge given in Protocol P below allows one to prove either $\log_h(A) = \log_y(B/f^1)$ or $\log_h(A) = \log_y(B/f^0)$ without revealing which of two is the case. (In the following, symbol $x \in_R S$ means to pick x in S at random.)

Protocol P

Common input $[A,]\, B$, $[g,]\, y$, f; Prover's input $k \in_R Z_q$

What to prove? $k = \log_y(B)$ $[= \log_g(A)]$

or $k = \log_y(B/f)$ $[= \log_g(A)]$

Prover			Verifier
$B = y^k$	$B = y^k f$		
$w, r_1, c_1 \in_R Z_q$	$w, r_2, c_2 \in_R Z_q$		
$[a_1 = g^{r_1}/A^{c_1}]$	$[a_1 = g^w]$		
$b_1 = y^{r_1}/B^{c_1}$	$b_1 = y^w$		
$[a_2 = g^w]$	$[a_2 = g^{r_2}/A^{c_2}]$		
$b_2 = y^w$	$b_2 = y^{r_2}/(B/f)^{c_2}$	$\xrightarrow{[a_1,a_2,]\, b_1,b_2}$	
			$c \in_R Z_q$
		$\xleftarrow{c}$	
$c_2 = c - c_1$	$c_1 = c - c_2$		
$r_2 = w + kc_2$	$r_1 = w + kc_1$	$\xrightarrow{r_1,r_2,c_1,c_2}$	$c \stackrel{?}{=} c_1 + c_2 (\bmod q)$
			$[g^{r_1} \stackrel{?}{=} a_1 A^{c_1}]$ $y^{r_1} \stackrel{?}{=} b_1 B^{c_1}$
			$[g^{r_2} \stackrel{?}{=} a_2 A^{c_2}]$ $y^{r_2} \stackrel{?}{=} b_2 (B/f)^{c_2}$

We have specified Protocol P in two variations with respect to the optional inclusion/omission of the information in the square parentheses. When the protocol is run with the square parentheses unselected, part of the ElGamal ciphertext block A will not be available and thus the result will not allow decryption of the encrypted information in the block B. On the other hand when the protocol is run with the optional material selected, the ElGamal ciphertext pair (A, B) will be available to the verifier and therefore it is possible for the shareholders to perform decryption and reveal the one-bit information f^0 or f^1.

3 Publicly Verifiable Partial Key Escrow

We know that for f^X computing the dis-log X can use a square-root method (also known as the Shanks' baby-step-giant-step algorithm, see [2], Section 5.4.1) in $O(2^{|X|/2})$ time ($|X|$ denotes the bit length of X); for small X this is the most efficient method known. In our partial key escrow scheme, the user (Alice) will construct a public key, escrow a portion of the matching private key and leaving the rest unescrowed. She must prove the bit length of the unescrowed portion. Setting the bit length of the unescrowed portion of the private key to be $l = 80$ seems to be compatable to the current resource available. The length setting may grow in the future in anticipation of the growing resources of computation.

When Alice requests for an escrowed key establishment, she chooses private key $X_A \in_R Z_q$ for $X_A = X_0 + X_1$ where she sets $|X_0| = l$. Her public key will be $Y_A = f^{X_A} = f^{X_0} f^{X_1} = Y_0 Y_1$. Let X_0 and X_1 be presented in the following binary forms (where $m = \lfloor \log_2(X_1) \rfloor$):

$$X_0 = x_0^{(0)} 2^0 + x_1^{(0)} 2^1 + \cdots + x_{l-1}^{(0)} 2^{l-1}, \; x_i^{(0)} \in \{0, 1\}, \; i = 0, 1, \cdots, l-1. \quad (2)$$

$$X_1 = x_0^{(1)} 2^0 + x_1^{(1)} 2^1 + \cdots + x_m^{(1)} 2^m, \; x_i^{(1)} \in \{0, 1\}, \; i = 0, 1, \cdots, m. \quad (3)$$

To publicly verifiable prove $|X_0|$, Alice will construct (for $0 \leq i \leq l-1$) $A_{(0,i)} = g^{k_i^{(0)}}$ and $B_{(0,i)} = y^{k_i^{(0)}} f^{x_i^{(0)}}$ where $k_i^{(0)} \in_R Z_q$. She shall send the ordered numbers $B_{(0,i)}$ to Bob. For each of these numbers, they will run Protocol P to prove the one-bit length of X_0. Note that the protocol is run without selecting the optional information inside the square parentheses. Setting $K_0 = \sum_{i=0}^{l-1} (k_i^{(0)} 2^i) (\mathrm{mod} q)$, and noticing equation (2), we can write

$$B_0 = \prod_{i=0}^{l-1} B_{(0,i)}^{2^i} = \prod_{i=0}^{l-1} [y^{k_i^{(0)}} f^{x_i^{(0)}}]^{2^i} = y^{(\sum_{i=0}^{l-1} k_i^{(0)} 2^i)} f^{(\sum_{i=0}^{l-1} x_i^{(0)} 2^i)} = y^{K_0} Y_0. \quad (4)$$

Bob can form B_0 using the numbers $B_{(0,i)}$ that he possesses. Upon completion of the protocol run for each bit, Alice will send the following triple to Bob: $A_0 = g^{K_0}$, $B_0 = y^{K_0} Y_0$, $V = \mathcal{O}(Y_0)$. Now using the Stadler's publicly verifiable encryption technique Alice can prove to Bob the proper encryption of the partial public key Y_0 under the pair (A_0, B_0). The following statement will hold.

Proposition *K_0 and Y_0 form the only method known to Alice for the publicly verifiable encryption proof.*

Proof We have shown a method to construct K_0 and Y_0. Suppose that Alice also knows some $K' \neq K_0$ and $Y' = f^{X'} \neq Y_0$ (hence $X' \neq X_0$) such that $B_0 = y^{K'} Y'$ holds. Then $\log_y(f) = \frac{K' - K_0}{X_0 - X'} (\mathrm{mod} q)$ which contradicts the assumption that for almost all $x \in G$, nobody knows $\log_x(f)$. □

Analogous to the above construction, Alice can encrypt the partial private key X_1 under the public key y. The resulting ciphertext pairs will be (for $0 \leq i \leq m = \lfloor \log_2(X_1) \rfloor$): $A_{(1,i)} = g^{k_i^{(1)}}$, $B_{(1,i)} = y^{k_i^{(1)}} f^{x_i^{(1)}}$ where $k_i^{(1)} \in_R Z_q$. Alice shall send the ordered pairs $(A_{(1,i)}, B_{(1,i)})$ to Bob. For each pair, they shall run Protocol P to prove proper encryption of each bit of the private key X_1 under the public key y. Here the protocol is run with the optional information in the square parentheses selected. For

$$A_1 = \prod_{i=0}^{m} A_{(1,i)}^{2^i}, \quad B_1 = \prod_{i=0}^{m} B_{(1,i)}^{2^i},$$

we know by analoging to the equation (4) that:

$$A_1 = g^{K_1}, \quad B_1 = y^{K_1} Y_1 \quad \text{where } K_1 = \sum_{i=0}^{m} (k_i^{(1)} 2^i) (\bmod q).$$

To this end, Alice shall disclose $(K_0 + K_1)(\bmod q)$ to Bob. This enables him to verify $A_0 A_1 = g^{K_0+K_1}$ and reveal

$$Y_A = y^{-(K_0+K_1)} B_0 B_1.$$

Bob shall certify Y_A and archive (A_0, B_0) and $(A_{(1,i)}, B_{(1,i)})$ $(i = 0, 1, ..., m)$ for future key recovery use. Before co-operational shareholders decrypt these data, neither of the partial public keys Y_0 or Y_1 is available, and so prematured key recovery is infeasible.

4 Concluding Remarks

We have devised a publicly verifiable partial key escrow cryptosystem. Although the encryption and the bit-length proof are in a bit-by-bit fashion, the scheme still gains efficiency from communication between the user and a single verification entity. The efficiency gain overwhelms the cost of bit-wise proof to many verifiers in the case of partial key escrow using VSS. Moreover, now there is no need to trust this single verification entity because it is well known that using a cryptographically secure hash function as commonly trusted challenge the proof steps can be turned into non-interactive with the data used become essentially signatures and therefore can be verified by the public. As a partial key escrow cryptosystem, the scheme defeats mass key recovery and early key recovery attacks, and is robust in guaranteeing recoverability of the escrowed key.

References

1. M. Bellare and S. Goldwasser. Verifiable partial key escrow. In Proceedings of *4th ACM Conference on Computer and Communications Security.* pages 78–91. ACM Press. April 1997.
2. H. Cohen. *A Course in Computational Algebraic Number Theory.* Springer-Verlag Graduate Texts in Mathematics 138. 1993.
3. R. Cramer, R. Gennaro and B. Schoenmakers. A secure and optimally efficient multi-authority election scheme. In *Advances in Cryptology — Proceedings of EUROCRYPT'97 (LNCS 1233)*, pages 103–118. Springer-Verlag, 1997.
4. T. ElGamal. A public-key cryptosystem and a signature scheme based on discrete logarithms. *IEEE Transactions on Information Theory*, IT-31(4):469–472, July 1985.
5. S. Micali. Fair public key cryptosystems. In *Advances in Cryptology — Proceedings of CRYPTO'92 (LNCS 740)*, pages 113–138. Springer-Verlag, 1993.
6. T. Okamoto. Threshold key-recovery system for RSA. In Proceedings of *1997 Security Protocols Workshop.* Paris. April, 1997.
7. T. Pedersen. Non-interactive and information-theoretic secure verifiable secret sharing. In *Advances in Cryptology — Proceedings of CRYPTO'91 (LNCS 576)*, pages 129–120. Springer-Verlag, 1992.
8. M. Stadler. Publicly Verifiable Secret Sharing. In *Advances in Cryptology — Proceedings of EUROCRYPT'96 (LNCS 1070)*, pages 190–199. Springer-Verlag, 1996.

A Secure Code for Recipient Watermarking against Conspiracy Attacks by All Users

Hajime Watanabe and Tadao Kasami
{hajime,kasami}@is.aist-nara.ac.jp

Graduate School of Information Science,
Nara Institute of Science and Technology
Ikoma, Nara 630-01 Japan

Abstract. To protect copyrights of data from illegal copying, a scheme called recipient watermarking which can specify dishonest users who got the data legally and copied it illegally has been studied. A recipient watermarking consists of two processes, called coding process and embedding process. Even though there exists a secure embedding process on which it is impossible to erase the codeword by conspiracy of users, there are some cases that they can make the data by which the server cannot specify even one of them as a illegal user because of the defect of the coding process of the information about users.
In this paper, we propose a secure code for recipient watermarking against any conspiracy attacks including the attack by all users. The proposed code can specify two dishonest users who are in conspiracy and this code is more fair than such code proposed previously. We analyze the relation between probabilities to succeed attacks and conspiracy patterns, and show the security of our code.

1 Introduction

In network services, digital data flows from a server to a user. It is important to protect digital data against illegal copying. But it is not sufficient that cryptographic system is used when a server send digital data to a client since once a legal user gets digital data, no one can prohibit the user from illegal copying the data. In the following, we call such user "dishonest user." To solve this problem, several schemes to embed a watermark (called recipient watermarking[4], which can be used to specify the user who received the data from the data) in secret into the digital data are proposed. A recipient watermarking consists of two processes(Fig. 1). Most researches about watermarking so far are how to embed information into digital data. However, even though there exists a secure embedding process, it is not sufficient to protect illegal copying. When legal users conspire, the places embedded watermarks are easily specified and, in those cases, the watermarks are also easily broken. In those researches, this conspiracy attack is rarely considered.

In [2], on the assumption that there exists a secure embedding process, for n users, it is proposed a code (called logarithmic length c-secure code, simply, ll-c-secure code) which is secure against conspiracy attacks to the extent of c users. The word "secure" means that the coding process allows the server who embedded watermarks, in very high probability, to specify "one" dishonest user in the conspiracy group. A code (called n-secure (l,n)-code) for any conspiracy attacks (including the attack by all users) is also discussed. This code is used to construct ll-c-secure code. The length of codewords of n-secure (l,n)-code is $O(nd)$.

In this paper, we propose a code which is secure against any conspiracy attacks and can specify "two" dishonest users in the conspiracy group. The probability to succeed in a conspiracy attack for n is shown and it is shown that this code is secure. The maximum probabilities to succeed in conspiracy attacks for some n are calculated and the relation between the probabilities and conspiracy attacks are discussed. It is also shown that the length of our code is equal to n-secure (l,n)-code on order.

The quality must be the same between each data embedded codewords. In most embedding processes proposed so far, more embedding data causes less quality of the data. For this reason, it is desirable that the weights of each user's codeword are the same for such embedding process. n-secure (l,n)-code, described in [2], does not satisfy this condition. On the other hand, our code satisfies this condition.

In n-secure (l,n)-code, an innocent user whose codeword is lower weight is specified incorrectly in relatively higher probability than users whose codeword are higher weight. This problem is also solved in our code because all codewords have the same weight.

2 Preliminaries

2.1 Recipient Watermarking

In recipient watermarkings, two processes, coding process and embedding process like Fig. 1:

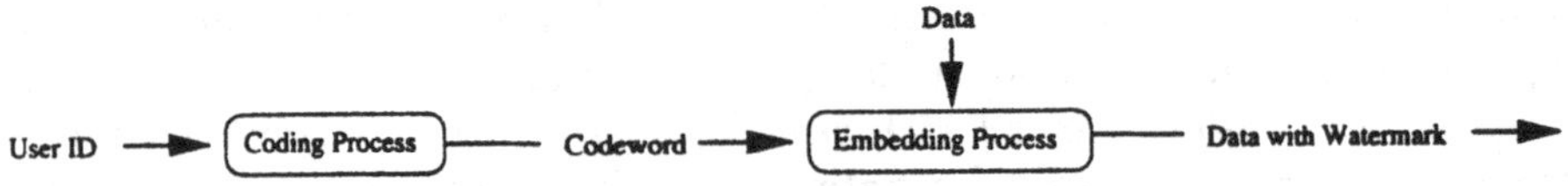

Fig. 1. Recipient Watermarking

1. a process to map a user's ID to a codeword (Coding process)
2. a process to embed a codeword into digital data (Embedding process)

Here, we focus our research interest on the coding process, and we propose a code which can specify (two) dishonest users in very high probability even though conspiracy attacks have been done.

In this paper, we consider the following sort of embedding process:

- The codewords embedded are presented in binary expression.
- Places where the server decide to embed a codeword are selected at random.
- Changing the value of a place where the server decide to embed a bit of a codeword indicates 1.
- Not changing the value of a place where the server decide to embed a bit of a codeword indicates 0.

We assume that there exists a "secure" embedding process, where "secure" means that the process has the following properties:

Property 1. Even when dishonest users have one value (which indicates 0 or 1) for a data place corresponding to a embedded bit, they cannot change the value to the value from which the server would fail to decide the bit value.

This means that, for a data place, they cannot create the value which indicates 0(or 1), and neither 0 and 1 from the value 1(or 0).

Property 2. Even when dishonest users have two value (each of which indicates 0 and 1) for a data place corresponding to a embedded bit, they cannot change the value to the value from which the server would fail to decide the bit value, that is, neither 0 nor 1.

In [2], an embedding process which has above properties for digital movie data is described briefly. In the following, we introduce this process.

2.2 Secure Embedding Process (Example)

This process includes the way of creating digital movie data. Digital movie data is created in the following algorithm. Let i-th frame of a movie data M be f_i^M.

Process for creating digital movie data

1. Create two movie data, A and B, using two cameras which are in slightly different view points and each of data are obtained from one camera.
2. By using A and B, create two data D and D' for watermarking: Choose f_i^A or f_i^B at random for i-th frame of D, and let i-th frame of D' be the frame not chosen. For example, $D = \{f_1^A, f_2^A, f_3^B, f_4^A, \cdots\}$ and $D' = \{f_1^B, f_2^B, f_3^A, f_4^B, \cdots\}$.
3. the watermarking server keeps two digital movie data, D is as original data for watermarking and D' is as embedding data which will be used in watermarking.

Let the codeword be w, j-th bit of w be w_j and the length of all codewords be l. For those movie data, the movie data embedded a codeword w is created in the following way:

Process for embedding a codeword

1. For a movie data D, the server chooses l frames, $\{f_{c_j}^D | 1 \le j \le l\}$, to embed l bits to the movie data.

2. For each frame $f_{c_j}^D$, do the following:

$$\text{If } w_j \text{ is} \begin{cases} \text{"1": change the frame to } c_j\text{-th frame of } D', \\ \text{"0": do nothing.} \end{cases}$$

For c_j-th frame of a movie data, if dishonest users only have one value, $f_{c_j}^A$ or $f_{c_j}^B$, they cannot create, of course, the other frame and they cannot create the frame from which the server would not decide the bit without very large loss of quality. Consequently, it is said that this embedding process satisfies Property 1.

For c_j-th frame of a movie data, even if dishonest users have both value, $f_{c_j}^A$ and $f_{c_j}^B$, they cannot create the frame from which the server would not decide the bit value, neither 0 nor 1. Consequently, it is said that this embedding process satisfies Property 2.

The code which we will propose in the next section does not depend on the sort of data, like digital movie. If a secure embedding process exists for a sort of data, our code can be used for the sort.

3 A Secure Code against Any Conspiracy Attacks

In [2], the n-secure (l, n)-code $(d = 1)$ is presented as Table 1. This code can

Table 1. n-secure (l, n)-code $(d = 1)$

user ID	codeword($n-1$ bits)								
	1	2	3	$\cdots$	$j-1$	j	$\cdots$	$n-2$	$n-1$
u_1	1	1	1	$\cdots$	1	1	$\cdots$	1	1
u_2	0	1	1	$\cdots$	1	1	$\cdots$	1	1
$\vdots$						$\vdots$			
u_j	0	0	0	$\cdots$	0	1	$\cdots$	1	1
$\vdots$						$\vdots$			
u_n	0	0	0	$\cdots$	0	0	$\cdots$	0	0

specify, in very high probability, one of dishonest users who are in conspiracy.

3.1 Structure of Proposed Secure Code

The code which we propose in this paper is constructed by using the same idea twice. Our code is presented as Table 2. The codeword w_{u_j} of our code is the concatenation of two subcodewords $w_{u_j}^R$ and $w_{u_j}^L$. The server only knows the

Table 2. Proposed secure code

user ID	subcodeword $w^L_{u_j}$ ($n-1$ bits)								
	1	2	3	$\cdots$	$j-1$	j	$\cdots$	$n-2$	$n-1$
u_1	0	0	0	$\cdots$	0	0	$\cdots$	0	0
u_2	1	0	0	$\cdots$	0	0	$\cdots$	0	0
$\vdots$						$\vdots$			
u_j	1	1	1	$\cdots$	1	0	$\cdots$	0	0
$\vdots$						$\vdots$			
u_n	1	1	1	$\cdots$	1	1	$\cdots$	1	1

user ID	subcodeword $w^R_{u_j}$ ($n-1$ bits)								
	1	2	3	$\cdots$	$j-1$	j	$\cdots$	$n-2$	$n-1$
u_1	1	1	1	$\cdots$	1	1	$\cdots$	1	1
u_2	0	1	1	$\cdots$	1	1	$\cdots$	1	1
$\vdots$						$\vdots$			
u_j	0	0	0	$\cdots$	0	1	$\cdots$	1	1
$\vdots$						$\vdots$			
u_n	0	0	0	$\cdots$	0	0	$\cdots$	0	0

correspondence between user IDs and users. For example, the codeword of a user whose ID is u_j, is

$$w_{u_j} = \overbrace{\underbrace{111\cdots1}_{j-1}\underbrace{000\cdots0}_{n-j}}^{w^L_{u_j}}\overbrace{\underbrace{000\cdots0}_{j-1}\underbrace{111\cdots1}_{n-j}}^{w^R_{u_j}}.$$

Note that our code does not have the parameter d. As described in the next section, this code can specify, in very high probability, two dishonest users since the idea of n-secure (l,n)-code are used twice. The weight of all codewords are $n-1$. The length of this code is $2(n-1)$.

3.2 Algorithm for Specifying Dishonest Users

In our recipient watermarking, the codeword of a user is embedded to a data for k times. When $k = d/2$ where d is that of n-secure (l,n)-code, the amount of embedded data is equal to that of n-secure (l,n)-code.

Since we assume that a secure embedding process like that described in the previous section is used in watermarking, places which are selected by the server to embed a codeword are not sensible by one dishonest user. The only attack for one dishonest user is to change the data but it is impossible to succeed because the embedding process has Property 1. Therefore, all codes using a secure embedding process are secure for all attacks by one user.

By the assumption about embedding process, the only possible conspiracy attack is to create a data from which the server would fail to specify dishonest users in the following way:

Attack

1. Compare each data and specify places whose values are different among users in conspiracy.
2. For each place, choose one of those values for that place.

Our specifying algorithm is slightly different from that of n-secure (l, n)-code because, in n-secure(l, n)-code, it is considered that one very large codeword is embedded to a data, and, in our code, it is considered that one (relatively) short codeword is embedded for k times. The specifying algorithm of our code which is very simple is given below.

Algorithm

1. Get k words, $w(i)(1 \leq i \leq k)$ from places embedded codewords of illegal copy data.
2. For each $w(i)$, do the following:
   (a) For the left subcodeword of $w(i)$, $w^L(i)$, check, from 1st bit to $(n-1)$-th bit, whether the embedded bit is 1 or not. If the p-th bit is the first place which indicates 0, let the dishonest candidate for $w^L(i)$, $U^L(i)$, be u_p.
   (b) For the right subcodeword of $w(i)$, $w^L(i)$, check, from $(n-1)$-th bit to 1st bit, whether the embedded bit is 1 or not. If the $(q-1)$-th bit is the first place which indicates 0, let the dishonest candidate for $w^L(i)$, $U^R(i)$, be u_q.
3. Specify u_a and u_b as dishonest users, where
   $u_a \in \{U^L(i) | 1 \leq i \leq k\}$ and u_a has the smallest index in $\{U^L(i) | 1 \leq i \leq k\}$,
   $u_b \in \{U^R(i) | 1 \leq i \leq k\}$ and u_b has the largest index in $\{U^R(i) | 1 \leq i \leq k\}$.

4 Security

Let the number of users in conspiracy be c and let the set of those users be $C = \{u_{i_j} | 1 \leq j \leq c, i_1 < i_2 < \cdots < i_c\}$. The codewords of u_{i_j} are presented as Table 3. On the assumption about the embedding process, by the attack described in the previous section, conspiracy users can create the words like: for left subcodewords of k words, $w^L(p)(1 \leq p \leq k)$,

bit position	1	$\cdots$	$i_1 - 1$	i_1	$\cdots$	$i_c - 1$	i_c	$\cdots$	$n-1$
possible words	1	$\cdots$	1	B	$\cdots$	B	0	$\cdots$	0

for right subcodewords of k words, $w^R(p)(1 \leq p \leq k)$,

bit position	1	$\cdots$	$i_1 - 1$	i_1	$\cdots$	$i_c - 1$	i_c	$\cdots$	$n-1$
possible words	0	$\cdots$	0	B	$\cdots$	B	1	$\cdots$	1

where B means that both 0 and 1 are possible.

From the assumption about embedding process and our specifying algorithm, the aim of dishonest c users in conspiracy is to create the word from which the server specify a innocent user $u_x(u_x \notin C, i_1 \leq i_j < x < i_{j+1} \leq i_c)$ in possible words for $w^L(p)$. The words by which they succeed in this conspiracy attack are the form

bit position	1	$\cdots$	$i_j - 1$	i_j	$i_j + 1$	$\cdots$	$i_{j+1} - 1$	i_{j+1}	$\cdots$	i_{c-1}	i_c	$\cdots$	$n-1$
success words	1	$\cdots$	1	1	B^*	$\cdots$	B^*	B	$\cdots$	B	0	$\cdots$	0

Table 3. Codewords of c users in conspiracy

user ID	subcodeword $w^L_{u_j}$ ($n-1$ bits)											
	1	$\cdots$	i_1-1	i_1	$\cdots$	i_j-1	i_j	$\cdots$	i_c-1	i_c	$\cdots$	$n-1$
u_{i_1}	1	$\cdots$	1	0	$\cdots$	0	0	$\cdots$	0	0	$\cdots$	0
$\vdots$							$\vdots$					
u_{i_j}	1	$\cdots$	1	1	$\cdots$	1	0	$\cdots$	0	0	$\cdots$	0
$\vdots$							$\vdots$					
u_{i_c}	1	$\cdots$	1	1	$\cdots$	1	1	$\cdots$	1	0	$\cdots$	0

user ID	subcodeword $w^R_{u_j}$ ($n-1$ bits)											
	1	$\cdots$	i_1-1	i_1	$\cdots$	i_j-1	i_j	$\cdots$	i_c-1	i_c	$\cdots$	$n-1$
u_{i_1}	0	$\cdots$	0	1	$\cdots$	1	1	$\cdots$	1	1	$\cdots$	1
$\vdots$							$\vdots$					
u_{i_j}	0	$\cdots$	0	0	$\cdots$	0	1	$\cdots$	1	1	$\cdots$	1
$\vdots$							$\vdots$					
u_{i_c}	0	$\cdots$	0	0	$\cdots$	0	0	$\cdots$	0	1	$\cdots$	1

where B means that both 0 and 1 are possible, and B^* means that both 0 and 1 are possible but at least one bit position of B^*s must be 0.

Consider the case that, for each bit position B, 0 or 1 are selected in the same probability, 1/2. This consideration is reasonable because this is fair for all users in C. The probability to create such word is

$$P_{i_j} = \left(\frac{1}{2}\right)^{i_j-(i_1-1)} \left(1 - \left(\frac{1}{2}\right)^{(i_{j+1}-1)-i_j}\right)$$

The probability to succeed in the conspiracy attack for $w^L(p)$ by users in C is

$$P(C) = \sum_{j=1}^{c-1} P_{i_j}.$$

We calculate maximum probabilities of $P(C)$ for some n and the results are shown in Fig. 2. This probability is almost equal to 1/2 for large n. This is the obvious upper bound since at least i_1-th bit must be set to 1. All maximum values are obtained from conspiracy attacks of two users, u_1 and u_n. The relationship between the probabilities $P(C)$ and C for $n = 5$ and 10 is shown in Fig. 3, where labels of horizontal axis express conspiracy patterns. In the binary expression of a label, i-th bit indicates whether or not u_i is in the conspiracy group C. For example, 19 means the conspiracy of users, u_1, u_2 and u_{16}.

To create the words from which the server would fail to specify the user u_{i_1}, users in C must succeed in this attack for all $w^L(p)(1 \le p \le k)$. Hence, its

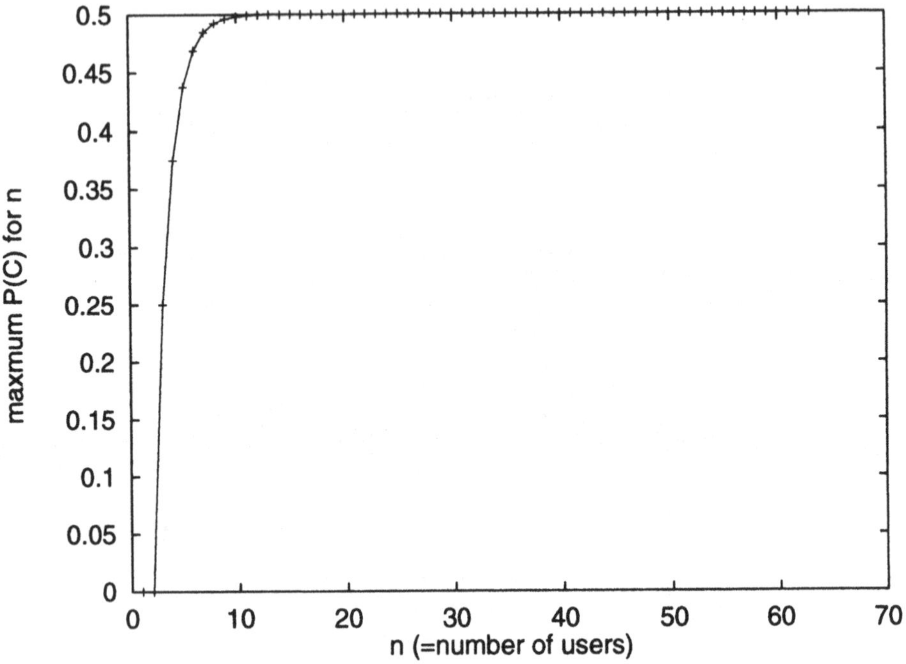

Fig. 2. Maximum probabilities of $P(C)$ for c users conspiracy attacks

probability is

$$(P(C))^k = \left(\sum_{j=1}^{c-1} P_{i_j} \right)^k .$$

For $w^R(p)(1 \le p \le k)$, the same result can be shown. Consequently, each dishonest user, u_{i_1} and u_{i_c}, is specified in

$$1 - (P(C))^k = 1 - \left(\sum_{j=1}^{c-1} P_{i_j} \right)^k .$$

If k is sufficiently large, for example, if the amount of embedded data is equal to that of n-secure (l, n)-code, that is, $k = d/2 = n^2 log(2n/\epsilon)$, this probability must be almost 0. From this result, we can conclude that our code is secure.

Our specifying algorithm is more simple than that of n-secure (l, n)-code because, on deciding a bit of codeword of our code, it is sufficient to check whether or not one "1" exists in k words, as we showed above. This check corresponds to the check on $weight(x|B_{s-1})$ of Algorithm 1 in section 4[2]. In step 3 of Algorithm 1, "for all $s = 2$ to $n - 1$ do;" must be "from $s = n - 1$ to 2 do;." If not so, this algorithm may return multiple innocent users except for one.

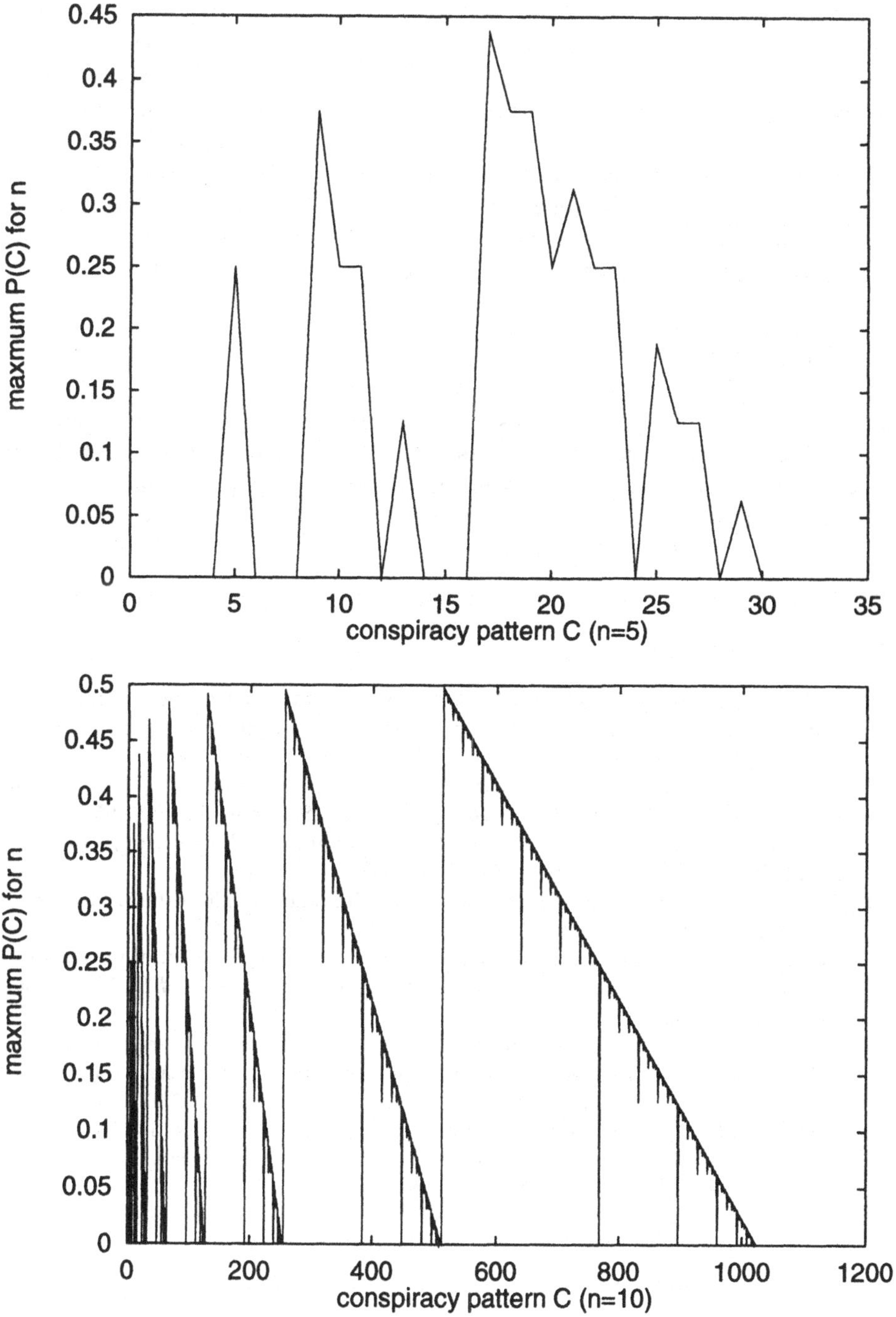

Fig. 3. probabilities $P(C)$ for C

For n-secure (l,n)-code whose specifying algorithm is described above, let the number of users in conspiracy be c and let the set of those users be $C = \{u_{i_j} | 1 \le j \le c, i_1 < i_2 < \cdots < i_c\}$ and let the correspondence between user ID and its codeword be given as Table 1. The probability that an innocent user whose index is larger and a little smaller than i_c is specified incorrectly is higher than users whose index is smaller than this user since the specifying algorithm tries to output U_{i_c} in high probability. On the other hand, this problem is solved in our code because our specifying algorithm tries to output U_{i_1} and U_{i_c}. This means that our code is more fair for users than n-secure (l,n)-code.

5 Conclusions

In this paper, we have proposed a secure code from which the server can specify two dishonest users even when all users conspire. We have shown the probability to succeed in a conspiracy attack and shown the security of our code.

Furthermore, this code also can be used to construct ll-c-secure code in the same manner of n-secure (l,n)-code.

References

1. Blakley, G.R., Meadows, C., Purdy, G.B.: Fingerprinting long forgiving messages, Advances in Cryptology- *CRYPTO '85*, LNCS 218, pp.180–189, Springer-Verlag, 1986.
2. Boneh, D., Shaw,J.: Collusion-secure fingerprinting for digital data, Advances in Cryptology- *CRYPTO '95*, LNCS 963, pp.452–465, Springer-Verlag, 1995.
3. Macq, B.M., Quisquater, J.-J.: Cryptology for digital TV broadcasting, *Proc. of the IEEE*, Vol.85, No.6, pp.944–957, June 1995.
4. Zhao, J.: A WWW service to embed and prove digital copyright watermarks, *Proc. of ECMAST '96*, pp.695–709, 1996.
5. Delaigle, J.-F., Boucqueau, J.-M., Quisquater, J.-J., Macq, B.: Digital images protection techniques in a broadcast framework: an overview, *Proc. of ECMAST '96*, pp.710–727, 1996.
6. Simon, C., Goray E., Vercken, G., Delivet, B., Delaigle, J.-F., Bouqueau, J-.M.: Digital images protection management in a broadcast framework: overview/TALISMAN solution," *Proc. of ECMAST '96*, pp.729–746, 1996.
7. Pfitzmann, B., Schunter, M.: Asymmetric fingerprinting, Advances in Cryptology- *EUROCRYPTO '96*, LNCS 1070, pp. 84–95, Springer-Verlag, 1996.

Protocols for Issuing Public-Key Certificates over the Internet

James W. Gray, III[1*] and Kin Fai Epsilon Ip[1**]

Department of Computer Science
Hong Kong University of Science and Technology
Clear Water Bay, Kowloon, HONG KONG
email: {gray, fai}@cs.ust.hk

Abstract. Until recently, public-key certificate issuance has involved verifying users' identities and public keys over a separate—and presumably secure—channel, such as in person or over the phone. However, the recent draft SET specification has changed that with a protocol for issuing public-key certificates to credit card holders in an interaction that takes place entirely over the Internet.
We describe the security concerns of protocols for public-key certificate issuance. These concerns include US export controls, weak DES encryption, and offline guessing attacks. We motivate and describe two protocols that have appeared in drafts of the SET specification. Plus, we describe a new protocol for public-key certification issuance.

1 Introduction

Due to the nature of standard Internet protocols (namely, TCP/IP), messages sent or received over the Internet may be read and/or modified by attackers in transit. This can be accomplished in many ways, including passive and active wiretapping, installation of packet sniffers at Internet service providers, and illicit modification of routing tables. Given that attackers are able to read and modify messages, requirements for secure communication are usually addressed using cryptography. Moreover, systems designed for the Internet most often use *public-key* cryptography. Some of the most prominent examples are PGP (Pretty Good Privacy) for secure email, ssh (Secure Shell) for secure remote login, and SET (Secure Electronic Transactions) for secure credit card transactions.

Whenever public-key cryptography is used, it is essential to use the correct public key. Consider, for example, a credit card payment application. Suppose an attacker tricks a merchant into using the wrong public key when verifying a bank's digital signature. In particular, suppose the merchant is tricked into using the attacker's public key. In this case, the attacker can convince the merchant that a payment has been authorized by the bank, when it was, in fact,

* Supported by grants HKUST 608/94E from the Hong Kong Research Grants Council and AF/253/95 from the Hong Kong Industrial Technology Development Committee.

** Supported by grant HKUST 608/94E from the Hong Kong Research Grants Council.

authorized by *the attacker.*[3] It is therefore vital—in all applications of public-key cryptography—to ensure the use of the correct public key. In the majority of Internet applications of public-key cryptography, this is accomplished by the use of public-key "certificates".

Generally speaking, a "public-key certificate" binds an identity with a public key. For email applications, a public-key certificate binds a user's *name* with his public key. For remote login, the user's *login id* is bound to his public key. In the SET specification, cardholder certificates bind the cardholder's *credit card account number* with his public key.

The binding in a public-key certificate is typically implemented by a digital signature produced by a trusted third party called a *Certificate Authority* (CA). That is, the CA digitally signs a certificate attesting to the relationship between the identity and the public key. Since the higher-level application (e.g., a credit card payment application) must trust the CA's certification, there must be some way for the CA to verify the relationship prior to signing the certificate.

Until recently, the CA's verification of the relationship between an identity and a public key has been done via a separate—presumably secure—channel. For example, the PGP User's Guide describes how this is expected to be done in person or possibly over the phone (given that the CA can recognize the user's voice). However, if public-key certificates are required by a large number of users, say all credit card holders, this approach is impractical. The SET specification therefore provides a protocol by which a cardholder can obtain a public-key certificate by communicating with a CA over the Internet [SET]. We will call such a protocol a *Certificate Issuance Protocol* (CIP).

In the remainder of this article, we discuss the security concerns and state of the art in CIP design. We will describe a model of certificate issuance which has a specific set of security requirements and assumptions. Then we describe and analyze the February 1996 draft SET CIP with respect to our model. Then we describe a flaw in that CIP that allows attackers with a $300,000 investment to obtain fake certificates. We go on to describe the correction that appears in Version 1.0 (the current version) of the SET CIP. Finally, we present a novel protocol that conforms to our model and assumptions, and is simpler, yet more secure than the SET CIP. Note, however, that the SET CIP has additional requirements that we do not address. (More on this below.) Hence, our protocol is not an immediate replacement of the SET CIP.

2 Security for Certificate Issuance Protocols

We describe the security requirements of a CIP by way of the attacker's goal. That is, we define the attacker's goal and then take the requirement to be "stop the attacker from achieving his goal". As discussed in the previous section, a public-key certificate binds an identity with a public key. We therefore take the attacker's goal to be the following.

[3] Analogous attacks against confidentiality are possible by tricking a party into using the wrong public key to encrypt a message.

Definition 1. For a given CIP, we say an attacker is *successful* if he can obtain a certificate that binds the identity of some user with a public key other than the one chosen by the user. We will call such a certificate a *fake certificate*.

In the context of SET, a principal's "identity" is the bankcard number and expiration date. Further, since such information can be used to make unauthorized purchases, this information must be kept *secret*. In some sense, every principal in the SET model has a "secret identity". Hence, in addition to preventing the attacker from obtaining fake certificates, the SET CIP must prevent the attacker from learning any principal's secret identity.

For the remainder of the paper, we focus on the requirement for CIPs to prevent attackers from obtaining fake certificates. A possible topic for future research is to additionally address the requirement to maintain secret identities.

3 The Draft SET CIP

In the February 1996 draft SET specification, a cardholder, C, obtains a public-key certificate from the certification authority, CA, as follows.

1. **The cardholder registration message is prepared.** The cardholder connects to CA over the Internet and obtains a blank registration form, F, and CA's public key, K_{CA}.[4]
   The cardholder fills in F, producing a completed registration form, which we'll denote F'. This completed form contains personal information about the cardholder such as home phone number and address. This information will be used later in the protocol as part of the verification of the cardholder's identity.
   The cardholder generates a public key/private key pair. The private key, say K_C^{-1}, is used to digitally sign the completed form, F', concatenated with the public key, say K_C. We'll denote this digital signature as $K_C^{-1}(F' \cdot K_C)$, where "$\cdot$" denotes string concatenation.
   The cardholder generates a DES encryption key[5], $K_{C,CA}$, and uses it to encrypt the completed registration form F', C's public key, and C's digital signature. We denote this encrypted component as $E_{K_{C,CA}}(F' \cdot K_C \cdot K_C^{-1}(F' \cdot K_C))$.
   The cardholder uses the CA's public key to encrypt the DES encryption key $K_{C,CA}$ and the cardholder's credit card number CN. We denote this encrypted component as $K_{CA}(K_{C,CA} \cdot CN)$.

---

[4] Actually, the CA has two public keys—one for use in encryption and one for signature verification. The cardholder gets both at this step. For simplicity, we refer to both keys as K_{CA}. Incidentally, the cardholder can verify that K_{CA} is the CA's true public key, since K_{CA} is contained in a certificate digitally signed by a higher-level certification authority. See [KPS95, Section 13.4] for more information on certification hierarchies.

[5] "DES" is the US Data Encryption Standard—an encryption algorithm that uses a 56-bit key.

2. **The cardholder sends the registration message.** The above two encrypted components are sent over the Internet to CA. This message is denoted as follows.

$$C \rightarrow CA : K_{CA}(K_{C,CA}\cdot CN))\cdot \mathrm{E}_{K_{C,CA}}(F'\cdot K_C\cdot K_C^{-1}(F'\cdot K_C))$$

3. **The CA decrypts the registration message.** Using its private key, the CA decrypts $K_{CA}(K_{C,CA}\cdot CN))$ to obtain $K_{C,CA}$ and CN. $K_{C,CA}$ is then used to decrypt the other encrypted component to obtain the completed registration form F', C's public key K_C, and C's digital signature, $K_C^{-1}(F'\cdot K_C)$. The CA uses C's public key to verify C's digital signature.[6]
4. **C's identity is verified and the certificate is issued.** In consultation with the bank that issued C's card, the CA uses CN and F' to verify C's identity, for example, by checking them against online bank records. If this verification succeeds, then CA digitally signs C's name and public key[7] thus producing the requested certificate. The CA then sends the certificate over the Internet to C.

3.1 Attacking the Draft SET CIP

Suppose the attacker intercepts a registration message, say,

$$C \rightarrow CA : K_{CA}(K_{C,CA}\cdot CN))\cdot \mathrm{E}_{K_{C,CA}}(F'\cdot K_C\cdot K_C^{-1}(F'\cdot K_C))$$

and then somehow determines the DES key $K_{C,CA}$ used to encrypt the second component. With this DES key, the attacker could decrypt the second component to obtain F' and use that to produce a replacement containing a different public key, say,

$$\mathrm{E}_{K_{C,CA}}(F'\cdot K_X\cdot K_X^{-1}(F'\cdot K_X))$$

This component could be concatenated with the first component from the intercepted message and sent to the CA. Since the recomposed message contains a valid credit card number CN and registration form F', the CA will issue a certificate binding C's name with the attacker's public key, K_X.

The above attack is qualitatively no different than, say, an attack where the attacker "somehow determines the CA's private key". However, there is an important *quantitative* difference. That is, it has been shown by Wiener [Wie94] that with a $1,000,000 investment, an attacker can build a DES key searching machine that takes a small amount of known plaintext (e.g., a fixed portion of the registration form) and the corresponding ciphertext (from F') and recovers the associated DES key in about 3.5 hours. More recently, an "ad hoc group of cryptographers and computer scientists" (including Wiener) reported that such a machine can be built with only a $300,000 investment [BDR+96]. Moreover,

[6] Although included in the current draft of the SET specification, it is unclear what benefit is gained by verifying C's digital signature.

[7] The CA also digitally signs the hash of C's bankcard number.

computing technology is expected to continue improving, so this investment and key-cracking time are likely to further decrease in the years to come.

Given these attacks, it is clearly unreasonable to assume it is infeasible for an attacker to determine the key used in a DES-encrypted component. In particular, we can say that with the February 1996 draft SET CIP, a well-financed attacker can obtain fake certificates.[8]

3.2 US Export Control Considerations

One approach to correcting the SET CIP is to replace DES with a stronger encryption algorithm. A rough measure of the strength of a symmetric encryption algorithm is the number of bits in its key. In fact, the short key—56 bits—used by DES is its main source of weakness. Further, it has been recommended that commercial applications use 90-bit keys at a *minimum* [BDR+96]. There are a number of encryption algorithms that could replace DES (see [Sch96]), if such a change were possible in the context of SET.

Unfortunately, replacing DES with an algorithm that uses a longer key would have the undesirable effect SET-enabled products would not be exportable out of the United States (US). One of the explicit objectives in the SET design was to use "exportable technology throughout" [SET]. As it applied to commercial encryption technology during the early stages of the SET development (1996), the following was a rough approximation of US export law.

> Any product that applies symmetric encryption to free-form messages must use keys with length 56 bits or shorter. Otherwise, that product cannot be exported from the US.

In the context of SET, credit card numbers (CN) have a limited format and, hence, are not considered to be free-form messages. In contrast, registration forms (F') are expected to contain fields for information such as the cardholder's home address, which can be used to send arbitrary free-form information. This is why the credit card number can be encrypted under the CA's public key, whereas F' can be encrypted only with DES.[9] Therefore, to remain consistent

---

[8] In the context of SET, this attack did not result in an immediate vulnerability. This is because the payment information sent by the cardholder during the payment protocol contains the cardholder's card number, encrypted with the public key of the authorizing bank. During payment authorization, this card number is decrypted and verified by the authorizing bank; thus, even though an attacker may obtain a fake certificate; he cannot use that certificate, unless he also knows the associated credit card number.

[9] Encryption under the CA's public key is considered more secure than DES encryption. However, a direct comparison in terms of the number of bits in the respective keys would be misleading, since the former is *public-key* encryption, whereas the later is *symmetric-key* encryption. Comparing the number of bits in the two keys would be like comparing apples and oranges. A better approach is to compare the amount of time needed to crack the respective keys, given a particular financial investment by the attacker. Such a comparison is beyond the scope of the present paper.

with the objectives of the SET design, F' cannot be encrypted with any algorithm stronger than DES.

Very recently, US controls on the export of cryptography have been somewhat relaxed. In particular, it is now possible to use strong encryption on free-form text, as long as the product developer agrees to migrate the product to an escrowed-key encryption algorithm in the near future. Thus, encrypting F' with a strong cipher becomes a viable option. Nevertheless, it introduces the necessity of moving toward an escrowed-key algorithm and the SET designers have avoided that complexity.

4 The Version 1.0 SET CIP

Taking into account US export control considerations, the version 1.0 SET specification takes an approach other than stronger encryption to address the attack described in Section 3.1. To understand their approach, note that the attack described in Section 3.1 involved replacing the component containing the cardholder's public key, say,

$$\mathrm{E}_{K_{C,CA}}(F' \cdot K_C \cdot K_C^{-1}(F' \cdot K_C))$$

with a substitute containing the attacker's public key, say,

$$\mathrm{E}_{K_{C,CA}}(F' \cdot K_X \cdot K_X^{-1}(F' \cdot K_X))$$

If such a substitution could be detected, the attack would be circumvented.

Fortunately, there is a cryptographic tool we can use to detect such substitutions. Namely, we can make use of the following property of cryptographic hash functions. (For a general introduction to cryptographic hashes, see [KPS95] or [Sch96].)

Definition 2. A hash function h is called *collision intractable* if it is infeasible for an attacker to find two values, say $x \neq y$, such that $h(x) = h(y)$. (If found, x and y are called a *collision*.)

To make use of a collision intractable hash function in the SET CIP, the registration message becomes

$$C \to CA : K_{CA}(K_{C,CA} \cdot CN \cdot h(K_C)) \cdot \mathrm{E}_{K_{C,CA}}(F' \cdot K_C \cdot K_C^{-1}(F' \cdot K_C))$$

where $h(K_C)$ is a cryptographic hash (e.g., the US standard SHA-1) applied to K_C.

Upon receipt of this message, the CA verifies the identity of C, as in the original SET CIP. In addition, the CA verifies the hash value contained in the first encrypted component. This additional verification ensures K_C was not modified in transit, even if the DES key is known to an attacker. That is, even if the attacker obtains the DES key ($K_{C,CA}$), he still cannot obtain a fake certificate, since, by the collision intractibility of h, he cannot find a public key, say $K_X \neq K_C$, such that $h(K_X) = h(K_C)$.

Although the version 1.0 SET CIP addresses the attack described in Section 3.1, this improvement has come at a cost. First of all, the modified SET CIP relies on an additional assumption, namely, the collision intractability of h.[10] In addition, the size of the first encrypted component, $K_{CA}(K_{C,CA} \cdot CN \cdot h(K_C))$ is larger than the original protocol by (in the case of SHA-1) 160 bits. This increase in size may lead to the necessity of dividing the message into two blocks and encrypting each blocks separately, which would have a significant performance impact.

With the above improvement, the primary remaining weakness in the SET CIP is the shared secret. In particular, note that if an attacker has sufficient funds to build a DES key cracking machine, the only remaining secret is the cardholder's card number. Moreover, in a typical 16-digit card number, perhaps as many as ten digits will be predictable by the attacker from knowledge of the issuing bank along with knowledge of that bank's practices in assigning numbers. If only six digits need to be guessed by the attacker, the attacker will succeed, on average, in 500,000 guesses. This is quite secure, since the CA can impose a limit on the number of times any given cardholder's certificate request will be processed. For example, imposing a limit of 100 would probably not be an inconvenience to legitimate cardholders, but would reduce the attacker's probability of success to 1/5000. Nevertheless, we will see in the following sections that it is possible to reduce this probability of success, even without relying on strong encryption of the completed registration form, F'.

5 Offline Guessing Attacks

In this section, we describe a straw man CIP which is much simpler than the SET CIPs. We use this protocol to introduce the concern of offline guessing attacks, as well as our general approach to CIP design.

The basic idea of the straw man CIP is to use the secret shared between the requesting user (C) and the certificate authority (CA) as a *shared key*. This shared key is used to authenticate the message containing C's public key. The following definition describes the basic cryptographic tool by which this is accomplished.

Definition 3. A *Message Authentication Code (MAC) Scheme* is an algorithm that takes a message, m, and a key, k, and produces a bitstring $\mathrm{MAC}_k(m)$. Typically, k is known only to two parties, say C and CA, and $\mathrm{MAC}_k(m)$ is appended to m when it is sent between the two. Since k is known to both parties, on receiving m, the receiving party can also compute $\mathrm{MAC}_k(m)$ and verify that m was sent by the other party. Further, it is infeasible for an attacker

[10] Actually, the SET CIP already relies on the collision-intractability of h, since h is used in the implementation of the CA's digital signature.

who does not know k to compute $\mathrm{MAC}_k(m)$ for any message m (that has not already been sent by C or CA appended with a valid MAC).

The purpose of a MAC is similar to a digital signature, except that with MACs, the sender and receiver perform the identical operation, whereas with digital signatures, the two parties perform different operations—namely, the sender signs using its private key and the receiver verifies using the corresponding public key.

In the case of SET, a suitable MAC key $K_{C,CA}$ can be computed from C's card number and personal information. For example, C's card number, expiration date, name, home phone number, home address, work phone number, etc. can be concatenated into a single string and then a cryptographic hash can be applied to the string to obtain the shared MAC key.[11]

Given that a suitable MAC key is shared between C and CA, the straw man CIP protocol proceeds as follows.

The cardholder generates a public key/private key pair, say (K_C^{-1}, K_C). Next, using the shared MAC key, $K_{C,CA}$, the cardholder computes the MAC for his public key; that is, the cardholder computes $\mathrm{MAC}_{K_{C,CA}}(K_C)$. Then, the cardholder sends a message containing his public key and this MAC to the CA, viz,

$$C \rightarrow CA : K_C \cdot \mathrm{MAC}_{K_{C,CA}}(K_C)$$

Upon receiving this message, the CA, in conjunction with the bank that issued C's card, uses $K_{C,CA}$ (which it can compute independently from C, using online bank records) to verify the MAC. If this verification succeeds, the CA generates the requested certificate (by signing C's name and K_C) and sends it back to C.

The above protocol may seem secure. However, as we now describe, even when using strong digital signatures and MACs, an attacker can obtain fake certificates. This is because the MAC key is guessable to some extent and the straw man CIP leaves the MAC key open to *offline guessing attacks*. See [GLN+93] for a general discussion of offline guessing attacks. In the following, we discuss how offline guessing attacks apply to our straw man CIP.

Suppose a cardholder registration message, say

$$C \rightarrow CA : K_C \cdot \mathrm{MAC}_{K_{C,CA}}(K_C)$$

is intercepted by an attacker. In this case, the attacker can search for the cardholder's secret as follows.

The attacker tries out a series of guesses. For each guess, g, the attacker generates the corresponding MAC key and applies it to K_C, which was obtained in the intercepted message. If the result is equal to $\mathrm{MAC}_{K_{C,CA}}(K_C)$, which

[11] Note that it is important for C and CA to use exactly the same information in exactly the same form as input to the cryptographic hash. Otherwise, their resulting keys will be different. Consistent formats can be ensured if the cardholder and bank software enforce identical conventions regarding upper/lower case, abbreviations, etc. For example, "St." can always be used in place of "Street".

was also obtained in the intercepted message, the attacker knows g is correct. The important point about this attack is that it can be done offline—that is, without sending any registration messages to the CA—so, there is no way to limit the number of guesses made by the attacker. The attacker is limited only by his computational resources; for example, the attacker can likely test millions of guesses per day on a standalone PC.

It may seem that an offline guessing attack is also possible with the SET CIP. That is, the attacker can intercept a registration message containing the public-key encrypted credit card number, say $K_{CA}(K_{C,CA} \cdot CN \cdot h(K_C))$. Assuming the attacker has the financial resources for a DES key-cracking machine, he can obtain $K_{C,CA}$. Also, since the adversary knows K_C, he can compute $h(K_C)$. What then, prevents the attacker from performing an offline attack, by taking a guess, g, computing $K_{CA}(K_{C,CA} \cdot g \cdot h(K_C))$ and comparing the result with the intercepted value?

In the case of SET, the technique that prevents this attack is known as "Optimal Assymmetric Encryption" [BR93], which is a form of probabilistic encryption. That is, the encryption of $K_{C,CA} \cdot g \cdot h(K_C)$ partly depends on a random number generated during the encryption process. This means that even if the attacker has guessed correctly (i.e., $g = CN$), the two encrypted packets will (with probability close to one) be different. Thus, an offline guessing attack is not possible with the SET CIP.

6 A New Certificate Issuance Protocol

In this section we present a CIP that is nearly as simple as the straw man CIP, yet satisfies our security requirements. It is unencumbered with US export controls; it does not depend on weak encryption (i.e., DES) to protect any part of the shared secret; and offline guessing attacks are infeasible.

The new CIP we describe below is based on the previous straw man CIP with a MAC, with additional mechanisms to prevent offline guessing.

The advantage of using a MAC is that US export controls place no restrictions on message authentication, so it is possible to use a MAC scheme utilizing a long key—say 128 bits. For example, the CBC MAC with a 128-bit block cipher would be appropriate. (For a description of the CBC MAC, see [KPS95], wherein it is called the "CBC residue".) Besides, the adversary must guess the cardholder's personal information at the same time as credit card number. That is, there is no way an attacker can obtain the personal information in a separate attack (such as DES key cracking); rather, the secret must be guessed in its entirety.

In the attack for the straw man CIP, eavesdropping on a registration message provides the attacker with both a message (i.e., C's public key) and the corresponding MAC. This provides sufficient information for the attacker to search for the MAC key. To protect the MAC key we need to produce a MAC that is not directly generated from the accompanying message body. One possible approach is to encrypt the message with the public key of the receiver, then append the MAC of the plaintext. However, in the straw man CIP, the message body is C's public key, which the attacker can presumably obtain by other means. Therefore, encrypting it is of no use.

The approach we use is for the cardholder to encrypt a random number with the CA's public key. The MAC is then applied to this random number in conjunction with the cardholder's public key. Since the attacker does not know the random number, offline guessing attacks are prevented.

Assuming the public key of CA is properly delivered to each eligible cardholder C, and a MAC key similar to the one described in the straw man CIP is shared between C and CA, our new CIP proceeds as follows.

The cardholder generates a public key/private key pair and random number, N. 128 bits should be sufficiently long for N. More on this below. The cardholder then encrypts N with CA's public key to produce $K_{CA}(N)$. Then, the cardholder produces a MAC for the *plaintext* of N concatenated with K_C. Finally, the cardholder sends the message containing the encrypted N, his public key K_C and the MAC.

$$C \rightarrow CA : K_{CA}(N) \cdot K_C \cdot \mathrm{MAC}_{K_{C,CA}}(N \cdot K_C)$$

CA retrieves the plaintext N with his own private key and the MAC is verified. As in other CIPs, C's identity is verified and the certificate is issued.

This protocol is at least as secure as the straw man CIP, since the same MAC mechanism is used to verify the identity of the cardholder. In addition, assuming the public key encryption is highly secure and the value of N is unpredictable, the MAC key is protected against offline guessing attacks. In particular, if N is 128 bits long, the attacker will need to test 2^{128} messages for each guess as to the value of the MAC key.

7 Conclusions

Certificate issuance protocols were first defined in the context of SET for the purpose of Internet bankcard payments. However, there is no reason why a variant of the SET CIP or the CIP introduced in this paper cannot be used to issue certificates for applications such as electronic banking, secure email, secure remote login, etc.

The major requirement on the use of the CIPs in this paper is that the requester and the Certification Authority (CA) initially share a secret that is difficult for an attacker to guess. In addition to credit card numbers, this secret can include personal information such as date of birth, address, phone number, social security number, driver's license number, bank account number, etc, as well as passwords or PINs. Roughly speaking, the more difficult-to-guess information included in the shared secret, the better.

On the other hand, the more information included in the shared secret, the higher the chance that a discrepency will arise between the requester's version and the CA's version, thus resulting in the certificate not being issued, even though a legitimate requester has applied. To avoid this problem, we recommend using only information that can be put in a consistent format.

As another general guideline, it is prudent to implement a maximum number of certificate requests from any single cardholder. That is, a legitimate cardholder is unlikely to apply for a certificate as many as 100 times, whereas an attacker may try a brute force search for the shared secret by applying tens of thousands of times. Such brute force attacks can be circumvented if the CA limits the number of applications by any single requester to 100.

The two most difficult design requirements for CIPs are (1) to avoid applying strong encryption technology to free-form information (in order to minimize US export control limitations); and (2) to protect the shared secret against offline guessing attacks. Our new CIP satisfies both requirements and we conclude that it is a strong candidate for general and practical use.

References

[BR93] Mihir Bellare and Phillip Rogaway. Optimal asymmetric encryption. In *Advances in Cryptology—Proc. Eurocrypt '94, (LNCS 950)*, pages 92–111. Springer-Verlag, 1994.

[BDR+96] Matt Blaze, Whitfield Diffie, Ronald Rivest, Bruce Schneier, Tsutomu Shimomura, Eric Thompson, and Michael Wiener. *Minimal key lengths for symmetric ciphers to provide adequate commercial security* `http://www.bsa.org/policy/encryption/~cryptographers.html` (January 1996).

[GLN+93] Li Gong and Mark A. Lomas and Roger M. Needham and Jerome H. Saltzer. Protecting Poorly Chosen Secrets from Guessing Attacks. *IEEE Journal on Selected Areas in Communications*, 11(5):648–656, June 1993.

[KPS95] Charlie Kaufman, Radia Perlman, and Mike Speciner. *Network Security, Private Communication in a Public World.* Prentice-Hall, Englewood Cliffs, New Jersey, 1995.

[Sch96] Bruce Schneier. *Applied cryptography: protocols, algorithms, and source code in C.* Wiley, New York, second edition, 1996.

[SET] Secure Electronic Transaction (SET) specification (Version 1.0) available on the Internet via `http://www.mastercard.com` or `http://www.visa.com`. Jointly developed and issued by Mastercard and Visa (May 1997)

[Wie94] Michael J. Wiener. Efficient DES key search. Technical Report TR-244, School of Computer Science, Carleton University, May 1994. Reprinted in *Practical Cryptography for Data Internetworks*, W. Stallings (ed.), IEEE Computer Society Press, pp 31–79, (1996).

Distributed Cryptographic Function Application Protocols

André Postma, Thijs Krol, Egbert Molenkamp
University of Twente, Department of Computer Science
P.O.Box 217, NL 7500 AE, Enschede, the Netherlands
e-mail: {postma, krol, molenkam}@cs.utwente.nl

Abstract. Recovery of data stored in a fault-tolerant and secure way requires encryption and / or decryption of data with secret cryptographic functions, for which a group of processors should be responsible. For this purpose, the here-introduced distributed cryptographic function application protocols (DCFAPs) can be applied. DCFAPs are executed on a set, $\mathcal{N}$, of N processors, in order to apply a secret cryptographic function $\mathcal{F}$, while satisfying the following properties:

- malicious behaviour of up to T processors from $\mathcal{N}$ does not inhibit application of $\mathcal{F}$
- any group of N-T or more processors from $\mathcal{N}$ is capable of applying $\mathcal{F}$
- T or fewer colluding processors from $\mathcal{N}$ are unable to compute or apply $\mathcal{F}$.

1. Introduction

In order to minimize the probability of permanent loss or corruption of data due to processor failures or undetected mutilations by malicious attackers, and to prevent data from being read by unauthorized users, data should be stored in a ***fault-tolerant and secure*** way. Recovery of data stored in such a way requires application of secret cryptographic functions, for which so-called ***distributed cryptographic function application protocols*** (***DCFAPs***) can be used. In the next section, we describe how fault-tolerant and secure data storage can be established. Section 3 introduces DCFAPs.

2. Fault-tolerant and secure data storage

2.1. Fault-tolerant data storage

In order to make a data storage system resilient to failures of one or more of its processors, redundancy in the form of extra processors can be employed. All data in such a system is stored in a fault-tolerant way as a collection of equal-length data fragments on the disks of N processors. The data is resilient to a number of arbitrary processor failures, provided that the collection of data fragments contain sufficient redundancy.

The required redundancy can be obtained by encoding the data into symbols of an error-correcting code. In this method, called ***data encoding***, every data fragment contains a number of symbols of an error-correcting code. Assume that every processor p_i ($1 \leq i \leq N$) possesses one data fragment D_i. Assume furthermore that D_i contains b symbols, numbered $s_{i,1}, \ldots s_{i,b}$. Then, for any j, with $1 \leq j \leq b$, the concatenation of the j-th symbol of every data fragment, i.e., $s_{1j} \ldots s_{N,j}$, forms a code word of the error-correcting code. Every processor p_i possesses a data encoding function C_i, which it may apply on a piece of data, B, to calculate its data fragment $C_i(B)$ of B. If a repetition code is used to encode data, then every data fragment is just a copy of the original data. This method, called ***data replication***, is widely used in existing fault-tolerant systems.

Provided that the collection of data fragments contains sufficient redundancy, data fragments that are lost or corrupted due to a processor failure, may be recovered with the help of the remaining data fragments. After recovery of all lost or corrupted data fragments, the system is able to survive a subsequent processor failure. Thus, it is advantageous to recover lost or corrupted data fragments as soon as possible.

2.2. Secure data storage

Secure data storage requires application of cryptographic techniques. In order to protect data from ***unauthorized reading***, the original data is encrypted. It is essential that decryption can only be done by authorized parties. In order to protect data against ***undetected mutilation***, processors should digitally sign their pieces of data. Digitally signing a piece of data, *B*, consists of concatenating an encrypted hash value of *B* to data *B*. The hash value of *B* is obtained by applying some strong one-way hash function on *B*. To prevent forgery of the hash value, the owner of *B* encrypts the hash value with a secret cryptographic function. In order to protect data against ***both unauthorized reading and undetected mutilation***, the above techniques should be combined.

It is assumed that the cryptographic functions used for encryption (decryption) and signing (verification) are different, since this enhances the security of the system. The ***encryption resp. decryption function*** of user *u* are needed to solve the *privacy* problem (i.e., they serve to protect data against unauthorized reading), and are denoted by $\mathcal{P}_{(u)}^{(-1)}$ resp. $\mathcal{P}_{(u)}$. The ***sign resp. verify function*** of a user *u* are needed to solve the *authentication* problem (i.e., they serve to protect data against undetected mutilation), and are denoted by $\mathcal{A}_{(u)}$ resp. $\mathcal{A}_{(u)}^{(-1)}$. If a *secret-key* cryptosystem (e.g., DES) is used for encryption and decryption (resp. signing and verification), then $\mathcal{P}_{(u)} = \mathcal{P}_{(u)}^{(-1)}$ (resp. $\mathcal{A}_{(u)} = \mathcal{A}_{(u)}^{(-1)}$), and both functions are kept secret. If a *public-key* cryptosystem (e.g., RSA) is used, the encryption (resp. verify) function $\mathcal{P}_{(u)}^{(-1)}$ (resp. $\mathcal{A}_{(u)}^{(-1)}$) is made public, whereas the decryption (resp. sign) function $\mathcal{P}_{(u)}$ (resp. $\mathcal{A}_{(u)}$) is kept secret.

Assume user *u* wants to protect a piece of data, *B*, against unauthorized reading and undetected mutilation by others. The above techniques can be combined in two ways:

Combination C1: Any piece of data is signed before it is encrypted. For user *u*, the result is equal to $\mathcal{P}_{(u)}^{(-1)}(B),\mathcal{A}_{(u)}(H(B))$ or $\mathcal{P}_{(u)}(B, \mathcal{A}_{(u)}(H(B)))$. ❑

Combination C2: Any piece of data is encrypted before it is signed. For user *u*, the result is equal to $\mathcal{P}_{(u)}^{(-1)}(B),\mathcal{A}_{(u)}(H(\mathcal{P}_{(u)}^{(-1)}(B)))$. ❑

Combination C2 is preferred to C1 for the following reasons. First, in C1, in order to check whether or not the data has been mutilated, in C1, both the (confidential) data and the hash value should be decrypted, whereas, in C2, it suffices to decrypt the hash value. Moreover, for this purpose, in C1, knowledge of both $\mathcal{A}_{(u)}^{(-1)}$ and $\mathcal{P}_{(u)}$ is required, whereas, in C2, knowledge of $\mathcal{A}_{(u)}^{(-1)}$ suffices.

2.3. Fault-tolerant and secure data storage

Establishing fault-tolerant and secure data storage requires combining (in any order) the above-mentioned techniques of encryption, signing, and encoding. We restrict ourselves to alternatives in which data is encrypted before it is signed, since these alternatives are preferred to alternatives in which data is signed before it is encrypted (see previous section). Assume that a user u wants to store a piece of data, B, in a fault-tolerant and secure way as a collection of N fragments R_i ($1 \leq i \leq N$). Then, the following three alternatives are possible:

Alternative A1: The original data is first encrypted, then signed, and finally, encoded. In this case, for any i (with $1 \leq i \leq N$), $R_i = C_i\,(\mathcal{P}_{(u)}{}^{(-1)}(B), \mathcal{A}_{(u)}(H(\mathcal{P}_{(u)}{}^{(-1)}(B))))$. ❑

Alternative A2: The original data is first encoded, then encrypted, and finally, signed. In this case, for any i (with $1 \leq i \leq N$), $R_i = \mathcal{P}_{(u)}{}^{(-1)}(C_i(B)), \mathcal{A}_{(u)}(H(\mathcal{P}_{(u)}{}^{(-1)}(C_i(B))))$ ❑

Alternative A3: The original data is first encrypted, then encoded, and finally, signed. In this case, for any i (with $1 \leq i \leq N$), $R_i = C_i(\mathcal{P}_{(u)}{}^{(-1)}(B)), \mathcal{A}_{(u)}(H(C_i(\mathcal{P}_{(u)}{}^{(-1)}(B)))$ ❑

In general, A1, A2, and A3 are different (unless the method of data replication is used to encode the data), and in this case, A2 and A3 are preferred to A1, because in A2 and A3 (in contrast to A1), the correctness of every data fragment can be verified (provided every data fragment is unique, i.e., its signed hash value carries a unique identification). This enables encoding of the data into symbols of a T-erasure-correcting code, since the ability to determine the locations of all mutilated symbols in a code word is a sufficient (but also necessary) requirement for application of a T-erasure-correcting code. The advantage of applying a T-erasure-correcting code over applying a T-error-correcting code is, that for $T > 0$, a T-erasure-correcting code requires much less redundancy[1]. Moreover, for $T \geq 2$, a T-erasure-correcting code is easier to decode [3].

Alternative A3 is preferred to A2, because, in A3, for recovery of lost or corrupted data fragments of a user u, knowledge of functions $\mathcal{A}_{(u)}{}^{(-1)}$ and $\mathcal{A}_{(u)}$ is sufficient (in order to be able to verify the correctness of the remaining data fragments and to sign recovered re-created data fragments), whereas, in A2, knowledge of $\mathcal{P}_{(u)}$ is also required. Hence, we assume that A3 is applied.

3. DCFAPs based on function sharing

As stated before, recovery of lost or corrupted data fragments should be done as soon as possible. Since recovery of data of a user u requires knowledge of the secret cryptographic function $\mathcal{A}_{(u)}$, it may seem straightforward to let user u be responsible for recovery of his own data. However, this would imply that *the owner's processor must*

---

1. Let k be the number of symbols in a data word, and n be the number of symbols in a code word. Then, from [3], we know that a ***T-error-correcting code*** can be constructed if and only if $n \geq k + 2T$, and a ***T-erasure-correcting code*** can be constructed if and only if $n \geq k + T$.

function correctly in order to make recovery possible. In other words, it would mean that the capability of the system to recover lost data fragments would depend on the correctness of a single processor (viz., the owner's processor). This is unacceptable, since the system should be resilient to a number of failures of *arbitrary* processors. Therefore, the responsibility of encryption and decryption of data with the secret cryptographic function of any user should be given to a **group** of processors.

Assume that a group, $\mathcal{N}$, of processors is responsible for encryption and decryption of data with a secret cryptographic function $\mathcal{F}$. In the presence of up to T faulty processors in $\mathcal{N}$, it should be possible for the correct processors in $\mathcal{N}$ to apply $\mathcal{F}$. Furthermore, the faulty processors in $\mathcal{N}$ may not be capable of unauthorized reading (and undetected mutilation, respectively) of data encrypted with $\mathcal{F}^{(-1)}$ (and $\mathcal{F}$ respectively).

The above-mentioned problem can be solved by means of a so-called ***distributed cryptographic function application protocol*** (***DCFAP***). A DCFAP is executed by a set, $\mathcal{N}$, of N processors p_i ($1 \le i \le N$), up to T of which may behave maliciously, in order to apply function $\mathcal{F}$ on data B, while satisfying the following properties:

P1. malicious behaviour of up to T processors from $\mathcal{N}$ does not inhibit application of $\mathcal{F}$.
P2. any group of N-T or more processors from $\mathcal{N}$ is capable of applying $\mathcal{F}$.
P3. T or fewer colluding processors from $\mathcal{N}$ are unable to compute or apply $\mathcal{F}$.

It is assumed that all correct processors in $\mathcal{N}$ have already reached agreement about data B **before** the start of the DCFAP. For any i, with $1 \le i \le N$, let $B(p_i)$ be the data that should be encrypted / decrypted according to processor p_i. If p_i is correct, $B(p_i)=B$.

A ***DCFAP*** resembles an interactive consistency algorithm (ICA) [4], and is defined as follows. Let $\mathcal{N}$ be a set of N processors p_i ($1 \le i \le N$). Every processor p_i possesses an initial value $B(p_i)$ (being the data on which $\mathcal{F}$ should be applied according to p_i). The initial values are encrypted / decrypted and distributed from all processors to all processors by means of a DCFAP. In the presence of up to T maliciously behaving processors, a DCFAP guarantees satisfaction of the following two conditions:

D1. All correct processors agree on the result of application of $\mathcal{F}$ on the initial values they think they have received from each of the processors.
D2. For every p_i ($1 \le i \le N$), it holds that if p_i is correct, the agreement mentioned in D1 equals the result of application of $\mathcal{F}$ on the initial value possessed by p_i.

Both a DCFAP and an ICA consist of N Byzantine Agreement Protocols (BAPs) [2], one per processor. However, while in an ICA, messages are simply distributed from all processors to all processors, in a DCFAP, messages are not only distributed but also encrypted (resp. decrypted).

From P2 and P3, it can be concluded that a certain minimal number of processors is both necessary and sufficient in order to apply $\mathcal{F}$. Thus, somehow, the secret information needed to apply $\mathcal{F}$ must be distributed among the processors in $\mathcal{N}$, such that **coop-**

eration of a certain minimal number of processors is both necessary and sufficient to apply $\mathcal{F}$. The technique of function sharing [1] can be used for this purpose.

A DCFAP based on function sharing is defined on a set, $\mathcal{N}$, of N processors p_i ($1 \leq i \leq N$), up to T of which may behave maliciously. We assume that $N = 2T+1$. We assume that by means of function sharing, cryptographic function $\mathcal{F}$ has been split into N shadow functions $\mathcal{F}_i$ ($1 \leq i \leq N$), that have been distributed among the N processors p_i. Each processor p_i is given one shadow function $\mathcal{F}_i$. A DCFAP consists of execution of N authenticated BAPs BAP_i ($1 \leq i \leq N$). Processor p_i acts as the source in BAP_i in order to have $\mathcal{F}_i(B(p_i))$ be communicated to all other processors in $\mathcal{N}$. After execution of all BAPs, every processor p_i combines the partially encrypted results from the BAPs in order to calculate $\mathcal{F}(B)$. If p_i has obtained at least $T+1$ out of N evaluations of different shadow functions $\mathcal{F}_j$ of data B, p_i may easily calculate $\mathcal{F}(B)$, otherwise, obtaining this result is computationally infeasible. We assume that if any processor $p_j \in \mathcal{N}$ functions correctly, then p_j correctly communicates $\mathcal{F}_j(B)$ to all other processors in $\mathcal{N}$. Since $N = 2T+1$, if at most T processors are faulty, the system contains at least $T+1$ correctly functioning processors, and all correct processors may easily calculate $\mathcal{F}(B)$.

A DCFAP based on function sharing can be described as follows:
For every processor $p_i \in \mathcal{N}$ ($1 \leq i \leq N$):

1. *Processor p_i communicates $\mathcal{F}_i(B(p_i))$ to all processors in $\mathcal{N}$ by means of BAP_i.*
2. *For any j ($1 \leq j \leq N$, $i \neq j$), processor p_i participates in BAP_j if it receives the first valid message of BAP_j started by processor p_j in order to communicate $\mathcal{F}_j(B(p_j))$.*
3. *For any j ($1 \leq j \leq N$), after p_i has terminated BAP_j, p_i decides on basis of the message values of the messages of BAP_j that it has accepted during BAP_j. If processor p_j is correct, the decision processor p_i takes in BAP_j equals $\mathcal{F}_j(B(p_j))$.*
4. *After p_i has terminated all the BAPs that it has started, p_i combines all decisions of the BAPs in order to calculate $\mathcal{F}(B)$.*

4. Conclusion

This paper describes DCFAPs based on function sharing. DCFAPs can be applied in order to recover lost or corrupted data fragments in a fault-tolerant and secure data storage system of processors, a number of which may behave maliciously.

5. References

[1] De Santis, A., et al., How to Share a Function Securely, in: **Proceedings of the 26th Annual ACM Symposium on the Theory of Computing**, Montréal, 1994, pp.522-533.

[2] Lamport, L., et al., The Byzantine Generals Problem, in: **ACM Transactions on Programming Languages and Systems**, Vol.4, No.3, July 1982, pp.382-401.

[3] MacWilliams, F.J. and Sloane, N.J.A., **The Theory of Error-Correcting Codes**, Amsterdam, the Netherlands, North Holland, 1978.

[4] Pease, M., Shostak, R., and Lamport, L., Reaching agreement in the presence of faults, in: **Journal of the ACM**, Vol. 27, No.2, April 1980, pp. 228-234.

Fault Tolerant Anonymous Channel

Wakaha OGATA[1], Kaoru KUROSAWA[2],
Kazue SAKO[3], Kazunori TAKATANI[1]

[1] Himeji Institute of Technology,
wakaha@comp.eng.himeji-tech.ac.jp
http://wwwj2.comp.eng.himeji-tech.ac.jp/home/wakaha/
[2] Tokyo Institute of Technology,
kurosawa@ss.titech.ac.jp, http://tsk-www.ss.titech.ac.jp/~kurosawa/
[3] NEC Corporation, sako@sbl.cl.nec.co.jp

Abstract. Previous anonymous channels, called MIX nets, do not work if one center stops. This paper shows new anonymous channels which allow less than a half of faulty centers. A *fault tolerant* multivalued election scheme is obtained automatically. A very efficient ZKIP for the centers is also presented.

1 Introduction

Chaum considered an anonymous channel which hides the correspondences between the senders and the messages [Ch81]. Suppose that there are l senders $P_1, \ldots, P_l$ such that each P_i has a message m_i. Assume a center called MIX which publicizes his RSA public key E. Each sender P_i sends $c_i = E(m_i)$ to the MIX. The MIX decrypts them and publicizes $m_1, \ldots, m_l$ in the lexicographical order. The MIX, however, knows who sent what message. Finally, Chaum proposed a *MIX net* in which n MIXes are sequentially connected [Ch81]. Anonymity is protected if at least one MIX is honest.

The *MIX net*, however, does not work if one MIX (center) stops. This paper shows new anonymous channels which allow less than a half of faulty centers, where faulty centers can stop or deviate from the protocol arbitrarily. *Fault tolerant* multivalued election schemes are obtained automatically. (Cohen and Fischer type election scheme realizes only yes/no votes [CF85, Be86].) A very efficient zero knowledge interactive proof system (ZKIP) for MIX is also presented.

2 Fault tolerant anonymous channels

This section presents two robust anonymous channels which allow less than a half of faulty centers. One is based on the hardness of factorization and the other is based on the difficulty of the discrete log problem. In both schemes, even if less than a half of centers are faulty, (1) randomly shuffled messages are output and (2) anonymity is protected.

2.1 Proposed scheme based on factoring

The r-th residue public key cryptosystem is defined as follows [CF85]. Let r be a prime.
Secret key: Two large prime numbers p, q such that $r|p-1$ and $r \not| q-1$.
Public key: $N(\stackrel{\triangle}{=} pq)$ and y such that $y \neq x^r \bmod N$ for $\forall x$.
Plaintext: m such that $0 \leq m < r$.
Encryption: $E(m, x) \stackrel{\triangle}{=} y^m x^r \bmod n$, where x is a random number.
Decryption: Let $c = E(m, x)$. Then $m = j_0$ if

$$c^{(p-1)/r} = (y^{(p-1)/p})^{j_0} \bmod p$$

This cryptosystem satisfies a homomorphic property such that

$$E(m_1, x_1)E(m_1, x_2) = E(m_1 + m_2 \bmod r, x_0), \text{ for some } x_0. \qquad (1)$$

Then our scheme is described as follows. Suppose that there are l senders $P_1, \ldots, P_l$ such that each P_i has a message m_i. Assume n centers $MIX_1, \ldots, MIX_n$ such that each MIX_j has a public key E_j of the r-th residue cryptosystem. Let

$$k \stackrel{\triangle}{=} \lfloor (n-1)/2 \rfloor + 1.$$

Definition 1. For a plaintext m, choose a random polynomial $R(x)$ of degree $k-1$ such that $R(0) = m$. Let

$$B(m, R) \stackrel{\triangle}{=} [E_1(R(1), x_1), \cdots, E_n(R(n), x_n)],$$

where $x_1, \cdots, x_n$ are random numbers. We say that $B(m, R)$ is an encrypted shares of m.

Definition 2. For $B(m, R)$, choose a random polynomial $U(x)$ of degree $k-1$ such that $U(0) = 0$. Let

$$\hat{B}(m) \stackrel{\triangle}{=} [E_1(R(1), x_1)E_1(U(1), w_1), \cdots, E_n(R(n), x_n)E_n(U(n), w_n)],$$

where $w_1, \cdots, w_n$ are random numbers. We say that $\hat{B}(m)$ is a reencryption of $B(m)$. ($\hat{B}(m)$ is again an encrypted shares of m because $\hat{B}(m) = B(m, R+U)$.)

(Sender's protocol)

Each P_i computes $B(m_i, R_i)$, an encrypted shares of his message m_i, and publicizes $B(m_i, R_i)$. He proves that he knows $R_i(x)$ by using a ZKIP of knowledge. He is then proving two things:

1. He knows $R_i(0)$ which is his message itself. This proof prevents the Pfitzmann's attack against Chaum's MIX net [PP89].
2. He indeed distributed the message correctly (verifiable secret sharing scheme).

(Center's protocol) Now

$$[B(m_1,R_1),\cdots,B(m_l,R_l)] \tag{2}$$

are publicized. MIX_1 randomly computes a reencryption of each $B(m_i,R_i)$ such that $B(m_i,R_i+U_i)$. MIX_1 chooses a random permutation π on $\{1,2,\ldots,l\}$ and publicizes

$$[B(m_{\pi(1)},R_{\pi(1)}+U_{\pi(1)}),\cdots,B(m_{\pi(l)},R_{\pi(l)}+U_{\pi(l)})).] \tag{3}$$

MIX_1 further proves that eq.(3) is computed from eq.(2) correctly by using a ZKIP. For $2 \le j \le n$, MIX_j executes the same process sequentially.

(Decryption)

At the end, MIX_n publicizes

$$[B(m_{\varphi(1)},\tilde{R}_{\varphi(1)}),\cdots,B(m_{\varphi(l)},\tilde{R}_{\varphi(l)})]$$

for some permutation φ, where $B(m_{\varphi(i)},\tilde{R}_{\varphi(i)})$ is an encrypted shares of $m_{\varphi(i)}$. Let

$$B(m_{\varphi(i)},\tilde{R}_{\varphi(i)}) = [c_{i,1},\ldots,c_{i,n}]$$

Each MIX_j decrypts $c_{i,j}$ and publicizes its plaintext $v_{i,j}$ for $i=1,\ldots,l$. Then everybody can recover $m_{\varphi(i)}$ from k or more $v_{i,j}$. Each MIX_j proves that he behaved correctly by using a ZKIP.

If some P_i or MIX_j is detected to be faulty, he is ignored from that time on.

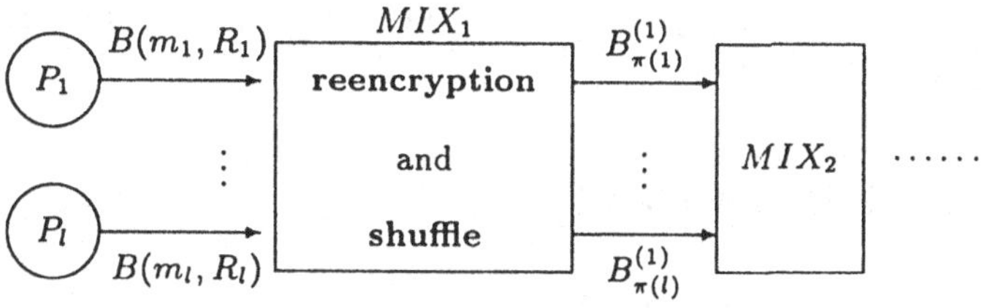

Fig. 1. Reencryption and shuffle, where $B^{(1)}_{\pi(i)} \triangleq B(m_{\pi(i)},R_{\pi(i)}+U_{\pi(i)})$.

2.2 Proposed scheme based on DLOG

This anonymous channel makes use of the scheme shown in Sec. 5.1 of [PIK93], which uses ElGamal public key cryptosystem. (It is called type 2 channel in [Pf94].)

Each MIX_j distributes his secret key x_j to all MIXes by using Shamir's (k,n)-threshold secret sharing scheme. Each sender proves that he knows his message by using a ZKIP to avoid the attack of [PP89]. Each MIX_j proves that he behaved correctly by using a ZKIP so that the attack of Sec.5.1 in [Pf94] does not work. Further, the attack of Sec.5.2 in [Pf94] does not work because x_j of a faulty MIX_j is revealed by using the (k,n)-threshold secret sharing scheme.

Let p and q be primes such that $q \mid p-1$. Let $g \in Z_p$ be a q th root of unity. Each MIX_j chooses a secret key $x_j \in Z_q^*$ and publicizes $y_j = g^{x_j} \bmod p$ as his public key. Further, he distributes x_j to all MIXes by using Shamir's (k,n)-threshold secret sharing scheme. He executes Feldman's non-interactive Verifiable Secret Sharing [Fe87].

(Sender's protocol) Each sender P_i encrypts his message m_i as

$$(G_i, M_i) = (g^r \bmod p, (y_1 y_2 \cdots y_n)^r m_i \bmod p)$$

He publicies (G_i, M_i) and proves that he knows m_i by using a ZKIP.

(Center's protocol) For $1 \le j \le n$, MIX_j reencrypts and shuffles $(G_1, M_1), \cdots, (G_l, M_l)$ sequentially. He proves that he behaved correctly by using a ZKIP.

(Decryption) At the end, suppose that MIX_n publicized $(\hat{G}_1, \hat{M}_1), \cdots, (\hat{G}_l, \hat{M}_l)$. For each $(\hat{G}, \hat{M})$, each MIX_j computes

$$G_j = \hat{G}^{x_j} \bmod p$$

and publicizes G_j. He proves the validity of G_j by using a ZKIP. Then the plaintext m is obtained by

$$m = \hat{M}/(G_1 \cdots G_n)$$

If there are some faulty MIXes and some G_j is not opend, the remained honest MIXes reveal the corresponding secret key x_j by using Shamir's (k,n)-threshold serect sharing scheme.

3 Efficient ZKIP for shuffle

In this section, we show a very efficient ZKIP for the center's protocol of Sec. 2.1. The communication complexity is $1/l$ of the standard ZKIP. A similar ZKIP can be obtained for that of Sec.2.2.

MIX_1 wants to prove that eq.(3) is computed from eq.(2) correctly in zero knowledge. In eq.(2) and eq.(3), let

$$B(m_i, R_i) = [a_{i,1}, \cdots, a_{i,n}]$$
$$B(m_{\pi(i)}, R_{\pi(i)} + U_{\pi(i)}) = [b_{i,1}, \cdots, b_{i,n}]$$

Let P denote a prover and V denote a verifier. Repeat the steps $1 \sim 4$ below $\log_2 N$ times.

(Step 1) P randomly computes a reencryption of each $B(m_i, R_i)$ such that $B(m_i, R_i + \hat{U}_i)$. Then P chooses a random permutation τ on $\{1, 2, \ldots, l\}$ and sends to V

$$[B(m_{\tau(1)}, R_{\tau(1)} + \hat{U}_{\tau(1)}), \cdots, B(m_{\tau(l)}, R_{\tau(l)} + \hat{U}_{\tau(l)})).] \tag{4}$$

Let
$$B(m_{\tau(i)}, R_{\tau(i)} + \hat{U}_{\tau(i)}) = [d_{i,1}, \cdots, d_{i,n}].$$
P also bit commits τ and $\phi \triangleq \tau\pi^{-1}$. He sends the commitals to V.
(Step 2) V sends to P a random bit e and random numbers t_i $(1 \leq i \leq l)$ such that $0 \leq t_i < r$.
(Step 3) P computes

$$U' \triangleq \begin{cases} \sum_{i=1}^{l} t_i \hat{U}_{\pi(i)} & \text{if } e = 0, \\ \sum_{i=1}^{l} t_i (\hat{U}_{\tau(i)} - U_{\phi(i)}) & \text{if } e = 1 \end{cases}$$

Let w_j $(1 \leq j \leq n)$ be the random number which satisfies

$$E_j(U'(j), w_j) = \begin{cases} \prod_{i=1}^{l} (d_{i,j}/a_{\tau(i),j})^{t_i} & \text{if } e = 0 \\ \prod_{i=1}^{l} (d_{i,j}/b_{\phi(i),j})^{t_i} & \text{if } e = 1 \end{cases} \qquad (5)$$

- If $e = 0$, P sends $U', w_1, \cdots, w_n$ and the decommmital of τ to V.
- If $e = 1$, P sends $U', w_1, \cdots, w_n$ and the decommmital of ϕ to V.

(Step 4) V accepts if

1. U' is a polynomial of degree at most $(k-1)$ such that $U'(0) = 0$.
2. Eq.(5) is satisfied for $1 \leq j \leq n$.

Theorem 3. *If eq.(3) is not computed from eq.(2) correctly, then*

$$\Pr(V \textit{ accepts in each round}) \leq 1/2 + 1/2r$$

for any (possibly cheating) prover.

References

[Be86] J. Benaloh, Secret sharing homomorphisms: Keeping a secret secret , in: *Proc. of Eurocrypt'86*, 251–260 (1986).

[CF85] J.D. Cohen and M.J. Fischer, A Robust and Verifiable Cryptographically Secure Election Scheme, in: *Proc. of 26th IEEE Symp. on Foundations of Computer Science*, 372–382 (1985).

[Ch81] D.L. Chaum, Untraceable Electronic Mail, Return Address, and Digital Pseudonyms, in: *Communications of the ACM, Vol.24, No.2*, 84-88 (1981).

[Fe87] P. Feldman, A Practical Scheme for Non-Interactive Verifiable Secret Sharing, in: *Proc. of 28th IEEE symposium on Foundations of Computer Science*, 427–437 (1987).

[MH96] M. Michels and P. Horster, Some Remarks on a Receipt-Free and Universally Verifiable Mix-Type Voting Scheme, in: *Proc. of Asiacrypt '96*, 125–132 (1996).

[Pf94] B. Pfitzmann, Breaking an Efficient Anonymous Channel, in: *Proc. of Eurocrypt '94*, 339–348 (1994).

[PIK93] C. Park, K. Itoh and K. Kurosawa, All/Nothing Election Scheme and Anonymous Channel, in: *Proc. of Eurocrypt '93*, (1993).

[PP89] B. Pfitzmann and A. Pfitzmann, How to Break the Direct RSA-implementation of Mixes, in: *Proc. of Eurocrypt '89*, 373-381 (1989).

An Implementable Scheme for Secure Delegation of Computing and Data

Josep Domingo-Ferrer[1] and Ricardo X. Sánchez del Castillo[2]

[1] Universitat Rovira i Virgili, Escola Tècnica Superior d'Enginyeria, Autovia de Salou s/n., E-43006 Tarragona, Catalonia, e-mail jdomingo@etse.urv.es
[2] Universitat de Barcelona, Servei d'Informàtica, Trav. Corts 131-159, E-08028 Barcelona, e-mail ricardo@dalila.ird.ub.es ***

Abstract. The need for delegating information arises when the data owner wants to have her data handled by an external party. If the external party is untrusted and data are confidential, delegation should be performed in a way that preserves security. Uses of delegation range from public administration to smart cards. In this paper, correctness and security requirements as well as protocols are specified for delegation of computing and data. A cryptographic solution to the secure delegation problem is described which provides data confidentiality and computation verifiability. Finally, an implementation allowing secure delegation of information over the Internet is briefly discussed.

1 Introduction

In many scenarios, data cannot be processed where they originate or belong to. In such cases, the data owner must transfer data to a remote environment (the handler) for processing. If data are confidential, a security problem arises. A legal solution to this problem is to require that the handler sign a non-disclosure agreement. For example, this is the procedure followed when some government agencies release data to universities for research purposes.

The above legal solution has a serious drawback: the data owner must *believe* that the handler is fair, since there is no technical means to prevent data misuse. In this paper, we describe an implementable solution that allows the data owner to control and limit the kind of operations performed by the handler. Depending on who is interested in the results of the data processing, two types of delegation scenarios can be distinguished:

Computing delegation A data owner (client) sends to a handler (server) a data set, some basic operations and an expression to be evaluated on the data. The handler subsequently returns to the owner the result of the evaluation of the expression on the set of data. The data owner is the client interested in the computation.

*** This work is partly supported by the Spanish CICYT under grant no. TIC95-0903-C02-02 and by the Statistical Institute of Catalonia under contract no. FBG-2577.

Data delegation A data owner (server) sends a data set to a handler (client), who thereafter performs some computation with those data. The handler is the client interested in the computation.

In the solution described below, *the work performed by the handler depends on the nature of the processing to be done, whereas the work done by the data owner is fairly independent of the processing* (in fact, the data owner functionality could be implemented in hardware). We illustrate next a few practical applications:

- A computing delegation problem happens whenever a (small) company wants to use external computing facilities to do some calculations on corporate confidential data. Think of a medical research team using a (insecure) university mainframe for processing confidential healthcare records. The reason for using external facilities may be the complexity of the calculations but also the huge volume of the data set.
- Data delegation problems appear when several lower-level organizations (municipalities, member states, etc.) cooperate with a higher-level organization (national agency, European Union, etc.) in data collection (census data, etc.). In return, the former organizations would like to analyze the whole collected data set, but only the latter organization is legally entitled to do so. Without secure data delegation, the higher-level organization will have to spend resources in (uninteresting) analyses requested by the cooperating organizations.
- In [4], availability of secure data delegation is relied on for increasing the multi-application capacity of smart cards. The basic idea is that if a very resource-demanding application is to be run on card-stored data then the card exports these data in encrypted form and the application is run on an external computing server.

Requirements and protocols for computing and data delegation are specified in section 2. Solving both problems relies on data confidentiality and computation verifiability. Encrypted data processing is sketched in section 3 as a way of preserving data confidentiality. A method achieving computation verifiability is described in section 4. Finally, section 5 briefly discusses the implementation of a prototype allowing secure delegation.

2 Requirements and Protocols for Delegation

The basic requirements of both computing and data delegation are *security* and *correctness*. In both types of delegation, security essentially translates to preservation of *data confidentiality*. This means that the handler should not know the computation input data.

In delegation of computing, correctness essentially translates to *verifiability of the handler computation* by the data owner. The handler may deviate from the claimed computation accidentally (overflow) or intentionally. In any verification procedure, there is a tradeoff between the confidence attained and the resources spent. If total confidence is desired, the owner must entirely repeat all computation (which makes delegation of computing useless); on the other hand, if the owner does not worry about the correctness of the computation, no check is needed. Intermediate solutions will be proposed in this paper which provide a reasonable confidence at a reasonable cost.

In data delegation, there is no correctness requirement from the data owner's viewpoint, since computation is done by the handler for his own use. From the handler's viewpoint, overflow detection is a serious problem when computing on encrypted data. The owner's verification can help as follows: if the owner detects fraud, she tells the handler. If the handler committed no fraud, he knows that overflow has occurred.

To meet the above requirements, the following protocols are proposed which rely on encryption

Protocol 1 (Computing delegation) *1. Prior to sending data to the handler, the owner encrypts them using an encryption transformation which allows some operations to be performed directly by the handler on encrypted data (homomorphical encryption).*
2. The handler returns a result which is still encrypted and must be decrypted by the data owner to recover the clear result.

Protocol 2 (Data delegation) *1. Data sent to the handler are encrypted by the owner using a homomorphical encryption transformation. The owner also supplies the handler with the operations on encrypted data supported by the encryption transformation used.*
2. Upon completing his computation, the handler sends the (encrypted) result to the owner for decryption. The handler also provides the expression used to obtain such a result, in order to allow some verification.
3. The owner verifies *the handler's computation. If no fraud can be detected and the claimed expression does not lead to evident disclosure of input data (inference controls should be applied here, but this is beyond the scope of this paper, see [2]), then the owner returns the decrypted result to the handler. Fraud occurs when the result does not correspond to the claimed expression.*

In summary, to solve both the secure computing and data delegation problems, we need to solve two more basic problems: data confidentiality and computation verifiability. We next propose solutions to such problems.

3 Data Confidentiality

As mentioned above, in both delegation problems the handler must be able to compute without being revealed the input data. This means that computation is to be carried out on encrypted data. Encryption transformations allowing some operations to be carried out directly on the encrypted data are known as privacy homomorphisms (PHs for short, see [8]).

Basically, such homomorphisms are encryption functions $E_k : T \longrightarrow T'$ allowing a set F' of operations on encrypted data without knowledge of the decryption function D_k. Knowledge of D_k allows the result of the corresponding set F of cleartext operations to be retrieved. As it has been shown above, the availability of secure PHs is central to the delegation of computation. RSA [9] is a well-known example of PH which allows multiplication and test for equality to be carried out on encrypted data. A summary of the state-of-the-art on PHs can be found in [4].

The choice of a particular PH depends on the type of the data to be dealt with. For qualitative (non-numerical) data, RSA resists chosen-plaintext attacks and provides a way for the handler to perform checks for equality on encrypted data. For numerical data, the PHs in [3] and [4] resist known-plaintext attacks and allow full arithmetic to be carried out by the handler on encrypted data.

4 Computation Verifiability

As noted above, absolute confidence can only be attained by the owner if she repeats the whole handler's claimed computation. But this makes delegation useless. Repetition of the claimed computation by the owner with low probability q, say 10%, only detects fraud or overflow with probability q. *Parity checking* is a more interesting alternative explained below, which can yield fraud or overflow detection probabilities as high as 50% (or more if intermediate results are verified).

Let the parity of an integer x be $Z(x) = x \bmod 2$. The parity of the addition or subtraction of two integers is obtained by XORing the parities of both integers. The parity of the product of two integers is obtained by ANDing the parities of both integers. If divisions are reduced to multiplications through use of fractions (as in the PHs of [3] and [4]), then any arithmetical expression that the handler claims to have computed on encrypted data can be mapped to a Boolean expression with XOR and AND gates yielding the parity of the claimed result.

Now, the following subprotocol can be embedded in protocols 1 and 2 to allow verification of the claimed expression:

Subprotocol 1 (Parity checking) *1. The handler supplies the Boolean parity formula associated to the claimed expression in SOP form (by a well-known theorem of switching theory, any Boolean expression can be rewritten in "sum of products" or SOP form,* i. e. *as the OR of several AND terms, see [5]).*

2. The data owner uses the supplied Boolean formula and her knowledge of clear input data to compute the claimed parity of the result (which is very fast). Then she decrypts the encrypted result computed by the handler and compares its actual parity against the claimed parity. If parities differ, fraud or overflow have been detected.

Notice that the handler never knows the parities of the data he is handling, because data are encrypted. The parity of the encrypted result is also unknown to the handler unless the computed expression is always even. Always-even expressions are those having the form $(d+d) \times e$, where d and e are subexpressions or input data (e is optional); as a side remark, notice that the handler cannot build always-odd expressions by just adding and multiplying encrypted data. If the claimed and the computed expressions differ but both are always even, they are indistinguishable using the parity checking subprotocol above. Therefore, always-even expressions should be forbidden by the data owner. This is not as restrictive as it may seem. If computing $(d+d) \times e$ is *really* needed, then compute instead $d \times e$. Once the clear outcome of $d \times e$ is known, just multiply by 2 in the clear. To recognize a (forbidden) always-even expression, the data owner must check whether the associated Boolean parity formula is equivalent to 0. This check is very easy for a Boolean formula in SOP form, where it amounts to checking that each AND term contains some input variable d and its complemented $\bar{d}$. Note that it is trivial for the owner to make sure that the Boolean formula provided by the handler is in SOP form.

The following properties of the parity checking method are easy to prove:

Lemma 1 Fraud detection probability. *If the claimed and the computed expressions differ, the probability of fraud detection by the owner is*

$$P(detect) = p(1-p') + p'(1-p)$$

where p and p' are, respectively, the probabilities of the claimed and computed expressions being odd.

Theorem 2. *For random input data, the probability of fraud detection by parity checking of the final result is tightly upper-bounded by 1/2.*

Theorem 3. *For random input data, the probability of fraud detection by parity checking of the final result is tightly lower-bounded by* $\max(p, p')$, *where p and p' are, respectively, the probabilities of the claimed and computed expressions being odd. If always-even claimed expressions are forbidden by the data owner, nonzero detection probability is guaranteed.*

Note. Even if always-even claimed expressions are forbidden, in data delegation the handler can fabricate for every $\epsilon > 0$ a claimed expression such that $0 < p < \epsilon$. But, for fraud to make sense, the computed expression is expected to be chosen on the basis of its usefulness to the handler, which means that p' takes a fixed value. Therefore, the lower bound $\max(p, p')$ —and consequently the detection probability— cannot be made arbitrarily small by the handler if the computed expression is to remain useful.

We conclude that the data owner can be assured of a nonzero detection probability but is *unable to compute $P(detect)$ nor the lower bound of theorem 3* because she does not know p'. On the other hand, theorem 2 gives an upper bound on the confidence attainable if verification of the final result passes. A way for the owner to increase the probability of detection (maybe above 1/2) is to verify several independent intermediate results. For example, if the claimed expression is regarded as a binary tree having input data as leaves and the result as root, then the owner may request from the handler the two encrypted intermediate results preceding the final result in the tree. If detection probabilities for such results are p_l and p_r, then the probability of fraud detection is $p_l + p_r - p_l p_r$. The procedure can be iterated by backtracking up the tree: if the owner requests also the four encrypted intermediate results preceding the previous two ones, then the probability of detection increases further. A trade-off is obvious: increasing the detection probability entails more verification work for the data owner.

5 Dikē: A Prototype for Secure Delegation of Computation

A prototype christened Dikē (blind Greek goddess of justice and social order, and also Delegation of Information without Knowledge Exposure) is about to be completed which allows secure delegation [1]. As a conclusion, the design rationale of Dikē is briefly discussed in this section.

It was argued in section 1 that delegation problems can be viewed as client-server problems. For computing delegation, the client is the data owner. For data delegation, the client is the handler. In this context, it becomes clear that Dikē must be a distributed application.

On the other hand, when operating on encrypted data, the way a given operation is implemented depends on the particular privacy homomorphism used for encryption. Polymorphism provided by object-oriented technology is very useful here, because it allows coding handler applications that are independent of the privacy homomorphisms used by the data owner. The code for an operation may depend on the data type in a transparent way. For instance, "multiply $E_k(a)$ and $E_k(b)$" will call a different code if $E_k()$ is RSA encryption or encryption using the PH of [4] (the code to be used is provided by the data owner, who chooses either privacy homomorphism).

So, we conclude that Dikē should have a distributed object-oriented architecture. Ideally, Dikē's architecture should also be language-independent and even platform-independent. CORBA [7][10] is a standard for distributed computing offering such independence and has been adopted for Dikē. More specifically:

- CORBA provides a universal notation for the software interfaces (Interface Definition Language or IDL), with standard maps for many languages.
- The CORBA abstraction simplifies the construction of distributed computing applications and this allows the developer to concentrate in the core of the problem.
- Reusability, interoperability and scalability are concepts central to CORBA, and are inherited by CORBA-based applications.

Even if the resulting application is platform and language-independent, a platform and a language were needed for development. Unix and C++ have been chosen. CORBA is available for Unix and translators from IDL to C++ exist. Moreover, C++ has the advantage of being object-oriented and widely used.

Finally, an arithmetical library had to be chosen to implement encryption, decryption and operations on encrypted data. Our requirements were: speed, wide range of number-theoretic primitives, easy multiprecision handling and an object-oriented interface. LiDIA [6] was our choice. Although no specific cryptographic primitives are included in LiDIA, they are very easy to derive from the available primitives.

References

1. J. Castilla, J. Domingo-Ferrer and R. X. Sánchez, *Dikē: Delegation of Information without Knowledge Exposure*, internal reports #1 (Nov. 1996) and #2 (Feb. 1997).
2. D. E. Denning, *Cryptography and Data Security*, Addison-Wesley, 1982 (chapter 6 on inference controls).
3. J. Domingo-Ferrer, "A new privacy homomorphism and applications", *Information Processing Letters*, vol. 60, no. 5, pp. 277-282, Dec. 1996.
4. J. Domingo-Ferrer, "Multi-application smart cards and encrypted data processing", *Future Generation Computer Systems*, vol. 13, no. 1, pp. 65-74, July 1997.
5. F. J. Hill and G. R. Peterson, *Introduction to Switching Theory and Logical Design*. New York: Wiley, 1981.
6. The LiDIA Group, *LiDIA Manual. A Library for Computational Number Theory*. Ver. 1.3, Feb. 1997. TH Darmstadt/Universität des Saarlandes. ftp://ftp.informatik.th-darmstadt.de/pub/TI/systems/LiDIA
7. Object Management Group, *OMG Common Request Broker Architecture: Architecture and Specification (CORBA)*, Revision 2.0. OMG Document Number 96.03.04, March 1996.
8. R. L. Rivest, L. Adleman and M. L. Dertouzos, "On data banks and privacy homomorphisms", in *Foundations of Secure Computation*, R. A. DeMillo *et al.*, Eds. New-York: Academic Press, 1978, pp. 169-179.
9. R. L. Rivest, A. Shamir and L. Adleman, "A method for obtaining digital signatures and public-key cryptosystems", *Communications of the ACM*, vol. 21, pp. 120-126, Feb. 1978.
10. J. Siegel, *CORBA Fundamentals and Programming*. New York: Wiley, 1996.

Electronic Commerce with Secure Intelligent Trade Agents

Jaco van der Merwe (vdmerwj@mwp.eskom.co.za) and
S.H. von Solms (basie@rkw.rau.ac.za)
Department of Computer Science, Rand Afrikaans University,
P.O. Box 524, Auckland Park, 2006, South-Africa.

Abstract
Electronic commerce on the Internet has the potential to generate billions of transactions but the number of merchants providing goods or services on the Internet will be so large, that it will become impossible for humans to visit each site and decide where it is best to buy or sell goods. In this paper we develop intelligent trade agents that roam a network, collect and analyse the data from servers on the network and make decisions to buy and sell goods on behalf of a user. The combination of distributed-object technology and single and public key encryption mechanisms makes these agents *secure intelligent trade agents*. We show that distributed-object technology is an enabling technology for intelligent trade agents.

Key words: Electronic commerce, intelligent agents, public key encryption, distributed-objects, transaction authorisation.

1. Introduction

The promise of high bandwidth at very low cost has conjured visions of an information highway that turns into the world's largest shopping mall. Billions of electronic transactions will be generated on a daily basis [4, 7, 12, 13].

The amount of information that is available on the Internet is so large that it becomes near impossible for humans to visit each site on the network, analyse this information and make sound business decisions as to the trade of goods. Businesses and individuals need electronic agents who can roam the network and trade on their behalf.

In this paper we develop *intelligent trade agents* that have the ability to roam a network, collect business related data and use this data to make decisions to buy or sell goods on its owner's behalf. We show how an authorisation server can be used to authorise transactions and provide payment for goods or services bought by intelligent trade agents.

For an intelligent trade agent to roam a network, it must visit many different server systems and execute at these server systems. When an agent executes at a foreign server system it is vulnerable to attacks from these servers. We show that intelligent trade agents, implemented as distributed-objects, can logically visit many different server systems, but physically remain secure inside the protective boundaries of an agent repository. The agents are further secured by encrypting all messages between the agents and other entities on the network with pubic key encryption combined with DES single-key encryption.

2. Intelligent Agents

Intelligent software agents are probably one of the fastest growing areas of information technology. They are being used, and touted for applications as diverse as personalised information management, electronic commerce, computer games, etc. An agent can be thought of as a computer program that simulates a human relationship by doing something that another person could do for you [6]. An agent is a self-contained program capable of controlling its own decision making and acting, based on its perception of its environment, in pursuit of one or more objectives [9].

More than one type of agent is possible. In its simplest form it is a software object that sifts through large amounts of data and present a subset of this data as useful information to another agent, user or system. An example of this is an agent that reads and analyse all incoming e-mail, and route it to an appropriate department or another agent for reply [14]. These types of agents are called *static agents*.

Mobile agents have the ability to migrate themselves across nodes of a network in order to perform their tasks and report back their findings. These agents would typically gather and analyse data from a multitude of nodes on the network, and present a subset of this data as information to a user, agent or system. An example of this is an agent, such as *Book Worms Bargainbot*, that visit all virtual bookstores, and present to its user a list of specific books available and the best available price on each [8]. Mobile agents can also act as brokers for users, for example a single sign-on agent can sign-on to many different systems relieving the user from typing in his/her password for every system [3, 14].

In this paper we examine mobile agents that have the ability to trade goods. One example that we describe in more detail in later sections, is an agent that roams the stock exchanges of the world and trade shares on a user's behalf. In this way the agent can, using fixed rules, build up a valuable portfolio of shares for a user.

3. Intelligent Trade Agents

An intelligent agent that roams a network to buy and sell goods on behalf of its owner has many advantages, for example it can allow businesses to respond quickly to market opportunities and give them the competitive edge that is required in today's business world.

An *Intelligent Trade Agent* (ITA) is a mobile agent with the ability to visit many different sites on a network, analyse the data at each site and make decisions as to whether goods should be traded at the specific site. An ITA is autonomous: it can decide where to travel on a network and what goods or services to buy without human intervention. An intelligent trade agent should be able to pay for goods or services bought. We show how an ITA can use credit cards to provide payment to merchants.

Figure 1 shows an ITA roaming a network, buying and selling goods or services at three different merchant's servers on a network.

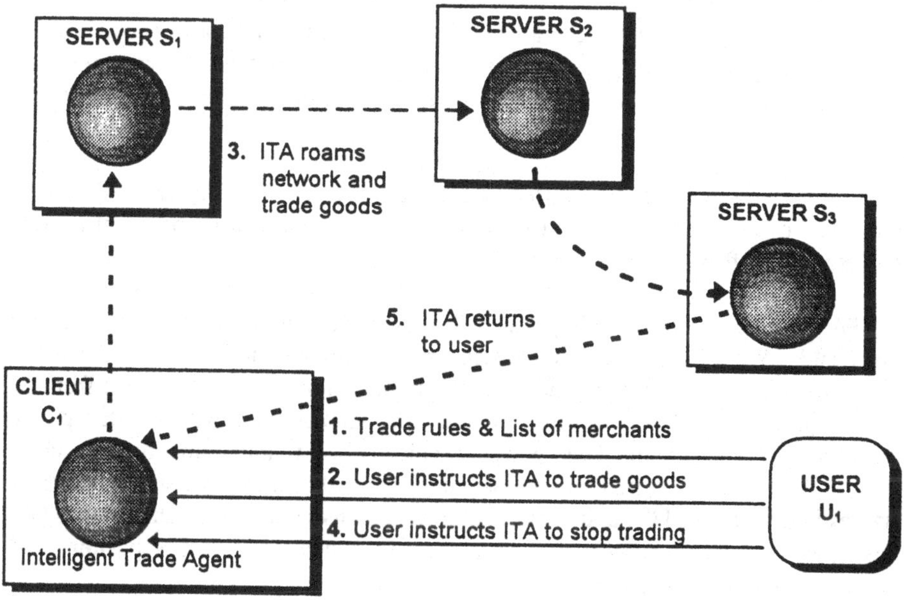

Fig. 1 An Intelligent Trade Agent Roaming a Network.

Steps in figure 1:

1. User U_1 sets-up its ITA by providing it with a list of trade rules (what to buy and sell and under what conditions) and a list of servers (merchants) to visit and trade with. Most of the set-up information will be done *only once* and thereafter the ITA is merely instructed to start trading and stop trading using the set-up information.
2. The user instructs its ITA to start roaming and trade. At this point the user could disconnect from its client computer and leave the agent to trade on its own.
3. The ITA now roams the network trading goods using the rules its user provided.
4. In this step the user instructs the ITA to stop trading and return home (see the section below).
5. The ITA returns to its user's client computer and the user could inspect the ITA to see what goods were traded.

It must be possible to stop agents from roaming the network and instruct them to occasionally report back to home. The question is where is an agent's home? It can't be the user's client computer since users don't always use the same client computer and might not even be logged-in while the agent travels the network. We introduce the notion of an *Agent Repository* where all agents are stored and centrally controlled and maintained. In section 4 we show how ITAs can logically roam a network but physically remain at the Agent Repository.

3.1 Agent Repository (AR) and Authorisation Server (AS)

The *Agent Repository (AR)* keeps all ITAs. ITAs are never kept at a user's client computer. The user of an ITA doesn't send its ITA directly to any server computer, it instructs the Agent Repository to activate the ITA which will instruct the ITA to roam the network.

b

The Agent Repository has a security advantage: it will only activate ITA's for its legitimate owner. The AR will first authenticate the ITA's owner before it accepts any instructions from the ITA's user.

An ITA must be able to provide payment for goods bought and receive payment for goods sold. We use a mechanism where an ITA provides to a merchant a user's credit card number. The merchant uses this information to claim payment from an *Authorisation Server (AS)*. The user's credit card number is encrypted with a public key encryption mechanism. The AS would typically be a financial institution such as a bank. The AS makes sure that merchant's servers only claim payment for goods or services bought by an ITA and ensures that merchants receive payment when ITAs buys goods or services from them. Figure 2 shows how an Agent Repository and Authorisation Server are used.

Note that there are many security issues involved in the process such as authentication of the user with its ITA and secure communication between the ITA and other ITAs, server systems and the user as well as secure payment. Figure 2 shows the logical steps in the process and describes the secure payment mechanism that is used. Section 4 describes the other security issues in the process such as authentication and secure communication.

Figure 2 indicates the *logical* steps involved in the mechanism, figure 3 in section 4 shows the *physical* steps in the mechanism and indicates that the ITA does not physically move to any servers but remains inside the Agent Repository.

Notation used:

A_{pub} is the public key of entity A and A_{priv} is the private key for entity A. E(M, A_{pub}) indicates that message M is encrypted with key A_{pub}.

Symbols used:

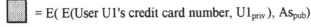

[light box] = E(E(User U1's credit card number, $U1_{priv}$), As_{pub})

[black box] = E(E(Description of goods/services bought, ITA_{priv}), As_{pub})

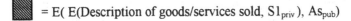

[hatched box] = E(E(Description of goods/services sold, $S1_{priv}$), As_{pub})

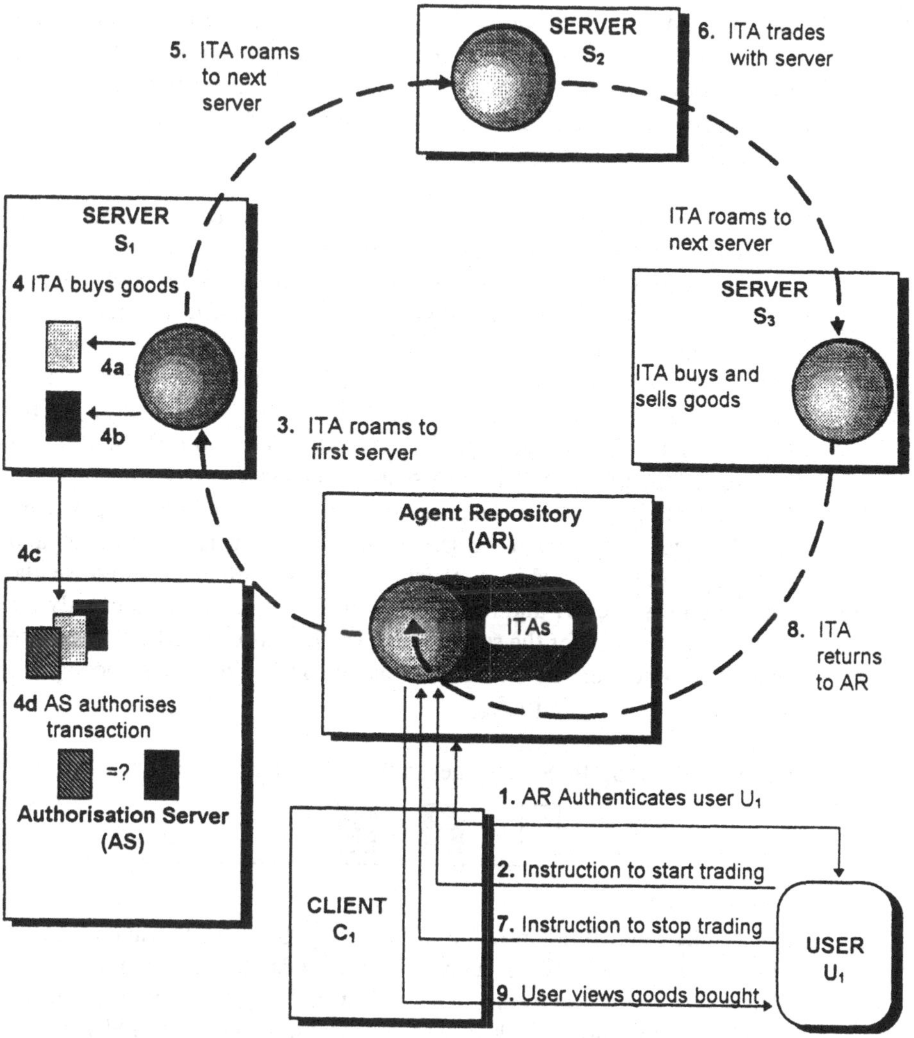

Fig. 2. An Intelligent Trade Agent Executing Transactions on a Network.

Steps in figure 2:

1. Before user U_1 at client computer C_1 can issue any instructions to its ITA, it must be authenticated by the Agent Repository. Any authentication mechanism such as a encrypted password could be used.
2. The user instructs the ITA to start trading.
3. The AR instructs the ITA to roam the network which causes the ITA to move to server S1.

4. In this step, the ITA trade with server S1. This means that the ITA buys or sells goods or services from the merchant at server S1. The ITA must be able to pay for goods or services bought. In this section we describe how the transaction is authorised and payment given. The payment method must be secure. We use a public key encryption mechanism.

4a. The ITA gives to server S1 its user U1's credit card number. This information is pre-encrypted (by the user U1) with AS_{pub} and $U1_{priv}$. The AS keeps a copy of the encrypted version of the user's credit card number. This means only the AS can decrypt the message and it can only be U1 that created the message.

4b. The ITA creates a message indicating what services or goods where bought. This information is encrypted with AS_{pub} and ITA_{priv}. This means only the AS can decrypt the message and it can only be the ITA that created the message. The ITA sends this message to the Server S1.

4c. Server S1 needs the AS to authorise the transaction and ensure payment. The ITA has already given the server user U1's financial information and a message indicating what it bought. The AS will only authorise the transaction if server S1 and the ITA can agree that they traded exactly the same goods/services for the same price. To do this server S1 generates the same description of the goods sold as the ITA did. This message is encrypted with $S1_{priv}$ and AS_{pub}.

Server S1 sends all three messages to the AS for authentication.

4d. The AS decrypts all the messages using AS_{priv}, $S1_{pub}$, ITA_{pub} and checks if S1's version and the ITA's version of the goods traded (and the time of transaction) are the same. The AS would typically be a financial institution which will check if user U1 has sufficient funds and if so the transaction is authorised and the AS will provide the merchant at server S1 with payment. The AS might be a financial institution and will use traditional fund transfer mechanisms to ensure that payment is transferred to the merchant at server S1's bank account. The details of how the AS transfers funds from user U1's bank account to the merchant at server S1's bank account is not relevant.

At this point the transaction is completed.

5,6 The ITA roams the network and trades with other servers. When required, the AS authorises transactions.

7. The owner instructs the ITA, via the AR, to stop trading. It is also possible to set-up the ITA in such a way that it automatically stops trading if certain conditions are met, for example a certain number of goods have been bought.
8. The AR waits for the ITA to report back and instructs it to immediately return to the AR.
9. The user may inspect the ITA to see what goods were bought or sold.

The above process shows how an intelligent agent can travel the intergalactic network and trade on its owner's behalf. The ITA is intelligent because it can decide when to buy/sell goods and when to roam to the next server without human intervention.

When an agent executes in the address space of a foreign server system, it (an specifically its private key) is vulnerable to attacks from these servers. The next section shows how ITAs are secured using distributed objects.

4. Securing Intelligent Trade Agents with Distributed Objects

There are several security issues to be considered in mobile agent-based computing: authentication of the user (the sender of the ITA), determination of the agent's integrity (ITAs must be protected from theft and unauthorised modifications) and determination of the agent's ability or willingness to pay for services or goods provided by a server [10].

In this section we describe *secure intelligent trade agents* and show how the communication between the user, the Agent Repository, the ITA and the merchant's server object are secured with a public key and DES single-key encryption mechanism.

4.1 Secure Intelligent Trade Agents

Figure 1 and figure 2 showed an ITA migrating itself from one server to another to trade with each server. The ITA is in fact *roaming* the network. The ability of an agent to roam a network has the advantage that the agent can visit many different servers to determine the best available supplier to trade goods or services with. If the ITA migrates itself from one server to another it means that it executes in the memory space of these servers. No matter how we protect this ITA, while a copy of the ITA is in the memory space of a foreign server system, it is very vulnerable to attacks from that server which means that its private key can be stolen and illegitimate transactions executed by the server system.

We need an architecture that will allow the ITA to visit foreign server systems to trade with them, but physically remain in the protective boundaries of the Agent Repository. This means that logically the ITA still roams form one server to the next, but physically it remains at one secure location. The architecture that makes this kind of roaming possible is *distributed-object technology*.

A C++ or Smalltalk object encapsulates code and data, and can be specialised by means of inheritance, but can't reach across compiled-language or address-space boundaries. In contrast, *distributed objects* are packaged as binary components accessible to remote clients by means of method invocations. Clients don't know which language or compiler built a server object, or where on the intergalactic network the object physically resides [4]. They need to know only its name and the interface it publishes.

If we implement an intelligent trade agent as a distributed object and have each merchant's server system present its goods or services to the outside world with another distributed object, then these distributed objects (the ITA and merchant object) can execute methods on each other without being in the same address space. Figure 3 shows an ITA, implemented as a distributed object, trading with a merchant object, also implemented as a distributed object.

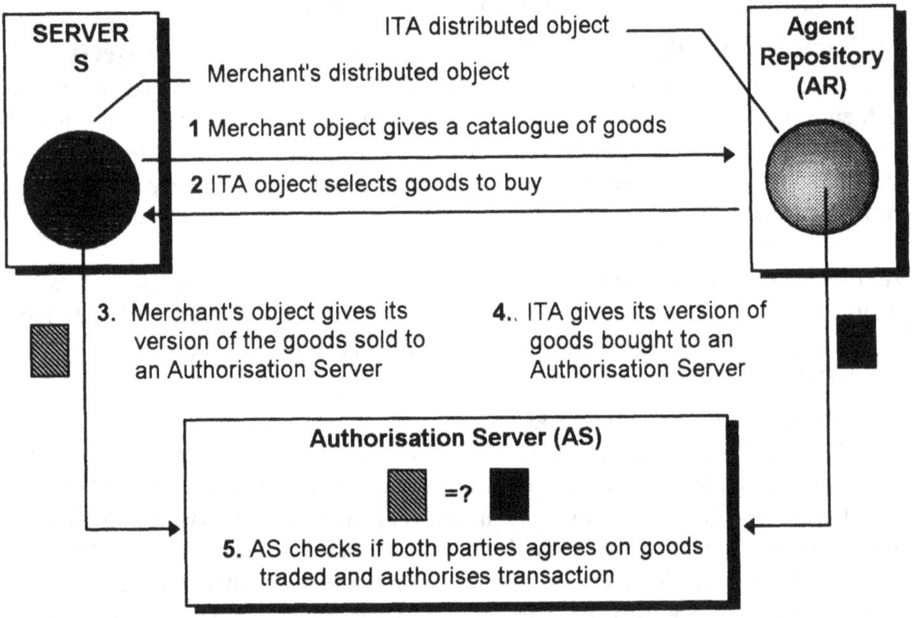

Fig. 3. An Intelligent Trade Agent Object and a Merchant Object Trading.

Figure 3 shows that an ITA, implemented as a distributed object, executing methods on a merchant distributed object (on a remote server system S) to obtain a catalogue of goods/services sold by this merchant. The ITA decides to buy these goods from the merchant which causes them to generate a signed message, indicating what they traded, and give this messages to an authorisation server for authorisation.

It is important to notice that all the signed messages are still encrypted with the private and public keys as described in the previous section. Previously, the ITA's private key moved with it to the server system which meant that the server system could have attacked the ITA and retrieved ITA_{priv}. With a distributed-object architecture, the ITA doesn't move to the server system, but rather remotely calls methods on the server object. This means that the ITA stays inside the secured boundaries of the Agent Repository and the ITA (and therefor ITA_{priv}) is not exposed to attacks from any server system.

4.2 Secure Communication

To create secured messages we use a public key encryption mechanism combined with the DES single-key cryptosystem.

Communication between the ITA, other ITA, merchant server systems, the Agent Repository and the Authorisation Server is encrypted using a combination of RSA public-key cryptography combined with the DES secret-key cryptosystem.

Suppose a server S sends a message M to an intelligent trade agent ITA. The message is encrypted as follow. S generates a DES secret-key K, encrypts M with K so that M2 = E(M, K). S then encrypts the secret-key K with the ITA's public key so that M3 = E(K, ITA_{pub}). The encrypted messages M2 and M3 together form a *digital envelope* [11] that is send to the ITA.

Upon receiving the digital envelope, the ITA decrypts the DES key with its private key and uses this secret-key to decrypt the message itself. The public keys of the ITA and the application servers must be published in some central directory. The Agent Repository could be used as a directory service.

5. Applications of ITAs Implemented as Distributed Objects

Intelligent agents travelling the intergalactic network buying and selling goods with the ability to securely and quickly pay or receive payment make a large number of applications possible. In this section we describe two applications that illustrate the use of intelligent trade agents.

5.1 Supply and Demand Agents

Jennings and Wooldrige [10] classify agents in commerce as either *user agents* acting on behalf of the consumer or *business agents* representing suppliers. There is no reason why an ITA, if configured and managed well, can't represent a user and business agent. Such an agent can analyse the market and as soon as it can identify a demand for a specific product, start roaming the network and buy these goods from merchants that can supply them at the best price. The ITA then goes back to the source of the demand and sells the goods at a profit.

These supply and demand ITAs generate income for its owner by identifying demands and supplying in those demands. The actual goods never reaches the ITA's owner, instead the ITA makes sure that any goods bought are sold immediately and delivered to the new owner.

5.2 ITAs on the International Stock Exchange

Stock brokers often buy and sell shares simply by analysing market indicators. If an ITA is given these market indicators and a set of rules associated with each indicator, then it is possible for such an agent to decide when and which shares to buy or sell.

If we equip an ITA with the above mentioned indicators, rules and a list of stock exchange servers to visit, then the ITA can roam the network, buying and selling shares and in this way build and manage a valuable portfolio of shares for a user.

A similar agent, *Mobile Trader* [13], is already active on the Internet. This agent keeps an eye on commodities, and speaks up when the time is right to buy or sell. This agent does not have the ability to buy without user intervention. Trade agents become truly useful when they can buy or sell before the opportunity is lost.

6. The Enabling Technology for Intelligent Trade Agents

Only a few months ago, it would have been near impossible to implement intelligent trade agents on a world-wide basis allowing agents to *securely* visit servers and trade goods. Proprietary solutions, such as General Magic [10], allows for roaming agents and make agent-based electronic commerce possible but it is a closed standard with a limited development environment [2]. Intelligent trade agents will only be useful in an open environment where they can roam freely to a large number of servers and interact with these servers.

Distributed-object technology promises the most flexible client/server systems. This is because it encapsulates data and logic in objects that can roam anywhere on networks, run on different platform, talk to legacy applications by way of object wrappers, and manage themselves and the resources they control [5]. This description of distributed-object technology shows that it is a very good candidate technology to implement intelligent trade agents with.

With intelligent trade agents implemented as distributed objects on an open ORB architecture, intelligent trade agents may soon become something no business can be without.

6.1 Existing Trade Agents

We emphasise that agents roaming a network, advising users when to buy or sell commodities is already a reality, for example *Mobile Trader* [13], but intelligent trade agents *securely* roaming across hundreds of servers all over the world with the ability to buy or sell commodities when they think it is appropriate, have only just become possible with the maturing of distributed-object technology.

7. Summary & Conclusions

Electronic agents can act as trade agents for users on a network. These intelligent trade agents have the ability to decide what goods to trade and with which merchants. ITAs make businesses more responsive to market changes and allow them to utilise business opportunities as soon as they appear.

With distributed-object technology maturing to a level were it is possible to implement intelligent trade agents that can roam to and from servers all over the world, intelligent trade agents can give businesses the ability to react quickly to market opportunities and maximise profits.

8. Further Research

This paper described how an intelligent trade agent can roam the markets of the world to trade on behalf of its owner. In further research these intelligent trade agents could be implemented in an existing network environment (such as the Internet) to investigate the actual performance of these agents in a physical environment. We are interested in the performance overhead these agents might cause (will they be able react quick enough to seize market opportunities). It is also necessary to investigate the maturity of current object request brokers as a means to implement the communication between servers and ITA distributed objects.

9. References

[1] R. Orfali and D. Harkey, *Client/Server Survival Guide with OS/2*, John Wiley & Sons, 1994.

[2] R. Orfali, D. Harkey and J. Edwards, *The Essential Distributed Objects Survival Guide*, John Wiley & Sons, 1996.

[3] J. van der Merwe and S. H. von Solms, *Authorisation and Authentication with Intelligent Agents*, submitted to Computers & Security, January 1997.

[4] R. Orfali and D. Harkey, *Client/Server with Distributed Objects*, BYTE Magazine, April 1995, p. 151-162.

[5] R. Orfali, D. Harkey and J. Edwards, *Intergalactic Client/Server Computing*, BYTE Magazine, April 1995, p. 108-122

[6] T. Selker, *A Teaching Agent that learns*, Communications of the ACM 37 (7), 1994, p 92-99.

[7] A. R. Immel, *It's Business as Unusual on the Information Superhighway*, Trends Magazine, October 1994.

[8] Netsurfer Digest, *BOOK' BOT A good Idea that needs work*, Vol 2 (17), 6 June 1996.

[9] N. Jennings and M. Wooldridge, *Software Agents*, IEEE Review, January 1996, p. 17 - 20.

[10] C. Harrison, D. M. Chess and A. Kershenbaum, *Mobile Agents: Are they a good idea?*, IBM Research Report (T.J. Watson Research Center), March 1995.

[11] P. Fahn, *About Today's Cryptography*, Answers to Frequently Asked Questions, RSA Laboratories, September 1993.

[12] Advanced Information Management Strategies, *Payment Systems for the Internet*, The Meta Group, August 1996.

[13] J. Horberg, *Talk to My Agent: Software Agents in Virtual Reality*, Computer-Mediated Communication Magazine, Vol 2 (2), 1 February 1995, p. 3

[14] L. Wirthman, *Gradient DCE has sign-on feature*, PC Week, March 1996, p. 31.

Efficient Scalable Fair Cash with Off-line Extortion Prevention

Holger Petersen[1] · Guillaume Poupard

École Normale Supérieure, DMI, 45, rue d'Ulm 75005 Paris, France
hpetersen@geocities.com Guillaume.Poupard@ens.fr

Abstract. There have been many proposals to realize anonymous electronic cash. Although these systems offer high privacy to the users, they have the disadvantage that the anonymity might be misused by criminals to commit *perfect crimes*. The recent research focuses therefore on the realization of fair electronic cash systems where the anonymity of the coins is revocable by a trustee in the case of fraudulent users.
In this paper, we propose a new efficient fair cash system which offers *scalable security* with respect to its efficiency. Our system prevents extortion attacks, like blackmailing or the use of blindfolding protocols under *off-line* payments and with the involvement of the trustee only at registration of the users. Another advantage is, that it is assembled from well studied cryptographic techniques, such that its security can easily be evaluated. The strength of this approach is clearly its simplicity. Although it might astonish the reader that the design matters little from existing schemes, it is nevertheless the first scheme offering these properties.

1 Introduction

The concept of *anonymous* digital cash was invented in 1982 by Chaum [Chau82]. Unfortunately, this perfect anonymity might be misused by criminals to commit a *perfect crime* [SoNa92]. Blackmailing of coins, money laundry, extortion or theft of secret keys or the use of blindfolded protocols with the bank or trustees have been considered as possible new attacks [SoNa92, BrGK95, JaYu96].

In order to prevent these threats the payment systems should provide *anonymity revocation* mechanisms, that allow the tracing of coins in any of the above scenarios by an authorized third party, the *trustee*, or a set of these parties. The first systems preventing blackmailing and money laundry were proposed by [BrGK95, StPC95]. Since then there have been several proposals [CaPS96, M'Raï96, FuOk96, JaYu96, JaYu97] to prevent these attacks. All schemes require the participation of the trustee in the opening of an account or even in the withdrawal of coins. The only two systems that don't require trustee participation except for initialization of the system and for anonymity

[1] The author's work was granted by a postdoctoral fellowship of the NATO Scientific Committee disseminated by the DAAD.

revocation have recently been proposed by [CaMS96, FrTY96]. However, they are unable to prevent extortion attacks and the use of blindfolding protocols. These attacks were only prevented in the systems of [JaYu96, JaYu97, FuOk96], which are therefore not as efficient as they require the trustee interaction in payment protocols. In case that one of these attacks is reported, they require an *on-line* payment protocol between user, shop and trustee in order to prevent the spending of illegal coins.

Our results
We propose a secure payment system that allows anonymity revocation by trustees in the case of any extortion attack under an *off-line* payment protocol. This is achieved by registering pseudonymous user–keypairs at a trustee before withdrawal and by the use of securized revocation lists. Thus our system is more efficient than the one of [JaYu96], which achieves the same extortion prevention at the cost of an interaction with the trustee at each withdrawal and under on-line payments. Although different payments under the same pseudonym are linkable, the *privacy* of payments is *scalable* by the user by increasing their number up to a maximum of one pseudonym per payment to satisfy total unlinkability. Remarkable *benefits* of this approach are a modular, simple design that is easy to understand, to implement and to analyze with respect to security requirements, as different cryptographic tools interact as little as possible (in contrast to systems like e.g. [Bran93, JaYu96, CaMS96]). The design results from a careful analysis of the attacks and requirements described in section 2. Finally, our system is versatile, as it allows the integration of multi-spendable and divisible coins and supports the *challenge semantics* [JaYu96]. Due to space limitations, we give only a brief description of our system. A more detailed view can be found in the full version [PePo97].

2 Electronic Payment Systems

In this section we review the main properties of electronic payment systems.

Events
We assume a simplified electronic cash system with just one bank, one user and one shop. Extensions to many of each are straightforward. In this setting seven main events are distinguishable:

1. *Initialization:* Choice of system parameters and keypairs of all entities.
2. *Opening account:* Bank opens user's account and registers personal data.
3. *Registration:* In the pseudonymous systems, the user registers at the trustee.
4. *Withdrawal:* The user withdraws coins from his account onto his device.
5. *Payment:* The user pays at the shop using the coins stored on his device.
6. *Deposit:* The shop deposits digital coins at the bank and is credited for them.
7. *Revocation:* The trustee computes either the shape of the coin from the withdrawal transcript or computes the user's identity from the payment transcript.

Models
In an electronic payment system there participate eight types of entities: These

are users, banks, shops, trustees, judges, certification authorities, key directories and finally the attackers. All entities have different security requirements and also a different amount of trust in each other. We refer to [JaYu96] for a detailed discussion of the underlying security and trust model.

Attacker model: In the *traditional world* exist many different kind of attackers: thieves, bank robbers, blackmailers, kidnappers or terrorists. Their interests and methods are different but they all have to be physically involved in their crime at some moment or even for a longer period (as for kidnapping). In the setting of *electronic cash* systems some new attacks have been considered, as the *perfect crimes* first mentioned in [SoNa92] and the *ultimate crimes* introduced in [JaYu96]. These attacks are of short duration and without physical involvement of the attacker. After their appearance, the entity acts again by his own will. Observe, that this model *excludes* crimes like kidnapping or robbery of goods, as it is obvious that cryptography cannot help to prevent them.

Communication model: We assume for the *internet payment* scheme, that all communication is transmitted via a network and thus interceptable and vulnerable to eavesdropping attacks. For the *electronic purse* scheme, we assume that the communication between the bank and user and also between the user and shop is protected against these attacks and thus confidential, as the user is physically present.

Technical requirements

For an electronic payment system a secure user device is needed in order to store the user's secret information. For example, in the setting of an *internet payment* system the device can be integrated in the user's computer as a black box. In the case of an *electronic purse* it might be a tamper resistant hardware[2] integrated on a smart card. In this context the use of a secure and trustworthy device under control of the user, the *electronic wallet*, is discussed [ChPe92].

Attacks

Attacks against electronic cash systems can be classified by who is *attacking*, who is *attacked* or their *strength*. We give a classification by the attacking entity.

Fraudulent user

- *Overspending:* A user spends coins for a value exceeding their allowed value.

Fraudulent shop

- *Impersonation:* A shop respends or deposits a user's coin several times.
- *Money laundry:* A shop obtains digital coins by an illegal action. To conceal the origin of the money he issues a fictitious bill.

Fraudulent bank

- *Tracing a user:* The bank traces the relationship between a coin and a user.
- *Tracing a user with help of trustee:* False conviction of trustee that a coin has been overspent. As a result a honest user is incriminated after tracing his identity for this coin.

[2] The existence of tamper resistant devices is questioned, see e.g. [AnKu96].

- *Framing a user:* False accuse of overspending a coin.
- *Framing a shop:* False accuse of double deposit of a valid coin.
- *Coin forgery after overspending:* The bank generates fictitious payment transcripts for an already overspent coin in order to be reimbursed immoderately.

Fraudulent trustee

- *Framing a user:* The trustee falsely identifies a honest user. Thereby the bank might incriminate this user without justification.

Fraudulent outsider

A fraudulent outsider is an entity which might be – but is not necessarily – registered to the trustee or have a bank account. Banks and trustees are not considered to be fraudulent outsiders.

- *Universal coin forgery:* An entity knowing public parameters (and old payment transcripts) forges the bank's signature scheme and obtains valid coins.
- *One-more coin forgery:* An entity, that participates in n (parallel) withdrawal protocols obtains $n+1$ valid coins.
- *Overspending coin forgery:* An entity knowing several payment transcripts of an overspent coin generates transcripts for spendable fresh coins.
- *Eavesdropping of coins or pseudonyms:* A passive attacker eavesdrops the communication at withdrawal, payment or deposit in order to obtain spendable coins. An active eavesdropper might also act as *man-in-the-middle* and modify the protocol data. The same strategy is applicable at registration to obtain signed pseudonyms.
- *Theft or extortion of coins from the user:* An attacker either steals coin transcripts from the user's device or forces him to withdraw coins from his account and to transfer them to the attacker's device.
- *Theft or extortion of coins from the shop:* An attacker either steals transcripts of coins from the shop's device or forces the shop to reveal them.
- *Theft or extortion of secret keys:* The attacker either steals the secret keys of the bank (user/trustee), e.g. by hacking into its system or forces to reveal them. In the second case, the bank (user/trustee) is aware that the attack happened, which might not be the case after a theft. If this attack is used against a user, the attacker can also steal his electronic purse. Extortion might be impossible, if the user's secret keys are only stored in the tamper proof hardware of his device.
- *Blindfolding:* The attacker forces the bank (trustee) to engage in blindfolded protocols in order to obtain digital coins (certified pseudonyms), that he can spend successfully (use to withdraw untraceable coins).

These attacks might also be performed by a coalition of several entities, e.g. the bank and shop or the user and shop. A *fraudulent outsider* might attack several entities at once, e.g. bank and user or bank and trustee, in order to execute one of the attacks.

Security requirements

In order to resist the above attacks, untraceable electronic payment systems should fulfill many security requirements. We first describe those, that are imposed to be satisfied by all systems.

Unforgeability: Only authorized entities are able to issue valid digital coins.

Untraceability: The relationship between a digital coin and a user is untraceable for the bank, except in the case of authorized revocation.

Unlinkability: Different coins spent by the same user are unlinkable.

Framing: No user or shop can be falsely incriminated by the bank or trustee.

Secondly, there are additional requirements in order to obtain fair electronic cash systems with revocable anonymity [CaMS96, FuOk96, FrTY96]:

Overspent-tracing: The bank can determine the identity of the user who overspends a coin. It is either realized by a separate mechanism or in the same way as user-tracing.

User-tracing: Bank and trustee cooperate to match a spent coin to the user.

Coin-tracing: Bank and trustee cooperate to compute information that allows the matching of a coin when it is spent or deposited. In some systems it might be possible, that the user himself announces the necessary information after an extortion of digital coins.

Extortion-tracing: Bank and trustee cooperate in order to compute information that allows the matching of a coin when it is spent or deposited. In some systems it might be necessary, that the trustee participates in an online payment protocol in order to prevent the undetectable spending of an extorted coin [FuOk96, JaYu96].

Notice that overspent-tracing is achieved in all recent payment systems but extortion-tracing was only possible using an *on-line* payment protocol [JaYu96, FuOk96]. We are the first to present a solution that allows an *off-line* prevention of this attack.

Versatility

The basic concept of electronic coins might be extended in order to obtain versatile and efficient systems. Additional properties are the *transferability* of coins between users, the *k-spendability* of a coin of value v or more generally its *divisibility* in fractions, such that the value of all fractions is equal to v. Other useful properties might be the *loss tolerance* of unspent coins to credit their value if the device is lost or *multi-currency* coins [JaYu96, WaPf89].

3 An efficient fair cash system

We focus the description on the payment of coins for a *unique* coin value and a *single* time of validity. Generally, a combination of these two aspects is achieved either by a corresponding keypair of the bank or by using advanced techniques (e.g. [Bran93]).

3.1 Basic design

The design of most payment systems is based on the approach of Chaum [Chau82]. We also take use of it but apply the following modifications: The user opens an account at the bank and registers at the trustee. He obtains a certified pseudonymous keypair (PS_x, PS_y). Then he uses a blind signature scheme in order to withdraw anonymous digital coins from the bank. To prevent various attacks, the public part of the user's pseudonym PS_y is embedded into the coin. At payment the coin is spent using a "challenge-and-response" protocol between the user and shop, where PS_x is used to generate a signature σ_C on a random challenge m, which is verifiable using the corresponding public key PS_y.

3.2 Cryptographic tools

We need six cryptographic tools as building blocks. These are

1. a collision resistant cryptographic hash function h,
2. three (different) signature schemes $(\mathcal{G}_T, \mathcal{S}_T, \mathcal{V}_T)$, $(\mathcal{G}_U, \mathcal{S}_U, \mathcal{V}_U)$, $(\mathcal{G}_C, \mathcal{S}_C, \mathcal{V}_C)$ used by the trustee, user and for signing coins resp. (see [GoMR88]),
3. a blind signature scheme $(\mathcal{G}_\mathcal{B}, \mathcal{S}_\mathcal{B}, \mathcal{V}_\mathcal{B})$ used by the bank to sign coins,
4. an interactive authentic key exchange protocol $\mathcal{K}$ with mutual user authentication that generates a fresh authentic session key $K_{A,B} := \mathcal{K}(ID_A, ID_B)$ for the authentic communication between A and B (for a survey see [RuOo94]),
5. a probabilistic encryption scheme $(\mathcal{E}_\mathcal{Z}, \mathcal{D}_\mathcal{Z})$ used with the trustee and shops. The encryption of a message m for $\mathcal{Z}$, $\mathcal{Z} \in \{T, S\}$ is denoted as $e := \mathcal{E}_\mathcal{Z}(m)$ and decryption as $m := \mathcal{D}_\mathcal{Z}(e)$,
6. a symmetric cryptosystem $(\mathcal{E}_K, \mathcal{D}_K)$, where K is the session key.

These tools are defined as usually (see e.g. [MeOV97]). The three signature schemes might be chosen independently by the entities. $(\mathcal{S}_T, \mathcal{V}_T)$ must be an existentially unforgeable signature scheme in order to prevent transparently blindfolding attacks [JaYu96]. The key generation algorithms $\mathcal{G}_\mathcal{Z}$ with $\mathcal{Z} \in \{B, T, U, C\}$ return keypairs $(x_\mathcal{Z}, y_\mathcal{Z})$ for the bank, trustee, user and signature generation for the coins respectively. The generation of a signature by entity $\mathcal{Z}$ for message m is described as $\sigma_\mathcal{Z} := \mathcal{S}_\mathcal{Z}(x_\mathcal{Z}, m)$ and the verification of this signature by $\mathcal{V}_\mathcal{Z}(y_\mathcal{Z}, m, \sigma_\mathcal{Z}) \in \{true, false\}$. $Ind(x)$ is a pointer to the stored secret key x.

3.3 Databases and revocation lists

The databases (DB), black- and whitelists (BL/WL) used in our scheme are resumed with their content and security requirements in figure 1.

We assume – as it was implicitely assumend in all other proposals – that the databases are *confidential/authentic* to their owners and that the black- and whitelists are *integer* and *authentic* even after an attack. This is the only strong assumption needed by our scheme to guarantee its security. Furthermore the blacklists are assumed to be updated faster than an illegal withdrawn or extorted coin might be spend, i.e. there is a delay associated with each payment. Remark, that only a supplement to the existing revocation lists has to be distributed, which consists usually only of a little amount of data.

owner	abbre-viation	access at/after	stored information	authenticity guaranteed by	confiden-tiality
databases					
trustee	PS$_y$-DB	registration	$ID_U, PS_y, \sigma_U, Ind(x_T)$	σ_U	yes
bank	U-DB	opening account	ID_U, acc_U	–	no
	W-DB	withdrawal	ID_U,C,$Ind(x_B)$, $\mathcal{E}_T(h(PS_y, C))^3$	$\tilde{\sigma}_B$	no
	D-DB	deposit	$C, PS_y, ID_s, \sigma_T, \sigma_C$	σ_T, σ_C	no
user	C-DB	withdrawal and	$C, \sigma_B, Ind(PS_x)$	σ_B	no
	PS$_x$-DB	payment	$PS_x, (PS_y), \sigma_T$	σ_T	yes
shop	P-DB	payment	$C, PS_y, \sigma_B, \sigma_T, \sigma_C$	σ_B, σ_C	no
revocation lists					
shop	U-BL	user extortion	voided PS_y	$\mathcal{S}_T(x_T, PS_y)$	no
	C-WL	bank extortion	valid $h(PS_y, C)$'s	$\mathcal{S}_T(x_T, C)$	no
	B-BL	bank extortion	voided y_B	$\mathcal{S}_T(x_T, y_B)$	no
	PS-WL	trustee extortion	valid PS_y	$\mathcal{S}_T(\tilde{x}_T, PS_y)^4$	no
	T-BL	trustee extortion	voided y_T	$\mathcal{S}_{CA}(x_{CA}, y_T)$	no

Fig. 1. Databases and revocation lists

3.4 General protocol

Opening an account

1. The user U registers at the bank and obtains an account number acc_U.
2. The bank stores (ID_U, acc_U) in its U-DB.

Registration at trustee

1. To identify himself to the trustee and to obtain an authentic session key $K_{U,T}$, the user and trustee participate in the authentic key exchange protocol $\mathcal{K}$ and obtain $K_{U,T} := \mathcal{K}(ID_U, ID_T)$. To obtain a confidential communication $K_{U,T}$ is used to encrypt all communication in steps 2. and 3.
2. The user generates a pseudonymous keypair $(PS_x, PS_y) := \mathcal{G}_C$. He computes $\sigma_U := \mathcal{S}_U(x_U, (ID_U, PS_y))$ and sends $\mathcal{E}_{K_{U,T}}(ID_U, PS_y, \sigma_U)$ to the trustee.
3. The trustee verifies $\mathcal{V}_U(y_U, (ID_U, PS_y), \sigma_U) \stackrel{?}{=} true$, calculates $\sigma_T := \mathcal{S}_T(x_T, PS_y)$ and transmits $\mathcal{E}_{K_{U,T}}(\sigma_T)$ to the user U. He stores $(ID_U, PS_y, \sigma_U, Ind(x_T))$ in his PS$_y$-DB.
4. U verifies $\mathcal{V}_T(y_T, PS_y, \sigma_T) \stackrel{?}{=} true$ and stores all values.

These steps might be processed several times to obtain several pseudonymous keypairs (PS_x, PS_y). For an efficient electronic purse protocol, it is possible to make the modification that the trustee generates $(PS_x, PS_y) := \mathcal{G}_C$ himself, signs PS_y, transmits all values confidentially to the user and keeps PS_x secret for a later use in a symmetric cryptosystem.

[3] In case of electronic purse, this is the symmetric encrypted value $\mathcal{E}_{PS_x}(h(PS_y, C))$.

[4] The signature is generated using a fresh trustee's key $\tilde{x}_T$ generated after the attack.

Withdrawal protocol

1. In order to identify himself to the bank and to obtain an authentic session key $K_{U,B}$ for the communication, the user and bank participate in the authentic key exchange protocol $\mathcal{K}$ and obtain $K_{U,B} := \mathcal{K}(ID_U, ID_B)$. $K_{U,B}$ is used to authenticate the communication in steps 2. and 3. as during registration.
2. The user U generates random coin C (possibly containing some redundancy). He computes $\tilde{C} := blind(h(C, PS_y))$, $e_T := \mathcal{E}_T(h(PS_y, C))$[5] and transmits these values to the bank. In case of an extortion attack against the bank, it sends all stored values e_T to the trustee (as described below).
3. The bank computes $\tilde{\sigma}_B := \mathcal{S}_B(x_B, \tilde{C})$ and sends it to the user U. It subtracts the value of the coin from the user's account acc_U and stores $(ID_U, Ind(x_B), \tilde{C}, e_T)$ in his W-DB.
4. The user computes $\sigma_B := unblind(\tilde{\sigma}_B)$ and verifies $\mathcal{V}_B(y_B, h(C, PS_y), \sigma_B) \stackrel{?}{=} true$. If the verification fails, he asks the bank to resend the signature for the blinded coin $\tilde{C}$. He keeps (C, σ_B) as his coin and notices the relation to the tuple (PS_x, σ_T).

Payment protocol

1. The shop sends the user a uniquely generated message *mess*.
2. The user generates $\sigma_C := \mathcal{S}_C(PS_x, (C, ID_S, mess), PS_y)$ and sends the payment transcript $(C, PS_y, \sigma_B, \sigma_T, \sigma_C)$ to the shop. In the case of internet payments, the value σ_C is encrypted as $e_S := \mathcal{E}_S(\sigma_C)$ in order to prevent its eavesdropping.

3a. If *no extortion* attack was reported, the shop verifies $\mathcal{V}_T(y_T, PS_y, \sigma_T) \stackrel{?}{=} true$, $\mathcal{V}_B(y_B, h(C, PS_y), \sigma_B) \stackrel{?}{=} true$ and also $\mathcal{V}_C(PS_y, (C, ID_s, mess), \sigma_C) \stackrel{?}{=} true$. He stores the payment transcript together with *mess* in his P-DB.

3b. After a *user extortion* attack the shop receives an actualized U-BL from the trustee. If $PS_y \in$ U-BL, he rejects the coin. Otherwise he accepts it.

3c. After a *bank extortion* attack the shop receives an actualized B-BL and C-WL from the trustee. If ($y_B \in$ B-BL and $h(PS_y, C) \notin$ C-WL) he rejects the coin. Otherwise he accepts it as in step 3a.

3d. After a *trustee extortion* attack, the shop receives an actualized T-BL and PS-WL from the trustee. If σ_T was generated under $y_T \in$ T-BL and $PS_y \notin$ PS-WL, he rejects the coin. Otherwise he accepts it as in step 3a.

Step 1 is omitted if the user computes a fresh, unique value *mess*, e.g. as a function of *time*, *contract* etc. In the case of divisible coins *mess* must contain the fraction C_f of coin C that is spent (see section 3.5).

Deposit protocol

1. The shop sends the tuple $(C, PS_y, \sigma_B, \sigma_T)$ to the bank.
2. The bank verifies $\mathcal{V}_T(y_T, PS_y, \sigma_T) \stackrel{?}{=} true$, $\mathcal{V}_B(y_B, h(C, PS_y), \sigma_B) \stackrel{?}{=} true$ and checks, whether the coin was already deposited under the same pseudonym PS_y. In this case, it finds a tuple (C, σ_C') in D-DB and sends

[5] $e_T := \mathcal{E}_{PS_x}(h(PS_y, C))$ in the case of electronic purse payments,

σ'_C to the shop as a proof. If $\sigma_C = \sigma'_C$ the shop is accused of double deposit and cannot deposit the coin. If $\sigma_C \neq \sigma'_C$ or if the coin C was not already deposited, the bank sends the shop a signed acknowledgment herefor.

3. The shop sends the tuple $(ID_S, acc_S, mess, \sigma_C)$ to the bank.
4. The bank verifies $\mathcal{V}_C(PS_y, (C, ID_s, mess), \sigma_C) \stackrel{?}{=} true$ and checks if C has been overspent under PS_y. In this case it initiates the *user-tracing* protocol described below with the trustee to identify the cheating user. As a proof, it sends the trustee the payment transcripts $(C, PS_y, mess_1, \sigma_B, \sigma_T, \sigma_{C,1}), \ldots, (C, PS_y, mess_k, \sigma_B, \sigma_T, \sigma_{C,k})$.
5. If every thing is okay the bank stores the payment transcript in its P-DB and credits the shop's account acc_S by the value of C.

Overspent-tracing and user-tracing

1. To prove that a coin C was overspent the bank sends the trustee several transcripts $(C, PS_y, mess_1, \sigma_B, \sigma_T, \sigma_{C,1}), \ldots, (C, PS_y, mess_k, \sigma_B, \sigma_T, \sigma_{C,k})$.
2. The trustee verifies all transcripts. If they are correct, he looks for the tuple (ID_U, PS_y, σ_U) in his PS_y-DB and sends (ID_U, σ_U) to the bank.
3. The bank checks $\mathcal{V}_U(y_U, (PS_y, ID_U), \sigma_U) \stackrel{?}{=} true$.

Coin-tracing

Usually coin-tracing is considered to prevent extortion or theft of coins. Here, this attack is impossible, as the knowledge of C without the corresponding secret key PS_x doesn't allow to spend the coin. Thus this attack should be considered only if the pseudonymous keypair (PS_x, PS_y) is compromized at the same time. In this case, the *extortion-tracing* described below applies as countermeasure.

Extortion-tracing

We focus on the actions taken by the attacked party to guarantee that the extorted values are put on a revocation list. The treatment by the shops has already been described in steps 3b-d. of the payment protocol.

1. *Extortion of user's secret key PS_x:* After finding (PS_y, ID_U) in his PS_y-DB, the trustee puts PS_y on U-BL and distributes this list immediately among the shops. He issues $\mathcal{S}_T(x_T, (ID_U, PS_y))$ to the user which allows him to exchange unspent coins withdrawn under PS_y at the bank, after he has identified as their legal owner.
2. *Extortion of bank's secret key x_B:* To avoid the forgery of electronic coins in case of an extortion of x_B, the bank sends the encrypted values $\mathcal{E}_T(h(PS_y, C_i))$ of all legally withdrawn coins C_i in his W-DB to the trustee. The trustee decrypts the $h(PS_y, C_i)$'s and puts them on C-WL. Additionally, the public key y_B corresponding to x_B is put on B-BL. Both lists are immediately distributed among the shops. The bank generates a fresh keypair $(\tilde{x}_B, \tilde{y}_B) := \mathcal{G}_B$ for the issue of new coins.
3. *Extortion of trustee's secret key x_T:* After an extortion of x_T the trustee puts all pseudonyms PS_y in his PS_y-DB legally signed under this key on PS-WL and at the same time the public key y_T that corresponds to x_T on T-BL. Both lists are distributed immediately among the shops. The trustee generates a fresh key pair $(\tilde{x}_T, \tilde{y}_T) := \mathcal{G}_T$ for the certification of user pseudonyms.

4. *Blindfolding coins under bank's secret key x_B:* Transparent blindfolding of coins is not possible, because of the authentic key exchange protocol $\mathcal{K}$ with user authentication between the user and the bank. If the user enforces the blindfolded withdrawal of a coin without proper identification, the bank is aware that the attack happens and takes the same countermeasures as in case of extortion of x_B.
5. *Blindfolding pseudonyms PS_x, PS_y under trustee's secret key x_T:* As transparent blindfolding is impossible due to the use of an existentially unforgeable signature scheme [JaYu96] the trustee is aware of the attack and takes the same countermeasures as under point 3. in the case of extortion of x_T.

Reducing trust in trustees
In the above protocol, the user has to trust the trustee that

- he doesn't reveal his pseudonym to anyone, such that his payments will become traceable (*anonymity w.r.t. to trustee* as requested in [JaYu96] is not yet satisfied).
- he interacts correctly for decryption of $e_i := \mathcal{E}_T(PS_y, C_i)$ in case of an extortion of the bank's secret key x_B as otherwise the user is unable to spend valid coins C_i signed under x_B afterwards.

It depends on the *trust model* if the users accept these conditions. Otherwise it is possible to reduce trust in the trustee by sharing PS_y and e_i among several trustees [FuOk96]. Then the signature scheme $(\mathcal{G}_T, \mathcal{S}_T, \mathcal{V}_T)$ is substituted by a multisignature scheme and the probabilistic encryption scheme $(\mathcal{E}_T, \mathcal{D}_T)$ by an encryption scheme with threshold-decryption (for a survey see [Desm94]).

3.5 Versatility of system

The *k-spendability* or more general *divisibility* of coins is obtained as in the schemes [JaYu96, M'Raï96] by including the spend fraction C_f of the coin C into the message at payment. The overspent-tracing is modified accordingly. The system also supports the use of *challenge semantics* introduced by Jakobsson and Yung. The issue of *receipts* can be added to all transactions in order to minimize the trust between the entities.

3.6 Scalability of cryptographic tools

The choice of concrete cryptographic tools in our general scheme allows to obtain scalability w.r.t. to security and efficiency. We discuss possible choices of those tools, that influence the efficient implementation on the user's device.

Choice of blind signature scheme $(\mathcal{S}_\mathcal{B}, \mathcal{V}_\mathcal{B})$: Tradeoff between efficient blinding & storage of σ_B on the user's device and provable security. The most *efficient* scheme w.r.t. *blinding* is the blind RSA signature scheme with small public exponent. It allows message-recovery and therefore signed values don't have to be stored. The most *efficient* scheme w.r.t. signature *storage* is the blind Schnorr signature [ChPe92, Okam92]. A *provably computational secure* choice is the blind Okamoto signature [PoS96b]. Also other schemes, like restrictive blind signatures

[Bran93] that allow to prevent overspending without the help of the trustee, might be considered.

Choice of user and coin signature schemes ($\mathcal{S}_\mathcal{C}, \mathcal{V}_\mathcal{C}$) and ($\mathcal{S}_\mathcal{U}, \mathcal{V}_\mathcal{U}$): Tradeoff between efficient generation and provable security. PKP [Sham89] or CLE [Ster94] are *original* choices well suited for electronic purse implementation. An *efficient* and at the same time *provably computational secure* scheme is the Schnorr signature [Schn89, PoS96a]. Besides these, any signature scheme from the ElGamal-family (e.g. [ElGa85, HoMP94]) or RSA might be used.

Coin C: Tradeoff between *efficiency*, like a multi-spendable or divisible coin and *privacy*, i.e. the linkability of such payments.

4 Conclusion

We presented an efficient payment system with anonymity revocation. It is one of the first schemes, that achieves off-line prevention of all kind of extortion attacks. Due to its scalable security w.r.t. efficiency, it can be scaled for the two main applications of secure internet payment and efficient electronic purse, where the security and efficiency requirements are quite different. A detailed description of these schemes including a concrete realization is given in the full version of this paper [PePo97]. Furthermore a classification of fair payment systems and a detailed security analysis of the scheme can be found there.

References

[AnKu96] R.Anderson, M.Kuhn, "Tamper Resistance - A Cautionary Note", Usenix Electronic Commerce Workshop, (1996), 11 pages.

[Bran93] S.Brands, "Untraceable Off-Line Cash in Wallets with Observers", LNCS 773, Advances in Cryptology – Crypto '93, Springer, (1994), pp. 302–318.

[BrGK95] E.Brickell, P.Gemmell, D.Kravitz, "Trustee-based Tracing Extensions to Anonymous Cash and the Making of Anonymous Exchange", Proc. 6.ACM-SIAM SODA, (1995), pp. 457–466.

[CaMS96] J.Camenisch, U.Maurer, M.Stadler, "Digital Payment Systems with Passive Anonymity-Revoking Trustees", LNCS 1146, Proc. ESORICS'96, Springer, (1996), pp. 31–43.

[CaPS96] J.Camenisch, J.-M.Piveteau, M.Stadler, "An efficient Fair Payment System", Proc. 3rd ACM-CCS, ACM Press, (1996), pp. 88–94.

[Chau82] D.Chaum, "Blind signatures for untraceable payments", Advances in Cryptology – Crypto '82, Plenum Press, (1983), pp. 199–203.

[ChPe92] D.Chaum, T.P.Pedersen, "Wallet databases with observers", LNCS 740, Advances in Cryptology – Crypto '92, Springer, (1993), pp. 89–105.

[Desm94] Y.Desmedt, "Threshold Cryptography", European Trans. on Telecommunications, Vol. 5, No. 4, (1994), pp. 35–43.

[ElGa85] T.ElGamal, "A public key cryptosystem and a signature scheme based on discrete logarithms", IEEE Transactions on Information Theory, Vol. IT-30, No. 4, July, (1985), pp. 469–472.

[FrTY96] Y.Frankel, Y.Tsiounis, M.Yung, " "Indirect discourse Proofs": Achieving Efficient Fair Off-Line E-Cash", LNCS 1163, Advances in Cryptology – Asiacrypt'96, Springer, (1996), pp. 286–300.

[FuOk96] E.Fujisaki, T.Okamoto, "Practical Escrow Cash System", LNCS 1189, Proc. 1996 Security Protocols Workshop, Springer, (1997), pp. 33–48.

[GoMR88] S.Goldwasser, S.Micali, R.Rivest, "A secure digital signature scheme", SIAM Journal on Computing, Vol. 17, 2, (1988), pp. 281–308.

[HoMP94] P.Horster, M.Michels, H.Petersen, "'Meta-ElGamal signature schemes"', Proc. 2. ACM conference on Computer and Communications security, ACM Press, November, (1994), pp. 96–107.

[JaYu96] M.Jakobsson, M.Yung, "Revokable and Versatile Electronic Money", Proc. 3rd ACM-CCS, ACM Press, (1996), pp. 76–87.

[JaYu97] M.Jakobsson, M.Yung, "Applying Anti-Trust Policies to Increase Trust in a Versatile E-Money System", Proc. Financial Cryptography Workshop, (1997), 21 pages.

[MeOV97] A.Menezes, P.C.van Oorshot, S.Vanstone, "Handbook of Applied Cryptography", CRC Press, (1997).

[M'Raï96] D.M'Raïhi, "Cost Effective Payment Schemes with Privacy Regulations", LNCS 1163, Advances in Cryptology - Asiacrypt '96, Springer, (1996), pp. 266–275.

[NIST95] National Institute of Standards and Technology, Federal Information Processing Standards Publication, FIPS Pub 180-1: Secure Hash Standard (SHA-1), April 17, (1995), 14 pages.

[Okam92] T.Okamoto,"Provable secure and practical identification schemes and corresponding signature schemes", LNCS 740, Advances in Cryptology - Crypto'92, Springer, (1993), pp. 31–53.

[PePo97] H.Petersen, G.Poupard, "Efficient scalable fair cash with off-line extortion prevention", full version of this paper, Technical Report LIENS–97–07, May, (1997), 33 pages, `http://www.dmi.ens.fr/EDITION/preprints/`.

[PoS96a] D.Pointcheval, J.Stern, "Security Proofs for Signatures", LNCS 1070, Advances in Cryptology - Eurocrypt'96, Springer, (1996), pp. 387–398.

[PoS96b] D.Pointcheval, J.Stern, "Provably Secure Blind Signature scheme", LNCS 1163, Advances in Cryptology - Asiacrypt'96, Springer, (1996), pp. 252–265.

[RiSA78] R.L.Rivest, A.Shamir, L.Adleman, "A method for obtaining digital signatures and public-key cryptosystems", Comm. of the ACM, Vol. 21, (1978), pp. 120–126.

[RuOo94] R.Rueppel, P.C.van Oorschot, "Modern key agreement techniques", Computer Communications, Vol. 17, Vol. 7, (1994), pp. 458 - 465.

[Schn89] C.P.Schnorr, "Efficient identification and signatures for smart cards", LNCS 435, Advances in Cryptology - Crypto '89, Springer, (1990), pp. 239–251.

[Sham89] A.Shamir, "An efficient Identification Scheme Based on Permuted Kernels", LNCS 435, Adv. in Cryptology - Crypto '89, Springer, (1990), pp. 606–609.

[SoNa92] S.von Solms, D.Naccache, "Blind signatures and perfect crimes", Computers & Security, Vol. 11 , (1992), pp. 581–583.

[StPC95] M.Stadler, J.-M.Piveteau, J.Camenisch, "Fair-Blind Signatures", LNCS 921, Advances in Cryptology - Eurocrypt '95, Springer, (1995), pp. 209–219.

[Ster94] J.Stern, "Designing identification schemes with keys of short size", LNCS 839, Advances in Cryptology - Crypto '94, Springer, (1995), pp. 164–173.

[WaPf89] M. Waidner, B.Pfitzmann, "Loss-tolerant Electronic Wallet", Proc. Smart Card 2000, North-Holland, (1991), pp. 127–150.

A Security of the scheme

For the security analysis we benefit from the modular design of our system using well known cryptographic primitives. Although all attacks mentioned in section 2 are proved to be prevented, the security analysis is clearly structured as we avoided interaction between the mechanisms as much as possible. Theorems 1 – 6 analyse the influence of the cryptographic tools described in section 3.2 for security and theorems 7 – 9 with the importance of the different revocation lists, introduced in section 3.3.

Definition 1 *A signature scheme* $(\mathcal{S}, \mathcal{V})$ *is called* provably computational secure, *if existential forgery of a signature with respect to an adaptively chosen message attack is proved to be equivalent to a known computational hard problem (e.g. factorization or discrete log).*

Theorem 1 *Assuming* $(\mathcal{S}_\mathcal{B}, \mathcal{V}_\mathcal{B})$ *is a provably computational secure* blind *signature scheme (in the sense of [PoS96b]) and the function h is a collision intractable hash function, the system achieves unforgeability and therefore is secure against* coin forgery.

Proof 1: Since $\mathcal{S}_\mathcal{B}$ is secure against existential forgery, all coin signatures σ_B on the message $h(C, PS_y)$ must have been (blindly) generated by the bank. As the hash function h is collision intractable by the user, it is impossible for him to find a coin value $C' \neq C$ (or $PS'_y \neq PS_y$) with $h(C', PS_y) = h(C, PS_y)$ ($h(C, PS'_y) = h(C, PS_y)$ respectively). Thus, the system achieves unforgeability of coins. □

Theorem 2 *Assuming* $(\mathcal{S}_\mathcal{B}, \mathcal{V}_\mathcal{B})$ *is a provably computational secure* blind *signature scheme (in the sense of [PoS96b]) and* $(\mathcal{E}_T, \mathcal{D}_T)$ *is a strong probabilistic encryption scheme, the system achieves untraceability and therefore is secure against* tracing a user by the bank.

Proof 2: As $(\mathcal{S}_\mathcal{B}, \mathcal{V}_\mathcal{B})$ is a perfect blind signature scheme, the signed coin (C, σ_B) is untraceable to the view $(\tilde{C}, \tilde{\sigma}_B)$. Furthermore, the ciphertext e_T delivered with the blind coin, doesn't allow the bank to match it to a given coin later-on, as $(\mathcal{E}_T, \mathcal{D}_T)$ is a probabilistic encryption scheme. □

Theorem 3 *Assuming* $(\mathcal{S}_C, \mathcal{V}_C)$ *is a provably computational secure signature scheme, the system is secure against* (1) impersonation, (2) framing a user by the bank, (3) tracing a user with help of trustee *and* (4) coin forgery after overspending.

Proof 3: Since anybody who does not know the secret key PS_x cannot produce a signature σ_C of a tuple $(C, ID_S, mess)$ such that $\mathcal{V}_C(PS_y, \sigma_C, (C, ID_S, mess)) =$ *true*, impersonation by the bank is impossible. If the bank wants to frame an honest user who did not overspent a coin, she must show at least two tuples $(C, PS_y, mess_1, \sigma_B, \sigma_T, \sigma_{C,1})$ and $(C, PS_y, mess_2, \sigma_B, \sigma_T, \sigma_{C,2})$. But since $(\mathcal{S}_C, \mathcal{V}_C)$ is a provably computational secure signature scheme, the bank cannot produce more signature $\sigma_{C,i}$ than she received. So the attack (2) is prevented. The two last ones are also avoided for the same reason, the knowledge of some signatures $\sigma_{C,1}, \ldots \sigma_{C,k}$ on the same coin by the bank does not help to generate new signatures of the user on this coin. □

Theorem 4 *Assuming* $(\mathcal{S}_\mathcal{U}, \mathcal{V}_\mathcal{U})$ *is a provably computational secure signature scheme, the system is secure against* framing a user by the trustee.

Proof 4: Since the trustee needs to generate a valid signature $\mathcal{S}_\mathcal{U}(x_U, (ID_U, PS_y))$ for a given PS_y and any known ID_U in order to frame user U, this attack is not possible assuming $(\mathcal{S}_\mathcal{U}, \mathcal{V}_\mathcal{U})$ is a provably computational secure signature scheme. □

Theorem 5 *Assuming* $(\mathcal{S}_\mathcal{T}, \mathcal{V}_\mathcal{T})$ *is a provably computational secure signature scheme, the system achieves* overspent- *and* user-tracing *and is therefore secure against* overspending, money laundry *and* transparently blindfolding of the trustee.

Proof 5: Since $\mathcal{S}_\mathcal{T}$ is unforgeable, a signature σ_T is a proof that the trustee knows a relation between a real user identity and a signed pseudonyms PS_y. In the mechanism of user tracing, the bank sends a tuple $(C, PS_y, mess, \sigma_B, \sigma_T, \sigma_C)$ and the trustee verifies all signatures. Since it was already proved, that they were necessarily generated by the correct entities, σ_B authenticates a coin $h(C, PS_y)$, then σ_C proves that C was in fact spent under pseudonym PS_y and finally σ_T proves that the trustee knows a pair (PS_y, ID_U) to relate C to ID_U. So *overspending* and *money laundry* are prevented. Moreover, engaging the trustee in a transparently blindfolding protocol for σ_T is impossible, as assuming the contrary the user would know two valid signatures after only one interaction with signer (the blinded one and the unblinded one), which contradicts the existential unforgeability of $\mathcal{S}_\mathcal{T}$. □

Theorem 6 *Assuming, the key exchange protocol* $\mathcal{K}$ *is secure against any active and passive attacks and the symmetric cryptosystem* $(\mathcal{E}_\mathcal{K}, \mathcal{D}_\mathcal{K})$ *is not breakable without knowing the proper key* K*, the system is secure against* eavesdropping of coins and pseudonyms *and* framing of a shop by a bank.

Proof 6: The user communication at registration and withdrawal is protected under an authentic session key obtained from the authentic key exchange protocol $\mathcal{K}$. As it is resistant to eavesdropping and man-in-the-middle attacks, the communication protocols inherit this property. Since transmission of σ_C between the user and the shop is encrypted under y_S, the bank does not know σ_C and thus can't frame the shop. □

We are now going to study the security of our system against extortion of coins or secret keys. The case of their theft is similar if it is immediately discovered (e.g. by frequent audit). If the theft is not remarked, there are no cryptographic ways to protect the system.

Theorem 7 *Assuming that the user blacklist (U-BL) is properly used*[6]*, the system achieves all kinds of extortion-tracing and is thus secure against* extortion of coins.

Proof 7: The *extortion of coins from the user* in order to spend or deposit them is impossible, as the relation between C and PS_y is embedded in the structure of the coin and protected by σ_B. Thus C has to be signed with the corresponding

PS_x at payment, which appears in the U-BL, if it was stolen together with C. The *extortion of coins from the shop* in order to deposit them at the bank on the attackers account is impossible, as the shop's identity is included in the signed message as payment, which prevents to credit them to another account. □

Theorem 8 *Assuming that all blacklists described in section 3.3 (U-BL, B-BL and T-BL) are properly used* [6], *the system achieves all kinds of extortion-tracing and is thus secure against* extortion of secret keys.

Proof 8:

- In case of *extortion of the user's secret key* PS_x, PS_y is blacklisted immediately and therefore any coin withdrawn under this pseudonym is rejected by any shop.
- In case of an *extortion* of the bank's secret key x_B, it is put immediately on B-BL, which prevents his further use. The hash values of all legally obtained coins signed under x_B are listed in an authentified whitelist, which prevents the use of others, that have been signed hereafter using x_B.
- In case of *extortion* of the trustee's key, the mechanism is the same as before, replacing x_B by x_T and coins by pseudonyms. □

Theorem 9 *Assuming, that the coin and pseudonym whitelists (C-WL and PS-WL) are properly used, no money that hasn't been already spent, is lost for the honest user.*

Proof 9:

- If a pseudonym PS_x was extorted and therefore blacklisted, the legal user of this pseudonym gets a signature $\mathcal{S}_T(x_T, (PS_y, ID_U))$ from the trustee. This allows him to exchange withdrawn coins under this pseudonym at the bank for fresh coins, after he has authentified himself as their legal owner.
- If the bank's secret key x_B was extorted and therefore blacklisted, the users who possess legally withdrawn coins C under this secret key can still spend them, as their hashes $h(PS_y, C)$ appear on the whitelist C-WL. Nobody else can forge these coins, as he doesn't know C and the pre-image of PS_y, which are both necessary to spend it.
- If the trustee's secret key x_T was extorted and therefore blacklisted, the users who possess legally obtained pseudonyms PS_y certified under this secret key, can still use them, as they appear on the whitelist PS-WL. Nobody else can reuse them, as he is not able to obtain the corresponding secret key PS_x. □

Acknowledgements

We are greatful to the editors Yongfei Han, Tatsuaki Okamoto and Sihan Qing that they allowed us three additional pages for the security considerations of our scheme.

[6] i.e. it is immediately updated and distributed after fraud has been reported

An Anonymous and Undeniable Payment Scheme

Liqun Chen[†1] and Chris J. Mitchell[‡]
[†] Hewlett-Packard Labs. Bristol, UK. liqun@hplb.hpl.hp.com
[‡] Royal Holloway, University of London, UK. cjm@dcs.rhbnc.ac.uk

Abstract. At Asiacrypt 1996, M'Raihi presented an electronic payment scheme using a blinding office to achieve anonymity. This scheme allows both a bank and a blinding office to impersonate a user without being detected. It may result in a denial problem where the user can deny his bad behaviour by suggesting that either the bank or the blinding office did wrong. This paper proposes a variant of the M'Raihi scheme to prevent the bank and blinding office from impersonating the user, so that the user cannot deny it if he abuses a coin himself.

1 Introduction

In electronic payment systems there are likely to be two contradictory requirements. On the one hand users want to have *anonymous* electronic cash, and on the other hand banks have requirements to ensure the electronic cash cannot be misused. For instance, if abuse (e.g. over-spending) is suspected, the related user's identity should be *traceable*. Some recent research has made use of blind signature and escrow-like techniques to design a payment scheme, which meets the requirements of both the users and banks, e.g. [4, 5].

M'Raihi [7], at Asiacrypt 1996, presented a payment scheme using a Blinding Office (BO) as a pseudo-identity escrow agency. This protocol relies on the assumption of strong trust relationships among a User (U), Bank (BA) and BO, because both BA and BO can impersonate U without being detected. A potential problem resulting from these trust relationships is that U can deny his bad behaviour by claiming no longer to trust either BA or BO. In this case it is difficult for an impartial Judge (J) to arbitrate amongst the three parties.

In this paper we introduce an extra requirement in the M'Raihi scheme: *non-denial*, which, if it holds, will enable J to determine who is lying: U, BA or BO. This requirement has been met in a number of other payment schemes, e.g. [3]. We then present a modification of the M'Raihi scheme to offer the three properties of *anonymity*, *traceability* and *non-denial*. The main advantage of the new scheme is in preventing BA and BO from impersonating U, so that U cannot deny that he has misused a coin himself.

2 An electronic payment model

In this section we describe a general electronic payment model which is suitable for both the M'Raihi scheme and our modified one. It consists of five participants, i.e., U, BA, BO, a Shop (S) and J.

Briefly, this electronic payment model works as follows. U gets a coin (C) blindly signed by BA. BA retains a relationship-proof between U's real identifier

[1] Part of the first author's work was funded by the European Commission under ACTS project AC095 (ASPeCT) when she worked in Royal Holloway, University of London.

(ID) and pseudo identifier (PID). BO, involved in the blind signature, maintains another relationship-proof between PID and C. To spend C, U proves to S that he has knowledge of a private key x corresponding to C. If C is abused, e.g. if it is over-spent, BA and BO will collaborate to build a link between ID and C. J will be involved in this tracing procedure to arbitrate.

We suppose that both BA and U have public-key based signature schemes, respectively named (S_{BA}, V_{BA}) and (S_U, V_U), where V_{BA} is known to U, BO and S, and V_U is known to BA. We also suppose that BO has a public-key based encryption scheme, named (E_{BO}, D_{BO}), where E_{BO} is known to U and BA. All the schemes can be verified by J. A possible implementation for these asymmetric cryptographic schemes is RSA [8].

We assume that the coin C consists of three components. The first is the public verification key y for a public-key based signature scheme, where the corresponding private signature key is denoted by x. The second is a data field D containing certain relevant information about C, such as its expiry date and value. The third is BA's signature on both y and D. There are two different ways of including D in BA's signature. Firstly D can be concatenated with y by BO, prior to computing a blinded public key denoted by $\hat{y}$. Secondly D can be 'added' by BA by using a different signature key depending on the data which is to be indicated. In the subsequent discussion in this paper we ignore this distinction, and assume that either may be used.

We actually require the signature scheme of the bank to have certain special property, i.e., $S_{BA}(z_1)S_{BA}(z_2) = S_{BA}(z_1 z_2)$, which holds for RSA. Of course this is normally a most undesirable feature for a signature scheme, and is one reason why RSA should, in normal circumstances, always be used in conjunction with a one-way hash-function or a special 'redundancy' function (such as that specified in ISO/IEC 9796 [2]). In our case, we either explicitly specify the use of a one-way hash-function (denoted by $H(z)$ for message z) with S_{BA}, or prevent frauds resulting from the use of 'straight' RSA by other means. We denote a blinding function by F, and let it be an 'inverse' of the signature function, so that $S_{BA}(F(z_1)z_2) = z_1 S_{BA}(z_2)$ for any z_1, z_2. If BA's signature scheme is RSA, then F is simply exponentiation using the public verification exponent.

3 Outline of the M'Raihi scheme

The M'Raihi scheme works as follows. U and BA first establish a shared secret s. BA then signs a collision-free one-way function of s, i.e. $S_{BA}(H(s))$, which is used to construct PID by concatenating it with $E_{BO}(s)$. BA also retains a relationship-proof between ID and s, which we denote by $\{ID, s\}$. It, for example, is a signature on $H(s)$ using S_U. To withdraw C, U shows BO both PID and x which is generated by U. BO computes a corresponding y and a set of pre-computed values (which is used for reducing the computation of U's signature used in spending C). BO then blinds y with a random blinding factor v to obtain $\hat{y} = F(v)y$. BA signs $\hat{y}$ without knowing y and withdraws a real coin from U's account. BO derives C from BA's signature on $\hat{y}$ and gives it to U. BO maintains a relationship-proof between PID and C, which we denote by $\{PID, C\}$. To spend C, U signs a message, which is generated by S as a challenge, to prove U knows x. S claims a real coin back from BA later. If C is over-spent, BA will ask for a tracing procedure in which BA and BO collaborate to build a link between C and ID, based on $\{ID, s\}$ and $\{PID, C\}$.

As mentioned earlier, this scheme relies on strong trust relationships amongst U, BA and BO. Both BA and BO must be trusted not to impersonate U to obtain and spend C, since they are capable of doing so if they wish. During a tracing procedure, U can make one or more of the following claims to J to suggest that BA and/or BO has been impersonating U.

Claim 1. *BA can impersonate U to BO to obtain C, and is then able to impersonate U to S to spend C.* This holds because BA knows s.

Claim 2. *BO can impersonate U to spend C.* It holds because BO knows x.

Claim 3. *BO can cheat BA and U to obtain more than one coin from a single delegating blind signature.* This can be seen from the following implementation example.

Suppose that BA's signature scheme is RSA, in which BA's private signature key is d and public verification key is (e, n). That is, BA's signature on a message z is defined as $S_{BA}(z) = z^d \bmod n$, and the corresponding blinding function of the blinding factor v is $F(v) = v^e \bmod n$. Suppose also that the signature scheme used in spending C is DSA, in which the private signature key is x and its corresponding public verification key is $y = g^x \bmod p$. The use of d, e and n in RSA and x, y, g and p in DSA follows [8] and [1] respectively.

If BO wants to obtain m different coins from one blind signature, he chooses $v = v_1(g^{m-1} \bmod p)$, where v_1 is a random number, and computes

$$\hat{y} = (v_1^e \bmod n)((g^{m-1} \bmod p)^e \bmod n)y.$$

After obtaining $S_{BA}(\hat{y})$, he unblinds it by using different blinding factors $(v_1 g^{m-1-i} \bmod p,\ i = 0, 1, ..., m-1)$ as follows:

$$\begin{aligned} S_{BA}(\hat{y}) &= v_1(g^{m-1} \bmod p)((g^x \bmod p)^d \bmod n) \\ &= v_1(g^{m-1-i} \bmod p)((g^{(x+ie)} \bmod p)^d \bmod n). \end{aligned}$$

Following the M'Raihi scheme, U obtains one pair of private and public keys $(x, y = g^x \bmod p)$, and BO obtains a set of key pairs $(x_i = x + ie,\ y_i = g^{x+ie} \bmod p,\ 0 \leq i \leq m-1)$. Each key pair is relevant to a valid coin.

4 A new scheme

The new scheme has the following three differences from the M'Raihi one.

- ◇ BO does not know x, and hence is not able to spend C.
- ◇ U and BO jointly generate a random v. It ensures that neither U nor BO can individually control the value of v, and hence neither U nor BO can obtain more than one coin from a single blind signature.
- ◇ BA retains U's signature on $\hat{y}$ as a relationship-proof between ID and $\hat{y}$, denoted by $\{ID, \hat{y}\}$. It ensures that BA and U cannot dispute whom x was issued by. Note that in the M'Raihi scheme, although BA can record ID with $\hat{y}$, it is still possible for U to refuse responsibility for C because he has no idea about the relationship between y and $\hat{y}$ when making a contribution to this record, in other words, because he was blinded as well.

In the new scheme, the procedure for withdrawing a coin works as follows, where all messages exchanged between U and BO are assumed to be encrypted with s if the communication channels between them are unprotected.

1. U randomly chooses x, computes y, and then sends $E_{BO}(y)$ to BO.
2. U and BO establish a shared v, e.g. using the Diffie-Hellman algorithm [6].
3. BO then computes $\hat{y} = F(v)y$ and sends it to U.
4. U verifies $\hat{y}$, and then sends BA a message signed using S_U. This message is made up of s, $\hat{y}$, BO's name and (possibly) other application data, such as a nonce and/or a time-stamp, to ensure that the uniqueness and freshness of the signature is verifiable.
5. BA retains U's signature, withdraws a real coin from U's account, and then replies to U with $T = E_{BO}(S_{BA}(\hat{y}))$. U passes T to BO.
6. BO decrypts T to obtain $S_{BA}(\hat{y})$, unblinds $S_{BA}(y)$ to construct C, and then sends C to U. After that, BO stores $\{PID, C\}$, which consists of a record of PID and two encryption values, i.e. the encryption of $E_{BO}(y)$ with s and the encryption of T with s.

We now sketch the proofs that the new scheme holds the following security properties. Unless giving a specific indication, we suppose that all participants, U, BA, BO, S and J, do not collude with each other.

Proposition 1. *U cannot obtain C without the involvement of BA and BO.*

Proof sketch. In order to obtain C without BA and/or BO being involved, U must be able to compute $S_{BA}(\hat{y})$ from either $\hat{y}$ or $E_{BO}(S_{BA}(\hat{y}))$, both of which are assumed to be infeasible. □

Proposition 2. *A valid coin relevant to a private key x can only be spent by the participant who is the issuer of x.*

Proof sketch. Assume that given y and other related public information, it is computationally infeasible to recover x. Following the scheme, x is known only to its issuer and is not revealed to anyone else. Thus, since knowledge of x is required to spend C, the result follows. □

Proposition 3. *BO cannot impersonate U to either BA or S.*

Proof sketch. To impersonate U to BA, BO must obtain U's signature on a $\hat{y}$ chosen by BO, which is assumed to be infeasible. To impersonate U to S for spending C, BO must know both C and x. Since U is able to verify $\hat{y}$, BO cannot blind U and then obtain BA's signature on $\hat{y}$ corresponding to his own x. Therefore, it is infeasible for BO to obtain C and its corresponding x. □

Proposition 4. *If BA impersonates U to obtain and to spend a coin, he cannot claim that the coin was issued by U.*

Proof sketch. To prove $\{ID, \hat{y}\}$, BA needs U's signature on $\hat{y}$. Such a signature cannot be obtained, even with BO's collaboration, if U was not involved in the coin generation. □

We conclude this analysis with the following result.

Proposition 5. *If a coin, with a relationship-proof of $\{ID, C\}$, is misused, U cannot deny responsibility for this abuse.*

Proof sketch. Based on Propositions 2, 3, and 4, a valid coin, C, with a relationship-proof of $\{ID, C\}$, maintained jointly by BA and BO, must be related to a private key, x, issued by U, and thus U is the only person able to spend the coin. The theorem follows. □

5 Conclusions

This paper discussed a potential denial problem in the M'Raihi payment scheme and proposed a variant scheme to overcome the problem. The main advantage of the new scheme is that neither BA nor BO can impersonate U to obtain and spend a coin without being detected, so that if U abuses the coin, he cannot deny it by suggesting that it was done by either BA or BO. This advantage is at the cost that the user computational requirements are more onerous than for the M'Raihi scheme, because the user himself needs to do precomputations of the signature (if using DSA) for spending a coin.

Acknowledgment

The authors would like to thank Wenbo Mao for pointing out a weakness in an earlier version of the paper.

References

1. U.S. Department of Commerce/National Institute of Standards and Technology, *Digital Signature Standard.* Federal Information Processing Standard Publication (FIPS PUB) 186, May 1994.
2. ISO/IEC 9796: 1991. *Information technology — Security techniques — Digital signature scheme giving message recovery.*
3. S. Brands. Untraceable off-line cash in wallet with observers. In *Advances in Cryptology - CRYPTO '93, Lecture Notes in Computer Science 773*, pages 302–318. Springer-Verlag, Berlin, 1993.
4. E. Brickell, P. Gemmel, and D. Kravitz. Trustee-based tracing extensions to anonymous cash and the making of anonymous change. In *Proceedings of 6th Annual Symposium on Discrete Algorithm (SODA)*, pages 457–466. ACM Press, 1995.
5. J. Camenisch, U. Maurer, and M. Stadler. Digital payment systems with passive anonymity-revoking trustees. In *Computer Security - ESORICS 96, Lecture Notes in Computer Science 1146*, pages 33–43. Springer-Verlag, Berlin, 1996.
6. W. Diffie and M.E. Hellman. New directions in cryptography. *IEEE Transactions on Information Theory*, 22:644–654, November 1976.
7. D. M'Raihi. Cost-effective payment schemes with privacy regulation. In *Advances in Cryptology - ASIACRYPT '96, Lecture Notes in Computer Science 1163*, pages 266–275. Springer-Verlag, Berlin, 1996.
8. R.L. Rivest, A. Shamir, and L. Adleman. A method for obtaining digital signatures and public key cryptosystems. *Communications of the ACM*, 21:294–299, 1978.

Author Index

Lecture Notes in Computer Science

For information about Vols. 1–1265

please contact your bookseller or Springer-Verlag

Vol. 1266: D.B. Leake, E. Plaza (Eds.), Case-Based Reasoning Research and Development. Proceedings, 1997. XIII, 648 pages. 1997 (Subseries LNAI).

Vol. 1267: E. Biham (Ed.), Fast Software Encryption. Proceedings, 1997. VIII, 289 pages. 1997.

Vol. 1268: W. Kluge (Ed.), Implementation of Functional Languages. Proceedings, 1996. XI, 284 pages. 1997.

Vol. 1269: J. Rolim (Ed.), Randomization and Approximation Techniques in Computer Science. Proceedings, 1997. VIII, 227 pages. 1997.

Vol. 1270: V. Varadharajan, J. Pieprzyk, Y. Mu (Eds.), Information Security and Privacy. Proceedings, 1997. XI, 337 pages. 1997.

Vol. 1271: C. Small, P. Douglas, R. Johnson, P. King, N. Martin (Eds.), Advances in Databases. Proceedings, 1997. XI, 233 pages. 1997.

Vol. 1272: F. Dehne, A. Rau-Chaplin, J.-R. Sack, R. Tamassia (Eds.), Algorithms and Data Structures. Proceedings, 1997. X, 476 pages. 1997.

Vol. 1273: P. Antsaklis, W. Kohn, A. Nerode, S. Sastry (Eds.), Hybrid Systems IV. X, 405 pages. 1997.

Vol. 1274: T. Masuda, Y. Masunaga, M. Tsukamoto (Eds.), Worldwide Computing and Its Applications. Proceedings, 1997. XVI, 443 pages. 1997.

Vol. 1275: E.L. Gunter, A. Felty (Eds.), Theorem Proving in Higher Order Logics. Proceedings, 1997. VIII, 339 pages. 1997.

Vol. 1276: T. Jiang, D.T. Lee (Eds.), Computing and Combinatorics. Proceedings, 1997. XI, 522 pages. 1997.

Vol. 1277: V. Malyshkin (Ed.), Parallel Computing Technologies. Proceedings, 1997. XII, 455 pages. 1997.

Vol. 1278: R. Hofestädt, T. Lengauer, M. Löffler, D. Schomburg (Eds.), Bioinformatics. Proceedings, 1996. XI, 222 pages. 1997.

Vol. 1279: B. S. Chlebus, L. Czaja (Eds.), Fundamentals of Computation Theory. Proceedings, 1997. XI, 475 pages. 1997.

Vol. 1280: X. Liu, P. Cohen, M. Berthold (Eds.), Advances in Intelligent Data Analysis. Proceedings, 1997. XII, 621 pages. 1997.

Vol. 1281: M. Abadi, T. Ito (Eds.), Theoretical Aspects of Computer Software. Proceedings, 1997. XI, 639 pages. 1997.

Vol. 1282: D. Garlan, D. Le Métayer (Eds.), Coordination Languages and Models. Proceedings, 1997. X, 435 pages. 1997.

Vol. 1283: M. Müller-Olm, Modular Compiler Verification. XV, 250 pages. 1997.

Vol. 1284: R. Burkard, G. Woeginger (Eds.), Algorithms — ESA '97. Proceedings, 1997. XI, 515 pages. 1997.

Vol. 1285: X. Jao, J.-H. Kim, T. Furuhashi (Eds.), Simulated Evolution and Learning. Proceedings, 1996. VIII, 231 pages. 1997. (Subseries LNAI).

Vol. 1286: C. Zhang, D. Lukose (Eds.), Multi-Agent Systems. Proceedings, 1996. VII, 195 pages. 1997. (Subseries LNAI).

Vol. 1287: T. Kropf (Ed.), Formal Hardware Verification. XII, 367 pages. 1997.

Vol. 1288: M. Schneider, Spatial Data Types for Database Systems. XIII, 275 pages. 1997.

Vol. 1289: G. Gottlob, A. Leitsch, D. Mundici (Eds.), Computational Logic and Proof Theory. Proceedings, 1997. VIII, 348 pages. 1997.

Vol. 1290: E. Moggi, G. Rosolini (Eds.), Category Theory and Computer Science. Proceedings, 1997. VII, 313 pages. 1997.

Vol. 1291: D.G. Feitelson, L. Rudolph (Eds.), Job Scheduling Strategies for Parallel Processing. Proceedings, 1997. VII, 299 pages. 1997.

Vol. 1292: H. Glaser, P. Hartel, H. Kuchen (Eds.), Programming Languages: Implementations, Logigs, and Programs. Proceedings, 1997. XI, 425 pages. 1997.

Vol. 1293: C. Nicholas, D. Wood (Eds.), Principles of Document Processing. Proceedings, 1996. XI, 195 pages. 1997.

Vol. 1294: B.S. Kaliski Jr. (Ed.), Advances in Cryptology — CRYPTO '97. Proceedings, 1997. XII, 539 pages. 1997.

Vol. 1295: I. Prívara, P. Ružička (Eds.), Mathematical Foundations of Computer Science 1997. Proceedings, 1997. X, 519 pages. 1997.

Vol. 1296: G. Sommer, K. Daniilidis, J. Pauli (Eds.), Computer Analysis of Images and Patterns. Proceedings, 1997. XIII, 737 pages. 1997.

Vol. 1297: N. Lavrač, S. Džeroski (Eds.), Inductive Logic Programming. Proceedings, 1997. VIII, 309 pages. 1997. (Subseries LNAI).

Vol. 1298: M. Hanus, J. Heering, K. Meinke (Eds.), Algebraic and Logic Programming. Proceedings, 1997. X, 286 pages. 1997.

Vol. 1299: M.T. Pazienza (Ed.), Information Extraction. Proceedings, 1997. IX, 213 pages. 1997. (Subseries LNAI).

Vol. 1300: C. Lengauer, M. Griebl, S. Gorlatch (Eds.), Euro-Par'97 Parallel Processing. Proceedings, 1997. XXX, 1379 pages. 1997.

Vol. 1301: M. Jazayeri, H. Schauer (Eds.), Software Engineering - ESEC/FSE'97. Proceedings, 1997. XIII, 532 pages. 1997.

Vol. 1302: P. Van Hentenryck (Ed.), Static Analysis. Proceedings, 1997. X, 413 pages. 1997.

Vol. 1303: G. Brewka, C. Habel, B. Nebel (Eds.), KI-97: Advances in Artificial Intelligence. Proceedings, 1997. XI, 413 pages. 1997. (Subseries LNAI).

Vol. 1304: W. Luk, P.Y.K. Cheung, M. Glesner (Eds.), Field-Programmable Logic and Applications. Proceedings, 1997. XI, 503 pages. 1997.

Vol. 1305: D. Corne, J.L. Shapiro (Eds.), Evolutionary Computing. Proceedings, 1997. X, 307 pages. 1997.

Vol. 1306: C. Leung (Ed.), Visual Information Systems. X, 274 pages. 1997.

Vol. 1307: R. Kompe, Prosody in Speech Understanding Systems. XIX, 357 pages. 1997. (Subseries LNAI).

Vol. 1308: A. Hameurlain, A M. Tjoa (Eds.), Database and Expert Systems Applications. Proceedings, 1997. XVII, 688 pages. 1997.

Vol. 1309: R. Steinmetz, L.C. Wolf (Eds.), Interactive Distributed Multimedia Systems and Telecommunication Services. Proceedings, 1997. XIII, 466 pages. 1997.

Vol. 1310: A. Del Bimbo (Ed.), Image Analysis and Processing. Proceedings, 1997. Volume I. XXII, 722 pages. 1997.

Vol. 1311: A. Del Bimbo (Ed.), Image Analysis and Processing. Proceedings, 1997. Volume II. XXII, 794 pages. 1997.

Vol. 1312: A. Geppert, M. Berndtsson (Eds.), Rules in Database Systems. Proceedings, 1997. VII, 214 pages. 1997.

Vol. 1313: J. Fitzgerald, C.B. Jones, P. Lucas (Eds.), FME '97: Industrial Applications and Strengthened Foundations of Formal Methods. Proceedings, 1997. XIII, 685 pages. 1997.

Vol. 1314: S. Muggleton (Ed.), Inductive Logic Programming. Proceedings, 1996. VIII, 397 pages. 1997. (Subseries LNAI).

Vol. 1315: G. Sommer, J.J. Koenderink (Eds.), Algebraic Frames for the Perception-Action Cycle. Proceedings, 1997. VIII, 395 pages. 1997.

Vol. 1316: M. Li, A. Maruoka (Eds.), Algorithmic Learning Theory. Proceedings, 1997. XI, 461 pages. 1997. (Subseries LNAI).

Vol. 1317: M. Leman (Ed.), Music, Gestalt, and Computing. IX, 524 pages. 1997. (Subseries LNAI).

Vol. 1318: R. Hirschfeld (Ed.), Financial Cryptography. Proceedings, 1997. XI, 409 pages. 1997.

Vol. 1319: E. Plaza, R. Benjamins (Eds.), Knowledge Acquisition, Modeling and Management. Proceedings, 1997. XI, 389 pages. 1997. (Subseries LNAI).

Vol. 1320: M. Mavronicolas, P. Tsigas (Eds.), Distributed Algorithms. Proceedings, 1997. X, 333 pages. 1997.

Vol. 1321: M. Lenzerini (Ed.), AI*IA 97: Advances in Artificial Intelligence. Proceedings, 1997. XII, 459 pages. 1997. (Subseries LNAI).

Vol. 1322: H. Hußmann, Formal Foundations for Software Engineering Methods. X, 286 pages. 1997.

Vol. 1323: E. Costa, A. Cardoso (Eds.), Progress in Artificial Intelligence. Proceedings, 1997. XIV, 393 pages. 1997. (Subseries LNAI).

Vol. 1324: C. Peters, C. Thanos (Eds.), Research and Advanced Technology for Digital Libraries. Proceedings, 1997. X, 423 pages. 1997.

Vol. 1325: Z.W. Raś, A. Skowron (Eds.), Foundations of Intelligent Systems. Proceedings, 1997. XI, 630 pages. 1997. (Subseries LNAI).

Vol. 1326: C. Nicholas, J. Mayfield (Eds.), Intelligent Hypertext. XIV, 182 pages. 1997.

Vol. 1327: W. Gerstner, A. Germond, M. Hasler, J.-D. Nicoud (Eds.), Artificial Neural Networks – ICANN '97. Proceedings, 1997. XIX, 1274 pages. 1997.

Vol. 1328: C. Retoré (Ed.), Logical Aspects of Computational Linguistics. Proceedings, 1996. VIII, 435 pages. 1997. (Subseries LNAI).

Vol. 1329: S.C. Hirtle, A.U. Frank (Eds.), Spatial Information Theory. Proceedings, 1997. XIV, 511 pages. 1997.

Vol. 1330: G. Smolka (Ed.), Principles and Practice of Constraint Programming – CP 97. Proceedings, 1997. XII, 563 pages. 1997.

Vol. 1331: D. W. Embley, R. C. Goldstein (Eds.), Conceptual Modeling – ER '97. Proceedings, 1997. XV, 479 pages. 1997.

Vol. 1332: M. Bubak, J. Dongarra, J. Waśniewski (Eds.), Recent Advances in Parallel Virtual Machine and Message Passing Interface. Proceedings, 1997. XV, 518 pages. 1997.

Vol. 1333: F. Pichler. R.Moreno Díaz (Eds.), Computer Aided Systems Theory – EUROCAST'97. Proceedings, 1997. XI, 626 pages. 1997.

Vol. 1334: Y. Han, T. Okamoto, S. Qing (Eds.), Information and Communications Security. Proceedings, 1997. X, 484 pages. 1997.

Vol. 1335: R.H. Möhring (Ed.), Graph-Theoretic Concepts in Computer Science. Proceedings, 1997. X, 376 pages. 1997.

Vol. 1336: C. Polychronopoulos, K. Joe, K. Araki, M. Amamiya (Eds.), High Performance Computing. Proceedings, 1997. XII, 416 pages. 1997.

Vol. 1337: C. Freksa, M. Jantzen, R. Valk (Eds.), Foundations of Computer Science. XII, 515 pages. 1997.

Vol. 1338: F. Plášil, K.G. Jeffery (Eds.), SOFSEM'97: Theory and Practice of Informatics. Proceedings, 1997. XIV, 571 pages. 1997.

Vol. 1339: N.A. Murshed, F. Bortolozzi (Eds.), Advances in Document Image Analysis. Proceedings, 1997. IX, 345 pages. 1997.

Vol. 1340: M. van Kreveld, J. Nievergelt, T. Roos, P. Widmayer (Eds.), Algorithmic Foundations of Geographic Information Systems. XIV, 287 pages. 1997.

Vol. 1341: F. Bry, R. Ramakrishnan, K. Ramamohanarao (Eds.), Deductive and Object-Oriented Databases. Proceedings, 1997. XIV, 430 pages. 1997.

Vol. 1342: A. Sattar (Ed.), Advanced Topics in Artificial Intelligence. Proceedings, 1997. XVIII, 516 pages. 1997. (Subseries LNAI).

Vol. 1344: C. Ausnit-Hood, K.A. Johnson, R.G. Pettit, IV, S.B. Opdahl (Eds.), Ada 95 – Quality and Style. XV, 292 pages. 1997.